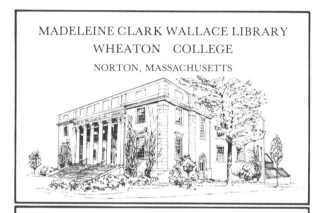

Primate Taxonomy

Primate Taxonomy

Colin Groves

Smithsonian Institution Press • Washington and London

editor: Eileen D'Araujo
production editors: Deborah L. Sanders and
    Ruth Spiegel
designer: Janice Wheeler

Library of Congress Cataloging-in-Publication Data
    Groves, Colin P.
    Primate taxonomy / Colin P. Groves.
    p. cm. — (Smithsonian series in comparative
        evolutionary biology)
    Includes bibliographic references (p.).
    ISBN 1-56098-872-X (alk. paper)
        1.Primates—Classification. I. Title. II. Series.
    QL737.P9 G86 2001
    599.8′01′2-dc21   00-045017

British Library Cataloguing-in-Publication Data available

Manufactured in the United States of America
08 07 06 05 04 03 02 01   5 4 3 2 1

Contents

Preface

The feeling that there is a need for a book like this has grown over the past 10 years or so; the certainty that there is became apparent when I subscribed to an Internet group called *primate-talk* (now discontinued). Primatologists ask questions like:

—What is the official taxonomy of [name your group]?

—Why have the names for [name your taxon] changed?

—Wouldn't it be simpler to recognize just one species of [name your polytypic group]?

And, at one level higher in the scale of sophistication:

—Why not keep humans in a family by themselves? They are so different from apes.

—Taxa A and B can interbreed in captivity: Doesn't this mean that they are the same species?

And, at one level higher still:

—Taxa A and B sometimes interbreed in the wild: Aren't they the same species?

Primatologists on the whole don't understand taxonomy, but they need to, and, in the main, they want to.

Taxonomy is one of the last areas where the gifted amateur can still make significant contributions, and primatologists know this; what they don't know is that the amateur must be *very* gifted. So often nonspecialists—eminent ethologists, ecologists, anatomists, geneticists—feel at liberty to have a go and proceed to make a mess of it. With honorable exceptions, of course!

This is not meant as an insult to my colleagues. I would make a mess of ethology or ecology (some would say I already have). Taxonomy is a speciality just as ethology and ecology (and anatomy and genetics) are; there is absolutely no reason why a primatologist whose speciality is, say, ethology should be expected to be expert in taxonomy.

Taxonomy is, however, fairly fundamental: it is the basic means by which we organize information about the living world. This is probably the reason why people expect there to be an "official" taxonomy. The notion that a taxonomic scheme is a hypothesis, an interpretation, like an equivalent interpretation of a behavioral trait or an ecological relationship, and so is absolutely bound to change as understanding advances, is a hard lesson to learn. That's why the names "keep changing"!

But taxonomy is, in today's tragic biodiversity crisis, the very starting point for conservation efforts. It makes a difference to the aye-aye's "conservation status rating" whether it rates its own infraorder, superfamily, or just family. It matters whether gold and golden-headed lion tamarins are different species, or just subspecies.

Thinking about all this, trying to classify my operational taxonomic units as objectively as possible, I found I had to focus more and more on the basic unit of taxonomy, the species. It is thus that I reluctantly, perhaps, came to recommend the phylogenetic species concept. Readers may be dismayed at the large number of species that I have proposed in this book (including a few that are not yet formally described). I hope they will think about this and at least discuss the argument for objectivity, as I see it.

I hope that the classifications in the book will be taken as a starting point, not as a solution to problems. I hope that others will take up some of the propositions and test them. I have, in many cases, deliberately asked questions and insisted that the way I have approached them is not definitive, nor meant to be.

Finally, I point out the obvious fact that primatology does not stand on its own. Part One of this book could be for mammalogists in general, ornithologists, and any other student of sexually reproducing animals. But Part Two is about Primates, the group I know best. I did a good deal of museum research for it, both more recently, specifically for the book, and at earlier periods for other purposes. That's why it is a book about Primate taxonomy.

I am ever grateful to colleagues all over the world for allowing me access to anything from their collections to their thoughts. Among so many, I may single out the late Prue Napier and the late John Hill; Mitsuru Aimi, Peter Andrews, Ignacio Avila, Simon Bearder, Boeadi, Doug Brandon-Jones, Judith Caton, Anwaruddin Choudhury, Adelmar Coimbra-Filho, Andrew Collins, Pierre Dandelot, the late Dao Van Tien, Jack Fooden, Laurent Granjon, Peter Grubb, Sandy Harcourt, Paula Jenkins, Jonathan Kingdon, Jean-Marc Lernould, Romari Martinez, Guy Musser, Tilo Nadler, John Oates, Anthony Rylands, Esteban Sarmiento, Myron Shekelle, Jatna Supriatna, Wang Yingxiang, and Roland Wirth. Thanks to Sue Fraser and Susan Lindsay, who have helped me enormously in formatting, organizing, and printing the manuscript. Thanks above all to Phyll, who as always helped in all those little ways and very big ones, not least that in the midst of writing her Ph.D. thesis she nonetheless found time to help me with this book. Thanks to everyone—don't be too dismayed by the result!

1

The Theory of Primate Taxonomy

1

What Taxonomy Is Meant to Do and How It Should Do It

Taxonomy was defined by Simpson in 1961 (p. 11) as "the theoretical study of classification, including its bases, principles, procedures, and rules," and **classification** (p. 9) was "the ordering of [organisms] into groups (or sets) on the basis of their relationships." Since that time there has been a growing tendency to extend the term taxonomy to cover classification in practice, such that "a taxonomy" means much the same as a classification, with the added proviso that it incorporates something of the rationale for it, which a classification need not.

A related term is **systematics**. Simpson (1961:7) defined systematics as "the scientific study of the kinds and diversity of organisms and of any and all relationships among them." For Simpson, the implications of the term were much wider than for taxonomy: more or less equivalent, in fact, to the modern "biodiversity." Today, systematics has become a virtual synonym for taxonomy, and I make no apology for using the two words synonymously here, but we must realize that there has been a subtle shift in terminology over the past 30 or 40 years, so that we are in some respects no longer speaking the same language as Simpson.

The basis of taxonomy is the taxon. A **taxon** (plural, **taxa**) is "a group of real organisms recognised as a formal unit at any level of a hierarchic classification" (Simpson 1961:19). Thus the species *Cercopithecus mitis* is a taxon, and so is the genus *Ateles,* and the family Lemuridae, and the order Primates, and so on.

The value (one might insist, indispensability) of a taxonomy lies, as I have hinted in the title of this section, in its role as an information retrieval system. Mayr and Ashlock (1991) listed the following as uses of a taxonomy:

—An index to stored information. It orders otherwise unwieldy masses of information, as the Dewey system does in libraries; more than that, it tells us whether a taxon is speciose or not, its degree of isolation, and its general position in the living world.

—Heuristic properties. It functions, like any scientific hypothesis, as a source for predictions and for future research directions.

—Generalizability. It enables one to make statements about particular organisms, with the expectation that they can be generalized to wider groups.

—Explanatory powers. It can be used to deduce the former existence of common ancestors.

LEVELS OF TAXONOMY

There are, as Mayr (1969) put it, three "stages of maturation" in the classification of a given taxonomic group, dubbed alpha, beta, and gamma taxonomy.

Alpha taxonomy is concerned with basic description, mainly of species (and subspecies) and perhaps genera and subgenera. Because it is basically descriptive, there is little analysis beyond "these specimens represent a new species, which is hereby named . . . " or "species B is hereby shown to be a synonym of species A, and is therefore formally sunk."

A paper describing the taxa within a group (a series of related genera with their contained species and subspecies, perhaps) is called a **revision** (literally a "new look" at that group). The genera, the subgenera if any, the species, and the subspecies if any should be listed one by one using the following sequence, which is helpful although not, unfortunately, obligatory:

1. The heading should be the name of the species (or other taxon) that is being considered. Usually the name is followed by the author and date, or, if the taxon is new, by some indication such as "new species," "n. sp.," or "sp. nov." (short for "species novum").

2. A synonymy or a bibliographic list should follow the heading. A **synonym** means literally "a name that means the same," so every synonymy is a listing, in chronological order, of every taxon that its author thought at the time was distinct but that the reviser thinks actually refers to the taxon he or she is dealing with. I will say more about the synonym business later on. A bibliographic list is generally as brief as possible, although some especially thorough taxonomists (such as the late Philip Hershkovitz) listed every publication in which the taxon had ever been mentioned, under whatever name. If one is describing a new taxon (a new species, for example), it sometimes happens that the species has been mentioned before in a publication, but the author thought that he or she was mentioning an already known species; references to

such publications are especially important to list.

3. Then, under one or more subheadings, the identity of the **type specimen** (for a new species or subspecies) or (for a new genus or subgenus) the **type species** (i.e., the species that is being described to typify the new genus/subgenus) must be given. This is crucial if a new taxon is being described (and useful, but optional, if a known taxon is simply being redescribed). The most vital piece of information for a new species/subspecies is the **type locality** (i.e., the locality from which the type specimen was obtained).

4. Give the paradigm (for a species or subspecies): other specimens that have been examined for the purposes of this description.

5. The diagnosis appears next: the features by which the species (or genus or subgenus) can be infallibly recognized or, for a subspecies (which cannot strictly speaking be diagnosed), the features by which it can most reliably be distinguished.

6. The description follows: other features that characterize the taxon, but are not infallibly diagnostic of it.

7. Include the distribution (for a species or subspecies): other localities where the taxon has been recorded.

8. Include notes, or comments next: anything else that is worth saying—what its nearest relative appears to be; who in the past may have notably misidentified it; who cleared up certain confusions about it; whether there are age or sex or individual variations that can confuse identification.

9. If one is describing a new taxon, including the heading "Etymology" is a nice touch: what the name means, and why it was picked; if it is named after someone, include why you think this person deserves to be perpetuated in this context.

Beta taxonomy is the more detailed, considered elucidation of relationships, including sometimes the construction of a more encompassing classification.

Some authorities differentiate a third level, **gamma taxonomy**. Mayr (1969) included under this heading the study of detailed intraspecific variation, such as by multivariate analysis; evolutionary studies; and functional or ecological studies. Actually, gamma grades into beta taxonomy; for myself, I am never certain which of the two I am doing at any one time, but I always know when I am doing alpha taxonomy.

CHANGING THE NAMES

The relationship between taxonomy and **nomenclature** is often misunderstood. If systematics, in a sense, may be said to be a reflection of the real world, nomenclature (the naming of organisms) is a system invented by the human mind, to be of service in systematics. We may grumble about taxonomists "always changing the names," but there are two good reasons for doing so.

Because of new discoveries or hypotheses about interrelationships, alterations in classifications are necessary. Taxonomy, like other fields of biology (ecology, ethology, physiology, genetics), is a dynamic science. Classifications are not engraved in stone, nor should they be; it is unfortunate that advances in the taxonomic field, unlike those in ecology and other disciplines, often require changing the names we give to species. If *Callithrix pygmaea* turns out to reflect present-day understanding of the relationships of the pygmy marmoset more accurately than does *Cebuella pygmaea,* then it is annoying for a while (especially for a field-worker publishing on the species), but that is the way it must be, and the irritation felt by the field-worker will pass quickly. Indeed, new predictions, to be tested in the field, may well emerge from the reclassification.

But, in addition, new discoveries about the way our predecessors played the nomenclature game can cause changes. The rules of nomenclature are as objective as any legal system; they were drawn up to make things easier for us, and on the whole they do. Where they do not, anyone may make an application to the International Commission on Zoological No-menclature to suspend the rules in a given instance, and the case will then be judged on its merits.

THE MEANING OF "RELATEDNESS"

So a classification, as Simpson (1961:9) defined the term, should reflect "relationships," or relatedness.

What is it, then, this relatedness? Simpson's definition of classification was quoted at the beginning of this chapter, but left incomplete; it concludes by explaining that "relationships" means "associations by contiguity, similarity, or both." The mention of "contiguity" is puzzling, because earlier on the same page Simpson had implied that he meant synecology, which is indeed a part of what Simpson called systematics and what we today call biodiversity, but to my mind it has nothing to do with classification as such.

To bring it into a more familiar field: what does it mean when we say that two people are related? Brothers and sisters are more closely related than cousins; first cousins are more closely related than second cousins. What we are saying, of course, is that being more closely related means having a more recent common ancestor. When we move across generations the question becomes more difficult: Are parents and children more closely related than brothers and sisters? Here we appeal to genetics: What proportion of their genome do they share?

Can we transfer these criteria to the natural world? Surely we can. That species A and B share a more recent common ancestor than they do with species C should mean that A and B are more closely related to each other than to C; they share DNA sequences, inherited from their common ancestor, that are missing from C. If we think "relatedness" in this sense (recency of common ancestry) is worth celebrating, we can place A and B in a taxonomic group that excludes C. A taxonomic schema in which a group AB is contrasted with a group C immediately predicts that members of group AB will have more in common genetically than they do with group C or any other.

Is this sufficient? The proposition has been put

(see Ashlock [1971] for example) that one lineage, A for example, may have changed so rapidly since its separation from its sister lineage (B in this case) that B is left sharing more of its genome with C. A good example is the position of *Homo* among the Hominoidea. Although *Homo* (human) and *Pan* (chimpanzee) share a common ancestor subsequent to the separation of the ancestry of *Pongo* (orangutan), it is obvious that far more morphological change occurred during human evolution, so that chimpanzees and orangutans seem to resemble each other more than chimpanzees resemble humans. It was on this basis that chimpanzees and orangutans (and gorillas, and sometimes gibbons) used to be placed together in the family Pongidae, leaving humans and their fossil relatives alone in the family Hominidae. Well, why not?

The reason is that, when we measure total genomic resemblance, chimpanzees and humans really do share more, significantly more, of their genetic equipment than they do with orangutans (humans and chimpanzees share about 98.5% of their genome with each other and only 95% with orangutans). The explanation is as follows: First, the proportion of the genome that consists of actual genes (that code for phenotypic characters) is very small (under 10%), so that rapid change in genes coding for gross morphological features gets swamped by the stochastic change occurring in the noncoding 90+%. Second, gross morphological features themselves account for only a small fraction of the phenotype, which also includes immunology, biochemistry, and other less obvious morphological features (patterns of blood vessels, musculature, details of skin and hair structure). Finally, there is the likelihood that, the "grosser" these morphological features are, the more probable it is that they are not separately coded characters as much as functional epiphenomena of a few general trends. Any way we look at it, the "rapid change argument" falls away as applying to only a tiny fraction of the total genome, and we are returned to the equation between relatedness (total similarity) and common ancestry.

Obviously, it is not usually possible to measure total genomic similarity. Most DNA-based studies of phylogeny involve sequencing relatively short segments in a few related taxa, and the assumption is made that these have diverged stochastically. This assumption is more likely to be justified in noncoding segments of DNA ("molecular clock" theory) than in genes themselves, and in general relative genetic distances found in these studies do reflect those found in measures of total genomic distance.

As an aside, why bother with morphology at all? Why not go straight to the DNA, which can these days be extracted from preserved skin, bone, even feces? There are many reasons why not. DNA extraction and sequencing is still quite an expensive business and is a skill that is not within everyone's competence. DNA sequences may not survive in fossils beyond a few tens of thousands of years (though this is currently the subject of a ferocious controversy!), but we still need to be able to compare fossils with extant taxa. Taxonomy has come to be highly valued in conservation, and when a monkey is glimpsed in the treetops we generally want to identify it on the spot and not after collecting its feces and taking them back to the laboratory. Finally, I would not want to miss the excitement of finding out something of the why and how of evolutionary change, its functional correlates, its geography, its ecology, its demography, and perhaps contributing to the current ferment in evolutionary theory. The importance of getting back to the genome cannot be overestimated, but when we have done that a vast amount is still left unsaid.

There is another argument, put by such authors as Martin (1990), that taxonomic stability is important and as such may sometimes require us to override its phylogenetic content. As an overall philosophy, this cannot be accepted: a classification that has been stable for 20 years was preceded by a different one that was stable for 20 years before that, and so on. The Hominidae vs. Pongidae classification, which some authors might like to preserve (on the grounds of stability), represented in its time an advance in knowledge over one that separated humans as a distinct suborder, Bimana, from all other Primates (Quadrumana). Each taxonomic scheme represents an advance in understanding on what went before; stability, as such, is not a virtue. Yet before we dismiss the plea for stability out of hand, let us admit that some taxonomists have hastened to revise "accepted" classifications on the evidence of little more than this week's cladistic analysis or (perish the thought!) of comparisons of DNA sequences of just a few hundred base pairs. Taxonomists do owe their colleagues the courtesy of a thought-out, careful analysis; a slenderly based revision is as misleading as a spurious orthodoxy.

THE EMERGENCE OF DEFINED TAXONOMIC PHILOSOPHIES

Mayr, Cain, Simpson, Ashlock, and Martin describe (or described) themselves as **evolutionary taxonomists**. They recognize(d) the primacy of common descent (indeed, Simpson [1961:59–60] came close to accepting it as the leading criterion of "relationship") while incorporating that extra dimension, "evolutionary distance." And this proved to be the stumbling block: there seemed no objective way to define "evolutionary distance." Two schemes have been proposed to objectify taxonomy.

Phenetics

The first definitive statement of dissatisfaction with the way taxonomy was being conducted was the seminal *Principles of Numerical Taxonomy* by Robert Sokal and Peter Sneath, published in 1963. This was the culmination of work by two noted biologists/biometricians (an economic entomologist and a bacteriologist, respectively), who for a number of years had been devising more objective, repeatable methods of measuring taxonomic distance. They insisted that characters differentiating taxa should be investigated for their validity (in particular, their intercorrelation) and, after being accordingly "cleaned up," should be used to quantify the relative distances between the taxa concerned. At its simplest, this amounts to no more than counting up how many

characters differ between A and B, between B and C, and between A and C and devising a taxonomy to reflect this.

In fact, their system is often much more sophisticated than this. The discussions on character correlation, the weighting and coding of characters, distance measures and clustering techniques, size and shape, and so on, are well worth reading even today, both because they are still used at lower taxonomic levels and because they are highly relevant in a phylogenetic scheme as well as in a phenetic one.

A term introduced by Sokal and Sneath (1963) that is useful in all taxonomic work is **operational taxonomic unit** (**OTU**). OTUs are the units of comparison in taxonomic studies, whether morphometric or cladistic. They may be populations, subspecies, species, genera, families, or (most likely) a mixture of these; the advantage of the term is that it does not prejudge rankings or interrelationships, or in fact anything except, ideally, monophyly.

Therefore, although phenetics was almost immediately superseded by cladistics, its influence continues to be felt; in effect, it was only after Sokal and Sneath's insights had been incorporated into Hennig's grand edifice that modern cladistics took shape. Perhaps that is why phenetics was still thought important enough to be the subject of a swingeing attack nearly 30 years after Sokal and Sneath's book (Mayr and Ashlock 1991, chapter 8). Mayr and Ashlock deemed phenetics to be inductivist, hence sterile (p. 196); subjective in its application (pp. 197, 200); laborious and too dependent on large data sets (p. 199); lacking methodological objectivity and repeatability (p. 200); replete with unsubstantiated assumptions (pp. 200–202); and in many ways atheoretical (pp. 202–204). They regarded the concept of OTU as too heterogeneous to be meaningful. Some of these criticisms are justified, some are unfair; especially, I would deem the very heterogeneity of the OTU idea to be a strength, in permitting a data set to be analyzed without first organizing it into taxa and so potentially begging the question.

Let us acknowledge, too, that phenetics is alive and well in the form of multivariate morphometrics.

It would be difficult to conceive of a more useful tool of analysis at the species level and below.

Cladistics

Sokal and Sneath (1963) were aware of the work of Hennig (1950) and discussed it. In this, they were way ahead of most of their Anglophone colleagues, so when Hennig's book, updated, was translated into English and published as *Phylogenetic Systematics* in 1966, it was a bombshell. This is an excellent example of a scientific revolution in the sense of Thomas Kuhn (1962).

Hennig (1950, 1966) maintained that one could classify only by phylogenetic relatedness. His arguments struck an immediate chord, and his new terminology rapidly achieved widespread use. He argued (1966:22–23) that making phylogeny the basis for taxonomy has the following advantages:

—Evolutionary relationships are at the basis of all biology and interrelate systems that are not directly related to each other.

—The interrelationships between organisms are, at least in principle, exactly measurable in a phylogenetic system, but not in any other taxonomic system.

—In many cases, though not all by any means, general similarities correspond with phylogenetic kinship and, in cases where they do not, "suitable evaluation" of the similarities enables the relationships to be recognized; whereas a system based merely on morphological similarities would group together ontogenetic stages, morphs, and so on.

The first point Hennig considered to be "decisive." His second point is one that has taken on a new lease of life since the DNA revolution and the discovery that genetic distance measures, as discussed above, can be used to represent degrees of phylogenetic separation. The third point is perhaps a bit cheeky: early objections to phylogenetic systematics (such as Ashlock [1979]) emphasized precisely that under this system taxa that seem very *dis*similar are often classified together; to discover genuine phylogenetic relationships one has to undertake a particular and often rather tortuous process of analysis (see "The Cladistic Method" later in this chapter).

Nearly all modern systematists have, nonetheless, agreed with Hennig: a logical taxonomic system has to be phylogenetic. Hennig thus irrevocably changed the entire basis of taxonomy. A colleague of mine, at a mammalogy congress, spoke in public discussion of "reading the words of the saints—Linnaeus, Darwin, and Hennig." As serious as that!

It is interesting to note that Sokal and Sneath (1963) considered the desirability of phylogenetically based taxonomy and rejected it for several reasons. The most important was that one commonly finds that a phylogenetically derived character state unique to OTUs A and B, which Hennig would use as evidence linking them to the exclusion of C, is balanced by a different derived state linking A and C. This circumstance, called **homoplasy**, is a very real problem with which cladists have to deal, and there are methods for sorting it out; but phylogeny is not thereby rendered unknowable, as Sokal and Sneath maintained. A subsidiary problem that they saw was that classifying by phylogeny alone often excludes some important information, if not as homoplasy then as uniquely derived states that might be, for example, crucial biochemical reactions in some bacterium that, in a phenetic system, would tend to distance that species from its relatives. The accumulation of uniquely derived states along a single lineage does not, of course, affect assessments of relationships, nor should it. Otherwise equally important phylogenetic information would be lost, whereas these derived states are recorded in the data set and so are available for inspection after the phylogenetic analysis has been made.

A curious, short-lived aberration of cladism was **transformed** or **pattern cladism**. As described by one of its initiators (Patterson 1980), it is an attempt to free cladistics from evolutionary theory; the pattern of character distribution thus reflects not phylogeny but a sort of neo-Cuvierian "natural order." Charig (1983) showed that, if thus divorced from its phylogenetic underpinnings, cladistics is indistinguishable from phenetics, and Ridley (1989) described the pattern cladist philosophy as "incoherent." Needless to say, the rise of pattern cladism has been used by every smart creationist to claim that eminent biologists "have doubts" about evolution; because in creationist philosophy (if one may be excused an oxymoron) a phrase once said cannot be unsaid as long as it can be interpreted as supporting their cause, the rapid demise of pattern cladism has remained unremarked, and in retrospect the incautious phraseology of Patterson and others was poorly chosen.

THE IMPORTANCE OF MONOPHYLY

Hennig changed forever the way we think of relatedness, the basis of taxonomy. He insisted that relatedness requires that taxa be monophyletic.

The definition of **monophyly** and **monophyletic** has changed considerably over the course of this century. The derivation seems simple enough: "mono," single; "phylum," stem. Evidently, what one regarded as this "single stem" was at stake.

Simpson (1945, 1961) defined a monophyletic taxon as one that was descended from a taxon of equal or lesser rank. He was especially interested in the origin of the mammals and thought it possible that the Prototheria (monotremes) and Theria (marsupials and placentals) were descended from different groups of mammal-like reptiles. The Mammalia rank as a **class** (within the phylum Chordata); the mammal-like reptiles are (or were) ranked as a **subclass** of the class Reptilia. The mammals therefore derive from a taxon of lesser rank, and so are monophyletic. Simpson referred in similar vein to the origins of the mammalian order Edentata, the genus *Equus,* and other examples; he considered that their origins were in multiple stocks outside the groups themselves. It is difficult to escape the conclusion that Simpson's definition of monophyly was a semantic sleight-of-hand specifically designed to preserve these taxa as monophyletic groups.

Some of Simpson's contemporaries were unhappy with this argument and the definition that seemed to flow from it. Reed (1960), for example, pointed out that reclassifying the mammal-like reptiles within the Mammalia would go at least some way toward reflecting their actual relationships.

Hennig's definition is uncompromising. **Monophyletic** groups must be descended from a common ancestor that was itself a member of that taxonomic group and must include all the descendants of that ancestor. Groups that include only descendants of a common ancestor but not all its descendants are called **paraphyletic**. Groups descended from a common ancestor that is not itself part of that group are called **polyphyletic**.

Ashlock (1971) deplored Hennig's restriction of monophyly to the case of only and all descendants of the common ancestor. He proposed that a group be deemed monophyletic if it consisted only of descendants from a common ancestor and that monophyletic groups could be either **holophyletic** if they included all its descendants or paraphyletic if they did not. Although it is true that, as Mayr and Ashlock (1991) have bewailed, cladists have done a good deal of redefining of terms to fit better into the cladistic program, in this case it must be pointed out that Hennig's redefinition had been with respect not to the old, Haeckelian meaning of monophyly, but to Simpson's unfortunate ad hoc usage. The term holophyletic is commonly now regarded as a well-meaning but ultimately unnecessary attempt to dichotomize phylogenetic concepts.

By Hennig's definitions the class Mammalia (sensu Simpson) is polyphyletic: descended from a common ancestor that belonged to the Reptilia. The family Pongidae, as recognized by Simpson and as "traditionally" recognized (to include orangutans, gorillas, and chimpanzees) is paraphyletic: descended from a common ancestor that was itself a pongid, but not including humans (or australopithecines), which descended from that same common ancestor. Because the only natural groups (groups that reflect relatedness) are those that are monophyletic, neither Pongidae nor Mammalia, at least as defined by Simpson, can take their place in a meaningful natural classification.

Each of these natural groups should, in principle, be a taxon. The aim of taxonomy is to make, as far as possible, a list of monophyletic taxa, from top (kingdom) to bottom, or nearly so (genus). Taxa of descending rank are therefore nested groups of increasing relatedness. "Nested" means that they are exclusive, nonoverlapping; the families fit within the orders, the orders within the classes. "Increasing relatedness" means that, as their ranks descend, the common ancestors become more recent. So taxa, down to genus/subgenus level anyway, are evolutionary branches, lineages, clades.

But species: should they be monophyletic? This is quite a controversial matter and will be considered in the next chapter.

PROBLEMS WITH CLADISTICS

Biologists using the cladistic method and cladistically based taxonomy are under no illusions about the difficulties that this may involve. In no case, however, have these problems been deemed so insuperable as to require that the whole system be junked. For the cladist, the advantages are so obvious, the logic so clear, that no further discussion is required. It may be this attitude of infallibility that has infuriated some critics, to the extent that much of chapter 9 of Mayr and Ashlock (1991) reads like a creationist's denunciation of demon "evolutionists." As in their forthright (and equally scathing) attack on phenetics in their previous chapter, Mayr and Ashlock certainly put their finger on weaknesses, but their bathwaterist approach has led to a refusal by cladists to take their strictures seriously.

Mayr and Ashlock draw attention to the universality of homoplasy (parallelism or reversal in evolution). This is perfectly true; in fact, the degree to which homoplasy is rampant has become clear only since the rise of cladism. The choice of characters is, as they say, absolutely crucial in any cladistic analysis; it remains equally true, however, that as the number of characters increases, so does the convergence of morphological and molecular analysis (Shoshani et al. 1996).

Classifying fossil taxa with living members of their lineages is equivalent to considering Charlemagne more closely related to his living descendants than to

his mother or his siblings; and it requires the abandonment of "readily recognisable" taxa such as Reptilia, because some reptiles are phylogenetically more closely related to birds or mammals than to other reptiles (Mayr and Ashlock 1991:226). It requires a vast increase in the number of taxonomic ranks, or some special arrangement to avoid this, and there are no objective criteria for ranking anyway (pp. 225–226, 230–231). Cladists ignore the potential information content of primitive character states and of uniquely derived states (p. 229). Ancestors are neglected; indeed it is sometimes denied that they can be recognized (pp. 231–232). Cladistic classifications are inherently unstable (pp. 233–234).

All these criticisms are doubtless true, and different ways of approaching the problems have been proposed. They are treated briefly in "The Cladistic Method" later in this chapter.

THE DNA REVOLUTION

The coming of "molecular methods" has revolutionized our understanding of evolution more than it has taxonomy. First came immunology, then amino-acid sequencing, followed by RFLP (restriction fragment length polymorphism), and finally DNA sequencing; but from the very start there was the concept of a **molecular clock**.

The molecular clock depends on the idea that much mutation is neutral: it has neither beneficial nor detrimental effect on the phenotype, or perhaps (because of the redundancy of the genetic code) no effect at all. The discovery that in general well over 90% of a species' genome is not transcribed ("junk DNA") has augmented the theoretical basis for widespread molecular clocks. These have been tested and retested since the early 1960s, with different results under different theoretical and practical regimes but always with some degree of regularity, even though the clocks are commonly fairly "sloppy" and work much better locally than globally. In theory, molecular findings have reinforced cladistic principles, refuting the idea that there can be enormous (pangenomic) variations in evolutionary rates, while in a practical sense, in cases where reliable clocks have appeared to work, it has enabled simpler genetic distance methods to be used in place of more exacting cladistic methods of analyzing relationships.

A particular application of molecular genetics to taxonomy is **phylogeography** (Avise 1994), which has shed enormous light on problems of relationships at the species and subspecies level, but stands to be misinterpreted unless care is taken. This question will be discussed farther on.

THE CLADISTIC METHOD

There are several good surveys of cladistic methodology available, such as Forey et al. (1992) and Maddison and Maddison (1992).

As discussed earlier, taxa (except perhaps at and below the species level: see next section) are meaningful only if they are monophyletic. The cladistic method aims to discover what the monophyletic groups are; it does this by analyzing the polarity of characters.

A **character** is any attribute of an organism. It might be an anatomical feature, a behavior, or a DNA base pair, just as long as it can be plausibly assumed to have a genetic basis. The relationship of phenotypic characters (anatomy being most discussed) to their genetic basis is often difficult to find out: is tail development a character, or does it consist of several characters, say, coccyx development and free caudal development? Studies of character covariance can help to elucidate this.

Characters vary among organisms; the forms their variations take are called their **states** (or conditions). Thus tail development might have two states: present and absent. Related to whether tail development is a single character or not is the question of how much genetic change (how many mutations) gets it from present to absent or perhaps vice versa. Where there is any information on this, it often turns out that the genetic basis of the difference between two character states is simple.

Character states change during evolution. Those that are unchanged from an ancestral condition are called **plesiomorphic** (crudely speaking, "primitive"); those that have changed are called **apomorphic** (or "derived"). Tail in catarrhines was primitively present, so the absent state is the apomorph (-ous, -ic) one. Plesiomorph and apomorph (or primitive and derived) states are known as the **polarities** of a character.

Many catarrhines (Old World monkeys and extinct forms like *Aegyptopithecus* and *Pliopithecus*) possess a tail. Is this evidence for their relatedness? No, because this possession is inherited from a precatarrhine ancestor (i.e., it is plesiomorphic). A shared primitive character state like this is called **symplesiomorphic** and of course can never be evidence for relatedness. On the other hand the Hominoidea lack a tail. Is this, then, evidence for their relatedness? Yes, because the most plausible, parsimonious explanation is that they share a common ancestor that had lost the tail. A shared derived character state like this is called **synapomorphic**. Two taxa sharing synapomorphies, representing the branching of an evolutionary lineage, are called **sister groups**, and only synapomorphies are acceptable evidence for relatedness (sister-group status).

Humans, gorillas, and chimpanzees, then, share the derived state of taillessness; but humans virtually lack body hair, whereas gorillas and chimpanzees are hairy. Hairiness, however, is a primitive condition, a symplesiomorph state, and so cannot be evidence that gorillas and chimpanzees are more closely related to each other than to humans. Human hairlessness is a uniquely derived state; such a state is called **autapomorphic**, and is of course not evidence for any relationship hypothesis.

Cladistic analysis, therefore, consists of the discovery of synapomorphies. The character states that differentiate the OTUs are listed, and their polarity determined. Several methods have been used: outgroup comparison, ontogeny, ingroup commonality, a priori methods, and paleontology.

Outgroup comparison is by far the most widely used, and productive, method, although it involves some circularity. When the character states differentiating the OTUs of interest (the **ingroup**, whose interrelationships one is interested in analyzing) are listed, the corresponding states in one or more **outgroups** are listed as well. Character states shared by the outgroup with one or more members of the ingroup are most parsimoniously interpreted as symplesiomorphic, and those not found in the outgroup as apomorphic: either autapomorphic if only one OTU of the ingroup exhibits them or synapomorphic if more than one does so. Those sharing most synapomorphies are deemed to be most closely related, those sharing fewer are less so, and so on.

The outgroup should be selected sensibly. When analyzing the Hominoidea, elephants would not be a very informative outgroup; Old World monkeys would seem a good choice, being closest to the Hominoidea, but it is often a good idea to have two or more outgroups in case of unresolved polarity, so perhaps New World monkeys could be used as well. The circularity of outgroup choice comes in here: if one is laying aside all previous assumptions to work out relationships from scratch, how to be sure that the chosen outgroup, too, is not actually part of the ingroup (i.e., that the chosen ingroup is not paraphyletic)?

The ontogeny-based method relies on Von Baer's Law: that primitive features appear first in **ontogeny**, so that those appearing later are the more derived ones. Apart from the fact that organizing polarity by the ontogenetic criterion relies on information that is often quite unavailable, both **heterochronic** processes (developmental dissociations) and the insertion of novelties into ontogeny are well-established occurrences, and thus the ontogenetic method should be used with the greatest circumspection.

Ingroup commonality relies on the assumption that a plesiomorphic condition is more likely to be widespread within the group under study; it is clearly incorrect, and this method is rarely used nowadays.

A priori methods, as Forey et al. (1992) referred to them, rely on assumptions about evolutionary change: that low chromosome numbers are usually derived from high (that fusion is more common than fission); that dominant genes are more likely to be fixed quickly, so recessives are likely to be plesiomorphic; that small brains are more primitive than large;

and so on. These sometimes have validity, but again must be used cautiously.

The paleontology-based method posits that, by definition, primitive character states appeared earlier, in geologic time, than derived ones, so the chances are that earlier fossils have more primitive states than later ones. But the vagaries of fossil discoveries make this rather dicey, and this method is best used as a fallback criterion unless the pattern of character distribution in ingroup fossils is very highly consistent.

So, once polarities have been satisfactorily decided, all that remains is to group the OTUs that share the most synapomorphies, group those with the OTUs with which they share just slightly fewer, and so on. The result is a **cladogram**, which to the uninitiated looks like an evolutionary tree, but is not: it depicts sister-group relationships, although in actuality one of the sister groups at a dichotomous split could have been the ancestral form (this has to do with evolutionary theory, not with taxonomy per se).

But it is not quite as straightforward as this, because of the existence of a confounding factor called **homoplasy** (character incongruence). Homoplasy may be a result of any of three occurrences: convergence, parallelism, or reversal.

Convergence is the development of apparently similar structures from quite different ancestral conditions. **Parallelism** is the development of similar structures independently but from similar bases. **Reversal** is the loss of a derived condition, reverting to the primitive. Homoplasy, it appears, is universal; the cladist's main defense against it is **parsimony**, the principle that evolutionary change is a statistically improbable event anyway, so the phylogenetic tree requiring fewest changes (= least homoplasy) is the one most likely to be correct. The identification of homoplasy by cladistic analysis can, in fact, produce patterns that yield important clues to the adaptive significance of morphological characters (Faith 1989).

Because of homoplasy, most cladistic analyses are too complicated to do by hand. A number of excellent computer packages are available. Perhaps the most widely used, at least for morphological analyses, is Phylogenetic Analysis Using Parsimony (PAUP) (Swofford 1993). Phylogeny Inference Package (PHYLIP) (Felsenstein 1991) is also very popular and is especially ideal for DNA analyses, in which **maximum likelihood** methods are often preferred (Felsenstein 1981; Bishop and Friday 1987) to simple parsimony. MacClade (Maddison and Maddison 1992) is simply brilliant at a different level, being designed to enable one to manipulate phylogenetic trees by hand and test different assumptions. To use any of these packages, all that has to be done is to convert the character states into codes (1, 2, 3 . . .) and list them in columns, the rows being the OTUs. Using such packages one can calculate **tree length** (the number of "steps" [= evolutionary changes] that must have occurred) and the **consistency index** and **retention index** (the proportion of these steps that are not due to homoplasy). One can test different assumptions about evolutionary mechanisms: whether reversal is more or less common than parallelism, for example.

Characters may be **ordered** or **unordered**. In a character "size" there might be three states: small, medium, and large. Is it a single evolutionary change to go from small to medium and another to go from medium to large (i.e., two steps in total [ordered]), or can we get from small to large in one step (unordered)?

Often several shortest (most-parsimonious) trees are reached, or a single shortest but several nearly as short. How to decide between them? There are several ways, which are now routinely built into most cladistic packages:

— Consensus trees: These depict what relationships are in common between all the shortest trees (**strict consensus**) or between more than half of them (**majority-rule consensus**).
— Bootstrap (Felsenstein 1985): This facility randomly samples (usually 100 times or more) the character columns in the data set, replacing those omitted by dummy columns or by repeats of those sampled. The percentage of times each clade reappears is calculated. This gives an estimate of the degree of character support for each clade.

— Permutation tail probability (PTP) (Faith and Cranston 1990): The entire data set is randomized, 100 times or more, to see whether a tree as short as the calculated one could occur by chance alone. This reveals, sometimes quite disconcertingly, whether there actually is any cladistic information in the original data set! A refinement of this test is the topology-dependent PTP (t-PTP), which tests the significance of each clade in turn.

There was presumably only one true phylogeny, so strictly speaking none of these can be regarded as tests of probability. The question they are answering is, with your choice of characters and the way you have handled them, have you discovered the true phylogeny or at least parts of it?

Cladistic analysis takes practice, but can be so revealing if done cannily that it is well worth acquiring the knack.

THE FOSSIL RECORD

How should we classify fossils?

The case for giving primacy to monophyly in this case is not as open-and-shut as for living organisms. They lived long ago, and they are all dead now, so their molecular clocks, were they recoverable, had not ticked as much and the amount of DNA sequence they share with their living relatives is consequently not as much as if they had survived till today. Yet the arguments are still compelling; especially the argument from taxonomy as an information retrieval system, because one can predict with confidence that even a fossil member of a lineage will share significant similarities with the other members.

When drawing an evolutionary tree it is customary, and helpful, to distinguish on any one branch the lineages terminating in modern taxa, as far back as their last common ancestor (the **crown-group**), and the shorter lineages along the stem from that ancestor back to the next major branch (the **stem-group**); the two together constitute the **total group**. So the crown-group of the Catarrhini includes all the taxa right back to the last common ancestor, 20+ million years ago, of the Cercopithecoidea and Hominoidea; the stem-group includes the taxa along the catarrhine branch from this common ancestor back as far as the separation of the Platyrrhini. *Proconsul* is part of the catarrhine crown-group, and *Aegyptopithecus* is part of the stem-group. Within the Cercopithecoidea, the crown-group includes Cercopithecinae and Colobinae, back as far as their last common ancestor, whereas *Prohylobates* and *Victoriapithecus*, which preceded this ancestor but are on the cercopithecoid stem subsequent to its separation from the Hominoidea, are stem-group forms.

As can be imagined, there is some disagreement whether a taxon originally based on modern forms should, when expanded to include fossils, be restricted to the crown-group or extended to include the stem-group as well, and whether it should begin at the basal node itself or along the stem leading to one or other reference node. Most recently Sereno (1999) has reviewed this problem and introduced the concept of the **node-stem triplet** (NST) to clarify taxon content (at a phylogenetic node, the more inclusive taxon equals the two subordinate taxa plus their most recent common ancestor). I will argue, later on, that the plesion concept should be modified to answer this problem.

2

Taxonomic Ranking and Nomenclature

The taxonomic placement of any organism consists of placing it in more and more inclusive, nonoverlapping ranked categories. Not only does each group of organisms have its own name at each level, but each rank itself characteristically has its own designation. The system of ranks was first proposed in the eighteenth century by Carl Linnaeus.

THE LINNAEAN HIERARCHY

Living organisms are initially classified into **kingdoms**. Originally, it was customary to recognize just two kingdoms, Animalia and Plantae, but gradually the Fungi, the Protista, and two or more kingdoms of bacteria (Eubacteria and Archaebacteria, and perhaps others) came to be recognized as separate kingdoms.

Within each kingdom there are a number of **phyla** (singular, **phylum**). In the Animalia, there are some 25 to 30 phyla. The largest, in terms of both species and individuals, is the phylum Arthropoda ("jointed-legs"), which contains insects, crustaceans, centipedes, millipedes, and the spiders and their relatives. Other groups, especially the Onychophora (velvet worms), are sometimes included in this phylum, ac-cording to whether a given specialist feels that the Arthropoda would be paraphyletic without them. Another proposed solution to the problem has been to break up the Arthropoda into three or four different phyla, on the hypothesis that some of their main diagnostic features, the segmented limbs and hard exoskeleton, have evolved in parallel in the different arthropod groups.

This short diversion reminds us that there is some instability in taxonomy even at the very highest level (phylum). There is bound to be: a taxonomic arrangement is a testable hypothesis, and the hypothesis of arthropod monophyly, on which the whole validity of the phylum Arthropoda depends, is under test at this moment.

Some phyla are very small and inconspicuous. Most of us will go through life without ever setting eyes on any members of the phyla Priapulida or Phoronida, which consist of marine burrowing worms.

Phyla are divided into **classes.** The phylum Chordata (broadly speaking, backboned animals) is divided into many classes, of which the most conspicuous are Mammalia and Aves (birds). Formerly, a class Reptilia was recognized, but it is now clear that this is paraphyletic: some groups (crocodiles and dinosaurs) are related to the bird lineage, others (the fossil "mammal-like reptiles") to the mammals, others (the turtles, and the lizard/snake group) form their own clades as distinct as the mammal-like and bird-like clades. Different ways of classifying the land chordates have been proposed to reflect these relationships. Similar problems arise when classifying the various classes of fish. In all classifications, however, the Mammalia survive as a class (but perhaps incorporating the so-called "mammal-like reptiles").

The Mammalia are divided into a number of **orders**, including Primates, Chiroptera (bats), Carnivora (dogs, cats, bears, seals), Proboscidea (elephants): about 20 in all.

The orders are divided into **families.** Whereas the names of the higher categories (phyla, classes, orders) can take any form as long as it sounds like a plausible Latin or Greek word in the plural, families always end in "-idae." Primates families include Lemuridae, Loridae, Cercopithecidae, Hominidae.

Families are divided into **genera** (singular, **genus**). A generic name is a Latin-sounding word in the singular: *Cercopithecus, Lemur, Homo*. Genera contain **species,** and the species name is a **binomial** (has two words). The first word of the binomial is the generic name, the second is the specific name: *Cercopithecus mitis, Cercopithecus diana, Lemur catta, Homo sapiens*.

These seven categories—kingdom, phylum, class, order, family, genus, species—are referred to as the Linnaean hierarchy because they are only a slight modification of the categories first proposed by Carl Linnaeus in the eighteenth century. Every organism is classified according to this system. The classification of the blue monkey is as follows:

Kingdom Animalia
  Phylum Chordata
    Class Mammalia
      Order Primates
        Family Cercopithecidae
          Genus *Cercopithecus*
            Species *Cercopithecus mitis*

Note that generic and species names (the ones that are part of the binomial) are italicized, names of higher categories are not; the specific name begins with a small letter, the others begin with capitals. It is customary to indent each successive level; this is very useful when listing the classification of an entire group, for example the Cercopithecidae, when one would want to list other genera as well as *Cercopithecus,* each with its one or more species.

ADDING MORE CATEGORIES

Of course, these seven categories are hardly going to be sufficient to classify all 5 million (or whatever the number is) species of living organisms. Finer and finer degrees of subdivision are constantly being required.

The first level of division is to make **sub-** categories: subphyla, subclasses, and so on down to subgenera and subspecies. As before, the names of the groups from family down are subject to rules; the names of higher categories and their subcategories are not. **Subfamilies** end in "-inae" instead of the families' "-idae." Subgeneric names, like generic ones, consist of a single word: they tend to be used rarely. The name of a subspecies is a trinomial: the species' binomial plus one extra word. *Cercopithecus mitis* is divided into several subspecies, including *Cercopithecus mitis mitis, Cercopithecus mitis maesi, Cercopithecus mitis stuhlmanni,* and others.

This is as far as subdivision of the family, genus, and species group goes, but in some groups a further subdivision, **infra-**, is used (infraorder and so on). Above each category can be inserted **super-**, and this includes superfamily, which (like family and subfamily) has its own termination, "-oidea." So a superfamily containing the Cercopithecidae is called Cercopithecoidea. Mirorder, grandorder, and so on have also been dreamed up when required.

Finally, entire new ranks can be invented. These are often quite ad hoc: section, cohort, and others, wherever one needs them. But one that seems to have stuck is **tribe**, below subfamily. Tribes have been given the termination "-ini" (Cercopithecini). Subtribes, ending in "-ina," can also be inserted.

These non-Linnaean ranks are not obligatory. The mammal with which all dictionaries begin, the aardvark *(Orycteropus afer)*, has no close relatives. It is placed in an order by itself, Tubulidentata, which has one family, Orycteropidae, and this has one genus, *Orycteropus,* which has one species, *Orycteropus afer.* Those are all the classificatory levels needed, from order on down. But there are so many monkeys and lemurs that within the order Primates we need suborder, infraorder, superfamily, family, subfamily, tribe, subtribe, genus, subgenus (probably), species, and subspecies, and a few invented ones as well.

AGES OF THE DIFFERENT RANKS

Yet, when we have made our phylogenetic analysis, shown who is related to whom, and named our resulting groups, one step remains before we can construct a classification. How do we know when a particular group is an order? A family? A genus, even?

This is a reasonable question. Historically, the phyla were named and constituted, and then as knowledge of the fossil record progressed it was discovered that where we have enough evidence, they are all there at the beginning of the Cambrian. So there is some kind of time linkage, although a billion-plus years of pre-Cambrian time leaves a large window for them to have evolved in. Again, the orders of mammals have mostly been shown to have originated in the latest Cretaceous.

Hennig (1950, 1966) suggested that we should regularize these links and say that if it arose in the early Paleozoic it is a class, if in the late Paleozoic it is a subclass, and so on. This proposal was rejected by the biological community as being totally impracticable. One problem is that the fossil record is not good enough to make it work. Another problem is that those age values that are well known are quite different in different kingdoms and in different phyla of animals. Thus the plant genus *Eucalyptus* is known from the Cretaceous: it is as old as whole orders of mammals! The insect genus *Drosophila* is even older. For consistency, the insects and the plants would have to be split much more finely, and the mammals lumped much more, than at present; this might ultimately improve communication between botanists, entomologists, and mammalogists, although the classification of the entire living world would be in chaos in the meantime.

Hennig did in fact realize this and discussed the problem at some length (1966:154–193). He suggested that we should actually link particular ranks to particular geological time periods, but adjust them differently for different zoological groups. So in insects any groups that arose in the Carboniferous or Permian must be designated as orders, but in mammals, rather than reducing the entire class of Mammalia (including "mammal-like reptiles") to a single order, we could set the period of Late Cretaceous to Oligocene as that within which orders arose.

Is this the argument for stability, which I refused to contemplate in the first chapter? In a way it is, but in a way it is much more. Replacing paraphyletic groups is a gradual process, proceeding as understanding increases; replacing the whole taxonomy of the living world at one go (because the general outlines, if not the full details, of the age links are known) is a different matter. It may well come, but if so it must be a gradual, consensual change.

Goodman et al. (1998) pointed out that some idea of the ages of groups without a fossil record can be gained with molecular clock theory, so that in a restricted group like the Primates the ages of different groups can be compared. Using local (not global!) clocks, they have calculated approximate times of separation of various Primate clades and showed what a time-ranked classification of Primates would look like. Some groups turned out to be (in conventional terms) oversplit, others overlumped.

The times and geological periods of origin for different ranks (total groups [i.e., time of separation of lineages, not merely crown-groups]) proposed by Goodman et al. (1998) are given in Table 1.

As one of the coauthors of the Goodman et al. (1998) scheme, I can hardly disagree with the concept. By 1998 we had the molecular clock by which

Table 1

Times and Geological Periods of Origin for Different Ranks of Primates

Ranks	ma	Geological periods
Semiorders	63	Early Paleocene
Suborders	58–50	Late Paleocene to early Eocene
Infraorders	45–40	Middle Eocene
Superfamilies	39–29	Middle Eocene to middle Oligocene
Families	28–25	Middle to late Oligocene
Subfamilies	23–22	Early Miocene
Tribes	20–14	Early to middle Miocene
Subtribes	14–10	Middle to late Miocene
Genera	11–7	Late Miocene
Subgenera	6–4	Late Miocene to early Pliocene

to calculate lineage splitting times, whereas Hennig in 1966 had only the fossil record. The molecular clock must be tested and retested, inside the Primates and out, and adjustments will be made continually. As always, we should avoid being too radical on insufficient evidence. Yet the Primates are becoming better and better known in a molecular sense, and the latest phylogenies (Shoshani et al. 1996; Schneider et al. 1997; Goodman et al. 1998) have very little to disagree about. I think the time has come at last when we should resurrect the time/rank correlation seriously, and see where it gets us.

The specific time/rank correlations of Goodman et al. (1998, tables 5, 6) are proposals only, and because some of them, as listed in Table 1, are very uneven, we might reconsider them here. The calculations of Takahata and Satta (1997) differed somewhat from those of Goodman et al.: thus, the catarrhine/platyrrhine split was put at 57.5 million years ago, which in Goodman et al.'s ranking is high in the subordinal level, on the borderline of semiorder. We should concern ourselves just with the obligatory ranks, which as far as Primates are concerned means order, family, and genus. In line with the "principle of least violence," enunciated several times already, we should try to preserve these with the content with which they are widely recognized (in so far as they are). The nonobligatory ranks (the super-, sub-, and infra- categories, and tribes and subtribes) are

not considered here because their function is more to split up extensively polytypic taxa of the obligatory ranks; it is the obligatory ranks to which we should consider giving some sort of objective reality.

The approximate age of the order Primates is the Cretaceous/Paleocene boundary; no conflict there. The families Lemuridae, Cheirogaleidae, Indridae, and Hylobatidae have achieved wide currency, whereas loriform and platyrrhine families are in a state of flux. According to Goodman et al. (1998, table 5), the subsidiary groups of the Lemuroidea (Cheirogaleidae, Lemuridae, Indridae) are about equally old and have time depths of about 28 million years; the Hylobatidae, however, have a separate existence only from 18 million years ago (ma), and the Cercopithecoidea and Hominoidea separated only 25 ma (i.e., more recently than the three Malagasy families) (Takahata and Satta [1997] put this split at 31 ma).

The genus *Callithrix* began to split up 5 ma, so it must be older than that; *Callicebus* began to split up 6 ma. On the other hand *Chiropotes* and *Cacajao* separated only 6 ma; *Cercocebus* and *Mandrillus* separated only 4 ma, as did *Papio* and *Theropithecus;* and *Homo* and *Pan* separated 6 ma (Goodman et al. 1998). So, in principle, if we wish to preserve the unity of the first two genera, we have to unite those of the second group of genera into three pairs.

Outside the Primates, we can find the following time relations:

1. In the Carnivora, according to Martin (1989), the earliest member of the Felidae, *Dinaelurus,* is from the early or middle Miocene, and so is the earliest known member of the Mustelidae, *Miomustela;* the earliest appearance records for Viverridae, Herpestidae, Hyaenidae, and Ursidae are late Oligocene; that for Canidae is early Oligocene (*Hesperocyon*). Molecular clock data (Wayne et al. 1989) put the four-way split leading to Felidae, Hyaenidae, Viverridae, and Herpestidae nearer to 40 ma, but the well-known fossil record of the Felidae shows that other, now extinct, families (Nimravidae, Smilodontidae) branched off the felid stem, so the four families are not themselves

sister taxa. The molecular separation of the Ursidae is some 18–25 ma, and the separation of Otariidae and Phocidae is only 15 ma (Wayne et al. 1989).

2. In the Artiodactyla (= Cetartiodactyla), most families with a good fossil record (Bovidae, Giraffidae, Antilocapridae, Cervidae) are younger than those in the Carnivora and go back to the early or middle Miocene (Scott and Janis 1987; Gentry 1994). First appearance dates for the Cervidae are MN3, so about 20 ma *(Procervulus, Lagomeryx),* and for the Bovidae, MN4, about 18 ma *(Eotragus);* in the bovid case diversification seems already to have begun. The two families can be supposed to have separated from their sister families about 20 ma or perhaps slightly more.

These may be the best-studied orders, in this context, outside the Primates. The families in these two orders go back as fossils to the late Oligocene or early Miocene, and the molecular data are not inconsistent with this conclusion. In fact in the case of the Ursidae the two data sets are in very good agreement. Under these circumstances Hylobatidae is acceptable as a family, although on the young side.

The fossil records of well-studied mammals, mainly African, suggest that between about 4 and 7 million years is the normal range of time depths for living genera:

1. In the Ursidae, both *Ursus* (brown, polar, sloth, and sun bears) and *Tremarctos* (spectacled bear) are present in the middle/late Pliocene (Martin 1989); the initial diversification in *Ursus* began about 5 ma (Wayne et al. 1989).

2. In the Canidae, *Canis* (wolf, dog, jackal, coyote) is already present in the late Miocene, but other modern genera such as *Vulpes* (fox) are Villafranchian in first appearance (Martin 1989). The common ancestor of *Canis* is put by molecular data at 6 ma; that of *Vulpes,* at 5.5 ma (Wayne et al. 1989).

3. In the Elephantidae, already well-differentiated representatives of *Loxodonta,* the African elephant, and its sister clade of *Elephas* plus *Mammuthus* (Asian elephant and mammoth) are known from the early Pliocene. *Loxodonta adaurora* is known from Kanapoi, Lothagam-3, and Kubi Algi, all about 4 ma. *Elephas ekorensis* is first recorded from Kubi Algi and (slightly later) Ekora, and *Mammuthus subplanifrons* from Langebaanweg, about 5 ma (Coppens et al. 1978); the latter is poorly differentiated and difficult to distinguish from *Elephas,* suggesting that the split had occurred not long before, or was still occurring.

4. In the Rhinocerotidae, the stem-species of white rhino, *Ceratotherium praecox,* difficult to distinguish from the sister genus *Diceros,* is known from Kanapoi, Ekora, Lothagam-1, Langebaanweg, Omo-Mursi (Yellow Sands), Aterir, Chemeron, and elsewhere (Hooijer 1972), all latest Miocene or early/middle Pliocene, suggesting that the two African rhinoceros genera separated around 6 ma.

5. In the Hippopotamidae, the sister genera *Hippopotamus* (common hippo) and *Hexaprotodon* (pygmy hippo) separated in the late Miocene. The first appearance data (FAD) are (presumed early) Pliocene of Kaiso, Uganda *(Hippopotamus kaisensis),* and latest Miocene, about 5.5 ma, of Lothagam-1 for *Hexaprotodon harvardi* (Coryndon 1977).

6. Gentry (1978) and Gentry and Gentry (1978) gave the following data for genera in the Bovidae: *Tragelaphus* (bushbuck, kudu, eland) first occurred at Kaiso and had already begun differentiating into species-groups. *Kobus* appeared first at Langebaanweg, but was again already partly diversified. The genera of the Alcelaphini (*Alcelaphus, Damaliscus, Beatragus, Connochaetes,* the hartebeest/wildebeest group) were well separated by 2 ma. *Antidorcas* (springbok) was first present at Laetoli, 3.75 ma.

So carnivore and ungulate genera tend to date from about 6 to 4 ma. Old World monkey genera of only 4 million years seem acceptable, if only marginally.

We should note that subordinate categories in at least two New World monkey genera go deeper than the split between some catarrhine genera (*Mandrillus* and *Cercocebus; Papio* and *Theropithecus)*, and one might possibly consider revising their ranks upward; I will not do it here, but rather leave to it future revisers after more rigorous analysis.

Finally, note that Easteal and Herbert (1997), arguing for a global molecular clock, timed some of the lineage separations much later than Goodman et al. (1998). In particular, they placed the catarrhine/platyrrhine split at about 30 ma, too late for *Aegyptopithecus* to be a catarrhine, and the human-chimpanzee split at 3.6 to 4.0 ma, too late for *Ardipithecus* and some of the australopithecines to be ancestral to *Homo*. Something is obviously wrong! Nonetheless, their arguments, if accepted, would have important consequences for taxonomy, among other things, and must be carefully considered by other molecular workers.

SHOULD TAXA BE RANKED AT ALL?

There have been several proposals to modify the system of taxonomic ranking. Hennig (1969) suggested a system of numbering: the first division would be 1, 2, 3, and so on; the next would be 1.1, 1.2, 2.1, 2.2, and so on; then 1.1.1, 1.1.2, and so on until the finest possible division is achieved. The names do not disappear; only their ranks do. A system like this can become very irritating, not to say confusing: as the number of subdivisions increases, so does the potential for 3.3.2.2.2.2.4 to become just 3.3.2.2.2.4, and slight human carelessness can destroy entire meanings.

Nelson (1972) proposed sequencing, with or without ranks. Each taxon in a classification is the sister group of the one succeeding it. Some form of sequencing and indenting, a closely allied concept, is in fact quite widely adopted.

A problem with monotypic taxa has long been recognized. Gregg's Paradox is that as we go down the ranks of, for example, the aye-aye, we meet it first as an infraorder, Chiromyiformes (or a superfamily, Daubentonioidea, according to different viewpoints); then as a family, Daubentoniidae; then there it is again as a genus, *Daubentonia*. And this is only

the obligatory categories. This has its illogical side, especially in mathematical applications. Thus, how many Primate infraorders has Madagascar? Why, two: Lemuriformes and Chiromyiformes; the second of these is the aye-aye. How many families? Five: Cheirogaleidae, Lemuridae, Megaladapidae, Indridae, and Daubentoniidae; and look, one of them is the aye-aye again. And so it goes on. Farris (1976) suggested simply omitting the intermediate ranks and just listing *Daubentonia* as a genus alongside Lemuriformes; in the case of fossils, their rank would be determined by the time between their origination and extinction. This is a rather unexpected proposal coming from a cladist, flouting as it does the Hennigian principle that sister groups must have the same taxonomic rank, and it has not been widely adopted.

A problem with traditional ranking, numbering, sequencing, and indenting alike is that, as new discoveries are made (usually new fossils are found, although there can also be radical new interpretations of known living or fossil forms), there is a risk that the whole scheme has to be ratcheted up one rank, or number, or the sequence altered, or the indent pushed over, to accommodate the new sister-group relationship. There are, again, a number of proposals to deal with this.

The most radical proposal is to drop ranking altogether. Given Gregg's Paradox, the different phylogenetic implications of a rank, such as order, in different phyla (referred to in the previous section), and the unavoidable proliferation of categories as knowledge and understanding improves, there should be no orders, suborders, and so on—just names, sequenced and indented in a classification as in Nelson's (1972) scheme (Queiroz and Gauthier 1992, 1994). In my view, such a scheme may suffice—just—as long as a full classification is being presented, but isolated reference to any suprageneric group would no longer be meaningful; and within a restricted group the categorical ranks do, in fact, carry a great deal of meaning.

In 1989 I suggested that limits be placed on the requirement to recognize each sister-group relationship in a classification, at least as far as fossils are concerned (Groves 1989). Thus, if the Plesiadapiformes are regarded as Primates, and if we rank them dichotomously, thus:

Suborder Plesiadapiformes
Suborder Euprimates
   Infraorder Strepsirrhini
   Infraorder Haplorrhini

we have already run out of ranks and will have to invent new ones to fit the Tarsiiformes and Simiiformes within the Haplorrhini, and the Platyrrhini and Catarrhini among the Simiiformes. But it now seems increasingly unlikely that the Plesiadapiformes really are Primates; if they are removed, the Strepsirrhini and Haplorrhini revert to suborders and we can now fit the Tarsiiformes and Simiiformes in as infraorders, although the Platyrrhini and Catarrhini are still left in limbo. It is much more flexible, surely, to rank Plesiadapiformes, Strepsirrhini, and Haplorrhini as three suborders so that no undue disturbance is created if the Plesiadapiformes get expelled from the Primates. In such a scheme, monophyly is preserved even though not every single evolutionary dichotomy is depicted.

A further step would be to leave a few (the more unstable) levels unranked while ranking most of them. Plesiadapiformes and Euprimates would appear as sisters below order Primates, but not be assigned ranks (Strepsirrhini and Haplorrhini would be suborders under Euprimates), so that if the Plesiadapiformes disappear the now-unnecessary Euprimates can be discarded, but no wholesale reranking is needed.

One proposal for fossils that has great value is the concept of **plesion** (Patterson and Rosen 1977). Many fossil taxa have very limited diversity (a genus with a single species, often), yet, by cladistic rules, they would have to be classified at the same rank as their diverse sister group (unless Farris's [1976] solution is adopted). So the Oligocene *Aegyptopithecus,* with one species, with high probability the cladistic sister group (read: phylogenetic ancestor) of the living Catarrhini, would have to be given equal rank to a group containing Cercopithecoidea plus Hominoidea:

Section Catarrhini
   Subsection Aegyptopitheci
   Subsection Eucatarrhini
      Superfamily Cercopithecoidea
      Superfamily Hominoidea

which seems overblown, and something similar would have to be done with every such case, requiring a further proliferation of ranks. Instead, in the proposal of Patterson and Rosen (1977) we would have:

Section Catarrhini
   Plesion *Aegyptopithecus*
   Superfamily Cercopithecoidea
   Superfamily Hominoidea

Strictly speaking, the term plesion can be used for a taxon of any rank: genus, family, order, and so on. Some authors even use it to designate every group known only as fossils. Used like this, the term loses its value; if it is particularly desired to indicate that a taxon consists entirely of fossils, the convention is to put a dagger in front of it. Patterson and Rosen (1977) designated the groups without known autapomorphies (putative ancestors, as argued earlier) confusingly as incertae sedis. In 1989 I proposed to restrict the term plesion to this latter case (Groves 1989), and I will continue to do this here.

RULES OF NOMENCLATURE

Questions of nomenclature in animals are determined by the International Commission on Zoological Nomenclature, who every so often issues a new edition of the *International Code of Zoological Nomenclature* (we are now, as of January 2000, up to the fourth). The *Code* gives the rules of naming, which are aimed at increasing the facility with which we communicate information about taxa. The *Code* is very clearly written, in English and French, and is aimed at stability of nomenclature. In those (relatively few) cases in which its strict application might actually threaten stability, anyone is at liberty to apply to the commission for an ad hoc ruling; the commission then votes on the application, and publishes (in the *Bulletin of Zoological Nomenclature*) an **Opinion**, which becomes binding, like a precedent in law, although in this case there are no sanctions (except the scorn of one's colleagues).

The rules aim to preserve the freedom of taxonomic thought: they don't tell you how to classify something, but what to call it after you have classified it.

For purposes of the *Code,* names are considered in three groups: the species-group (species and subspecies), the genus-group (genera and subgenera), and the family-group (superfamilies, families, subfamilies, tribes, subtribes, and any other categories that are inserted above genus). I will first consider the rules for the species- and genus-groups.

Priority and Availability

The guiding principle of nomenclature is **priority**. All too often a taxon (let us say, in this instance, a species) has been given several names in the past. In the eighteenth and early nineteenth centuries, quite often the same species became known to two or more taxonomists unbeknownst to each other, and so was given two different names. It even happened that two or more workers examined the selfsame specimen and independently described it as a new species. Then there is the much more common occurrence that someone thought he (almost invariably he, until quite recently) was describing a new species, but subsequent study has shown that his specimen(s) actually fitted into the variation of a previously described species. Someone performing a taxonomic **revision** (an exercise of alpha taxonomy) has to try to sort out what names are synonyms of each other and what names are not: this of course is classification and has a subjective element.

Names are (or pretend to be) in Latin, including latinized Greek, so that a generic name has a gender and the specific/subspecific names, if they are adjectives, have to agree with that gender. That is where a classical education comes in handy. If a generic name is not "classical," the describer may award it a gender of choice.

Different names given to what a reviser concludes is the same taxon are called **synonyms**. Names applied to the species on the basis of the same specimen are **objective** synonyms; those awarded on the evidence of different specimens are **subjective** synonyms. The earliest name given to a taxon is the **senior** synonym; the others are **junior** synonyms.

To be usable, a name has, of course, to have been published and published properly. Publication simply means "in numerous identical copies . . . for purposes of permanent record." So proofs (of typeset texts) are out, and so are theses and dissertations; the compilers of the fourth edition of the *Code* are currently discussing whether dissemination on the internet constitutes publication. A name that was published in a work whose author did not use the binomial system of nomenclature correctly cannot be used. A name must not, after 1950, have been published anonymously; or, after 1960, have been published "conditionally" ("if this turns out to be a valid taxon, I would name it as follows . . .": you must believe in what you are doing, which seems a reasonable stipulation). A name must have been published in conjunction with a description (features purporting to differentiate the taxon from its relatives) or a bibliographic reference to a description; if it was not, it is a **nomen nudum** (a "naked name") and cannot be used. Finally, a name must be unique: no two species (or subspecies) in a genus may bear the same name, and no two genera (or subgenera) in the entire animal kingdom may bear the same name; two names that are the same in these contexts are called **homonyms**, and the junior one must be replaced by the next synonym or, if there are no synonyms, by a new name.

A name that is usable is said to be **available**. Nomina nuda, homonyms, and other names that are improperly published are **unavailable**. Other synonyms are available; they just may not have priority, unless someone revises the classification and decides that one of the junior synonyms does actually designate a valid taxon after all, in which case the name is revived.

And that's it. Really extremely simple.

Author and Date

The author of a name is the person who first published it in an available form. The date is the year in which the publication appeared. In the case of a name of a genus-group, there is nothing more to be said.

In the case of a name of a species-group, it is important to note that the author and date refer to the specific/subspecific name itself, in whatever genus it was published. If the name was at a later date transferred to a different genus, this does not make it a new name, merely a **new combination**. In such a case, the author and date of the specific/subspecific name are placed in parentheses, indicating that the species/subspecies is not now placed in the genus in which the original describer placed it.

Examples of a Synonymy

Here are examples of a synonymy. I have not given the full bibliographic references for the names, and I will not in the synonymies later in the book; these can be looked up in easily available sources such as the series *Catalogue of Primates in the British Museum (Natural History) and elsewhere in the British Isles* (vols. 1–3 [1976–1985] by P. H. Napier, vols. 4–5 [1987, 1990] by P. D. Jenkins). The first example is for *Colobus guereza guereza*, a West Ethiopian black-and-white colobus; it illustrates several of the points I have mentioned.

Colobus guereza guereza Rüppell, 1835

1816 *Lemur abyssinicus* Oken. Abyssinia. Unavailable.
1835 *Colobus guereza* Rüppell. Gojjam Province, Abyssinia (Ethiopia).
1870 *Colobus rüppellii* Gray. Replacement for *guereza*.
1901 *Colobus abyssinicus poliurus* Thomas. Omo River, S. Ethiopia.
1913 *Colobus (Guereza) poliurus managaschae* Matschie. Managasha Forest, west of Addis Ababa, Ethiopia.

Let us go through these names one by one. Oken was completely uncaring about how he used scientific names. Because he did not use the binomial system of nomenclature with any great consistency (to put it mildly), the International Commission on Zoological Nomenclature has declared them un-

available, and the name *abyssinicus* can no longer be used, even though *Colobus abyssinicus* is used in some of the old literature for what we now call *C. guereza*. But *abyssinicus* is the senior synonym of the taxon.

Rüppell was an early explorer of Northeast Africa. He used the binomial system correctly, his descriptions are good, and his type localities are given with exactitude. His types are in Frankfurt. His name *guereza* is a junior synonym, but because of the unavailability of *abyssinicus* it is the first available name for the subspecies.

Gray thought that you could "up" a species to generic rank and then give a new species name. Even in his day, this was rare. His name *rüppelli* (which, in any case, should have been spelled *rueppelli,* because there are no diacritics in Latin, nor in zoological nomenclature) is an objective synonym of *guereza.*

Thomas thought that his *poliurus,* from the Omo River, really was subspecifically different from *guereza* from the Gojjam Plains. Matschie thought that, indeed, *poliurus* was so different that it ought to be a full species and he described *managaschae* as a subspecies of it. Both names have type specimens, which are in London (BM[NH]) and Berlin (ZMB), respectively, and can be examined by colobologists. Various folk have examined them in the past and concluded that they actually fall into the range of variation of *C. g. guereza,* but it is possible that someone someday may demur and decide that there is after all a valid subspecies *Colobus guereza poliurus* (for example). So the two names are currently junior synonyms of *C. g. guereza,* but subjective synonyms.

Here is another synonymy, for *Papio ursinus ursinus,* the chacma baboon of southernmost Africa. In this case, I have cited only the objective synonyms: a full synonymy, giving subjective synonyms as well, is given in the taxonomic section later in the book.

Papio ursinus ursinus (Kerr, 1792)
Chacma Baboon

1787 *Simia porcaria* Boddaert. Africa. Not of Brünnich, 1782 (= *Macaca nemestrina*).
1792 *Simia (Cercopithecus) hamadryas ursinus* Kerr. Cape of Good Hope.
1812 *Papio comatus* E. Geoffroy. South Africa: Cape of Good Hope.

In this shortened synonymy (which excludes a couple of names that may or may not be junior subjective synonyms) we see three names that all refer to the same population (from the Cape of Good Hope) and so are objective synonyms. As with the colobus example just given, the senior synonym, *porcaria*, cannot be used; in this case, because it is a homonym of the name *Simia porcaria* given by Brünnich five years earlier to the Sunda pig-tailed macaque (even though Brünnich's name is itself an objective synonym of *Simia nemestrina* [now *Macaca nemestrina*] of Linnaeus, 1766). So we go to the next synonym, *ursinus*, which is available, even though it was described as a subspecies.

Because Kerr described *ursinus* as a taxon of *Simia*, whereas today we put it in a different genus, *Papio*, the author and date are placed in parentheses when we cite the current combination of the name.

Names of the Family-Group

The name of a family is derived from that of one of its included genera. Take the name of the chosen genus, pretend it is Latin (many of them actually are), find what the stem is (or deduce what it should be), and add -idae to it. Thus *Homo*, stem *Homin-*: so Hominidae; *Cercopithecus*, stem *Cercopithec-*: so Cercopithecidae.

For a subfamily, make the suffix -inae; for a tribe, -ini; and so on.

When first published, a family-group name must have been based on a name that was at that time used for a valid genus. It does not matter whether the genus is today regarded as valid, or if that generic name is now a junior synonym.

Family-group names have their own priority; they do not take the priority of the genus-group names on which they are based.

The author and date of a family-group name is whoever named the family or subfamily or superfamily and when. This is consistent with the principle of coordination (see the following section), despite the slight change of termination that occurs when a family name is altered to become that of a subfamily (for example).

Coordination

Within, but not between, the three levels (family, genus, species) there is coordination. What is good for the species is good for the subspecies, and vice versa, but has no effect on the genus. A subspecies can be upgraded to a species, or a species downgraded to a subspecies, and it keeps the same name, but in binomial instead of trinomial combination (or vice versa). A genus can be downgraded to a subgenus, or a subgenus upgraded to a genus, and it keeps the same name. Within the family-group, the suffix must change, that's all.

But the name of one of the subspecies must be the same, in trinomial form, as that of the species. If *Cercopithecus mitis* is divided into subspecies, one of them must be called *Cercopithecus mitis mitis*. Which one? It must be the one found at the species's type locality (see the following section). It is called the **nominotypical** subspecies.

If a genus is divided into subgenera, one of them must have the same name as the genus: it must be the nominotypical subgenus, the one containing the type species.

If a family is divided into subfamilies, again there must be a nominotypical subfamily. Finally, if a number of families are united into a superfamily, this must have the same stem as one of the families, but with the -oidea suffix designation for superfamilies.

Types

When a species (or subspecies) is described, a **type specimen** should be designated. This is not necessarily an especially typical specimen, it is just the specimen on which the name is hung. And its locality of origin should be known: the **type locality**.

When a taxonomist revises a group, he/she may find that where there was thought to be one species/subspecies A, there are actually two (A and B). What names to apply? Look at the type specimens of names hitherto deemed to be synonyms of A. Perhaps some are clearly members of species A, others of species B. The names are thus automatically allocated to A and to B, and the senior available synonym, in each case, is the one that should be used. Perhaps, however, all the type specimens belong to species A. Species B, then, needs a new name.

Sometimes the types cannot be examined: they are in some far-flung museum, or there may not be any (they have been lost). Then try to allocate the names by type localities. Second best but, except in the case of two **sympatric** species, a good second best.

A specimen that was designated as type at the time of description, by the describer, is called a **holo-type**. If the describer designated more than one specimen to be joint types, these are called **syntypes**. If one of the syntypes is chosen by a later reviser to be the one and only type, it is called a **lectotype**. Finally, if no type exists, but it is felt desirable that there be one to fix the nomenclature, some suitable specimen may be chosen as **neotype**.

Sometimes, too, a describer may chose subsidiary specimens to serve as "backups" to the holotype. These are called **paratypes**.

3

The Species-Group

Even in the popular imagination there is something special about the species. Species are "kinds of animals" (and other organisms). They are what the public responds to: they are lions and tigers, they are chimpanzees and gorillas; for the more sophisticated they are black rhinos and white rhinos. For most lay persons with a smattering of biology, species are groups that do not interbreed with each other (which is presumably why mules are regarded, in the Bible and elsewhere, with some suspicion).

Before the revolution in evolutionary biology of the late 1930s, working taxonomists had their own criteria for recognizing species. Thus Miller (1934: 126) classified Southeast Asian langurs of the genus *Presbytis* into numerous species, rather than subspecies, although many of their differences were very slight, because "among the 72 specimens . . . I have found no one that is intermediate or of doubtful status." Even well after the promulgation of the biological species concept (defined in the next section), the same sort of operational definitions were still being given. Deraniyagala (1955:39) defined a species as "a community of similar individuals in which each is distinguishable from another of any other community."

There is currently a lively debate in taxonomic circles about what a species actually is or ought to be: what is special about these "kinds of organisms"?

Brief surveys of species concepts have been given recently by Groves (1996a), Crisp and Chandler (1996), and Lee (1997). Essentially, there is a split between theoretical concepts (what a species is in essence) and operational concepts (how you can recognize one when you meet one).

THEORETICAL SPECIES CONCEPTS

The new evolutionary synthesis, the long-overdue meshing of genetics and natural history, took shape in the late 1930s and was finally crystallized by Huxley (1942). Systematics could not escape being caught up in the synthesis. Populations were now seen as basal to evolution; species, long recognized as basal to taxonomy (being "kinds" of organisms), came to represent circumscribed populations. Hence if population genetics is the key to evolution, as per the new synthesis, the species becomes the largest unit within which population genetics is a meaningful field of concern (see contributions to Huxley [1940]).

The Biological Species Concept

It was in this context that Ernst Mayr (1940:256, elaborated in 1942 and further in 1963) proposed the biological species concept. He defined species as "groups of actually or potentially interbreeding natural populations which are reproductively isolated from other such groups" (Mayr 1963:19). The biological species concept (BSC) has served as the dominant paradigm ever since, and anyone wishing to propose an alternative way of looking at species has had first to demonstrate why the BSC is incomplete or in some other way inadequate.

Species, in the BSC, are thus defined by having reproductive gaps between them, and phenotypic differences between them take second place. Indeed there is a category of species, called sibling species, that differ so little that only very detailed examination, for example of karyotypes, can find any differences at all, yet they are still reproductively isolated.

Despite its predominance, the BSC is widely misunderstood, even (one may say, especially) by its supposed adherents, who seem often to have a rather superficial understanding of what exactly this idea is to which they are paying lip service. Such misapprehensions were flourishing less than a quarter of a century after the initial formulation of the BSC, such that Mayr felt it necessary to devote in effect two chapters (5 and 6) of his 1963 compendium to correcting them. "In spite of everything that has been written in the past 25 years," he wrote on p. 90, "there are still some authors who seem to think that reproductive isolation and sterility are synonymous terms. Nothing could be further from the truth." On p. 92 one finds a table (table 5–1) called "Classification of isolating mechanisms." Reproductive isolating mechanisms (RIMs) may, in Mayr's view, be either pre- or postmating: the latter include failure of hybridization, hybrid inviability, and hybrid sterility; but, as he noted, these are uneconomical, because the organisms have in the meantime wasted a lot of time and effort in mating and perhaps even bringing a hybrid offspring to term, whereas the premating RIMs prevent them from coming together in the first place. These are ecological separation, ethological incompatibility, mechanical barriers to fertilization. Later on, Mayr even acknowledged that there may be cases where premating RIMs break down, partially or completely.

The consequences for Primate taxonomy are clear. *Macaca nemestrina* and *M. fascicularis* share a wide geographic range, from the southern half of the Indochina Peninsula through the Malay Peninsula, Sumatra, Bangka, and Borneo, yet when placed together in captivity they hybridize, and their hybrids are fully fertile. They even occasionally hybridize in the wild when habitat alteration by human interference breaks down their ecological RIMs. They qualify in every respect as biological species in Mayrian terms. Indeed, if taxonomy is to serve as an information retrieval system, it would be nonsense to call them the same species and unite them under a single specific name.

The museum taxonomist in the BSC era seems to have argued in the following way: if two samples can be distinguished absolutely, they must be reproductively isolated; therefore they are distinct species. There is very little to distinguish this from the way people like Miller (quoted early in this chapter) actually went about things; it can be claimed that the BSC had made very little difference in how practicing taxonomists actually practiced. In fact, however, a new factor had entered into consideration: geographic distribution; sympatry is ipso facto evidence for reproductive isolation, but allopatry is not.

The trend in mammalian taxonomy was set in the 1950s by Ellerman and Morrison-Scott (1951) and to some extent has persisted until the present day: vicarious taxa are conspecific, unless there is good evidence that they are not. Throughout their work phrases recur such as "There is no evidence that these two taxa occur together [i.e., are sympatric], so we combine them into one species." The example of this influential volume ushered in an era of taxonomic "lumping."

The Evolutionary Species Concept

Simpson, as a paleontologist, was interested in what species mean in evolutionary terms and devised the evolutionary species, "a lineage (an ancestral-descendant sequence of populations) evolving separately from others and with its own unitary evolutionary role and tendencies." This was essentially the

BSC given a time dimension, with one important but only barely explicit addition (Simpson 1961:153–154): it "omits the criterion of interbreeding. Interbreeding promotes the unity of role involved and is also evidence that the criterion of unity is met. It is not, however, the only way in which unity is maintained or the only evidence that it exists." Simpson went on to talk about uniparental populations, but the way in which he phrased himself suggests very strongly that this was not all that was on his mind.

Simpson's concept is not so much an alternative to the BSC as an expansion of it or, one might say, an explanation of it. It is even less practicable than the BSC, but it suggests why the species is important and stands at the base of evolutionary as well as taxonomic theory.

Species as Fuzzy Sets Concept

The idea that set theory can be applied to the species concept is not new. Sets are simply a collection of objects, ideas, etc. They have essences, the differentia of which are peculiar to a given set; properties, shared attributes not part of the essence; and accidents, attributes present only in some members. In the view of Van Valen (1988:56), species can be viewed as sets; moreover they are fuzzy sets: "sets in which for each potential member there is a number which expresses the degree to which it is a member of the set." The idea has recently been revived, in the context of species in fossil *Homo,* by Willermet and Hill (1997).

Simpson (1961) was opposed to this idea as inadequate to the description of natural entities or taxa, but it is interesting that this is even today the traditional way of organizing alpha taxonomic descriptions: diagnosis (the taxon's differentia), description (the taxon's properties), and finally the heading called "notes" or some such, giving the accidents.

The Species-as-Individuals Concept

A question in logical philosophy is whether species, like other levels in the taxonomic hierarchy, are classes or, uniquely, are individuals. Ghiselin (1966) registered an objection to numerical taxonomy on the grounds that it treated species as collections of traits rather than as populations; collections of traits characterize classes, whereas species have a bounded existence whatever traits happen to characterize them. Later Ghiselin (1974:536) greatly elaborated this idea: species are individuals within classes, an individual in logic being a "thing," with its own internal integration, whereas a class is a universal. Species are "the most extensive units in the natural economy such that reproductive competition occurs among their parts," whereas between species competition is on a different level. Species are thus analogous to individual nations, firms, persons, and so on, whereas classes have no such reality.

Such a concept, as argued by Ghiselin, would circumvent the objections to the BSC and related concepts. Species may indeed change over time but, as individuals, they retain their real existence. Species may be only "potentially interbreeding," rather than actually, at any given time, but the tightness of their integration does not compromise their individuality.

The part about "potentially interbreeding" shows up the flaw in the argument: Ghiselin is, in reality, talking about populations, not about species. Two populations of a species may well share much the same assortment of genes, but if they are isolated (say, on different islands) they have nothing to do with one another, no integration, and the species has no individuality in the logical sense. The definition of "population" is fully as problematic as is that of "species" (Nelson and Platnick 1981), but that is a different matter.

The Ecological Species Concept

Van Valen (1976) considered a species as a population that occupies a distinct ecological niche, whence its reproductive and general biological cohesion over time. This concept was never widely adopted, and indeed it would be very hard to operationalize, although the ecological aspect of a species was favorably commented upon by Ridley (1989) as an essential component of its cohesion. We might say that ecological niche is an attribute of a species, but it can never be taken as a defining attribute.

The Recognition Species Concept

Paterson (1986:63) argued that reproductive isolation is a by-product of the divergence of two populations, not intrinsic to it, and that we must return to what is basic to species status: its reproductive inclusiveness. So a species is "that most inclusive population of biparental organisms which share a common fertilization system." The exact meaning of this last phrase is a bit obscure and can be misunderstood: is it not just the BSC turned on its head? The BSC says that members of a species do not interbreed with members of other species; isn't the recognition concept just saying that, instead, they interbreed with each other? More than this, however: they have specific mate recognition systems.

The way many primatologists, in particular, have interpreted the meaning of specific mate recognition system (SMRS for short) has been extraordinarily productive and insightful. In Part Two of this book I describe something of how such workers as Bearder, Masters, Zimmermann, and Honess have used the SMRS concept to disentangle the taxonomy of the bushbabies.

The recognition concept is, of course, essentially a tool for field-workers. It is of little use, except in imagination, to a museum worker, let alone to a molecular geneticist.

The Cohesion Species Concept

Templeton (1989:13) was one of many authors to worry that "species" was meaning different things in sexual and asexual organisms. The cohesion concept was his attempt to rectify this: a species is "the most inclusive population of individuals having the potential for phenotypic cohesion through intrinsic cohesion mechanisms," such mechanisms being either genetic or demographic. Genetic mechanisms include interbreeding; demographic mechanisms are either

evolutionary or ecological, and these are the ones that apply to asexual as well as sexual organisms.

OPERATIONAL SPECIES CONCEPTS

Morphological definitions of species ("the morphospecies") lapsed after the triumph of the evolutionary synthesis, but survived in the theoretical literature as a kind of straw man whose deficiencies threw the virtues of the BSC into relief. Thus Cain (1954:52) characterized morphospecies as "static, with no reference to changes in space and time," while acknowledging that it is an indispensable tool for cataloging museum material; and Simpson (1961) defended it as the only concept that paleontologists have to work with.

An early, little-noticed critique of the BSC came from Sokal and Crovello (1970). The basis of their argument consists of a flowchart for determining species (fig. 1, p. 136). One might do this for museum specimens of, say, bushbabies or mouse-lemurs. The steps are as follows:

1. Lay out specimens that are broadly similar to one another;
2. Divide them up geographically, as finely as possible into "localized population samples" (LPS);
3. Determine whether individuals within each LPS interbreed, and if they do not, partition the LPS into "localized biological population samples" (LBPS);
4. Organize the LBPSs into phenetically similar groups;
5. Determine whether these groups of LBPSs are "actually or potentially interbreeding";
6. Divide the groups into geographically connected samples and add to them iteratively until homogeneous species are obtained.

Now this is all very well in theory, but in practice the information necessary for steps 3 and 5 are usually unknown and unknowable. To deem two LPSs not interbreeding, as Charles Kingsley might have said, you must see them not interbreeding. Laboratory interbreeding experiments can perhaps be tried, but of course these will at best reveal only postmating RIMs. Steps 3 and 5 would therefore be, in Sokal and Crovello's (1970) terminology, "phenetic bottle-necks." If and when the requisite information becomes available, there might even be equivocal results: partial RIMs. Thus the authors accused the BSC of being (p. 145) "imprecise in its formulation and inapplicable in practice."

The first accusation is a little unfair. As we have seen, Mayr (1963) was well aware of the existence of incomplete RIMs: the very purpose of the BSC was to usher systematics into the dynamic world of evolution, and RIMs, like everything else, have evolved. The second accusation is more cogent. We can partition our LPSs into similarity groups, but to go on to assert that these are LBPSs is inference, no more; very reasonable inference, perhaps, but hypothesis, not observation.

Sokal and Crovello concluded that it would be more realistic to drop steps 3 and 5 altogether and define species purely phenetically, as in fact had been previously proposed by Sokal and Sneath (1963). Yet at no time did Sokal and Crovello or Sokal and Sneath actually define "phenetic species," leaving the reader to infer that for the pheneticist they are no more than a rung in the shifting stepladder of the taxonomic hierarchy.

The Phylogenetic Species Concept

Despite the criticisms of Sokal and his colleagues, most taxonomists seemed willing to continue subscribing to the BSC and, as was mentioned earlier, new species definitions up to the 1990s were generally modifications of it, or at least based on it. They were endoheresies (i.e., heresies from within the system). The phylogenetic species concept (PSC) of Cracraft (1983) was an exoheresy, a proposal to junk the system altogether and start afresh. At first, the PSC was largely ignored, but since about 1990 it has garnered more and more serious discussion, perhaps assisted by the widespread acceptance of the cladistic perspective and by its elaboration by Nixon and Wheeler (1990).

A criticism to which the BSC had always been prone was the indeterminable status of geographically isolated populations. Would they interbreed if they came into contact, or not? In the end, this comes down to mere speculation or even weight of authority. Surely we can be a bit more objective than this?

Then there is the little matter of what reproductive isolation actually is. As noted earlier, Mayr (especially Mayr [1963]) was well aware that it is often not clear-cut: we are dealing with evolution, after all, a dynamic state of affairs. Already Bigelow (1965) had questioned whether interbreeding really does necessarily imply gene flow; but the degree to which the concept of reproductive isolation is not clear-cut has become clear only with the advent of DNA sequencing, particularly mitochondrial DNA that does not recombine and so survives intact in a "foreign" population as a kind of fossilized relic of past hybridization. One of the earliest described examples of this was in deer: the mtDNA of *Odocoileus virginianus,* the white-tailed deer, has penetrated deep into the sympatric mule deer *(O. hemionus)* population where the two are sympatric, but yet on purely morphological evidence one would not suspect anything of the kind (Carr et al. 1986). Since then, cases of this nature have accumulated, documenting abundant evidence of introgression, generally asymmetrical, between what had hitherto been assumed on morphological evidence to be reproductively isolated species (Avise 1994). Evidence of interbreeding between distinct species has long been available in Primates. This has generally been put down to habitat alteration by human agency, bringing ecologically separated species into contact, and this may be true although the evidence of this in some described instances, such as the hybridization between generically distinct (!) *Theropithecus gelada* and *Papio anubis* (Dunbar and Dunbar 1974) is admittedly tenuous. Recently the whole matter of interspecific hybridization, using another *gelada/anubis* example as a basis, has been discussed in great detail by Jolly et al. (1997).

Again, if reproductive isolation is made the main criterion of species status, this automatically becomes the focus of attention when accounting for taxonomic diversification. It is one such problem but, Cracraft maintained, not the only one nor even necessarily the most important. For example, geographically separated populations often exhibit new, fixed character states differentiating them from close relatives, and these are surely "units of evolution" in all meaningful senses, yet the BSC commonly denies them this status (if they "don't differ enough") by denying them species status.

The "biological species," Cracraft (1983:170) concluded, obscures more than it illuminates. Instead, the phylogenetic species concept is based on the results of evolution (on pattern), not on the processes by which these results may or may not have come about: a species is "the smallest diagnosable cluster of individual organisms within which there is a parental pattern of ancestry and descent."

Two of the corollaries of this may cause some dismay. The first is that the capacity or propensity to interbreed becomes a character like any other, in fact a plesiomorphic one. It follows from this that there ought to be cases where two sister species are reproductively isolated but one or both of them can or do hybridize with a third, more distantly related species. Such cases, where application of the BSC would actually obscure evolutionary relationships, do actually exist among Primates, the genus *Eulemur* for example.

The second corollary is that more taxa would merit species status than traditionally recognized, perhaps many more. An example in birds is Cracraft's own revision (1992) of birds-of-paradise, in which 40–43 previously recognized species became 90; in Primates, application of the PSC resulted in the number of species of *Callithrix* increasing from 3 to 13 (Vivo 1991).

Avise (1994; Avise and Ball 1990) attempted to tame these perceived excesses of the PSC by means of a **concordance principle**: that if several independent lines of genetic evidence converge upon a single partition, then species status for the populations in question is suggested if the inferred reproductive barrier is intrinsic, but subspecies status if they are solely extrinsic (geographic). Although claiming to infuse the PSC with BSC principles to effect a synthesis, this principle does not seem to be much help when seeking an objective way of deciding when (or whether) allopatric yet diagnosable populations should be downgraded to subspecies status—the same crucial stumbling block identified by Cracraft (1983).

What might be the drawbacks of the PSC?

Ford (1994b) suggested that it has the potential for researcher bias: according to what features are selected by a given researcher, a taxon may be diagnosable, or it may not. This is perfectly true: a taxonomist must just be conscientious and spread the net as widely as possible, taking osteodental, soft anatomy, external, ethological, karyological, and molecular features into account to the extent that they are available. The bias attendant on the selection of features for study is not confined to PSC analyses, but is the more crucial because species-status decisions depend on it.

Avise and Ball (1990) listed other drawbacks, as follows:

1. The number of species to be recognized "depends on the resolving power of the analytic tools available" (pp. 46–47).
2. Different gene genealogies will usually be discordant (p. 48).
3. The fact of shared ancestry is implicit in the PSC, "a parental pattern of ancestry and descent" (p. 48) in Cracraft's definition.

All these criticisms have force, but still leave the PSC more valuable for working taxonomists than the alternatives. Point 1, of course, is a valid criticism of any species, indeed any taxonomic, definition whatever; yet the criterion is applicable at a far earlier stage of analysis than, say the BSC. Point 2, the discordance of different gene genealogies, is a peculiarly molecular geneticist's concern: the conclusion that there are fixed differences between OTUs (i.e., that they are diagnosable) operates at any level, be it morphological, behavioral, or genetic, and nonconcordance is not a fundamental concern. It is point 3, that the PSC does not after all free taxonomic decision making from concern with reproductive status, that is of most concern. We have no actual knowledge of the place of specimens in reproductive communities and generally can have no such knowledge; the reference to "patterns of ancestry and descent" is perhaps best dropped from the definition. A species, then, is simply a diagnosable entity: it has fixed heritable differences from other species, and whether they really are fixed and really are heritable is, as rightly noted by Avise and Ball (1990), dependent on the level of knowledge we have in any given instance. We just have to do the best we can with what information is available to us.

The Autapomorphic Species Concept

There was no stipulation in the PSC as to what kind of character state is necessary to diagnose a species, whether plesiomorphic or apomorphic. This was slightly surprising in an increasing cladistic environment, and Donoghue (1985) proposed to update it, in the form of the monophyletic or autapomorphic species concept, under which a species must possess at least one autapomorphy.

This concept has been discussed by Crisp and Chandler (1996:815), who pointed out that "the smallest autapomorphic unit may have as its sister-group an unresolved symplesiomorphic cluster of organisms," and these cannot be ignored taxonomically. Crisp and Chandler estimated that in higher plants fully half of all species may be nonmonophyletic.

Freudenstein (1998:100) has taken exception to Crisp and Chandler's conclusion, on what seem to me to be purely semantic grounds. His argument is as follows: (1) because a species is the lowest level in the taxonomic hierarchy (individuals being related to each other in a reticulating, not a hierarchical, pattern of genealogical relationships); and (2) because, in his definition, a monophyletic group is "an ancestor (taxon) and all of its descendants (which are also taxa)," therefore concepts of monophyly and nonmonophyly cannot apply to species.

I want to make two points here. First, the relationships among subspecies are reticulating rather than hierarchical, yet they are by definition taxa (unless one wishes to define them out of existence). So some species (those divided into subspecies) would have the potentiality for monophyly or nonmonophyly. More seriously, to add "which are also taxa" to the

definition of a monophyletic group is unduly narrow and unwarranted; it is essentially a revival of Hennig's (1966) own definition, although it does not seem to have achieved very wide acceptance since then. I see no reason why a group of individuals should not also be potentially monophyletic. What Crisp and Chandler (1996) were saying is that it is quite common to have species with no (detectable) autapomorphies, and this to most taxonomists means that such species are not, or not demonstrably, monophyletic.

THE SPECIES IN PALEONTOLOGY

If there has been such a diversity of views on how best to define species in the present-day world, how much more difficult, one might suppose, would be their theoretical delimitation in the fossil record. Actually, yes and no.

The debate about what the concept "species" should mean in the fossil record began with Simpson's evolutionary species concept (discussed earlier in this chapter). This theoretical concept is of such fundamental importance as an explanation for the pivotal status of the species, present-day or fossil, that it was necessary to describe it earlier. Yet, as I have argued, it is of no great help when we come to the problem of simply recognizing them.

As far as paleoanthropology is concerned, I briefly explored the meaning and recognition of species in this fraught field, with an admittedly rather simplistic example of how it might be done in practice (Groves 1997). Work by Cope and Lacy (1992) and Plavcan (1993) shows why it is likely that the number of fossil species will always be underestimated, whether within a single time-slice or across a time span. Henneberg and Brush (1994) and Henneberg (1997), from a similar theoretical starting point, have proposed that one should give up trying to recognize fossil species altogether and simply call a definable fossil sample a **similum**, without taking up a perhaps purely subjective position on whether it is a sex-specific sample, a morph, a species, a collection of closely allied species, or whatever. On the face of it, this seems to be simply a counsel of despair, but I believe the concept has some value as a stage of analysis in the taxonomic assessment of both modern and fossil samples (see the section on Protocol for Alpha Taxonomy in the next chapter).

The Cladistic (or Internodal) Species Concept

When Hennig (1950, 1966) proposed that monophyly is the only logical basis for taxonomy, he promoted a new view of what a species is: essentially, it is the space between a node and a terminal point on a cladogram, or between two nodes.

The first of these alternatives proposes that a species represented by a living population should be regarded as extending back in time to the point where its lineage became separate from others, and that the same should hold in the case of a species that became extinct without issue.

The second alternative has been vigorously defended by Ridley (1989), but has otherwise been received with dismay, even embarrassment, by cladists and noncladists alike. It says, in effect, that a species ceases to exist as soon as a lineage split occurs. This may be justified in a case where a population separates into two (or more) sizeable segments, each of which commences to diverge along its own path; but speciation theory holds that it is more usual that a small segment of a species becomes isolated from the rest and rapidly diverges, leaving the main body of the species genetically unaffected.

Hennig gave his concept no separate label; Ridley (1989:2) dubbed it the cladistic species concept. The natural objection, that in many cases we would have two separate species differing in no respect whatever, was swept aside by Ridley as "phenetic in inspiration": for him, "The virtue of the cladistic definition is its perfect objectivity. Species are defined unambiguously as branches" (p. 4). Having taken this extreme stance, Ridley then very interestingly built links to the BSC and to Van Valen's ecological species concept. The cladistic species may be complete in itself (theoretically prior), he said, but reproductive and ecological parameters are necessary to explain precisely why species should exist. This is perhaps the most explicit attempt to produce a combined view of species.

Another interesting aspect of Ridley's definition is that any split whatever qualifies as the end of one species and the beginning of two new ones, even if the split later closes up again (in which case a new species has been formed by hybridization). There is a reductio ad absurdum in this view: short-lived splits in a population happen all the time, "even between siblings that do not immediately interbreed" (Kornet and McAllister 1993:63).

Nixon and Wheeler (1990) referred to a lineage between two splitting points or nodes (including between a split and a termination) as an internodon; hence the cladistic species concept of Ridley, they suggested, would be better called the internodal concept.

The fundamental drawback to the cladistic or internodal concept is its sheer impracticality. Overwhelmingly, we have no information whatever about lineage splits: the fossil record simply is not that good. When a new discovery shows us that there has been a branching event between what we had thought were simply two time-successive samples, A and B, of a species, suddenly we would be required to place A and B in different species, presumably describing B as new ("Diagnosis: morphologically indistinguishable from species A").

The Composite Species Concept

Because species must be recognizable and assuming that we must delimit them in the fossil record as well as in the living theatre, how is this best done? Kornet and McAllister (1993:78) noted that new character states can be seen arising, spreading, and becoming fixed in fossil lineages as they have been in living populations; in "rectangular speciation" models (punctuated equilibria), these processes are conceived as occurring rapidly after a lineage split, but often they can be demonstrated to occur anagenetically. They referred to an internodon in which a character state becomes fixed as an originator internodon, and proposed the composite species concept (CSC): "the set of all organisms belonging to an originator internodon, and all organisms belonging to any of its descendant internodons, excluding further originator internodons and their descendant internodons."

That is to say, we may define a new species as commencing at the base of an internodon in which a new character state becomes fixed, and as terminating at the base of the next internodon in which there is such a fixation.

This concept marries the PSC to the internodal concept. The internodal species concept maintains that a species ceases to exist, giving place to two (or more?) daughter species at the moment a lineage splits (i.e., at every node); it ignores morphology, relying exclusively on cladistic patterning and assumed biological and ecological cohesion. The phylogenetic species concept, however, maintains only that species must be diagnosable; it ignores evolution, relying exclusively on morphology. The CSC agrees that species must be diagnosable (i.e., must have unique fixed character states), while acknowledging that new character states arise and become fixed progressively along a lineage, and cuts the evolutionary tree into species at the most convenient point. Although one is never going to be in a position to falsify a claim that the plesiomorphic state perhaps still exists (in low frequency) in a given lineage, Kornet and McAllister's (1993) definition is probably as close as one can come to operational applicability. Indeed, their claim that the PSC is, in a sense, a special case of the CSC, restricted to the present day, seems well justified.

RESOLUTION?

Of course there are different species concepts: they are trying to do different things. Theoretical concepts try to explain what species should be: why they are the basis for evolution, and why they occupy a justly special place as the foundation stone of taxonomy. Operational concepts try to explain why it is that working taxonomists know when they are in the presence of species: how they recognize them and, once again, why they occupy a special place as the foundation of taxonomy!

There may be a resolution within each of the two types of concepts, but it is difficult to see how there could ever be a resolution between the theoretical

and the operational approaches. Christoffersen (1995: 447) recognized this quite clearly and offered two definitions. His "theoretical (ontological) species concept" defines a species as "a single lineage of ancestor-descendant sexual populations, genetically integrated by historically contingent events of interbreeding," whereas his "operational (epistemological) concept" defines it as "an irreducible cluster of sexual organisms within which there is a parental pattern of ancestry and descent and that is diagnosably distinct from other such clusters by a unique combination of fixed characters."

The first definition is more or less Simpson's evolutionary concept; the second is Cracraft's PSC. Christoffersen is not having a bet each way; he is saying, in essence, that the first definition is why a species is important, and the second is how we recognize it. This is about as far as one can go; we have an operational criterion and a reason why it is important. The difference corresponds to the distinction between pattern and process; we can observe the pattern, but can only infer the process.

In this book I employ the phylogenetic species concept for living Primates, because there is no alternative, and the composite species concept for fossils, again because there is no alternative. But the reader will know that the meaning of species runs deeper than just whether it can be diagnosed or not; that it has an importance in the realm of evolutionary theory, more or less as the evolutionary species concept envisages it, and that very likely it is this idea that lies behind the diagnosability by which we recognize it.

NEED SPECIES BE MONOPHYLETIC?

Crisp and Chandler (1996) found widespread nonmonophyly among irreducible clusters in higher plants and concluded that nonmonophyletic species are an inescapable consequence of trying to divide the natural world into species. Whereas in the case of higher categories groups that are not monophyletic can be either divided or amalgamated until they are, in the case of species this is not possible: the buck stops here.

As far as their evolutionary role is concerned, monophyletic groups are those that are descended from a common ancestor that was itself a member of that taxonomic group and must include all the descendants of that ancestor (see the section on The Importance of Monophyly in Chapter 1). Therefore, if a species gives rise in evolution to other species, it automatically becomes paraphyletic.

The terms paraspecies and metaspecies have been coined for these contingencies (Donoghue 1985). A **paraspecies** is a species whose monophyly has been tested and refuted: some of its members are, genealogically, more closely related to other species than they are to other members of the same paraspecies. A paraspecies would be one in which a synapomorphy occurs in some but not all individuals, which is shared with another species. A **metaspecies** is one that lacks any autapomorphic states; this could be a living species that lacks distinguishing autapomorphies or a putative ancestral species: indeed, Lee (1997) has argued that lack of autapomorphies is how one recognizes potential ancestors.

It is simply not possible to insist on a monophyly criterion for species (or indeed for subspecies). As Crisp and Chandler (1996:833) said, "Species are different from higher taxa because they are basal, so a special criterion is justified . . . the appropriate species concept in a phylogenetic system is either the phylogenetic species concept or the related composite species concept. Both predict that some species will be monophyletic and others paraphyletic or metaphyletic."

SUBSPECIES AND POPULATIONS

The lowest rank in the classificatory hierarchy is the subspecies. Some species are pretty much uniform across their geographic ranges; such species, which have no subspecies, are called **monotypic**. Others have well-defined geographic variation and may be divided into subspecies; they are said to be **polytypic**. Note that a species either has two or more subspecies or it has none: there is no such thing as a species with one subspecies.

The subspecies, it must be firmly understood from the start, is a geographic segment of a species. By definition, subspecies do not inhabit the same region; they are population-based. Well-marked morphs do not qualify as subspecies. Polytypism is not the same as polymorphism, although one way of putting it would be to say that polytypism is a kind of geographical patterning of polymorphism.

The subspecies concept, with its trinomial nomenclature, was established during the latter half of the nineteenth century; it crept in, rather than being formally proposed, and achieved early a wide acceptance among American ornithologists. In those days there was a kind of "cultural cringe" in America, rather like that pervading Australia in the 1970s (and still persisting in some quarters), and it mattered deeply in some quarters that the concept was slow to gain acceptance in Europe; J. A. Allen (1884) was forced to cite several prominent European ornithologists who had adopted the concept, to placate some worried American amateur colleagues. Among these Europeans was Schlegel, who was even said to be "perhaps . . . the father of the system." If this is so, then Schlegel did not use it much in his work on Primates; the only instance in his monograph (1876) is the name *Lemur collaris rufus,* a trinomial whose significance he left unexplained (by *Lemur collaris* he meant what we now call *Eulemur fulvus*).

Mayr (1963:348; Mayr and Ashlock 1991:43) defined a subspecies as "an aggregation of phenotypically similar populations of a species inhabiting a geographic subdivision of the range of that species and differing taxonomically from other populations of that species." Although this definition was made by the founder of the New Systematics, it compares closely with that of Deraniyagala (1955:39), whose operational (non-BSC) species definition was quoted earlier. His definition of a subspecies was "one of several populations within a species which differ from one another as a whole but possess types of individuals that are common to some or all of these populations." The difference is that for Deraniyagala all populations, as long as the average frequency for some character varied among themselves, were regarded as subspecies; so he divided up *Elephas max-*

imus, the Asian elephant, into about a dozen subspecies, differing from one another in average size, in the frequency of depigmentation on the ears, and in the percentage of males bearing tusks.

It is interesting to go back to the earliest users to find what they meant. J. A. Allen (1884:102–103) noted that "varieties" had long been recognized in zoology, but that term had an unacceptably elastic usage, whereas terms like subspecies, conspecies, incipient species, imperfectly segregated species, geographical races, local forms, etc. . . . all imply the character of the forms thus designated, namely, that they are *intergrading* [italics in original], which, while characterized by differences easily recognized in their well-developed phases, yet so coalesce through intermediate stages of differentiation that they run the one into the other and cannot be sharply defined." Coues (1884:198), in much the same vein, specified that "If the links still exist, the differentiation is still incomplete." It would be difficult to find more lucid modern characterizations of subspecies than these.

For Mayr different populations had to differ from one another "taxonomically," so subspecies may often be clusters of adjacent populations. He quoted with approval the 75% rule: 75% of individuals of one population (or cluster of populations) must differ from all other individuals in the species to qualify as a separate subspecies. Statistically, this corresponds to a 90% joint nonoverlap, so metrical features can be tested by the coefficient of difference (C.D.), which is the sum of the standard deviations divided by the difference between the means (Mayr 1969). If C.D. > 1.28, the 90% joint nonoverlap requirement is fulfilled, and in theory two subspecies can be recognized. Of course, just as one cannot run a large number of *t*-tests sequentially and triumphantly point to a few $P < 0.05$ results as indicating a significant difference, so a large number of C. D. tests is bound to generate a few spurious positive results. The C. D. is just a rule of thumb in any case.

Much geographic variation is smoothly clinal, and there is no point along the cline where one can logically break it and make a different subspecies. One sometimes finds that two physically identical populations are separated geographically by one that is different (as in *Saimiri sciureus*). In addition, much geographic variation is nonconcordant (i.e., different characters vary independently across the species's range). All these problems were brought forward by Wilson and Brown (1953 and subsequent papers) and used as a reason to reject the subspecies concept altogether.

Cracraft (1992) discarded the subspecies concept on the grounds that in some cases fully diagnosable taxa had been regarded as subspecies, whereas in other cases they had been used merely to demarcate points on a cline. For him, they are species or they are nothing. Nixon and Wheeler (1990) were not so dismissive: subspecies are groups that are "centric and not monothetic."

The subspecies concept has problems, for sure, and can be used with different evolutionary meanings (Marroig 1995), but mammalogists, including (perhaps especially) primatologists, find it useful. It is a worthwhile exercise, especially, in today's world where conservation of named entities is required, to describe and characterize population groups in which the majority of individuals, but not every single one, differ from other such groups. Subspecies are the point along the scale of differentiation at which it becomes worth giving names. It follows, of course, that subspecies, unlike species, cannot truly be diagnosed. Yet it is traditional, in taxonomic revisions, to "diagnose" subspecies, and I have done so in this book simply to maintain custom; but let us realize that we are using the term in what is probably too flexible a way, as it is in the very nature of subspecies that their characters overlap.

PHYLOGEOGRAPHY AND SUBSPECIES

Phylogeography is "the study of the principles and processes governing the geographic distributions of genealogical lineages, including those at the intraspecific level" (Avise 1994:233). Avise (1994) and Avise and Ball (1990) noted that geographic sorting of genomes is actually what subspecies are all about, but they went on to deduce that where they cannot find such sorting there are no subspecies. This does not follow.

Typically what are studied are mitochondrial DNA (mtDNA) lineages (i.e., those inherited solely [or almost solely] through the female line). Presumably this is mainly because mitochondrial DNA, being orders of magnitude more abundant in any cell than nuclear DNA, is much easier to extract and clone by the polymerase chain reaction (PCR) technique; but its use in this context has important consequences.

Diagrams of the lineage sorting characteristic of mtDNA (and, incidentally, of Y chromosome DNA, the other kind of DNA that is uniparentally inherited and does not recombine) are given in sources such as Avise and Ball (1990; and reproduced in Avise [1994]), who depicted the assorting of lineages in two populations of organisms that separate from a common ancestor. In each generation, a certain proportion of females have only male offspring, so that although their nuclear DNA is perpetuated their mtDNA is terminated; at the same time, of course, mutations in other females' mtDNA maintain the genetic diversity in this system. So at the time of the separation of the two populations A and B, there are multiple mtDNA lineages in each of them: some shared, some unique. Some of these lineages, shared and unique alike, became differentiated only a few generations earlier, but others have been separated much longer; so as far as their mtDNA is concerned, populations A and B are polyphyletic. Time moves on, and some of the lineages become extinct while new ones are generated, and each of the two populations slowly achieves genetic monophyly. Given time, therefore, lineage exclusiveness will occur, but in between there will be a phase when the populations may be para- or even polyphyletic. This intermediate period is precisely when the two will be prime candidates for subspecies status.

This, then, is a theoretical problem for equating DNA lineages with subspecies: lineage sorting is not instantaneous, and gene trees are not population trees (or even, often enough, species trees). Maddison (1997) showed that the probability of what he referred to as deep coalescence in any given lineage

is a function of the relation between effective population size and number of generations, and he gave guidelines for deriving the "true" phylogeny from gene phylogenies.

There is also a practical problem. Time and again Avise (1994 and elsewhere) examined subspecies in particular species (mainly birds and fish) by the lineage sorting criterion; if sorting had occurred, the subspecies were validated, if not they were sunk. The subspecies in question were originally described on the basis of phenotypic features, typically (in the bird examples) color and pattern; if these are confirmed as characterizing particular populations, then surely this is in itself prima facie evidence that they are to some extent genetically different, whether the mtDNA lineages have become sorted or not. The phylogeographic method tells us a great deal about evolutionary events, especially demographic, but is not informative about the validity of subspecies.

Like other new methods that burst upon the scene (multivariate analysis and chromosome studies come to mind), DNA sequencing, including phylogeography, has exciting new perspectives to offer, but these augment, not replace, other, older methods.

THE CONSEQUENCES FOR CONSERVATION

Science is a system dedicated to the task of Finding Out. It is dynamic and progressive; scientific hypotheses and models rise and fall according to whether or not they survive the process of testing (Popper's criterion) and whether or not they generate new research programs and directions (Lakatos's criterion), not according to what nonscientists think of them. Of course, science also serves society and cannot ignore social issues; that is why it is such a cause for celebration when some scientific model slots unmodified into some category of social utility. Such a case is the value of the PSC to conservation.

Adoption of a phylogenetic concept can have only beneficial effects on conservation law and planning. Subspecies have a tendency to simply get lost in listings: see, for example, Wolfheim (1983), whose

painstaking compendium was marred somewhat because she worked only at species level. This is, in
context, no criticism; it is, in fact, entirely usual in
conservation writings. In legal processes, "species" is
all-important: if it is not a species, it is legally invis-
ible in most polities. As far as conservation is concerned, it is absolutely vital that we, in Brockelman's
(1991:7) words, "when in doubt, trot it out," and the
PSC does just that.

4

A Brief History of Primate Taxonomy

In antiquity, the Western world knew just a few non-human Primates. The Greek words *pithekos* (Latinized as *pithecus*) and *kephos* or *kebos* or *keipon (cephus, cebus, ceiphon)* meant any monkey, probably mainly the Barbary macaque, *Macaca sylvanus,* known to the Romans as *simia,* because the few long-tailed monkeys that reached the Mediterranean region tended to be specified as *kerkopithekos* (Latinized form *cercopithecus* = tailed monkey). Baboons, which were known to the Greek and Roman worlds through Egypt, were known to the Greeks as *kynokephalos (cynocephalus),* to the Romans as *papio.*

The Egyptian god Thoth was depicted in tomb paintings and in statuettes as a male hamadryas baboon. The ancient Egyptians mummified everything they could lay their hands on. Many years ago I was called upon to identify some unwrapped mummies of "baboons" from the catacombs of Saqqara; I expected to find *Papio hamadryas,* but instead there were a dozen *P. anubis,* and two each of *Macaca sylvanus* and *Chlorocebus* sp. The mummies dated from the Ptolemaic era, probably third century B.C.; so by that time hamadryas baboons were evidently no longer living in Egypt, and monkeys for mummification were being obtained from farther south (Ethiopia or Sudan) and west (the Maghreb). This may also give a clue to what *kerkopithekoi* were, though

whether it has any bearing on the identification of the beautiful but highly stylized blue monkeys on frescoes at the Bronze Age Aegean site of Thera is another matter.

Occasionally something really unusual turned up in the classical world. Jolly and Ucko (1969) discussed the identity of the "sphinx-monkey" that was brought to Egypt from an expedition to the south; it was clearly a female gelada *(Theropithecus gelada).* We must not jump to the conclusion, however, that the gelada was the origin of the Sphinx either in Greek mythology or the statue at Giza.

And this is how matters remained until not much more than 300 years ago, when European science freed itself from its utter dependence on its classical past and began to make its own discoveries. Such innovators as John Ray (1627–1705) began to bring some order into the arrangement of the natural world, but modern taxonomy sprang almost full-blown from the remarkable mind of Carl Linnaeus (1707–1778), the first in our rogues' gallery of contributors to Primate taxonomy.

Linnaeus, a botanist, was from 1730 a lecturer in the university at Uppsala, Sweden (and later, occupant of the chair of botany). Although he did travel in Europe and took part in a famous expedition to Lapland, his lasting claim to fame is his book *Systema*

Naturae (The System of Nature), which was first published in 1735 and ran through 12 editions during his lifetime. He proposed that the natural world could be divided into three kingdoms—animal, vegetable, and mineral(!)—and that each of these could be divided into classes, these into orders, these into genera, and these finally into species. Each species he designated by two names: the first denoting the genus to which it belonged, the second peculiar to the species. He thus invented not only the taxonomic ranking method that we use today (only the ranks phylum and family remained to be inserted and the various sub- and infra- categories), but also the binomial system of nomenclature. Its success was immediate. The system achieved its mature form in the 10th edition of *Systema Naturae*, published in 1758, and this is therefore taken as the starting point of all modern nomenclature.

In this seminal 10th edition, Linnaeus recognized the Primates as one of his 20 or so orders of the class Mammalia. There are four genera in his Primates: *Homo, Simia, Lemur,* and *Vespertilio.* The last of these, the bats, was soon expelled from the Primates by his successors, and *Simia* and *Lemur* were split into several genera.

Linnaeus knew of a suprising number of species of monkeys and lemurs. Some of these were described purely from the literature, but for many he had actual specimens before him, sent back by his much-traveled students, who included such names as Thunberg, Osbeck, Sparrmann, Solander, Hoppe. These specimens are now in Uppsala (in the Zoology Department of the university, and in his house, kept as a heritage site) or in the Stockholm Natural History Museum; they are skulls, shriveled skins, or whole animals in bottles, and it is a fascinating experience to examine them and see exactly what it was that Linnaeus meant by, say, *Simia apedia, Simia sciurea,* or *Lemur tardigradus.*

Linnaeus wrote in Latin. His surname already sounds Latin-ish, but his first name, Carl, had to be Latinized (into Carolus). Later in life he was ennobled as Carl von Linné, but this was not his actual surname, and his name should always be cited as Linnaeus.

This was the age when the Great Apes were gradually becoming known. Linnaeus had, apart from *Homo sapiens,* other species in his genus *Homo.* One, *Homo monstrosus,* was admittedly an artificial conglomeration of giants, dwarfs, and famously unusual people; the other, *Homo troglodytes,* was an amalgam of travelers' tales and included some undoubted orangutans and possibly chimpanzees. Neither of these names is used nowadays; they are deemed undeterminable. But he did also collect what he thought were cases of genuine humanlike species and wrote them up as a special thesis called *Anthropomorpha.* In those days graduating students, instead of writing a thesis, had to learn a thesis written by the professor and be examined on it; *Anthropomorpha* was the thesis on which a student called Hoppe (Latinized as Hoppius) graduated. In it, Linnaeus described a juvenile orangutan that had been, all too briefly, the pet of an Englishman named Edwards; thinking that it was the species that the ancient Greeks had called "the pygmy," he gave it the name *Simia pygmaeus.* And this explains not only why the name should be ascribed not to "Hoppius" but to Linnaeus, but also the origin of the ridiculous circumstance that the second-largest living nonhuman Primate bears the specific name *pygmaeus.*

SUCCESSORS TO LINNAEUS

The immediate adoption of the Linnaean system, coupled with the rapid expansion of European travel and exploration, meant that the late eighteenth century saw the description of many new Primate species. Chief among Linnaeus's followers was Johann Friedrich Blumenbach (1752–1840), professor of medicine at Göttingen in Germany. Blumenbach has entered history as the father of anthropology, but he was also a notable natural historian, and his *Handbuch der Naturgeschichte* went through many editions. He was the first person clearly to describe the chimpanzee, giving it the name *Simia troglodytes.* Because Linnaeus's *troglodytes* was put in the genus *Homo,* the two names are not homonyms (although if, as some modern systems are beginning to advocate, chimpanzees ought to be included in *Homo,* then a situation of secondary homonymy will arise and the next available name for the chimpanzee may have to be used, though the *Code* permits us to "forget" unused names, like *Homo troglodytes* Linnaeus, under certain circumstances).

Mathurin-Jacques Brisson (1723–1806) was destined for Holy Orders and even tonsured, but during the ceremony of his ordination as deacon at St. Sulpice he got cold feet and walked out of the chapel for good! He became assistant to the naturalist Réamur, and during this period wrote a work, *Regnum Animale*, in which he tried to use the Linnaean system; it seems rather unfair that in this field, likewise, he was not a great success. New names, especially for new genera, appear in plenty in *Regnum Animale*, but alas, the work has been declared nonbinomial and so unavailable by the International Commission on Zoological Nomenclature, and none of them can be used. After Réamur's death, Brisson turned to chemistry.

More successful was Robert Kerr (1755–1813), who in 1792 published *The Animal Kingdom of the celebrated Sir Charles Linnaeus*, which began as a simple translation but became a work in its own right because of the enormous proliferation of newly discovered species that had occurred in the meantime. Notable is Kerr's occasional use of trinomials, anticipating their general adoption by almost a century.

The French contribution was meanwhile kept alive by the medical anatomist Louis-Jean-Marie D'Aubenton (1716–1800), first director of the National Museum of Natural History, and by the painter-naturalist Jean-Baptiste Audebert (1759–1802), whose beautiful paintings in his *Histoire Naturelle des Singes et Makis* are accompanied by a text of variable quality, but noteworthy for the introduction of several new names for newly discovered species, whose types are now in the Paris Natural History Museum.

THE EARLY NINETEENTH CENTURY: THE FRENCH SCHOOL

When a nation produces a towering genius, his contemporaries in other nations look nervously over their shoulders. So when France produced Georges Cuvier and the almost comparably eminent Étienne Geoffroy-Saint-Hilaire, for a while the progress of natural history, including Primate studies, was almost entirely French.

Léopold Chrétien Frédéric Dagobert Cuvier (1769–1832), early renamed Georges after an elder brother who died in childhood, had an early interest and training in natural history, including the Linnaean system. He first worked as a tutor for rural nobility, then (at the urging of Geoffroy-Saint-Hilaire, in fact) came to Paris in the 1790s and spent the rest of his life in the Natural History Museum there, except when he was presiding over this or that national scientific policymaking body. His ability to keep his political nose clean, as much as his undoubted genius, ensured him rapid advancement under successive royalist, republican, imperial, and restoration regimes. His name lives as the father of both paleontology and comparative anatomy, and even his model of geological catastrophism still creates waves today. Although he did describe new taxa of Primates, he was more a "big picture" man, and left the detailed descriptions to workhorses like de Blainville, Desmarest, Lesson, and above all to his younger brother, Frédéric Cuvier (1773–1838), a man of such a self-effacing nature that he wished for no more individual an epitaph for his tombstone than "Brother of Georges Cuvier."

The man who had brought Georges Cuvier to Paris, Étienne Geoffroy-Saint-Hilaire (1772–1844), had studied medicine under D'Aubenton. In 1793, at the astonishing age of 21, he became the inaugural occupant of the chair of zoology in the Natural History Museum, and in this position he risked his life to save several colleagues from execution in the Terror that succeeded the French Revolution. In 1809 he moved to the University of Paris. His studies of vertebrate comparative anatomy led him to support the early evolutionary theories of his senior colleague Lamarck, and in 1830 he conducted a famous public debate with Georges Cuvier on the matter. Unlike the latter he was a keen alpha taxonomist and described abundant new species of mammals, including Primates, that French explorers were bringing home from expeditions to, among other places, Madagascar; and, alas, he was not above looting specimens from museums (notably Lisbon and Leiden) in countries conquered by Napoleon, and he described these too.

French primatology, like French zoology in general, seems to have become burned out after this phase; though kept alive for a while by such figures as Isidore Geoffroy-Saint-Hilaire (1805–1861), son of Étienne, and Alphonse Milne-Edwards (1835–1900), it petered out. The one burst of renewed energy was the Madagascar fieldwork of Alfred Grandidier (1836–1921) published, if incompletely, in beautifully illustrated folio format as *Faune de Madagascar*. It is only since the 1960s that French primatology has once again emerged onto the world stage in a big way.

THE EARLY NINETEENTH CENTURY: THE GERMAN CONTRIBUTION

Some of the earliest contributors to Primate taxonomy in Germany after Blumenbach were Karl Illiger (1774–1813), professor of zoology at Braunschweig; Johann Gotthelf Fischer von Waldheim (1771–1853), who first worked in Mainz but in 1805 moved to Russia; and the truly bizarre Lorenz Oken (1779–1851), whose mystical *Naturphilosophie* has earned him a place in history but whose cavalier way with the Linnaean system has caused most of his work to be declared unavailable for purposes of nomenclature. But what very early distinguished the Germans in the field was their propensity to do their own fieldwork. And so Alexander von Humboldt (1769–1859), a student of Blumenbach, traveled through Spanish-speaking South America, while Johann Centurius Hoffmann von Hoffmannsegg (1766–1849) and Johann Baptist von Spix (1781–1826) went to Brazil. As one expects of people with "von" in their names, all of these three were independently wealthy (Hoffmannsegg was even a Graf [= Count]), which certainly helped, but they were competent both as field naturalists (and as collectors, which does not necessarily count in their favor these days) and as academic taxonomists. As many of the most important lemur species were described by E. Geoffroy-Saint-

Hilaire, so most of the most important platyrrhines were first introduced to the world and named by the aristocratic German trio.

PRIMATE TAXONOMY TAKES OFF IN THE ANGLOPHONE WORLD

There were a few small beginnings to the study of Primates in Britain. Sir Andrew Smith (1797–1872), a surgeon, led expeditions from Cape Town north into unknown Africa in 1834–1836 and made known several new Primates that he described himself. George Robert Waterhouse (1810–1888) was curator of the London Zoological Society's museum and later moved to the British Museum. But it was the appointment of the cantankerous John Edward Gray (1800–1875) to the natural history department of the British Museum that set Primate taxonomy going in Britain. Gray was astoundingly productive; his first publication, in 1821, was no less than a taxonomic synopsis of the entire Mammalia (introducing such innovations as the family rank with its -idae termination), and right up to his death he was producing fresh catalogs of the ever larger collections of the museum and still describing new species and genera (but with less and less critical judgement). Among his more eccentric claims were that he had invented the idea of the penny post and postage stamps, which was stolen from him by Roland Hill to whom he had explained it during a coach ride, and that he had discovered that Darwinism was really a cover for spiritualism.

Gray's equally odd contemporary was Edward Blyth (1810–1873), whose life and work has recently been researched in some detail by C. Brandon-Jones (1995 and elsewhere). Though of working-class origin, even in his twenties he began to earn himself some respect as both ornithologist and mammalogist (for example, he was chosen to rewrite the mammal section for the new English edition of an influential work by Cuvier, *The Animal Kingdom*), which was rapidly dissipated by his propensity to complain and to fall out with people. He was obliged to take the "hardship post" of curator of the Asiatic Society's museum in Calcutta, where despite his high productivity he was persistently exploited and underpaid. His unusual personality comes out even in the names he gave to the taxa he described: thus for some reason he made the bizarre decision to name numerous representatives of Indian langurs *(Semnopithecus)* after characters in the Trojan War: Ajax, Achilles, Hector, Anchises, Achates, Iulus, Thersites, and so on. When he retired to Britain he did not live long, though long enough to make some impression as a penniless, alcoholic, unpredictable character: "that odd fellow," as the generally sympathetic Charles Darwin referred to him.

Anyone living in Britain in the age of Darwin must perforce take sides: the mistake that St. George Jackson Mivart (1827–1900) made was to try to take a middle road. He was convinced about evolution, but his Catholic faith would not allow him to accept natural selection as its prime mover, and he confided his somewhat mystical alternative not only to Darwin but to Thomas Huxley, who was not of a temperament to put up with sloppy thinking. And so this bright young man, whose taxonomic arrangement of the Primates (whose very definition of the order Primates, indeed) was still being quoted with approval even in the latter half of the twentieth century, was sidelined and carried on as a mere "workhorse."

When the Americans enter the fray, one must expect that it will be with a bang. When the Reverend Thomas Staughton Savage (1804–1880) quit his missionary work in Liberia in 1847, he first of all visited a colleague in Gabon, where he was given skulls of a large unknown ape and regaled with stories of it. On his return to Boston he collaborated with the anatomist Jeffries Wyman to describe the largest living Primate, which, recalling his classical education, Savage named *Troglodytes gorilla.*

THE AGE OF PROLIXITY

The age of great discoveries was over; the age of little discoveries was beginning. The main species of Primates were now known; the genera described; the organization of the order more or less agreed upon.

But as exploration advanced and guns felled local faunas in ever larger quantities, it became evident that there was as much work to be done filling in the details as there had been making the outlines known.

The era was ushered in by the remarkable Honourable Lionel Walter Rothschild (1868–1932), a man of means (rather considerable means, actually) who founded his own museum. His life story was engagingly recounted by his niece, Miriam Rothschild (1983). To Rothschild's credit is that he was one of the first (stimulated by his employee, Ernst Hartert) in the English-speaking world to use subspecies, but set against this is that he had little idea of individual variation and described far too many of them, often with only the vaguest of geographical indication, despite the fact that he had sent many of the collectors off on their expeditions himself.

Rothschild's taxonomic prolixity pales in comparison with that of some of his contemporaries, especially Elliot in America and Matschie in Germany.

Daniel Giraud Elliot (1835–1915) had worked in both American and European museums. He was curator of zoology at the Field Museum, in Chicago, until 1906, when he retired—to the American Museum of Natural History in New York! His huge three-volume *A Review of the Primates* (1913) is indeed an immensely useful review, but one does not want to take any notice of the taxonomy. I regret to say that Elliot had even less idea of individual variation than Rothschild, and no idea of age variation at all. He went to a great deal of trouble to examine all the type specimens that he could lay hands on, but in many cases there is no indication that he studied anything but types. He described a perfectly normal subadult male gorilla as a pygmy species, even making a new genus *(Pseudogorilla)* for it, and it persisted in the literature almost up to the present (see Groves [1986]). In one case he misread a label on a skull, *Cercocebus hamlyni,* for *Cercopithecus hamlyni;* the former name is a synonym of *Lophocebus aterrimus,* a mangabey, whereas the latter name refers to a valid species of guenon. The skull, being a mangabey, had a hypoconulid on its third lower molars, which guenons do not, so Elliot placed the species in a new genus, *Rhinostigma.* And so it goes on. Pocock (see later in this chapter) got his teeth into Elliot in no uncertain fashion in many publications.

The most unusual (not to say eccentric and even bizarre) individual in the entire history of mammalian taxonomy was Paul Matschie. George Fridrich Paul Matschie (1861–1926) worked first as a private tutor, then in 1886 joined the Zoological Museum in Berlin, and during the course of his employment switched from ornithology to mammalogy. Even reading some of his scientific papers one gets the impression of someone who was, well, different; and his obituary by Schwarz (1927) tells us just how different he was. He had an encyclopedic knowledge of his subject, but was entirely without formal qualifications (and it showed). He was a believer in special creation, perhaps the last of the antievolutionists to hold a professional position. He held that species (and subspecies, a concept that he came to use only later on and with some evident reluctance) were created to live in river basins, so that specimens from adjoining river valleys simply had to be different species, and specimens from watersheds were hybrids, showing the characters of one parent on one side of the body, those of the other parent on the other side. The concept of *halbseitiger Bastärde* (it sounds so much better in German) was unfairly left to his pupil, Ludwig Zukowsky, to unleash onto an unsuspecting world in a paper in which he described two new (!) species of buffalo from the left and right sides of a single skull. By 1920 Matschie's river-basin model had evolved (if one may say so) into the mystical idea of "elementary areas of distribution": these areas, each of which had to have its own complement of species, were "delimited by the diagonals of the squares formed by the even degrees of longitude and latitude," as Schwarz explained it (p. 294). After that date the distributions of the new species that he was still describing, dozens every year, were indicated by a numerical shorthand.

Matschie was, as we might expect given his philosophy, much the most prolific of all the describers of Primate species and subspecies. But, probably because he lived at the right time, when genuine new discoveries were being made, he did sometimes get it right—almost by accident, one might say. Perhaps his crowning glory was *Gorilla beringei,* the mountain gorilla.

Matschie's opposite number in the Vienna Hofmus was Ludwig Lorenz von Liburnau (1856–1943). Lorenz was bit of a splitter, but hardly in the Matschie class. Had Matschie received a batch of red colobus skins, all from a single locality but bewilderingly variable, he would have found reason to suspect that really they were from 20 different localities and he would have described 20 different species or subspecies; Lorenz, though convinced that they represented a new taxon, accepted the fact of their variability, and named the species *Colobus variabilis!*

The only Swedish zoologist to have contributed conspicuously to Primate taxonomy was Einar Lönnberg (1865–1942). Appointed curator of zoology in the Naturhistoriska Riksmuseet, Stockholm, at the age of 39, Lönnberg took an early interest in conservation and was active in promulgating education about nature preservation and delimiting national park boundaries in Sweden. He continued to work on mammalian taxonomy, using the material collected for his museum in South America, East Africa, and elsewhere, right up until his death.

British and American contemporaries, Thomas and Miller, were also responsible for the description of numerous new species, but are regarded today in a very different light from Matschie. Their sheer competence, good judgment, and thoroughness mark them as being, perhaps more than any of their contemporaries, the founders of modern systematic mammalogy. It is all the more pity, then, that neither of them took much time to explain what they meant by this or that taxonomic action or their taxonomic philosophies in general. In the main we have to read between the lines, although Miller at least dropped hints several times about what his species concept amounted to, and I have referred to this earlier, at the beginning of Chapter 3.

M. R. Oldfield Thomas (1858–1929), of the British Museum (Natural History) in London, was academically undistinguished. In fact he began at the museum as a clerk, though he held hopes of joining the biological staff, inspired in this direction, it seems, by attending courses in biology given by T. H. Huxley.

In 1878 he succeeded in getting a transfer to the zoological staff and started working on echinoderms, but one day out of the blue the keeper of zoology, Albert Günther, sent for him and informed him, "Thomas, you will do the mammals." And the mammals is what he did, for 50 years. He realized that faunal surveys were the direction in which taxonomy must travel, and he stimulated explorers and big-game hunters to preserve and bring back all the specimens they had collected, in this way laying the foundations of the British Museum's present enormous collections. He worked on these collections and published on them as fast as they came into the museum. The rapidity with which he published never compromised his thoroughness, though to be sure he was not distracted by hobbies or outside interests; Hinton's (1929) and more especially Pocock's (1929) obituaries make clear the narrowness of his intellectual horizons, not to mention his growing hypochondria. Even after his retirement, in 1923, he continued working on the collections. But in 1928 Mary, his wife of 37 years, died, and he cracked up altogether, and the following year he committed suicide.

Gerritt S. Miller (1869–1956), of the United States National Museum (Smithsonian Institution) in Washington, unlike Thomas, was a well-trained biologist, but like him was an astute taxonomic practitioner. His species concept, at first unarticulated, was later explained in more than one paper (and was the forerunner of the phylogenetic species concept), and although advances in knowledge have shown that many of his species are more probably subspecies, they are at least valid in some sense. His papers are often disappointingly terse, and one has to read closely between the lines to detect that here was a man who, among all the taxonomists of his time, knew his stuff. His speciality, the mammals of Southeast Asia, was made possible by the vast collections of one Dr. W. L. Abbott, whose schooner *Terrapin* sailed the waterways, visiting all locales from Borneo to the tiniest islets and blazing away at the local large fauna on each. It is not given to everyone to write a paper entitled "Seventy new Malayan mammals" (Miller 1903).

It is worth noting that Miller was working alongside colleagues who were equally prolix but with much less justification, such as Merriam, who described 90 species of North American brown bears. This proliferation was the order of the day, and despite all appearances Miller was conspicuously apart from it.

THE AGE OF REVISIONS

So the early years of the twentieth century saw the description of more species and subspecies than any comparable period before or since. Probably a conjunction of circumstances was responsible: the arrogance of the times, which included Europe's colonial activities and the big-game hunters that followed in their wake; the acceptance of the subspecies concept, in which Rothschild himself played no small part; and the example of—well, who started it first? Merriam, Matschie, who? There is absolutely no doubt of Matschie's extraordinary influence in Europe, at any rate; by his example many mammalian taxonomists were set racing along the same path, at least until it finally dawned on them that behind all those species and subspecies there was a hidden agenda.

The age of prolixity did not come to an abrupt stop, but gradually gave place to a period of consolidation, when new taxa were still being described but in the context of a revision of a whole group. Leader in this trend was Reginald Innes Pocock (1863–1947), who started life as a British Museum arachnologist, but during his tenure as superintendent of the London Zoo from 1904 to 1923 turned more and more to mammals; after his retirement from the zoo he returned to the museum as "temporary scientific worker" in the Mammal Section. As Miller had been in the right place at the right time to receive new Southeast Asian collections, so Pocock benefited from new Indian collections, which he published in the context of revisions (culminating in his 1939–1940 two-volume work, *The Fauna of British India: Mammalia,* never completed). Pocock was a great splitter of genera as well as of species: he might take a genus of six species, which he would lump into three and then divide between three genera, so ending up with many monotypic genera and few, but highly polytypic, species.

Perhaps Pocock will be best remembered, among primatologists, as the person who originally proposed the division of Primates into the suborders Strepsirrhini and Haplorrhini. He did this during his period at the London Zoo, when he would carefully examine all the mammals that died there and write papers on their "external characters" (chiefly their noses, feet, and genitals). It was for this series of papers that he was made a Fellow of the Royal Society.

Angel Cabrera (1879–1960) was, as far as I know, the first Primate worker (and one of the first professional biologists, in fact) to have a Ph.D., which he obtained in Madrid in 1902. He worked in the National Museum of Natural Sciences in Madrid until 1925, when he migrated to Argentina to become simultaneously professor of paleontology in the Museum of La Plata and professor of zoology in the School of Agriculture and Veterinary Medicine of the University of Buenos Aires. His revisionary work on South American mammals, *Catalogo de los Mamíferos de America del Sur,* was published in two parts, in 1957 and posthumously in 1961. But he cannot quite be dubbed the initiator of the now-flourishing South American primatology school: that status belongs to the remarkable Eladio da Cruz Lima (1900–1943), a justice of the supreme court of the state of Pará in Brazil, who during his time as a law student had worked as a volunteer assistant to the zoologist Alipio de Miranda Ribeiro and during his legal career continued to work on Brazilian mammals. At the time of his death he was working on a book on mammals, of which only the first volume was published (in 1945) under the title *Mammals of Amazonia,* Volume 1: *General Introduction and Primates.* This astonishing work contains numerous original insights but is most remarkable for its beautiful color plates, executed by Cruz Lima himself and based on actual specimens in the Museu E. Goeldi. It should be added that Cabrera was also wont to illustrate his work with his own color paintings: what is it about the South American environment that promotes that wonderful marriage of science and art?

Ernst Schwarz (1889–1961) was another biologist with a Ph.D., which he obtained in Munich in 1912,

in both medicine and biology, two fields in which he alternated throughout his life. After working for three years in charge of the mammal collection in the Senckenberg Museum, Frankfurt, he practiced medicine from 1915 to 1925. He returned to biology, working briefly first in the Berlin Museum and then in the small Greifswald Natural History Museum; in 1933 Hitler came to power and Schwarz, a Jew, fled first to Britain, where he worked in the British Museum, and then to America in 1937, where he joined the Smithsonian. After the war he joined the World Health Organization and worked for them, on and off, until his death. His major work on Primates was done during the 1920s, when he published influential revisions of such groups as colobus monkeys, *Cercopithecus,* mangabeys, lemurs, and lorises, some in German, some in English. And, what he is best remembered for, his paper on "the subspecies of the chimpanzee," in which he made known to the world the pygmy chimpanzee, which he described as *Pan satyrus paniscus.*

Rather like Pocock, Schwarz was a splitter of subspecies but a lumper of species, to the extent that he even placed *Procolobus verus* among the subspecies of red colobus, *P. badius.* This classification was, fortunately, not followed by anyone else; but other cases of his overlumping (such as *Cercopithecus petaurista* and *C. ascanius* as subspecies of *C. nictitans,* and all black-and-white colobus as subspecies of *C. polykomos*) have often proved quite insidious, and these and his lemur revisions were more or less standard well into the 1970s. I was delighted to meet this famous man about a year before his death and, insolent young whippersnapper that I was, I asked him about these revisions: shouldn't subspecies of a species replace each other geographically? Courteously, in a thick German accent, he explained to me his philosophy: it was necessary only that they *approximately* replace each other, and there could be much overlap, depending on how similar they were.

Discoveries were still being made in Southeast Asia. Cecil Boden Kloss (1877–1949), after whom Miller named a notable gibbon, was an ornithologist who also wrote extensively on mammals during his tenure as director of museums for the Federated Malay States and Straits Settlements (corresponding to modern Singapore and part of West Malaysia), from which he retired in 1932. His subordinate colleague at the Raffles Museum in Singapore was Frederick Nutter Chasen (1896–1942), who wrote the first regional checklist: *A Handlist of Malaysian Mammals* (by which he meant what are now Malaysia and western Indonesia). Chasen revised the mammals, including the Primates, of the region, describing several new subspecies as he did so, as well as including some interesting discussions. The exact date of his death is unclear; all that is known is that, just a few days before the fall of Singapore, the ship on which he was escaping was sunk by enemy action and he was drowned.

Chasen's counterpart in what was then the Dutch East Indies (now Indonesia) was luckier. Henri Jacob Victor Sody (1892–1959) became a tea planter in Java in 1918, taught at a school in Buitenzorg (now Bogor) from 1920 to 1941, and then moved to the Museum Zoologici Bogoriense, an increasingly important regional collection. He was interned when the Japanese invaded and emerged intact after the war, to be pensioned and retired in 1949. His internment explains why he is best known for only a few (although very long) papers, including one revising many of Indonesia's Primates.

Unlike the previous phase of Primate and mammal taxonomy, the age of revisions came to an abrupt end in 1951, with the publication of a work that was so seminal and summed up the trends of the previous 30 years to such a degree that it marked a watershed: J. R. Ellerman and T. C. S. Morrison-Scott's *Checklist of Palaearctic and Indian Mammals, 1758–1947*. The general theme of this book, which set the tone for the next 30 years at least, was this: subspecies, every page said, replace each other geographically. If they are sympatric, then yes, they are different species; if allopatric, then unless there is some really major reason why not, they should be regarded as subspecies of the one species.

It must be admitted, too, that Ellerman and Morrison-Scott (1951) was regarded as the revision to end all revisions. The book quoted Chasen's *Handlist* approvingly as the counterpart for Southeast Asia and (less approvingly, because they quickly issued a partial revision [Ellerman et al. 1953]) G. M. Allen's

"A checklist of African mammals" (1939a). There was nothing left to be done. Mammalian taxonomy was over, at least for the Old World.

Taxonomy became the poor relation after that. It was only very gradually that people realized that not only was there more to be done after all and perhaps Ellerman and Morrison-Scott's theoretical bias was not the last word, but that taxonomy is actually a rather crucial umbrella science.

THE CHROMOSOME REVOLUTION

Although serious study of chromosome variation among living organisms is more than a century old, it was long hampered by technical problems; indeed it was not until 40 years ago (Tjio and Levan 1956) that the correct human chromosome number was established! As soon as the existence of substantial diversity in chromosome number and morphology in Primates became evident, new interpretations of interrelationships began to become widespread. It was almost as if a new era in understanding was opening up, and that gross anatomy could now be banished as being of no further relevance. Perhaps the grossest example of this "rush to judgment" was the proposal to take the Hylobatidae out of the Hominoidea and align them with the Colobinae (Chiarelli 1963). Another problem was fairly widespread misidentification of taxa being karyotyped, leading to long-standing errors in the literature, such as Bender and Chu's (1963) ascription of $2n = 44$ to *Hylobates hoolock*. I have long maintained that new karyotypes should routinely be accompanied by some evidence that the identification of the taxon concerned is correct, such as a photograph of the living animal from which the sample was taken. This has rarely been done.

Yet, as techniques have improved (such as the introduction of several different modes of banding), there does seem to have been an improvement in sophistication of interpretation, such that chromosomes have been used as an adjunct to taxonomy, not as the last word. Often the discovery of chromosome diversity has been itself the starting point of a train of fruitful research; thus differences between

the karyotypes of two morphotypes of greater galagos, used to argue that these represent two different species (Masters 1986, 1988), led to extensive
field studies on bushbabies that have added vastly to
our knowledge of this neglected group.

THE PROTEIN REVOLUTION

First by immunology, then by amino acid sequencing, comparison of proteins in different taxa has
opened up new avenues of comparison. Early methods were mainly distance methods: if a given protein
was more similar between taxa A and B than between either of them and taxon C, then it meant that
A and B formed a phylogenetic unit compared to C
and so should be classified closer together. From the
very start, therefore, molecular taxonomy involved
an assumption of regularity in evolutionary rates
(well before the molecular clock concept was articulated), and, interestingly, from the very start it assumed that taxonomy should depend on phylogeny
(well before Hennig).

Cladistic arguments have been rather slow in
invading molecular taxonomy, but are now widely
adopted, perhaps under the influence of packages
such as PHYLIP. In Primates, a good example is the
recent work on galagos; Masters et al. (1994) used
the inferred polarity of the distribution of alleles in
red cell enzymes to generate a hypothesis of the
relationships among various galagos specifically and
Loriformes in general. Where both distance and cladistic methods have been used, the same answers often emerge: in itself, a good confirmation of the general validity of the clock assumption.

THE DNA REVOLUTION

If looking at proteins takes us down a level from
gross morphology, studying the DNA itself obviously gets to the heart of the matter. At first, restriction enzymes were used, then the polymerase chain
reaction made it more and more easy to sequence
DNA itself. This has, unfortunately, come to be re-

garded as something of a panacea in some quarters;
less, I am happy to say, among its practitioners than
among some of the "consumers" of the findings. Apparently overwhelmed by the magic of it all, they
have sometimes taken the attitude that if a short
segment of DNA, a few hundred base pairs, fails to
show consistent differences between two putative
taxa then the fact that one of those taxa is black and
has horns and the other is red and has no horns is
somehow not relevant and they should be lumped
together in the one species or subspecies.

So what, if two taxa have not yet sorted out their
mitochondrial lineages? So what, if a particular pseu-
dogene does not differentiate them? The existence of
(heritable!) phenotypic differences between two taxa
is ipso facto evidence that there are genetic differences between them. Is it not so?

Comparing entire genomes in a way gets around
problems of the vagaries of local evolution, or non-
evolution, in small segments of the genome; but in
a way it does not, because so much of it consists of
repeated sequences (microsatellites) that change en
bloc. Yet even this fact is useful. So, Crovella et al.
(1994) used highly repeated sequences to study the
affinities of various Loriformes, and much the
same conclusions result as studying many shorter sequences—or gross morphology!

THE FIELDWORK REVOLUTION

Behavioral characters have long been used in the taxonomy of birds, but their use in Primates is only
some 20 years old. Bernstein (1971) drew attention
to species-specific behavioral differences in macaques in captivity, and Struhsaker (1975) described
many differences in behavior between red colobus in
different geographic areas. The amazing explosion
of fieldwork studies on Primates since about the
mid-1960s has yielded a vast mass of data, but going
through it all to digest it and extract taxonomically
informative characters has yet to be achieved.

Struhsaker (1970) showed that *Cercopithecus* species that were known to be related on other grounds
also tended to be similar in their vocalizations, especially their loud calls (advertisement calls). Groves

(1974) and Marshall and Marshall (1976) found the same for gibbons, the latter proposing a reshuffling of the gibbons on Borneo by use of vocalizations. An even more significant use of this source of information was the study on black-and-white colobus by Oates and Trocco (1983), who showed that not only were *Colobus polykomos* and *C. vellerosus* specifically distinct, they are not even sister taxa, despite the existence of a widespread hybrid population between them. This brought into question the meaning of the species category in Primates, and so in a way introduced primatologists to the flourishing debates on species concepts in the wider biological field.

Meanwhile the influence of Paterson's recognition species concept was spreading among fieldworkers on Primates. Nowhere has this influence been more pervasive than in the study of the Galagonidae. The apparently rather reasonable assumption has been made that vocalizations, particularly advertisement calls, serve as specific mate recognition systems; hence if two galagos, whether sympatric species or allopatric populations hitherto assigned to the same species, differ in these vocalizations they are ipso facto specifically distinct. Zimmermann et al. (1988) used vocalizations to confirm the specific separation of *Galago senegalensis* and *G. moholi,* and Masters (1991) used them to substantiate the specific distinctness of two species of greater galago (here *Otolemur* spp.), which had formerly been considered as one. Zimmermann (1990) and Bearder et al. (1994) have even proposed to make vocalizations the basis of a higher-level taxonomy, and on this basis Bearder et al. have divided the Galagonidae into six categories or species-groups. But the recent description of "dialects," not in any way of taxonomic significance, in mouse-lemurs (Hafen et al. 1998) reminds us to be cautious about overinterpreting vocalizations in the context of species differentiation.

LATE TWENTIETH-CENTURY SYNTHESIS

W. C. Osman Hill's monumental series *Primates: Comparative Anatomy and Taxonomy* began with the volume on Strepsirhini in 1952 and continued until his death in 1977, at which time eight volumes (covering all except the Colobinae and Hominoidea) had been published. One may criticize Hill's taxonomy, though this did achieve more and more coherence in his later volumes, but that he made extraordinary contributions and virtually revived primatology and kept it alive for quite a few years is undeniable.

The 1960s saw primatology become a worldwide field of interest under the leadership of John and Prue Napier. Comparative and functional anatomy, closely followed by behavior, were early beneficiaries of this renewed interest, but taxonomy was not far behind, especially after Napier and Napier's (1967) *A Handbook of Living Primates.* Later, under the rubric of a series of catalogs of Primate specimens in the British Museum (Natural History) and then elsewhere in the British Isles, Prue Napier began taxonomic revisions of all the Primates, pointing out the places where knowledge was incomplete, making changes where they appeared glaringly necessary, and stressing the need for new understandings. Three volumes were published, and after Prue's retirement the series was completed by the mammalogist Paula Jenkins in the same thorough and cautious yet forthright style.

Only one group of Primates benefited from continuing specialist attention almost throughout this period: the platyrrhines, subject of numerous revisionary monographs by Philip Hershkovitz. It is most unfortunate that, at the time of his death in 1997, the larger platyrrhines still remained unmonographed.

To call Osman Hill, the Napiers, and Hershkovitz giants is not to belittle the work of Albignac, Brandon-Jones, Coimbra-Filho, Colyn, Dandelot, Dao Van Tien, Froehlich, Geissmann, Jenkins, Jolly, Kingdon, Masters, Mittermeier, Olson, Petter, Rumpler, Rylands, Schwartz, Tattersall, Verheyen, Vivo, Wang Yingxiang, and others (to mention only those with experience in museum taxonomy) whose efforts have so increased our understanding of Primate taxonomy over the last half century. In fact, that all these people (with the exception of Dao Van Tien) are still alive and very much kicking serves to indicate how very recent the resurgence of Primate taxonomy really is.

WHERE THE MAIN
COLLECTIONS ARE

The enormous collections of Primates built up over the last couple of centuries are still, in the main, in the museums that first received them. Some museum collections have a virtually complete coverage of all modern Primate groups, down to species level; others have more regional bias, including large collections from particular regions but very little from elsewhere.

The value of museum collections should be evident: they are irreplaceable documents of biodiversity. As Tattersall (1992:38) emphasized, they are essential to conservation efforts: "we must be able to characterize the diversity we wish to protect." There is an irony in this: drawer after drawer of dead specimens, no longer able themselves to contribute to their gene pools, lie there waiting to be interpreted to inform us precisely what those gene pools actually are and what they signify. Without them, we would be helpless; biodiversity study would be impossible. The modern conservation ethic is a very recent phenomenon (Fitter and Scott 1978), and the idea that Primates deserve any special consideration is even more recent, so we must avoid judging our predecessors by today's standards. Nonetheless one cannot help but be appalled by the evidence of past massacres, and one hopes that at the very least the remains of these Primates may play their part in helping to avoid future massacres.

Each museum has its acronym, by which it is universally known, and when citing a particular specimen one includes the acronym of the collection.

The world's largest collection of dead animals is in the Natural History Museum in London, until the early 1990s called the British Museum (Natural History); until the 1870s it was the natural history collections of the one and only British Museum. The acronym remains BM(NH). In the days of Thatcherism/ Vandalism, the museum was subjected to huge involuntary staff redundancies, and one feared that the collections might go the same way. At any rate the great Mammal Section, which in its time housed such luminaries as Gray, Waterhouse, Gunther,

Thomas, Lydekker, Hinton, Chaworth-Musters, Pocock, Morrison-Scott, Laurie, Hayman, Crowcroft, Corbet, and John Edwards Hill, now grudgingly allows a few hours a week for the current incumbents to leave their collection-management duties and do a little research. The science that used to be public property is now under a user-pays regime: "bench fees," the very antithesis of intellectual freedom.

The other British institution with a major Primate collection is the Powell-Cotton Museum, Birchington, Kent (PCM). This consists almost entirely of the collections of Major P. H. G. Powell-Cotton, who from about 1900 to 1940 made numerous hunting trips to Africa (and a few elsewhere). He was just one of many bold white hunters in that era, but, appallingly, he was the only one who kept all his bag and made them available for scientific study (although, until the collection's "rediscovery" by John and Prue Napier, Matschie had been almost the only taxonomist to use it extensively). For monkeys and apes there are huge collections (and I mean huge) from Cameroon, slightly smaller ones from Congo-Brazzaville, eastern Congo-Zaire, Uganda, and Ethiopia, and still smaller ones from Guinea. He was not as interested in bushbabies or pottos, and there are only small collections of them.

The other great world museum with worldwide coverage is the United States National Museum at the Smithsonian Institution (USNM), in Washington, D.C., which has a special Natural History building. The collections are not quite as large as those in London, and in some groups the coverage is not as good, but in others it is better.

Not far behind the USNM is the American Museum of Natural History (AMNH), in New York, a privately funded institution that benefited enormously from the patronage of Theodore Roosevelt. There is very good overall coverage.

Other American institutions with good Primate collections include the Field Museum of Natural History, Chicago (FMNH); the Museum of Comparative Zoology, Harvard University (MCZ); and the Academy of National Sciences, Philadelphia (ANSP).

On the continent of Europe, the major collection

is the Zoologisches Museum A. Humboldt, Berlin (ZMB). It is in the former East Berlin, and it was neglected after World War II, with a single curator most of the time doing what a plethora of them had done before the war. The large collections amassed under Paul Matschie (with almost every specimen duly described by him as a distinct species or subspecies) were wrongly rumored to have been destroyed by a bomb during the war, but this is incorrect. A bomb did fall, but not much was destroyed (alas, the type of *Gorilla beringei* seems to have been), although when I first visited, in 1964, the broken glass was still lying around!

Other major European collections are the Rijksmuseum van Natuurlijke Historie, Leiden (RML); the Senckenberg Museum, Frankfurt (SMF); and the Zoologisches Staatssammlung, Munich (ZSM). The Koninklijk Museum voor Middenafrika or Musée Royale de l'Afrique Centrale, Tervuren (MRAC), near Brussels, has a vast African collection, almost entirely from Congo-Zaire, which was the former Belgian Congo. The Muséum Nationale d'Histoire Naturelle, Paris (MNP), has a wonderful collection of types, stuffed and mounted in lifelike poses, from the days of Geoffroy père and fils, Desmarest, Milne-Edwards, and so on, housed in an underground vault.

There are some smaller collections that are nonetheless quite useful. Two, overshadowed by their larger and more famous neighbors, are the Zoologisch Museum, Amsterdam (ZMA), and the Institut Royale d'Histoire Naturelle, Brussels (IRB); they have restricted but useful collections. Switzerland has the collections in the Universität Zürich-Irchel (UZ), both the Adolf Schultz collection in the Anthropology Department and a smaller collection in the Zoology Department; there are small collections also in Geneva (MSNG) and Basel (NHMB). Italy's only major collection, and that with few Primates, is the Museo Civico di Storia Naturale, Genoa (MSNG). Smaller collections in Germany include those of Karlsruhe, where there are lots of Primate skulls from Liberia, and Dresden, with an important Sulawesi collection (partially destroyed during the war however). In Sweden, the Naturhistoriska Riks-

museet, Stockholm (NRS), has fairly good collections and some of the Linnaean types; other Linnaean types are in nearby Uppsala, both in the Zoology Department of the university and in the Linnaeus House (the cottage where the great man lived).

In Africa, the Kenya National Museum, Nairobi (KNM), has a fairly good local collection, as (I am told) does the Institut Fondamental de l'Afrique Noire, Dakar (IFAN).

In Southeast Asia the Museum Zoologici Bogoriense, Bogor (MZB), has a very large Indonesian collection, and the Zoological Reference Collection, Singapore (ZRC), which is in the National University, has the big collections from Thailand, Laos, Malaysia, Singapore, and Indonesia that used to be in the Raffles Museum. In Vietnam there is an important collection in the Zoology Department, Hanoi University, Hanoi (UH). China has Primates in the Kunming Institute of Zoology, Kunming (KIZ), and other places that I have not visited. Finally, in India there are type specimens from Blyth and others in the Zoological Survey of India, Calcutta (ZSI); these were transferred in the twentieth century from the nearby Indian Museum and before that had been in the Museum of the Asiatic Society of Bengal, and the vicissitudes of transfer, as well as poor government funding, shows itself in the condition (or loss) of some of the skins and/or their labels.

PROTOCOL FOR ALPHA TAXONOMY

It may seem egotistical to tell anyone "how to do it," but for what it is worth it may be helpful to explain how I, at least, usually go about a taxonomic revision.

1. Decide which group to work on. There are often good pressing reasons for looking over the taxonomy of a group: you have noticed a variant that you think others have missed, for example, and you want to see whether it represents a new taxon; or there is already a lively controversy about a group; or there is a zoogeographic issue; or there is a conserva-

tion crisis. Or, often, there has been no revision for several decades, and this alone may seem an adequate reason.

2. Find out where there is a decent collection of specimens of this group. If possible, plan to travel to visit several collections, especially those holding type specimens.

3. In each of your chosen collections, go through the material. Read the labels and check them with the museum records. Note down the information on the label: sex (check this on the specimen if possible), collector, date, and especially the locality. Find the locality on a map if possible. Some collectors left maps or records of their routes. Does the locality, as recorded on the label, tally with where the collector was on that particular day?

4. Note down the taxonomic identification written on the label, and then ignore it. That, after all, is what you want to find out, isn't it? You want to form your own opinion of what species and subspecies it is; don't list a specimen under subspecies A or B just because the person who wrote the label said it is.

5. Take out all the skins (or as many as you can all at one time!) from a restricted (delimited) area and divide them into **simila** (groups of individuals that are mutually similar [Henneberg and Brush 1994]). Do the same with the skulls. Measure the skulls, and use the measurements to help determine what the simila are. Copy down any collector's measurements (head plus body, tail, hindfoot, and ear lengths, and weight, if available) from the labels, and use these too in assorting the specimens. Make sure that age changes and sex differences are well understood: you do not want to describe a special "small-sized" taxon only to find that all the specimens were subadult!

6. Compare the simila from different areas and see if any can be combined: similum A from locality 1 may be identical to similum D from locality 2, while similum B from locality 1 might be identical to similum C from lo-

cality 2. Combine them iteratively until all samples that are identical have been lumped. If they are alike, but not identical, lay them alongside one another, but do not actually combine them. You now have a number of nonoverlapping groups, formed by combining local simila that proved identical and, perhaps, within some of these, several groups that differ from each other on average but not 100%.

7. Try to diagnose the nonoverlapping groups and to characterize the overlapping ones. Your working hypothesis is that the nonoverlapping groups are different species and the overlapping ones are subspecies within species.

8. Test this hypothesis by looking for other characters that might support or refute it: hair banding, hand and foot pads, facial depigmentation, cusp formation, growth regularities, chromosomes. Put the measurements into a computer data file, and run a principal components analysis on them: do you get the same grouping tendencies? Run a series of discriminant analyses on them, with ages and sexes separate, then combined: do the relationships between the groups support those that emerged from the initial analysis? Finally, DNA sequences will not go amiss, but this should not be a substitute for a revision on traditional criteria.

9. Finally, examine the type specimens as far as is possible: which type can be submerged in which sample? In this way, a synonymy can be constructed for each combined similum. If the types of some of the described taxa are not available in the collections you have studied, look up the type descriptions and assess their affinities, taking all the evidence (geographical as well as descriptive and metrical) into account.

10. Now you have your species and subspecies, and you know what names apply to each. If a species or subspecies lacks a name, give it one.

By now, the crucial importance of step 4 has become obvious. One simply must not prejudge the taxonomy—it may all be different from what the books say!

And, of course, this protocol is all very well in principle, but the practice may be very different. For example, the material may be inadequate (or, more rarely, just too overwhelming). A common occurrence is that at least one similum may be represented by just one specimen or a very small number, but this need not necessarily be too much of a stumbling block: if other samples are large enough, then expected ranges of variation can be calculated, and one can assess the probability that the isolated specimen(s) can or cannot fit into one of the other simila.

Today there is a renewed burst of discoveries of new taxa. This is excellent, but with each new taxon there is more need to revise whole groups and fit the new ones in. The work of the taxonomist is never finished.

5

Taxonomy of Primates above the Family Level

Prosimii vs. Anthropoidea, Strepsirrhini vs. Haplorrhini, or what? The arrangement popularized by Simpson (1945), whereby the Primates are split into Prosimii and Anthropoidea, is universally admitted to be based on grades, not clades. Recommendations for its retention (Martin 1990) argue the case on the grounds of familiarity, not because it is in any way a "natural" arrangement; indeed, Martin more than almost any other author insists that prosimians do not form a clade but that tarsiers assort phylogenetically with "Anthropoidea." The tarsier is nocturnal, like many strepsirrhines but only one other haplorrhine *(Aotus);* it is a vertical clinger and has tarsal elongation, toilet claws, a short simple gut, prominent olfactory bulbs, cerebellum exposed in dorsal view, a large postorbital fissure, persistent metopic suture, and primitive tribosphenic molars, all features that are more characteristic of "prosimians" than of haplorrhines. These features have misled authors into classifying it in a clade with lemurs or into recommending that in just this one instance we could adopt a "grade" taxonomy; but the phylogenetic evidence is very clear (Luckett 1993).

The other challenge to the strepsirrhine-haplorrhine dichotomy is the idea that the strepsirrhines may not be monophyletic. The most serious study leading to this conclusion is that of Schreiber and Bauer (1997), who, in an analysis of plasma proteins by a method called comparative determinant analysis (CDA), concluded that the Loriformes, as represented by *Otolemur,* are the sister group of a clade incorporating Malagasy lemurs plus haplorrhines.

What is becoming more and more obvious is the convergence of molecular and morphological analyses of Primates (Goodman et al. 1998). This gives us confidence that a real phylogenetic signal is emerging and has even led to the joint analysis of the two kinds of data to give "total evidence" phylogenies (Yoder 1994).

DIVIDING THE STREPSIRRHINI

The taxonomy of what may broadly be called "lemurs" has gone through several phases. For Simpson (1945), the Lemuriformes formed an infraorder of Primates, under his suborder Prosimii, contrasting with the Lorisiformes and Tarsiiformes; moreover the tree-shrews (now excluded from the Primates altogether and assigned to a separate order, Scandentia) were part of the Lemuriformes and no more distinct from most of the lemurs than was the aye-aye. Once the tree-shrews had been removed, the Lemuriformes acquired the status of a purely Malagasy group.

Placement of the Cheirogaleidae

A suggestion that the Cheirogaleidae might actually be "Lorisiformes" (recte Loriformes) was made independently by me (Groves 1974) and by Szalay and Katz (1974) and acquired some currency, but was finally shown by Yoder (1994, 1997) to be without foundation. Today, it is the aye-aye (*Daubentonia*) whose affinities are in question; these are considered later in this chapter.

So is there any doubt remaining about the affinities of the Cheirogaleidae? I think not. One of the earliest molecular studies (Dene et al. 1976) placed the Cheirogaleidae firmly on the lemuriform stem, probably after the separation of *Daubentonia,* but well before the Lemuridae/Indridae separation. In the chromosomal phylogeny of Rumpler et al. (1989), the Cheirogaleidae share a short stem with other Lemuriformes before becoming separate. In Stanger-Hall's (1997) analysis, using behavioral characters as well as morphological, under the Lundberg rooting scheme Cheirogaleidae share a clade with *Daubentonia* (linked by the loss of chest glands, supposed to be primitive), as they do when Platyrrhini are used as the outgroup; but under schemes whereby tree-shrews or galagos are the outgroup there is no such linkage. Yoder (1997) found Cheirogaleidae to be the sister group to all other Malagasy lemurs except for *Daubentonia*.

Gebo (1985) found that the Lemuridae, together with Megaladapidae and Indridae, and unlike the Cheirogaleidae, exhibit a specialized type of pedal grasp, whereby the hallux and enlarged second toe form a pincer between them. The four lateral digits form one side of an arch; the hallux forms the other. In the Cheirogaleidae, as in the Daubentoniidae, Loriformes, and Haplorrhini, the articulated foot skeleton lies flat on a surface. The tarso-metatarsal joint, although arched transversely, forms a single continuous plane from ectocuneiform to cuboid between rays II and V, rather than being stepped between ectocuneiform and cuboid as in most other Primates. Gebo referred to this as the I–II adductor grasp and to the type of all other Primates as the I–V opposable grasp.

The upshot of all this is that, although the Cheirogaleidae are firmly lemuriform, they are clearly a distinct family and constitute the sister group to Lemuridae + Indridae + Megaladapidae.

Lemuridae, Indridae, and Megaladapidae

The makeup of these three families is not beyond dispute, particularly the correct placement of *Varecia* and *Hapalemur*.

Hagen (1978) proposed that *Varecia* is not a lemurid at all, but might be closer to the Indridae. Subsequent commentators in the main have overlooked this suggestion, but it was revived by Macedonia and Stanger (1994), on the basis of their study of communication signals. These latter authors, however, noted that their results could also be explained by postulating that *Varecia* is sister to a Lemuridae + Indridae clade.

The most emphatic conclusion that *Varecia* is not a lemurid at all is that of Stanger-Hall (1997). She found that the only derived condition that *Varecia* shares with the other Lemuridae is the presence of more than three vallate papillae (and that is, in fact, reversed to the primitive three in *Eulemur rubriventer*); on the other hand the remaining Lemuridae share with the Indridae and Megaladapidae the loss of interramal vibrissae and the reduction of functional mammae to one pair (*Varecia* has the primitive three pairs).

In the eye, however, the fundus is fully pigmented as is that of *Eulemur* (except for *E. mongoz*) (Pariente 1970); the presence of pigmentation in the fundus, whether homogeneous or patchy, appears to be unique to the Lemuridae and so almost certainly autapomorphic. Cones are claimed to be absent (Martin 1990). *Varecia* shares with *Lemur* alone the presence of a lacrimal sutural bone between maxilla and frontal (Yoder 1994).

The other genus whose affinities have been questioned is *Hapalemur* (here split into *Hapalemur* and *Prolemur*). Schwartz and Tattersall (1985) proposed that it was cladistically associated with *Lepilemur,* and on this basis Jenkins (1987) placed them together in Lepilemuridae. For a variety of reasons, including both morphology (Groves and Eaglen 1988) and karyology (Rumpler et al. 1989), this now appears to be incorrect: the gentle or bamboo lemurs, as they are called, are lemurids. Indeed, the real question is whether or not they form a clade with *Lemur* s.s.

It seems likely that Lepilemuridae Stephan and Bauchot, 1965, is a synonym of Megaladapidae Major, 1893. Like *Megaladapis, Lepilemur* has an adult dental formula of 2133/0133 (its deciduous formula is 113/213). It must be admitted, however, that Yoder et al. (1999) could find no similarities to *Lepilemur* in the mitochondrial DNA sequences they managed to extract from *Megaladapis* bone. Like the Lemuridae and Indridae, this family has the I–II adductor pedal grasp described by Gebo (1985).

Gebo and Dagosto (1988) discussed the question of whether the family is closer to the Lemuridae or to the Indridae; they found that the evidence is equivocal. It shares some features of the postcranial skeleton with Indridae, which may either be convergent or indicate that it is an early offshoot from the indrid clade; on balance, Gebo and Dagosto favored the former hypothesis. Yoder's (1997) detailed survey indicated that it is still impossible to say for certain whether the family is closer to Lemuridae or to Indridae, or even sister to a lemurid-indrid clade.

Gebo (1985) and Gebo and Dagosto (1988) summarized or cited the evidence that Indridae and Lemuridae (including Megaladapidae) form a clade, unique among the Primates in having the highly derived I–II adductor grasp of the foot, as opposed to the I–V opposable grasp of all other Primates. Within this clade the Indridae are the most derived in their foot structure, and the extant Indrinae are more derived than the subfossil Archaeolemurinae and Palaeopropithecinae.

In agreement with this anatomical assessment, Dene et al. (1976) have a lemurid/indrid branch after the separation of Cheirogaleidae, whereas Rumpler et al. (1989), on the basis of chromosome morphology, derived the Indridae from a common stem from which Megaladapidae and a Lemuridae + Cheirogaleidae branch emerge separately.

The Position of *Daubentonia*

I have suggested (Groves 1974) that *Daubentonia* is the sister group of all other lemurs and most likely of all other strepsirrhines. I argued that the following features in the aye-aye are plesiomorphic with respect to all other strepsirrhines:

1. Orbital plate of maxilla contacts frontal.
2. Possession of more turbinal bones.
3. Possession of claws (except on the hallux), which moreover retain a thick, deep stratum. Other strepsirrhines lack a deep stratum on claws and nails alike; among other Primates, only tarsiers and platyrrhines retain a (very thin) deep stratum.
4. Certain muscular dispositions.
5. A simpler carotid branching pattern.
6. Mammae (a single pair) inguinal in position.
7. Persistent oogenesis (in this case, shared with Loriformes).

These points have been overlooked by almost all subsequent commentators, but they all remain valid, with the apparent exception of no. 2: the primitive number of ethmoturbinals in Primates is apparently five, so the six reported for *Daubentonia* is a derived condition (Martin 1990).

Tattersall and Schwartz (1974) and Schwartz and Tattersall (1985) associated *Daubentonia* in a clade with the Indriidae (recte Indridae) on the basis of cranial, mandibular, and dental similarities. The cranial similarities are a rounded, globular cranium; deepened splanchnocranium and mandible, expanded gonial angles, and rounded mandibular condyles. These are, broadly speaking, shared by *Daubentonia* and some of the Indridae, especially *Propithecus,* but the cranial features are not, and the mandibular ones scarcely, exhibited by *Indri.* (On the other hand, *Prolemur* has a [superficially] very *Propithecus*-like cranium and mandible.) The mandible is shallow anteriorly but strikingly deepened posteriorly in the Indridae (and some other Lemuriformes, especially *Prolemur*); but it is uniformly deepened in *Daubentonia.* In *Daubentonia,* unlike the Indridae, the gonial angles are not in the least expanded. Indeed, they are the least expanded of any strepsirrhine!

Postcranially, too, *Daubentonia* lacks the derived

conditions of the indrid-lemurid clade (Gebo and Dagosto 1988).

The dental similarities are almost impossible to confirm or deny. Schwartz and Tattersall (1985) bravely tried to homologize the elements of the occlusal surfaces with those of other lemurs, but as they admitted (with reference to M³ in particular), "It is difficult to identify these units with certainty" (p. 34).

One of the points I listed in 1974 must lapse (no. 3). Recently Soligo and Müller (1999) took a fresh look at the structure of claws and nails in Primates and corrected what now appear to be some oversimplifications in the older literature. As far as the strepsirrhines are concerned, Loriformes (as represented by *Galago senegalensis* and *Nycticebus coucang*) have single-layered nails but two-layered toilet claws, as does *Microcebus murinus,* the only cheirogaleid studied, whereas *Lemur catta* has two layers in both nails and toilet claws, exactly like *Daubentonia.* Moreover, *Daubentonia* has a very clear toilet claw homologue (which they illustrated). Their findings abolished the distinction in cheiridial appendages between *Daubentonia* and other strepsirrhines. The divisions among the various taxa studied (in Haplorrhini also) make it essential to extend the study to other families and even genera.

One problem is that *Daubentonia* is so highly autapomorphic. Oxnard (1981) found its postcranial proportions to be absolutely unique in Primates, but could not determine whether this was because of apomorphic conditions or because it is the sister group of all other strepsirrhines. Its single maxillary and mandibular pairs of rootless, bilaterally compressed, sagittally expanded incisors are quite unique in the Primates, and there seems to be no way of working out whether its ancestors ever possessed a toothcomb. The deciduous dental formula is commonly cited as 212/202 or /112, but Luckett and Maier (1986) thought that there are actually three pairs of deciduous incisors in each half of each jaw, the second in each case being preserved as the hypertrophied, rootless "permanent" incisor. I recently corroborated this on museum material.

Cytochrome *b* sequencing placed the aye-aye as the most divergent taxon of the Malagasy group, al-though no cheirogaleids were studied (Pero et al. 1995). The chromosome evidence is equivocal. As reconstructed by Poorman-Allen and Izard (1990), the banded karyotype most resembles that of *Propithecus.* Rumpler et al. (1988, 1989) and Dutrillaux and Rumpler (1995), however, were emphatic that it is completely distinct from other lemurs, with only two possible shared derived conditions from the hypothetical primitive Primate karyotype.

The only evidence known to me that *Daubentonia* is a lemuriform at all are the bullar form, with the tympanic ring inside it, and the palmar and plantar pads. The polarity of the bulla is not clear. In the Lemuriformes, on both palm and sole the first interdigital pad is fused with the thenar, as shown by Rumpler and Rakotosamimanana (1972): a clearly derived condition. All of the pads are quite tiny in *Daubentonia;* the rudimentary little pad in that position is considered by Rumpler and Rakotosamimanana (1972) to represent the typical lemuriform fusion. In my opinion, even this is not quite clear-cut: it is questionable whether the pad is not the first interdigital alone, with the thenar having disappeared.

Most of the available evidence bearing on strepsirrhine phylogeny in general has recently been reviewed in detail by Yoder (1997). She placed *Daubentonia* as a lemuriform, but as unequivocally the sister to all other Malagasy lemurs, and noted that this is where almost all (not quite all) genetic studies place it.

The Fossil Strepsirrhines

The Adapiformes constitute a separate stem from the living strepsirrhines; it is currently not possible with any confidence to assign the living clade a sister group (Ross et al. 1998), although *Europolemur* appears to have had a toilet claw and may eventually be ranked as a plesion of a group containing the living taxa. Within the Adapiformes there are (at least?) three long-lived and divergent families: Adapidae, Notharctidae, and Sivaladapidae, the latter of which has recently been extended back to late Eocene levels (Qi and Beard 1998) and will doubtless be taken back still further in time. Godinot (1998) provided a very full summary of adapiform taxonomy and evolution, including some straightening out of the havoc wrought by previous commentators.

DIVIDING THE HAPLORRHINI

Pocock (1918) proposed the names Tarsioidea and Pithecoidea, the latter for what is still usually called Anthropoidea. He noted that "anthropoid" generally meant the Great Apes ("anthropoid apes"), so was inappropriate for, say, the marmosets. Within the Pithecoidea he recognized two informal groups, Platyrrhini and Catarrhini: an interesting anticipation of the new practice of unranked taxonomic groups.

The most recent and most comprehensive analysis of "anthropoid" relationships, including key fossils as well as living taxa, was by Ross et al. (1998), based on 291 morphological characters. They confirmed that the Adapiformes are strepsirrhines, not the basal stock to the "Anthropoidea" as sometimes argued, and that *Tarsius* forms a clade with the "Anthropoidea," not with the Omomyiformes, although certain genera sometimes considered omomyiforms might well be plesiomorphic sister groups to this clade. Successive splits within the "Anthropoidea" are *Eosimias* (and perhaps *Afrotarsius*), Oligopithecidae, Qatraniinae, Parapithecinae, and Platyrrhini and Catarrhini.

I rank *Eosimias* and *Afrotarsius* as plesia under Simiiformes. The other three groups each have considerable internal diversity and are given suprafamilial names but are left unranked in the classification presented later in this chapter (Table 3).

INTERRELATIONSHIPS OF PLATYRRHINES

The history of opinions on affinities within the Platyrrhini was reviewed by Rosenberger (1981). Early on, the marmosets were conceived as being the most distinctive group and separated as family Hapalidae or Callitrichidae from the rest (Cebidae). Rosenberger himself has since the 1970s advocated that the basic division is between Cebidae, including the marmosets as well as *Cebus* and *Saimiri,* and Atelidae, containing all the rest (Rosenberger and Kinzey 1976; Rosenberger 1977, 1981, 1983a).

Cronin and Sarich (1975) studied platyrrhine albumins and transferrins by immunological methods. They found all platyrrrhine stems to be well separated and could not make any further associations among the following: *Callicebus,* the atelines (including *Alouatta*), *Cebus,* the pitheciines, *Saimiri,* the callitrichines, and *Aotus.* Baba et al. (1979) affirmed the existence of a *Cebus* + *Saimiri* clade, a pitheciine + ateline clade, and a marmoset clade, which included *Aotus,* with *Callicebus* attaching near the base.

Schneider et al. (1996, 1997) sequenced intron 1 of the interstitial retinol-binding gene (IRBP), amounting to a 1.8-kb sequence, and combined this with previously determined sequences for the epsilon globin gene, using *Homo sapiens* as outgroup. The two major clades were (1) Atelinae and a *Callicebus*-Pitheciinae clade and (2) a clade consisting of *Aotus, Cebus* + *Saimiri,* and the Callitrichinae, forming a trifurcation as reported by Schneider et al. (1996), but, on reanalysis, two clades, *Aotus* vs. the rest. The first clade, however, was poorly supported on bootstrap analysis, and the authors proposed a three-family classification for the Platyrrhini: Atelidae, Pitheciidae (including *Callicebus*), and Cebidae (including *Aotus* and the Callitrichinae). Later, workers of the same school (Porter et al. 1997) added the 5' flanking sequences of the epsilon globin gene for some of the species they had previously studied, again concluding that an atelid-pitheciid linkage was possible but unsubstantiated.

Dutrillaux et al. (1986) constructed a phylogeny for the Platyrrhini based on chromosome banding patterns. There was a basal trifurcation between *Aotus,* the Callitrichinae, and a cebine/ateline/pitheciine group. Further evidence led Dutrillaux (1988) to remove the Cebinae from the ateline/pitheciine clade, so that it became a basal four-way split. The most plesiomorphic karyotype is probably that of *Cebus,* although the karyotype of *Lagothrix* also retains many primitive states (Dutrillaux and Couturier 1981).

Horovitz et al. (1998) used a "total-evidence" ap-

Table 2

Characters for a Cladistic Analysis of the Platyrrhini (Numbers in Parentheses Are from the Original Sources)

Source	Characters	Source	Characters
Kay and Williams (1994)	1. Mandibular incisor area relative to M_1 (i5)	Ford (1986)	22. Incisor occlusion (I8): Unordered
	2. I_1 crown width (i7)		23. Canine size (C1 and C2)
	3. I_1 crown shape (i8)		24. Maxillary buccal cingulum reduction (UM7)
	4. I_1 crown height (i9)		25. M_3 reduction (LM2)
	5. Mandibular premolars cristid obliqua (p11)		26. Mandibular buccal cingulum reduction (LM6)
	6. P_1 inflation (p29)		27. Intraparietal and sylvian sulci confluent (B3)
	7. Mandibular molar cusp relief (m17)		
	8. Mandibular molar entoconid (m25)		28. Orientation of rectus, intraparietal, and central sulci (B7)
	9. Mandibular molar posterior sulcus (m26)		29. Distinction between arcuate and rectus sulci (B8)
	10. M_2 hypoconulid reduction (m28)		30. Subcentral posterior sulcus (B9)
	11. Mandibular molar hypocristid (m38)		31. Sylvian and superior temporal sulci (B10)
	12. Mandibular molar distolingual fovea (m41)		32. Anterior orbital fissure (O11)
	13. Mandibular molars cristid obliqua (m43)		33. Dental eruption (E12)
	14. Molar cusp inflation (m44)	Conroy (1981)	34. Asymmetry of emissary foramen loss
	15. I^1 basal cusp (I12)		35. Postglenoid foramen size
	16. Maxillary canine shape (C1)	Sampaio et al. (1991)	36. CA II allele 13
	17. $P^1:M^1$ area (P5)	Rosenberger (1977)	37. P_2 hone
	18. Maxillary molar roots (M1)	Perkins (1975)	38. Epidermal melanin
	19. $M^1:M^2$ area (M4)		39. Dorsal apocrine glands reduction
	20. Maxillary molar cingulum form (M22)	Kay (1990)	40. P_2 projects, massive
	21. Body size (M23)		41. Maxillary canine lingual cingulum raised

proach, joining the 16S, 12S, epsilon globin, and IRBP sequences with 76 morphological characters. They found the same basal dichotomy as Schneider et al. (1996, 1997), but more decisively. The first clade divided into (1) *Aotus* alone and (2) Cebinae + the marmoset group, the second into (1) the pitheciines + *Callicebus* and (2) the atelines (including *Alouatta*).

Bauer and Schreiber (1997) studied a few platyrrhine samples by their comparative determinant analysis method, using serum proteins, and concluded that the platyrrhines are actually paraphyletic: *Cebus* is the sister genus to a clade including *Lagothrix* and the catarrhines.

I gathered together as many characters as possible

from the literature, checking them on specimens wherever possible (these are given in Table 2), and ran a PAUP analysis, with all characters ordered except one. The OTUs were as follows:

Cebus

Saimiri

Callitrichinae

Aotus

Callicebus

Pitheciinae

Atelinae

with *Tarsius, Apidium,* and *Aegyptopithecus* as outgroups. Many characters in the literature were unfortunately too polymorphic to be used; 41 resulted.

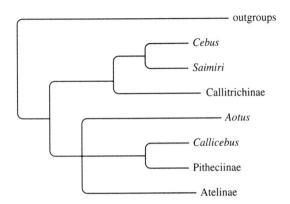

outgroups

Cebus

Saimiri

Callitrichinae

Aotus

Callicebus

Pitheciinae

Atelinae

FIGURE 1. Shortest platyrrhine tree: length 95, consistency index 0.529, retention index 0.360.

The PAUP run produced a shortest tree of length 95, consistency index 0.529, retention index 0.360 (Figure 1). *Aotus* is equivocal in position. The tree then divides into two clades: *Cebus/Saimiri* + Callitrichinae, and Pitheciinae/*Callicebus* + Atelinae. This is remarkably similar to the molecular trees (above), but before we get too enthusiastic it must be admitted that no clade has a bootstrap support of as much as 50%, and a trial PTP run of just 20 randomizations produced one tree of length 95 and one of 93 (i.e., the probability of this result is only 5%) Although the convergence of morphological and molecular phylogenies is noteworthy, much more morphological study of platyrrhines is clearly needed.

A CLASSIFICATION OF PRIMATES TO FAMILY LEVEL

From the discussions of the evidence presented in this chapter, and employing the concept of leaving some subordinate divisions unranked, we end up with the arrangement shown in Table 3. This outline is adhered to in the second part of the book.

Table 3
A Classification of Primates to Family Level

Suborder Strepsirrhini
    Adapiformes

                Family Adapidae
                Family Notharctidae
                Family Sivaladapidae

    Lemuriloriformes
        Infraorder Lemuriformes

            Superfamily Cheirogaleoidea
                Family Cheirogaleidae
            Superfamily Lemuroidea
                Family Lemuridae
                Family Megaladapidae
                Family Indridae

        Infraorder Chiromyiformes

                Family Daubentoniidae

        Infraorder Loriformes

                Family Loridae
                Family Galagonidae

Table 3

A Classification of Primates to Family Level (continued)

Suborder Haplorrhini
    Omomyiformes
                                            Family Omomyidae
    Tarsisimiiformes
        Infraorder Tarsiiformes
                                            Family Tarsiidae
        Infraorder Simiiformes
            Plesion *Afrotarsius*
            Plesion *Eosimias*
            Plesion *Amphipithecus*
            Plesion *Pondaungia*
            Plesion *Siamopithecus*
            Eusimiiformes
                Platyrrhini
                                  Family Cebidae
                                  Family Aotidae
                                  Family Pitheciidae
                                  Family Atelidae
              Catarrhini
                Eocatarrhini
                                  Family Parapithecidae
                Eucatarrhini
                    Plesion Oligopithecidae
                    Plesion *Propliopithecus*
                    Plesion *Aegyptopithecus*
                    Neocatarrhini
                        Plesion *Afropithecus*
                        Plesion *Heliopithecus*
                        Superfamily Pliopithecoidea
                                  Family Pliopithecidae
                        Superfamily Cercopithecoidea
                                  Plesion *Prohylobates*
                                  Plesion *Victoriapithecus*
                                  Family Cercopithecidae
                      Superfamily Hominoidea
                                  Plesion *Griphopithecus*
                                  Plesion *Otavipithecus*
                                  Plesion *Kenyapithecus*
                                  Family Proconsulidae
                                  Family Hylobatidae
                                  Family Hominidae

2

Putting Primate Taxonomy into Practice

Malagasy Lemurs

To a remarkably consistent degree, taxonomic diversity in lemurs is predictable from geographic patterns on Madagascar, which are partly (but not completely) climatically based. So within each widespread genus one can expect that there will be distinct taxa in the eastern rain forest, western dry forest, and southern xerophytic bush zones, and very likely in the Sambirano and Montagne d'Ambre regions as well (Tattersall 1992; Martin 1995). Within some of these zones, major (or even seemingly rather insignificant) rivers often separate presumed sister taxa. At the same time this regularity may even have tended to delay the recognition of the discovery or recognition of "out-of-place" taxa, such as the red mouse-lemur of the dry forest zone or the large hairy-eared dwarf lemur (perhaps *C. crossleyi?*) of the same region.

Family Cheirogaleidae Gray, 1873

Although most analyses find *Phaner* to be sister to other cheirogaleids, Stanger-Hall (1997) concluded that it is sister to *Mirza;* the two together form a clade with *Microcebus* as sister, and *Cheirogaleus* is sister to all these (*Allocebus* was not studied).

Genus *Cheirogaleus* E. Geoffroy, 1812
Dwarf Lemurs

Cheirogaleus major group:

1812 *Cheirogaleus* E. Geoffroy. *Cheirogaleus major* E. Geoffroy, 1812.
1840 *Cebugale* Lesson. *Lemur commersonii* Wolf, 1822.
1840 *Mioxicebus* Lesson. *Mioxicebus griseus* Lesson, 1840.
1841 *Chirogale* Gloger. (Unnecessary) emendation of *Cheirogaleus.*
1842 *Myspithecus* F. Cuvier. *Cheirogaleus major* E. Geoffroy, 1812. Not of de Blainville, 1839 (Daubentoniidae).
1846 *Myoxocebus* Agassiz. Emendation of *Mioxicebus* Lesson.
1873 *Opolemur* Gray. *Opolemur milii* Gray, 1873.

Cheirogaleus medius group:

1894 *Opolemur* Forsyth Major. Not of Gray, 1873.
1913 *Altililemur* Elliot. *Cheirogaleus medius* E. Geoffroy, 1812.

Mittermeier et al. (1994) pointed out that externally members of this genus are easily distinguishable by their prominent dark eye-rings and pink rhinarium.

Forsyth Major (1894) split this genus into two: *Chirogale* (for the *C. major* group, which he divided into *C. milii* [synonym *typicus*] and *C. melanotis*) and *Opolemur* (for the *C. medius* group, divided into *O. samati* and *O. thomasi*). He cited various cranial characters that he regarded as equivalent in value to those separating them from *Microcebus* (while at the same time retaining *furcifer* in *Microcebus!*). In *Opolemur* the skull profile is low and flat, whereas it is high-vaulted in *Chirogale* and *Microcebus;* the mastoid and surrounding regions are convex in *Opolemur* and *Microcebus,* but flat or slightly concave in *Chirogale;* the foramen magnum is higher than broad in *Chirogale* and *Microcebus,* but broader than high, or at least round, in *Chirogale;* and the coronoid is much higher than the condyle in *Opolemur* and in most *Microcebus,* but lower than the condyle in *Chirogale* and in *"Microcebus" furcifer.*

Petter-Rousseaux and Petter (1967) criticized this reliance on cranial characters, which they regarded as excessively variable (and this is the case with most of Forsyth Major's [1894] characters, with the conspicuous exception of the mastoid character, which is perfectly valid). Instead, they invoked diagnostic characters in the dentition: *Cheirogaleus* has the upper toothrows anteriorly convergent; the first premolar (P^2) is at least as well developed as the second (P^3); P^4 is somewhat molariform, with two well-developed small cusps; and the molars are large, with three rounded cusps and no hypocone. Schwartz and Tattersall (1985) cited additional dental distinctions, although some of these also are quite variable and (like some of Forsyth Major's cranial features) distinguish species within the *C. major* group as much as they distinguish the species-groups themselves.

The classification follows Groves (2000b).

Cheirogaleus medius group

Length of head plus body less than 200 mm; tail about the same length. Digits (and sometimes whole of extremities) tend to be white. Median facial strip always white; eye-rings vary. Color paler and grayer, the dorsal stripe narrower and usually sharper, and the white ventral zone more sharply demarcated than in the *C. major* group; there is a broken white "collar" on the sides of the neck; tail is more reddish, less bushy, and expands seasonally, acting as a nutrient store. Braincase is lower and flatter than in the *C. major* group; the maxillary incisors, especially the central incisor, even less forwardly projecting; palate is less extended behind third molars; auditory bullae

larger and mastoid region more convex; maxillary canine less back-curved and with less development of the lingual talon; P^2 higher-crowned, somewhat caniniform; P^3 only barely larger in mesiodistal and buccolingual dimensions than P^2, and more triangular in occlusal view, due to the much bigger lingual pillar; P^4 more triangular, less oblong in occlusal view; molar cusps higher, more pointed; upper molars more rounded lingually, with a relatively larger protocone; and M^3 less reduced distally.

Cheirogaleus medius E. Geoffroy, 1812
Western Fat-tailed Dwarf Lemur

1812 *Cheirogaleus medius* E. Geoffroy. Supposedly from Fort Dauphin; type locality fixed by Groves (2000b) as Tsidsibon River.
1812 *Cheirogaleus minor* E. Geoffroy. Supposedly from Fort Dauphin.
1867 *Chirogalus [sic] samati* Grandidier. Tsidsibon River.

DIAGNOSIS.   Light frosted fawn-gray above, with a broad but indistinct brown dorsal stripe; broadly creamy or yellowish below, yellower on the median strip; a partial white collar around throat, sharply marked, extending well up on sides of neck. A white median facial stripe; eye-rings dark brown, generally restricted to rims of eyes. Hands and feet white. Tail length 86–103% of head plus body length; ear short, 7.9–8.6% of head plus body length. A noticeable diastema between anterior and middle maxillary premolars. Auditory bulla length 8.3–9.7 mm.

DISTRIBUTION.   I have seen specimens from Beraboka, 65 km north of Morondava; Tsidsibon River, west coast; and Ampijoroa. The range therefore seems to be the dry forests of western Madagascar.

NOTES.   The three species described by Geoffroy (1812) were based on drawings; to fix the name, I nominated (Groves 2000b), as neotype for *C. medius*, the type of Grandidier's *C. samati*.

Cheirogaleus adipicaudatus Grandidier, 1868
Southern Fat-tailed Dwarf Lemur

1868 *Cheirogaleus adipicaudatus* Grandidier. Tuléar.
1894 *Opolemur thomasi* Forsyth Major. Fort Dauphin.

DIAGNOSIS. Usually darker above; dorsal stripe more vaguely expressed; more gray below, with traces of creamy yellow more toward midline; white collar much less sharply expressed on sides of neck. White median facial stripe shorter, not extending so far up beyond level of eyes; eye-rings black, thick, invading sides of nose. Only digits are white. Tail much longer than length of head plus body; the ear longer. Diastema poorly developed or absent. Auditory bulla length only 7.2–7.6 mm.

DISTRIBUTION. Apparently the xerophytic bush zone of southern Madagascar. I have seen specimens from Tabiky, west of Ankazoabo; 170 km east of Tuléar; and Fort Dauphin.

Cheirogaleus major group

Within this group, most specimens have pale digits, and pigmented ears. General color tone gray-brown (from wood brown to much grayer), with a reddish tone middorsally that forms a thick dorsal stripe of greater or lesser definition, light tone of underside grading more into upper side, and tail somewhat bushy. None has much white on the face; all have thick blackish eye-rings. There is no elbow stripe or neck collar. Braincase less low and flat than in the *C. medius* group, maxillary incisors more forwardly projecting; upper molars more rounded lingually, with a relatively larger protocone; and M³ less reduced distally (though there are differences among the different species of the group in all these characters). P² lower-crowned, P³ equal in size to P²; P⁴ more oblong in occlusal view; palate more extended behind third molars; auditory bullae smaller and mastoid region flatter; maxillary canine strongly back-curved with a lingual talon. Differences between the two groups are in many cases only exaggerations of the variations among the species within each group.

Cheirogaleus major E. Geoffroy, 1812

Greater Dwarf Lemur

1812 *Cheirogaleus major* E. Geoffroy. Supposedly from Fort Dauphin.
1822 *Lemur commersonii* Wolf. Renaming of *C. major.*

1828 *Cheirogaleus milii* E. Geoffroy. Madagascar.
1833 *Cheirogaleus typicus* Smith. Madagascar.
1840 *Mioxicebus griseus* Lesson. Madagascar.
1842 *Myspithecus typus* F. Cuvier. Madagascar.

DIAGNOSIS. Gray-brown above, creamy gray below, becoming yellower toward midline. Dorsal stripe, where present, vaguely expressed and red-brown. Midfacial zone pale but not white or yellow, does not extend above eye level; eye-rings relatively poorly marked. Feet not white. Ears poorly pigmented, naked or very sparsely haired. Tail much longer than the head plus body, 116% or more. Skull has a high braincase, which falls away steeply at the back, and a deeper facial skeleton. P² projects considerably above level of P³; upper incisors project strongly anteriorly. Skull and teeth very large: skull length 54–61 mm, toothrow length (C′ to M³) 20.3–22 mm.

DISTRIBUTION. The distribution extends down the east coast of Madagascar from (the western shore of?) Antongil Bay to nearly 23°S, corroborating Petter et al. (1977)'s field observations.

NOTES. Petter et al. (1977) described this as large with gray-brown pelage and rounded muzzle, and said that where it occurs sympatric with *C. crossleyi,* it is characteristic of the high forest.

Cheirogaleus crossleyi Grandidier, 1870

Furry-eared Dwarf Lemur

1870 *Chirogalus [sic] crossleyi* Grandidier. Antsianaka Forest.
1894 *Chirogale melanotis* Forsyth Major. Vohémar.

DIAGNOSIS. Warm red-brown, with golden tones, dorsally; no trace of a dorsal stripe; creamy gray below, becoming creamy toward midline. Yellow of midfacial zone invades the region above the eyes; eye-rings blackish, extending onto sides of nose. Feet not light. Ears heavily pigmented, with black fur inside and out. Tail length 85–110% of length of head plus body. Skull slightly smaller (51.5–52 mm), but much narrower than in *C. major;* facial skeleton and crown low; teeth smaller than in *C. major,* toothrow length 18–19 mm, P² lower-crowned, maxillary incisors project more forward.

DISTRIBUTION.   I have seen specimens from (south to north) Imerima, Lake Alaotra (48 km northeast, at 750 m), Périnet, Malewo, Antsianaka, and Vohima. This range is mainly inland of that of *C. major,* and extends farther north.

NOTES.   Petter et al. (1977) distinguished this dwarf lemur from true *C. major* (of which they regarded it as a subspecies) by its slightly smaller size, reddish color, and pointed nose, and its very marked black rings around the eyes. They recorded it along the edge of the plateau, inland from the coast, from north of Maroansetra south to Lake Alaotra, and wrote that it is abundant in plantations and "zones degradées" and in places is sympatric with *C. major.*

Cheirogaleus minusculus Groves, 2000
Lesser Iron Gray Dwarf Lemur

2000 *Cheirogaleus minusculus* Groves. Madagascar: Ambositra, on the central plateau at about 20°S, 47°E.

DIAGNOSIS.   Color iron gray with brownish tones, with a vaguely marked dorsal stripe; tail probably white-tipped; digits white; ears furred along rims. Size extremely small, at the upper end of the length range of *C. medius;* basal skull length 42 mm, toothrow length 16.3 mm; M^3 less reduced compared with other molars.

DISTRIBUTION.   Known only from the type locality, Ambositra.

NOTES.   Despite its more or less equivalent size, this species is distinguishable from *C. medius* by its much darker gray color, darker underside, more bushy tail, lack of white "collar," and by skull characters.

Cheirogaleus ravus Groves, 2000
Greater Iron Gray Dwarf Lemur

2000 *Cheirogaleus ravus* Groves. Madagascar: Tamatave (about 18°S, 14°E).

DIAGNOSIS.   Color and skull shape much the same as in *C. minusculus,* and tail often with a white tip, about equal in length to head plus body length.

Much larger in size; skull length 56–57 mm, with less flattened vault; toothrow longer (18.0–21.6 mm), molars broader even than in *C. major,* P^2 relatively higher-crowned, M^3 extremely reduced, and upper incisors less projecting forward.

DISTRIBUTION.   A small coastal range within that of *C. major,* from Tamatave (about 18°S) to Mahambo (about 17°S).

Cheirogaleus sibreei (Forsyth Major, 1896)
Sibree's Dwarf Lemur

1896 *Chirogale sibreei* Forsyth Major. Madagascar: Ankeramadinika, one day east of Antananarivo.

DIAGNOSIS.   Gray-fawn above, with abundant yellowish white frosting; a dark dorsal stripe; creamy below, this zone advancing farther up the flanks and on the outside of the thighs than in other species. Ears dark, nearly naked. Skull and teeth similar to those of *C. crossleyi* but smaller, length 46.6–48.0 mm, toothrow 17.0–17.8 mm, and M^3 still less reduced.

DISTRIBUTION.   Known from the type locality, Ankeramadinika; from Imerima; and from Pasandava.

Genus *Microcebus* E. Geoffroy, 1828
Mouse-lemurs

1828 *Microcebus* E. Geoffroy. "Le Microcèbe roux" = *Microcebus rufus* Wagner, 1840.
1835 *Scartes* Swainson. *Lemur murinus* J. Miller, 1777.
1840 *Myscebus* Lesson. *Myscebus palmarum* Lesson, 1840.
1840 *Gliscebus* Lesson. *Microcebus murinus* J. Miller, 1777.
1841 *Myocebus* Wagner. Error for *Myscebus.*
1871 *Azema* Gray. *Cheirogaleus smithii* Gray, 1842.
1871 *Murilemur* Gray. *Lemur murinus* J. Miller, 1777.

Maxillary toothrows are somewhat curved, anteriorly strongly convergent so that rostrum comes to a point in front; P^2 tends to be lower-crowned than P^3; P^4 is semimolariform as in *Cheirogaleus;* molars are broad, with pointed cusps including a small hypocone. Bullae are enlarged, reaching posteriorly behind level of basion.

Petter et al. (1977) first clearly distinguished the existence of three "forms" in this genus: a gray one in the west, and two red ones (eastern and western). Their western red species was reported from both Ankarafantsika and Analabe. It remained for Schmid and Kappeler (1994) to formally distinguish the Analabe red form as a species from the eastern red one, and for Zimmermann et al. (1997, 1998) to show that the Ankarafantsika red form is different yet again.

Tattersall (1982) detailed the extraordinarily complex nomenclature of the two widespread species, especially pertaining to the name *rufus*, which, as he showed, has been applied to both eastern and western forms, although fortunately priority belongs to its use by Lesson (1840) for the eastern rain forest species.

Table 4 gives measurements, from the literature, of the four species of *Microcebus* recognized to date. Skull measurements plotted by Martin (1995:555, fig. 9) for a single *M. myoxinus* specimen would be about 28.5 mm, and for two *M. ravelobensis* about 34.3 and 35.3 mm, hence quite consistent with those in Table 4.

Microcebus murinus (J. Miller, 1777)
Gray Mouse-lemur

1777 *Lemur murinus* J. Miller. Madagascar.
1785 *Prosimia minima* Boddaert. Madagascar.
1792 *Lemur prehensilis* Kerr. Madagascar (based on same reference as Boddaert's *Prosimia minima*).
1795 *Lemur pusillus* E. Geoffroy. Madagascar.
1812 *Galago madagascariensis* E. Geoffroy. Madagascar.
1812 *Cheirogaleus minor* E. Geoffroy. Fort Dauphin.

1840 *Myscebus palmarum* Lesson. Madagascar.
1840 *Microcebus rufus* Wagner. Not of Lesson, 1840.
1868 *Chirogaleus gliroides* Grandidier. Tulear.
1910 *M[icrocebus] minor griseorufus* Kollman. Southeast, south, and southwest coasts of Madagascar.

DIAGNOSIS. Gray dorsally with various reddish tones, off white ventrally. Ears long; tail barely longer than head plus body length. Jaw musculature well developed, with the temporal lines relatively close together; palate comparatively short (see Martin 1995). Rostrum does not project in front of incisors; cheekteeth cusps lower, more rounded; braincase flatter (Martin 1995).

DISTRIBUTION. Dry forests of west and south, from Fort Dauphin (= Tolagnaro) at least to the Sambirano River.

NOTES. Petter et al. (1977) signaled the existence of a probable subspecies from the Angavo escarpment, as far as Anjozorobe Forest, that is pure gray, lacking the reddish tones except on the head and around the eyes, giving it a very "contrasty" impression.

Microcebus rufus (Lesson, 1840)
Red Mouse-lemur

1840 *Gliscebus rufus* Lesson. "A few miles north of Fianarantsoa, central Betsileo" (Harper 1940).
1842 *Chirogaleus [sic] smithii* Gray. Madagascar.

Table 4

Measurements of Currently Recognized Species of *Microcebus,* from Forsyth Major (1894), Albrecht et al. (1990), Schmid and Kappeler (1994), and Zimmermann et al. (1998)

Species (n)	Head + body length (mm)	Tail length (mm)	Hindfoot length (mm)	Ear length (mm)	Skull length (mm)	Weight (g)
M. murinus (57)	100–140	107–152	19.9–25.5	18.8–28.0	29.6–34.3 (n = 247)	39–69
M. ravelobensis (27)	83–134	130–175	22.4–26.9	16.0–26.1	31.4–34.1 (n = 3)	38–68
M. rufus (15)	92–123	100–146.2	17.8–23.2	11.6–20.0	30.6–36.0 (n = 47)	35–62
M. myoxinus (32)	82–104	120–148	(mean 29.7)	14.4–19.0	27 (n = 1)	24.5–38

DIAGNOSIS. Slightly smaller than *M. murinus* (in head plus body length and weight measurements, but not in skull); tail longer. Reddish brown dorsally, yellowish ventrally. Ears smaller. Temporal lines much farther apart than in *M. murinus,* and palate longer; relative to skull size, the two do not overlap (Martin [1995]: see especially figures on p. 555); rostrum projects somewhat in front of incisors; cheekteeth with high, sharp cusps; braincase higher, more rounded (Martin 1995).

DISTRIBUTION. East coast from Tolagnaro to the Sambirano River.

NOTES. According to Petter et al. (1977), there are geographic variations especially in ear size: on Nosy Mangabe, there is a form with noticeably larger ears, whereas on the opposite mainland, 6 km away, the ears are very short. In museum specimens I have seen, specimens from the northeast are darker, with more gray, less yellow underparts, and shorter tails than those from farther south.

Microcebus ravelobensis Zimmermann et al., 1997
  Golden Mouse-lemur

1997 *Microcebus ravelobensis* Zimmermann, Ehresmann, Zietemann, Radespiel, Randrianambinina, and Rakotoarison. 2 km north of Ampijoroa, 16°35′S, 46°82′E.

DIAGNOSIS. Pelage golden brown above, yellow to whitish below; prominent white midfacial strip; dark brown eye-rings; tail brown, darker at tip; long yellow-brown ears. Two skulls of this species assort very clearly with *M. rufus* in the distance between temporal lines and in palate length, and other skull features resemble those of *M. rufus* (Martin 1995).

DISTRIBUTION. Apparently restricted to the Ankarafantsika region, where it is sympatric with *M. murinus.*

NOTES. Although a full description of this species, together with holotype, locality, metrical, and other data, was published by Zimmermann et al. (1998), the first introduction of the name (Zimmermann

et al. 1997) had "characters purporting to differentiate the taxon" (as specified in the *Code*) and so is available.

Microcebus myoxinus Peters, 1852
  Pygmy Mouse-lemur

1852 *Microcebus myoxinus* Peters. St. Augustine's Bay (Schwarz 1931c).

DIAGNOSIS. Size very small; pelage short, rufous-brown with orange tinge and a short median dorsal stripe; creamy white ventrally. White median facial strip well marked. Ears short. Tail relatively long, densely furred. Nonmetrical skull and tooth characters are said to resemble those of *M. rufus* (Martin 1995).

DISTRIBUTION. Known from scattered localities in the western forests. At Kirindy, north of Morondava, from where it was recently redescribed after a lapse of 140+ years (Schmid and Kappeler 1994), it is sympatric with *M. murinus.*

NOTES. The very small size of the skull makes it impossible to decide with confidence whether the relative distance between the temporal lines and the relative palate length resemble those of *M. murinus* or *M. rufus* plus *M. ravelobensis* (Martin 1995:555), although the other skull features are said to be like those of *M. rufus.* In a single museum specimen, the ear length falls into the range of that of *M. rufus* (Martin 1995:551).

Genus *Mirza* Gray, 1870
  Giant Mouse-lemur

1870 *Mirza* Gray. *Cheirogaleus coquereli* Grandidier, 1867.

Maxillary toothrows are convergent anteriorly, but differ from *Microcebus* in that they are straight, and the muzzle is blunt, not pointed anteriorly; on M^1 and M^2, cingulum is divided into two small cuspules. Bullae are reduced.

Although it is probable, according to most analyses (Stanger-Hall [1997] being an exception), that *Mirza* is sister to *Microcebus,* the differences are considerable (greater than among the four species of *Microcebus* now known), and generic status is best retained for the time being.

Mirza coquereli (Grandidier, 1867)
Giant Mouse-lemur or
Coquerel's Mouse-lemur

1867 *Cheirogaleus coquereli* Grandidier. Morondava.

DIAGNOSIS. As for genus.

DISTRIBUTION. Recorded sporadically down the west coast. There is no difference in size between western and northwestern (Sambirano) skulls, which average 50.9 and 50.4 mm, respectively (Albrecht et al. 1990).

NOTES. Mittermeier et al. (1994) gave length of head plus body for this species of 200 mm or more, tail 330 mm, weight about 300 g. Color is rich brown to gray brown; ventrally, reddish gray or yellowish gray. Tail thin, long-haired, dark toward tip.

Genus *Allocebus* Petter-Rousseaux and Petter, 1967
Hairy-eared Mouse-lemur

1967 *Allocebus* Petter-Rousseaux and Petter. *Chirogaleus trichotis* Günther, 1875.

Petter-Rousseaux and Petter (1967) pointed out that the species *trichotis,* formerly referred to *Cheirogaleus,* is at least as closely related to *Microcebus,* but quite distinct from both. The maxillary toothrows converge anteriorly only slightly, but the rostrum is somewhat pointed; P^1 and P^2 are both caniniform, the former almost as long as the canine; molars are small, with pointed cusps, including a small hypocone. P_1 and P_2 are elongated and slender. Auditory bullae are flattened, becoming lost posteriorly in the general mastoid inflation.

Allocebus trichotis (Günther, 1875)
Hairy-eared Mouse-lemur

1875 *Chirogaleus* [sic] *trichotis* Günther. Between Tamatave and Morondava.

DIAGNOSIS. As for genus.

DISTRIBUTION. Known localities are lowlands around Mananara, and the Masoala Peninsula.

NOTES. Mittermeier et al. (1994) gave length of head plus body for this species of about 140 mm, tail 168 mm, weight about 80 g. Color "rosy brownish-gray" dorsally, light gray ventrally. A detailed description is given by Meier and Albignac (1991).

Genus *Phaner* Gray, 1871
Fork-crowned Lemurs

1871 *Phaner* Gray. *Lemur furcifer* de Blainville, 1839.

Members of this taxon differ from all other members of the family in that the maxillary toothrows are parallel, and not anteriorly convergent; I^1 is elongated, and curved anteriorly; canine is extremely developed, and P^2 is caniniform; P^3 is very reduced, P^4 is somewhat so; molars are small.

A revision by Groves and Tattersall (1991) revealed the strong geographic differences between what are apparently population isolates in this species and described them as subspecies, but questioned whether they might not, in fact, represent distinct species. Because the available evidence indicates that these "subspecies" are in fact quite discrete, they are here raised to specific rank. The systematic position of another population, from the region of Iharana (Vohémar), remains to be studied, and an apparently distinctive one from the far south is known only from field observations.

Table 5 gives measurements of the species of *Phaner* and of the unallocated specimen from Vohémar.

Phaner furcifer (de Blainville, 1839)
Masoala Fork-crowned Lemur

1839 *Lemur furcifer* de Blainville. Fixed as Bay of Antongil by Groves and Tattersall (1991).

DIAGNOSIS. Dorsal color dark brown, ventrally creamy buff to buffy gray. Crown forks thick and black, fully continuous with dorsal stripe; dorsal stripe fails to reach base of tail, sometimes by a considerable distance. Tail dark on its terminal one-fifth to one-third. Hands and feet dark. Size large. Tail half as long again as head plus body length.

Table 5
Measurements of Taxa of *Phaner*, Based Partly on Groves and Tattersall (1991)

Species	Condylo-basal length (mm)	Bicanine breadth (mm)	Biorbital breadth (mm)	Toothrow length (mm)
P. pallescens	51.6–55.7	11.5–14.1	31.0–33.5	12.8–14.0
P. parienti	54.1–55.9	12.5–14.5	30.3–34.0	14.0–15.5
P. electromontis	55.8–56.3	13.0–13.5	34.3–35.7	13.9–14.5
P. furcifer	—	12.3–14.7	32.5–35.4	15.2–15.4
Vohémar	59.2	14.0	34.4	15.3

DISTRIBUTION. Known from the Masoala Peninsula.

Phaner pallescens Groves and Tattersall, 1991
Western Fork-crowned Lemur

1991 *Phaner furcifer pallescens* Groves and Tattersall. Tabiky (22°10′S, 44°15′E), ca. 20 km NW of Ankazoabo.

DIAGNOSIS. Dorsally light gray-fawn with silvery sheen; ventrally whitish to pale yellowish. Crown forks and dorsal stripe relatively poorly defined and only indistinctly continuous, but dorsal stripe reaches rump (though vaguely). Tail dark for its terminal half to three-fourths. Hands and feet only slightly darkened. Skull small with broad muzzle; teeth small.

DISTRIBUTION. Patchily found in the western forests, from latitude of Toliara north to Soalala; apparently absent between the Tsiribihina River and the Soalala District.

NOTES. The single available specimen from Soalala differs from those from farther south, being larger in size and with a shorter tail (equal to length of head plus body, instead of 137–164% of it). Further data are required to find out whether it represents a distinct taxon.

Phaner parienti Groves and Tattersall, 1991
Sambirano Fork-crowned Lemur

1991 *Phaner furcifer parienti* Groves and Tattersall. Sjangoi (= Djangoa, 13°50′S, 48°20′E), 20 km SW of Ambanja.

DIAGNOSIS. Pelage thick, dense; dorsal color dark brown; crown forks broadly continuous with dorsal stripe; dorsal stripe thick and black and extends without fading at least to base of tail; only distal one-third of tail dark; ventrally buffy, often with reddish tints. Occasionally a white tail tip. Size larger than *P. pallescens*, with larger teeth. Tail at least as long as head plus body length.

DISTRIBUTION. The Sambirano region.

Phaner electromontis Groves and Tattersall, 1991
Amber Mountain Fork-crowned Lemur

1991 *Phaner furcifer electromontis* Groves and Tattersall. Montagne d'Ambre (Ambohitra), approximately 12°40′S, 49°10′E.

DIAGNOSIS. Size very large; dorsally light gray, with thick, black crown fork and dorsal stripe, which extends onto rump; ventrally light gray. Hands and feet slightly darkened (especially hands). Tail considerably longer than head plus body length, darkened only on its distal one-third. Teeth relatively small.

DISTRIBUTION. Apparently confined to Montagne d'Ambre.

Family Lemuridae Gray, 1821

All members of the Lemuridae so far examined have at least spotty pigmentation of the ocular fundus, partially or completely obscuring the tapetum lucidum, which in all other strepsirrhines examined (including the diurnal *Propithecus*) shines brightly golden. The pigmentation is patchily distributed in *Lemur*, *Hapalemur*, and *Eulemur mongoz*, but homogeneously (totally obscuring the tapetum, assuming there is one) in *Varecia* and in other *Eulemur* species as far as known (Pariente 1970).

Dene et al. (1976), who combined *Varecia, Eulemur,* and *Lemur* in their genus *Lemur,* had *variegatus* branching off first, then *catta,* with *rufus* separating before a final divergence between *macaco* and *mongoz;* they did not examine *Hapalemur* or *Prolemur.* In their chromosomal phylogeny, Rumpler et al. (1989) showed *Varecia* branching off first, followed by a bifurcation between *Eulemur* and the other three genera. In a reconsideration, Dutrillaux and Rumpler (1995) made this *Varecia* vs. *Lemur/Hapalemur/Prolemur* vs. *Eulemur* split a trifurcation.

Genus *Lemur* Linnaeus, 1758

Ring-tailed Lemur

1758 *Lemur* Linnaeus. *Lemur catta* Linnaeus, 1758.
1762 *Prosimia* Brisson. *Lemur catta* Linnaeus, 1758. Unavailable.
1780 *Procebus* Storr. *Lemur catta* Linnaeus, 1758.
1785 *Prosimia* Boddaert. *Lemur catta* Linnaeus, 1758 (note that Tattersall's [1988] attempted designation of *mongoz* as the type of this name, following Hill (1953), Groves (1974), and Hagen (1978), is invalid [see Groves 1989]).
1806 *Catta* Link. *Lemur catta* Linnaeus, 1758.
1819 *Maki* Muirhead. *Maki mococo* Muirhead.
1878 *Mococo* Lesson in Trouessart. *Lemur catta* Linnaeus, 1758.
?1895 *Eulemur* Haeckel. Nomen nudum; no species mentioned.
1961 *Odorlemur* Bolwig. *Lemur catta* Linnaeus, 1758.

Brachial and antebrachial glands are present; perianal glands are present; scrotum is naked. The ocular fundus is only patchily pigmented, leaving a very bright tapetum mostly visible; there is a rudimentary foveal depression (Pariente 1970).

Lemur catta Linnaeus, 1758

Ring-tailed Lemur

1758 *Lemur catta* Linnaeus. Madagascar.
1819 *Maki mococo* Muirhead. No locality.

DIAGNOSIS.   As for genus.

DISTRIBUTION.   The likely existence of two subspecies of this species has been signaled by Goodman and Langrand (1996). Animals seen and photographed in the wild on the high (approximately 2,000 m), cold Andringitra Massif, on the eastern edge of the species's range, differed notably from those from the rest of the distribution. They are dark rufous brown on the body, dark brown on the rump, and dark gray-brown on the limbs, whereas "normal" *L. catta* are light reddish gray, with light grayish rump and limbs, and they have fewer tail rings.

Genus *Eulemur* Simons and Rumpler, 1988

Brown Lemurs

1988 (15 September) *Eulemur* Simons and Rumpler. *Lemur mongoz* Linnaeus, 1766.
1988 (6 October) *Petterus* Groves and Eaglen. *Eulemur fulvus* E. Geoffroy, 1796.

For discussion on the nomenclature of this genus see Groves (1992).

In the fundus of the eye in all known taxa except *Eulemur mongoz,* heavy pigmentation obscures a lighter base that may be the tapetum, which is otherwise universal in strepsirrhines; *E. coronatus* and *E. albocollaris* were not studied. *Varecia* is the only other lemur with homogeneous dark pigmentation of the fundus (Pariente 1970). As far as is known, only *E. mongoz* differs: as in *Lemur* and *Hapalemur,* the fundus is pigmented, but only patchily. Cones are certainly present in *E. mongoz* (as they are in *Lemur* and in the diurnal Indridae), but are claimed to be absent in the *E. fulvus* group (Martin 1990).

Hagen (1978) and Groves (1989) proposed the use of the principle of metachromism of Hershkovitz (1968), sketching preliminary phylogenetic schemes on this basis. Shedd and Macedonia (1991) elaborated this in some detail, refuting some of my (Groves 1989) tentative lineages and substituting their own (probably too few specimens [preserved skins] had been examined by Groves, whereas Shedd and Macedonia examined hairs from up to 30 living individuals for some taxa). They found that the primitive state in the Lemuridae has three phaeomelanin bands, and this is found in *Lemur* and *Hapalemur* (*Prolemur* and *Varecia* were not part of their study) and in some, but not all, females of *E. collaris* and *E. coronatus.* Males of *Eulemur* commonly have fewer bands than females and so are metachromatically more derived; hence Shedd and Macedonia (1991) constructed their phylogenetic sequences on the basis of females alone. *Eulemur rufus* and *E. albifrons* alone are sexually monomorphic; only males of *E. macaco* and some *E. fulvus* and *E. collaris* have unbanded (saturated) hairs.

Shedd and Macedonia (1991) proposed two metachromatic sequences. For what they regarded as the subspecies of *Eulemur fulvus* the sequence was *collaris-fulvus-sanfordi-rufus-albifrons;* no specimen of *albocollaris* was available to them, but they felt able to confirm that it would be a separate derivation from *collaris,* based on my (Groves 1989) data. The sequence of the other *Eulemur* taxa ran *coronatus-mongoz-rubriventer-flavifrons-macaco.* Broad phaeomelanin bands are more plesiomorphic than narrow ones; hence, of the two taxa with three phaeomelanin bands, *E. coronatus* is more primitive than *E. collaris,* and so stands at the base of the entire sequence.

It might be said that *E. rubriventer* makes an uncomfortable intermediate between *E. mongoz* and *E. macaco,* because its one or two phaeomelanin bands are much narrower than the single one of *E. macaco.* Likewise the double or single bands of *E. fulvus* and the single band of *E. rufus* are considerably broader than the bands of *E. collaris, E. sanfordi,* and *E. albifrons,* and it may be better to fork the sequence above the *collaris* level. In fact, Shedd and Macedonia (1991) remarked on the anomaly of a taxon with a more primitive facial pattern (*fulvus*) being higher in the metachromatic sequence than *E. collaris* with its strongly derived pattern.

Weakly developed antebrachial glands have been reported for *E. macaco* and *E. rufus,* but they are not structurally homologous with those of *Lemur* and *Hapalemur* (Macedonia and Stanger 1994). Those authors reported the presence of an interocular gland in *E. coronatus* and *E. rubriventer,* but not in other species.

Hypotheses of interrelationships among the species include the following:

—Groves and Trueman (1995), on the basis of morphological features modified after Groves and Eaglen (1988) and Tattersall and Schwartz (1991), placed *E. coronatus* as sister to the rest of the genus; then the members of the *fulvus* group separate in an unresolved polychotomy, with *albifrons* as sister to a clade containing *E. macaco, E. mongoz,* and *E. rubriventer.* Because this cladogram was based on an admittedly defective data set, it cannot be taken as definitive; but it is interesting that it sees the "species *E. fulvus*" as nonmonophyletic and thus emphasizes the need for its so-called subspecies to be taken separately in any analysis.

—Jung et al. (1992) and Crovella et al. (1992) saw *E. macaco* as sister to the other taxa, with *E. mongoz* and the *E. fulvus* group branching off successively, leaving a common *coronatus/rubriventer* terminal clade.

—Pero et al. (1995), sequencing the mitochondrial cytochrome *b* gene, found two clades: *E. macaco/coronatus,* and *E. fulvus/rubriventer/mongoz,* with *E. mongoz* being slightly farther from the other two than they were from each other.

—Macedonia and Stanger (1994), on the basis of communication signals (vocalizations and glandular), concluded that *E. mongoz* is sister species to the rest; there is then a dichotomy between a *coronatus/rubriventer* clade (in agreement with Crovella et al. [1992]) and a *macaco/fulvus* clade. All taxa of the *fulvus* group except *E. albocollaris* were studied, although lumped together for purposes of their analysis; there is some indication, however (see especially their fig. 6), that there are differences between them in vocalizations.

—Stanger-Hall (1997) found a trifurcation, which she could not resolve on available evidence, between *E. mongoz, E. fulvus* group plus *E. macaco,* and *E. coronatus* plus *E. rubriventer,* although she considered that it was mostly likely that the *E. coronatus/rubriventer* clade was sister to the rest.

In conclusion, although there seems to be some evidence that *E. mongoz* is the sister species to the rest, one cannot insist on this at the moment. There is also a persistent suggestion that *E. coronatus* and *E. rubriventer* are associated, but again this cannot be insisted upon. What one can insist on is full species status for what are currently regarded as subspecies of *E. fulvus.* These species are not only sharply distinct externally, but they also appear to differ consistently in craniodental characters (Tattersall and Schwartz 1991). Two of them, *collaris* and *albocollaris,* have unique DNA sequences and are already acknowledged as diagnosably distinct entities (Wyner et al. 1999). There is no evidence of overlap in phenotypic character states among the members of the group, so they qualify as species under the PSC; there is little or no evidence that they form a genetic continuum in the wild, so they even qualify under the BSC. There is also the distinct possibility (as already mentioned) that they form a nonmonophyletic group, with one (or more?) forming a clade with taxa that are acknowledged as full species of *Eulemur.*

Eulemur macaco (Linnaeus, 1766)
  Black Lemur

DIAGNOSIS.   Male entirely black, female reddish, varying from yellowish to dark brownish in tone. Dorsal hairs with one phaeomelanin band in female, none in male.

Eulemur macaco macaco (Linnaeus, 1766)

1766 *Lemur macaco* Linnaeus. Madagascar.
1775 *Lemur macaco niger* Schreber. Madagascar.
1863 *Lemur leucomystax* Bartlett. Madagascar.

DIAGNOSIS.   Hairs around face elongated, continuous with long ear tufts like those of male *E. sanfordi;* in female these are pale, whitish. Female has black muzzle and interocular stripe.

DISTRIBUTION.   From the Mahavavy River in the north to the Sambirano River in the south; south of that river is a zone of intergradation with *E. m. flavifrons.*

Eulemur macaco flavifrons (Gray, 1867)

1867 *Prosimia flavifrons* Gray. Madagascar.
1880 *Lemur nigerrimus* Sclater. Madagascar.

DIAGNOSIS.   Hairs around face and on ear rims not elongated; in female, pale cheek hair tends to continue around above eyes. Phaeomelanin band on dorsal hair of female is broader than in *E. m. macaco.*

Size is slightly larger (Albrecht et al. 1990). Iris, uniquely in the genus, is not brownish orange but pale blue or green.

DISTRIBUTION.   Between the Andranomalaza and Sandrakota Rivers. An intermediate population with *E. m. macaco,* between the Andranomalaza and Sambirano Rivers, is said to resemble *E. m. flavifrons,* but with light brown eyes instead of bluish (Mittermeier et al. 1994).

Eulemur fulvus (E. Geoffroy, 1812)
  Brown Lemur

1812 *Lemur fulvus* E. Geoffroy. Madagascar.
1840 *Prosimia macromongoz* Lesson. No locality.
1844 *Lemur bruneus* van der Hoeven. Replacement for *fulvus.*
1866 *Lemur mayottensis* Schlegel. Comoros: Mayotte.

DIAGNOSIS.   Dorsal hairs with one or two fairly broad phaeomelanin bands in female, one or none in male. Facial mask, crown, and temples black, somewhat more so in males than in females, with a full whitish cheek beard with remnants of agouti banding.

DISTRIBUTION.   In the west, from Ankarafantsika north to the Manongarivo (Andranomalaza) River and the Mahavavy River, in the Sambirano Massif, and probably the Tsaratanana Massif. In the east, northeast of Antananarivo and apparently "well into the suggested range of the white-fronted lemur" (Mittermeier et al. 1994:179); Evans et al. (1995) confirmed it for Ambatovaky (16°51′S) and cited evidence that the two are sympatric or parapatric along a wide contact zone.

NOTES.   Many authors (Jolly 1966; Petter et al. 1977; Mittermeier et al. 1994) have mentioned differences between northern/western and southern/eastern populations of this species. The northern ones (also found, presumably introduced, on Mayotte in the Comoros) have large, prominent light patches above the eyes; in the southern ones the patches are only slightly marked. Jolly (1966) and Petter et al. (1977) added that the northern/western form has a less thick pelage, whereas the thicker-coated eastern/southern form is also darker in color. The difference between eastern/southern and western/northern forms in size, as represented by skull length, is quite substantial (92.2 vs. 88.1 mm, respectively), although the standard deviations overlap; unexpectedly, perhaps, the Mayotte population is closer to the eastern one in size (Albrecht et al. 1990).

Museum skins in the London (BM[NH]) and Paris (MNP) collections actually assort into three regional groups, not just two:

1. Main eastern region (Lake Alaotra, Babaconte, Lakato, Lokosy): Olive-gray above, pale gingery below. Hands and feet redder. Head black, sending a short black line back to nape. Ear tufts white; cheek whiskers pale fawn ($n = 6$).

2. Northeastern region (Vohémar and Ambatoroa): Olive-gray above, but very pale below. Head with very little black; ears and cheeks with little white/fawn. Mayotte lemurs most resemble this variant. It is necessary to point out that, despite the localities, these are distinct from (female) *E. albifrons;* not only are there males in the sample, but the color of female *E. albifrons* is a nearly uniform dark brown. The ranges of the two species need to be closely delimited ($n = 4$).

3. Western region (Anerontsanga, Ampijoroa): More closely resembles the northeastern form, but much darker, grayer on average, and underside gray with less gingery tinge.

These differences are not those recorded in the literature (in Notes), and more attention needs to be paid by field workers to possible differences.

Eulemur sanfordi (Archbold, 1932)
   Sanford's Lemur

1932 *Lemur fulvus sanfordi* Archbold. Montagne d'Ambre.

DIAGNOSIS.   Dorsal hair with two, sometimes only one, narrow phaeomelanin bands in female, one in male. Male with long, not bushy, whitish cheek whiskers, almost continuous with long white ear tufts; facial mask (restricted to circumocular rim, interocular region, and muzzle) black. Female with en-

tirely gray face (except for blacker tones in midfacial region), without cheek whiskers or ear tufts or color contrasts.

DISTRIBUTION.   From the Ampasindava Peninsula south to the Mahavavy River in the west and the Manambato River in the east (Mittermeier et al. 1994). According to Mittermeier et al. (1994), populations south of the Manambato River are intermediate between *E. sanfordi* and *E. albifrons*.

Eulemur albifrons (E. Geoffroy, 1796)
   White-fronted Lemur

1796 *Lemur albifrons* E. Geoffroy. Madagascar.
1840 *Prosimia frederici* Lesson. Madagascar.

DIAGNOSIS.   Dorsal hairs with a single narrow phaeomelanin band in both sexes. Male with even more reduced black facial mask than in *E. sanfordi,* but pelage of rest of head (crown, temples, and cheeks) bushy and white. Face of female like that of *E. sanfordi*.

DISTRIBUTION.   From the Bemarivo River in the north to near Tamatava (= Toamasina) in the south (Mittermeier et al. 1994).

NOTES.   Tattersall (1982) drew attention to variation within this taxon: males that lack the white head, and females varying from gray-brown to gray, with light or dark gray head.

Eulemur rufus (Audebert, 1799)
   Red-fronted Lemur

1799 *Lemur rufus* Audebert. Madagascar.
1833 *Lemur rufifrons* Bennett. Madagascar.

DIAGNOSIS.   Dorsal hairs with a single, fairly broad phaeomelanin band in both male and female. Male with a broad, black midfacial stripe from anterior crown to nose; sides of nose and spots above eyes creamy white; cheek beard golden red, crown brick red. Female with smaller beard, black crown, and much larger, but grayish, supraorbital spots.

DISTRIBUTION. In the west, from the Betsiboka River south to the Fiherenana River and a little farther south; in the east, probably from the Mangoro River south to the Andringitra Massif and the Manampatrana River.

NOTES. I noted (Groves 1974) considerable variation in this taxon: males vary from gray with yellow underside to lighter, browner, with off-white underside; females vary from gray-red, even redder ventrally (and on rump), to more foxy red, yellower ventrally. Males average grayer in the west, darker and browner with strongly red crowns in the east; females too tend to be gray-red in the western part of the range, more yellowish red in the east.

As with *E. fulvus,* there are eastern and western populations, and they have been claimed to be somewhat different: in the west the pale eye patches are large and continue down around the eyes, ending in light cheeks in the females; in the east the eye patches are more notable in the males, and females tend to be redder in color (Mittermeier et al. 1994). The two populations are not much different in size: 88.4 mm in the western and 90.5 in the eastern (Albrecht et al. 1990).

As in the case of *E. fulvus,* I examined skins of this species by locality. Again, the museum specimens do not assort according to the literature descriptions. I found three regional forms, as follows:

1. Eastern (Ankafina, Ambohimanga, Manakara, Ampasinambe, Ankoja, Ivohibe): Males (*n* = 6) iron gray above; head dark red to brick red; tail black from about halfway along; digits reddish; underside gray-fawn. The type of *rufifrons* fits in this series. Females (*n* = 4) brownish olive-gray; head red, with virtually no gray above eyes; tail tip pale orange; underside whitish red.
2. Southwestern (Tabiky): Males (*n* = 4) similar to eastern form but less dark; more red on cheeks; the usual black line running through crown to nape is obsolescent; hands entirely red; tail less dark; underside more beige. Females (*n* = 3) similar to eastern form but with very little red on head.
3. Northwestern (Betsako, Namoroka, Tsiandro, Beroboka, Antsigny): I have seen no males from this region at all! Females (*n* = 10) very distinctive, gingery red, with large, light gray supraorbital spots; tail maroon-red; underside orange. The type of *rufus* fits in this series.

Eulemur ?cinereiceps (Grandidier and Milne-Edwards, 1890)

1890 *Lemur mongoz* var. *cinereiceps* Grandidier and Milne-Edwards. Madagascar; probably Farafangana.

DIAGNOSIS. Body bright red-brown or brown-red. Head completely gray, with sparsely haired, very light gray muzzle; the white skin of the face shows conspicuously through the sparse gray fur on the muzzle and around the eyes. No elongated hairs on cheeks, which are gray like crown; no adornments on the ears. Tail dark red-brown or gingery, with no variations in tone along its length. Underside orange. Arms and shanks paler, yellow-gray or gray; hands and feet dark gray or black.

DISTRIBUTION. Known only from Farafangana.

NOTES. I thought (Groves 1974) that this name designated the white-cheeked lemur, later described as *Lemur albocollaris* by Rumpler (1975). Tattersall (1982:63) disputed this, arguing that the illustration in Grandidier and Milne-Edwards's work (there was no verbal description) was more like *rufus,* though not precisely so.

If the plate were the only evidence, the name could be passed over as based on an artist's misinterpretation. But there are specimens. Schwarz (1931c) identified two mounts in Paris (MNP) as the specimens on which the illustration was based, and, though Tattersall (1982) doubted it, one of them really is remarkably like the illustration. They are here accepted as syntypes. One, a female, the holotype according to Schwarz's label (recte lectotype) and the one that resembles the illustration, is much lighter than the other, a sort of orange marmalade tone with pale yellow-gray shanks; the other, of uncertain sex (probably female), the paralectotype, is darker with gray shanks. They are stated to be from Farafangana.

Compared with females of *E. collaris* (and so presumably of *E. albocollaris*) they are much lighter and redder; the cheeks are light gray, with no trace of orange or white whiskers; the muzzle is very light, not black.

Compared with *E. rufus*, they are lighter, brighter red (and not the gingery of the northwestern females), and the head is all gray with pale skin and muzzle, and with just a slight dark mark on the anterior crown in the lectotype.

Compared with *E. mongoz*, which they resemble in the light gray muzzle, they are light red, with wholly light faces, and the hands and feet are not pale.

Eulemur collaris (E. Geoffroy, 1812)

Red-collared Lemur

1812 *Lemur collaris* E. Geoffroy. Madagascar.
1863 *Prosimia melanocephala* Gray. Madagascar.
1863 *Prosimia xanthomystax* Gray. Madagascar.

DIAGNOSIS.   Dorsal hairs with three, sometimes two, narrow phaeomelanin bands in female, occasionally only one; one or occasionally none in male. Male with orange ruff, more creamy around face; midfacial stripe present as in *E. rufus*. Female with less-developed ruff and entirely gray facial mask.

DISTRIBUTION.   From the Mananara River south to the limits of the rain forest near Tolagnaro (Mittermeier et al. 1994).

NOTES.   This form has always been outstanding because of its low chromosome number ($2n = 50$–52, polymorphic), and it has fixed DNA sequence differences from other members of the *E. fulvus* group (Wyner et al. 1999).

Eulemur albocollaris (Rumpler, 1975)

White-collared Lemur

1975 *L[emur] f[ulvus] albocollaris* Rumpler. No locality.

DIAGNOSIS.   Facial ruff of male white, not orange; female is more reddish in body tones.

DISTRIBUTION.   Restricted to a thin strip of forest between the Manampatrana and Mananara Rivers.

NOTES.   The karyotype ($2n = 48$) first set this taxon apart from other members of the *E. fulvus* group. Some DNA sequences are diagnosably different from both *E. collaris* and other members of the *E. fulvus* group taken as a whole (Wyner et al. 1999). Hybrids in captivity with *E. collaris* are sterile, despite their close external similarity. Tattersall (1982) drew attention to the anomaly that some females collected from within the range of this taxon resemble female *E. albifrons* and suspected an error in documentation.

I (Groves 1974) was the first to recognize that this was a distinct taxon, geographically separate from *collaris*, and used the name *cinereiceps* Milne-Edwards and Grandidier (1890) for it. Rumpler's *albocollaris* was named without reference to the possibility that *cinereiceps* could be an earlier name for the taxon. Tattersall (1979), however, doubted their equivalence, and *E. cinereiceps* is listed separately.

Eulemur mongoz (Linnaeus, 1766)

Mongoose Lemur

1766 *Lemur mongoz* Linnaeus. Madagascar.
1812 *Lemur nigrifrons* E. Geoffroy. Madagascar.
1812 *Lemur albimanus* E. Geoffroy. Comoros: Anjouan.
1834 *Lemur dubius* F. Cuvier. Comoros: Anjouan.
1840 *Prosimia brissonii* Lesson. Based on Brisson's (1762) *Prosimia pedibus albis*.
1840 *Prosimia micromongoz* Lesson. New name for *Lemur mongoz*.
1840 *Prosimia bugi* Lesson. Madagascar.
1840 *Prosimia ocularis* Lesson. Madagascar.
1870 *Lemur cuvieri* Fitzinger. Comoros: Anjouan.

DIAGNOSIS.   Unlike other taxa of the genus examined by Pariente (1970), this species was found to have only a spottily pigmented fundus; in this it resembles *Lemur* and *Hapalemur*, but unlike these no trace of a foveal depression was detected. Dorsal hairs with two or only one broad phaeomelanin bands in female (broader in redder than in grayer areas), only one in male. Male reddish-toned gray on body, with white muzzle, short reddish cheek beard, and gray face; female grayer, with gray-white muzzle, white cheek beard, and black face including forehead. Along with *E. coronatus*, this is the smallest species in the genus; skull length averages 79.7 mm on the mainland, 80.7 on Anjouan, and 81.1 on Mohéli (Albrecht et al. 1990).

DISTRIBUTION.   A small area around the Betsiboka River, north to Antsohihy (Mittermeier et al. 1994); introduced to Mohéli and Anjouan in the Comoros.

NOTES.   Groves (1974) and Petter et al. (1977) noted slight color differences between those on the mainland and those on the Comoros. The Comoro females have a dark gray crown, contrasting more strongly with the forehead than in the mainland females (Petter et al. 1977). Though sexual dichromatism is normal for the species, a few males have a femalelike coloration except that the muzzle is darker and the crown is blacker (Petter et al. 1977).

Eulemur coronatus (Gray, 1842)

Crowned Lemur

1842 *Lemur coronatus* Gray. Madagascar at 15°14′S.
1848 *Lemur chrysampyx* Schuermans. Madagascar.

DIAGNOSIS.   Ears very large, prominent, white. Males reddish gray, with black crown and white facial mask, partially surrounded by orange-red ring; females more grayish, including face, with a V-shaped orange-red mark on anterior crown. Dorsal phaeomelanin bands broad, three or two in number in female, two or one in male. Size small, as in *E. mongoz*: skull length averages 80.2 mm.

DISTRIBUTION.   From probably the Bemarivo River in the east and the Mahavavy River in the west to the northern tip of Madagascar (Mittermeier et al. 1994).

Eulemur rubriventer (I. Geoffroy, 1850)

Red-bellied Lemur

1850 *Lemur rubriventer* I. Geoffroy. Madagascar.
1850 *Lemur flaviventer* I. Geoffroy. Tamatave (Schwarz 1931c).
1870 *Lemur rufiventer* Gray. Error for *rubriventer*.
1871 *Prosimia rufipes* Gray. Betsimisaraka country.

DIAGNOSIS.   Ears very small, virtually hidden in fur. Dorsally deep chestnut to maroon red; ventral surface like dorsal in male, but yellow-white in female. Male with large white spot at and below inner corner of each eye. Dorsal phaeomelanin bands very narrow, two or one in female, only one in male. Size

markedly larger than *E. mongoz* and *E. coronatus,* but slightly smaller than *E. macaco* and the *fulvus* group: skull length averages 87.0 mm (Albrecht et al. 1990).

DISTRIBUTION.   Eastern rain forests at middle to high altitudes.

Genus *Hapalemur* I. Geoffroy, 1851

Lesser Gentle Lemurs or
Bamboo Lemurs

1851 *Hapalemur* I. Geoffroy. *Lemur griseus* Link, 1795.
1855 *Hapalolemur* Giebel. Emendation.
1913 *Myoxicebus* Elliot. *Lemur griseus* Link, 1795. Not of Lesson, 1840 (= *Galago*).

There have been persistent proposals that *Hapalemur* forms a clade with *Lemur,* to the exclusion of *Eulemur.* Highly repeated DNA sequences support the existence of such a clade (Montagnon et al. 1993); so does a study of communication signals (Macedonia and Stanger 1994).

As in *Lemur,* brachial and antebrachial glands are present, close to shoulder and wrist, respectively, in males; but whereas *Lemur* males have a spur associated with the antebrachial glands, in *Hapalemur* the structure is a brush (Macedonia and Stanger 1994). The X chromosome is acrocentric.

Again, exactly as in *Lemur,* the ocular fundus is only patchily pigmented, leaving a very bright tapetum mostly visible; there is a rudimentary foveal depression (Pariente 1970). No other strepsirrhine is known to have any trace of a fovea.

Groves and Eaglen (1988) could not confirm the monophyly of *Lemur* + *Hapalemur,* but considered it the favored option. Tattersall and Schwartz (1991) unfortunately used *Hapalemur* as one of their outgroups in their analysis of the Lemuridae, and Groves and Trueman (1995) could not exclude a basal position for it within the family, although neither could a clade with *Lemur* be rejected. Yoder (1997), considering all the evidence, especially that of genetics, very firmly supported a *Hapalemur/Lemur* clade.

Table 6 gives measurements from the literature for the four species of *Hapalemur* recognized here.

Table 6

Measurements of the Species of *Hapalemur,* from Vuillaume-Randriamanantena et al. (1985), Jenkins (1987), Groves (1988), and Albrecht et al. (1990)

Species	Head + body length (mm)	Tail length (mm)	Ear length as % head + body length	Skull length (mean) (mm)	Palate length (mean) (mm)	2n
H. griseus	250–310	320–400	avg. 11.3	65.5	26.1	54
H. occidentalis	250–300	260–400	10.3–11.3	62.6	24.9	58
H. alaotrensis	380–400	380–400	7.4	68.9	29.4	54
H. aureus	379–395	370–410	—	—	—	62

Note: The first three species are considered by the cited authors to be subspecies under *H. griseus.*

Hapalemur griseus (Link, 1795)
   Gray Gentle Lemur or
   Gray Bamboo Lemur

DIAGNOSIS.   Size small; gray with reddish head and often reddish tinges on dorsum; face lighter; no pale areas around the eyes; muzzle relatively short. Ears relatively large, but nearly hidden in fur. Chromosomes: $2n = 54$; 21 acrocentric pairs, 5 metacentric.

DISTRIBUTION.   Eastern forests from Tsaratanana Massif south to Tolagnaro, mainly in bamboo zones.

NOTES.   A specimen in Tananarive Zoo, from an unknown locality, was found by Rumpler and Albignac (1973) to have 56 chromosomes. It may represent an undescribed taxon or merely a polymorphism.

Hapalemur griseus griseus (Link, 1795)

1795 L[emur] griseus Link. Madagascar.
1820 Lemur cinereus Desmarest. Error for griseus.
1851 Hapalemur olivaceus I. Geoffroy. Madagascar.
1917 Hapalemur schlegeli Pocock. Madagascar.

DIAGNOSIS.   Patches on body and crown more olive-toned. Tail relatively long, 118–129% of length of head plus body.

Hapalemur griseus meridionalis Warter et al., 1987

1987 Hapalemur griseus meridionalis Warter, Randrianasolo, Dutrillaux, and Rumpler. Fort Dauphin region.

DIAGNOSIS.   Reddish patches on body, and red on head more accentuated (Mittermeier et al. 1994). Tail shorter, apparently 103–113% of length of head plus body (see Groves 1988).

Hapalemur occidentalis Rumpler, 1975
   Sambirano Gentle Lemur

1975 Hapalemur griseus occidentalis Rumpler. "The north and west" of Madagascar.

DIAGNOSIS.   More uniform gray-brown; face pale. Ears protruding from fur. Chromosomes: $2n = 58$, with 25 acrocentric pairs and only 3 metacentric.

DISTRIBUTION.   The Sambirano region, from south of Maromandia north through Beramanja, and other areas of the west: Lake Bemamba region between Mantirano and Belo-sur-Tsiribihina, and possibly the Ankarana Massif (Mittermeier et al. 1994).

Hapalemur alaotrensis Rumpler, 1975
   Alaotran Gentle Lemur

1975 Hapalemur alaotrensis Rumpler. Lake Alaotra.

DIAGNOSIS.   Size much larger than *H. griseus* and *H. occidentalis* (see diagram of maxillary dental metrics [Vuillaume-Randriamanantena et al. 1985], in which the range of measurements for this taxon does not overlap those of *griseus* and *occidentalis*); head rounder, pelage darker (Mittermeier et al. 1994). Ear relatively short, hidden in fur. Chromosomes: $2n =$

54 as in *H. griseus,* but with an accumulation of heterochromatin.

DISTRIBUTION. Known only from reed beds of Lake Alaotra.

Hapalemur aureus Meier et al., 1987
Golden Gentle Lemur or
Golden Bamboo Lemur

1987 *Hapalemur aureus* Meier, Albignac, Peyriéras, Rumpler, and Wright. 6.25 km from Ranomafana at 21°16′38S, 47°23′50E.

DIAGNOSIS. Size about the same as *H. alaotrensis.* Face black, with a golden yellow face-ring extending to cheeks and throat; gray-brown, with pale orange underfur; brachial gland less close to shoulder, antebrachial gland less close to wrist than in other species of *Hapalemur.* Ear short as in *H. alaotrensis.* Head appears like *H. alaotrensis* (Groves 1988). Diploid chromosomes: $2n = 62$, all acrocentric including the X chromosome; on this basis it was placed by Rumpler et al. (1991) as the sister group to other taxa of the genus (as here restricted).

Genus *Prolemur* Gray, 1871
Greater Bamboo Lemur

1871 *Prolemur* Gray. *Hapalemur simus* Gray, 1871.
1936 *Prohapalemur* Lamberton. *Hapalemur gallieni* Standing, 1905.

This genus is usually synonymized with *Hapalemur,* and the two do share the extremely broadened, shortened muzzle, a probable derived character state; in the chromosome phylogeny of Rumpler et al. (1991) it shares a short (but very short!) common stem with the species here assigned to *Hapalemur.* Vuillaume-Randriamanantena et al. (1985) listed the craniodental distinctions as follows:

—the muzzle is foreshortened, so that I^1 is nearly or quite hidden by the canine in lateral view;
—I^2 is reduced and situated lingual to the I^1– canine axis; its axis is rotated so that the lingual surface faces posteriorly;
—P^3 and P_3 are partially molarized and buccolingually expanded, not sharply contrasting with molarized P^4 and P_4 as is the case in *Hapalemur;*
—P^4 is less broadened than in *Hapalemur;*

—M^1 and M^2, and sometimes M^3, have hypocones, whereas in *Hapalemur* there is at most an "incipient hypocone" on the lingual cingulum;
—there are no mesostyles on the upper molars;
—all mandibular premolars and molars are greatly expanded buccolingually;
—buccal and lingual cusps on mandibular molars are set transversely;
—metaconids are poorly developed, especially on M_3.

There is an enormous brachial gland, situated just proximal to elbow; no antebrachial gland. Chromosomes: $2n = 60$, with two pairs of metacentrics.

In Rumpler et al. (1989), this taxon was depicted emerging from the *Hapalemur* stem just after the separation of *Lemur,* sharing only a single pericentric inversion with *Hapalemur.* On the other hand Macedonia and Stanger (1994) and Stanger-Hall (1997) found that it is less closely related to these latter two genera than they are to each other. Given these affinities, whether it is a sister genus to a clade containing *Lemur* and *Hapalemur* or to *Hapalemur* alone within such a clade, one predicts a similar retinal condition to these two genera, though it has not been investigated to date.

Prolemur simus (Gray, 1871)
Broad-nosed Gentle Lemur or
Greater Bamboo Lemur

1871 *Hapalemur simus* Gray. Madagascar.
1905 *Hapalemur gallieni* Standing. Ampasambazimba (subfossil).

DIAGNOSIS. As for genus.

DISTRIBUTION. Apparently confined to Ranomafana and nearby regions in southeastern Malagasy rain forests.

NOTES. Size is much larger than any *Hapalemur* species; mean skull length 83.6 mm (Albrecht et al. 1990). The subfossil specimens from Ampasambazimba and Andrafiabe, despite the large geographic gap separating them from extant (Ranomafana) and historic (Bay of Antongil) populations, show no sign of being any different (Vuillaume-Randriamanantena et al. 1985).

Genus *Varecia* Gray, 1863
Ruffed Lemurs

1863 *Varecia* Gray. *Lemur variegatus* Kerr, 1792.
1948 *Pachylemur* Lamberton. *Pachylemur insignis* Filhol, 1895.

In the eye the fundus is fully pigmented as is that of *Eulemur* (except for *E. mongoz*) (Pariente 1970); the presence of pigmentation in the fundus, whether homogeneous or patchy, appears to be unique to the Lemuridae and so is almost certainly autapomorphic. Cones are claimed to be absent (Martin 1990). *Varecia* shares with *Lemur* alone the presence of a lacrimal sutural bone between maxilla and frontal (Yoder 1994).

A subfossil species, *V. insignis* (Filhol, 1895), synonyms *Lemur jullyi* Standing, 1905, and *L. majori* Standing, 1908, is essentially part of the modern fauna, but was probably exterminated by human activity before the last two centuries.

Varecia variegata (Kerr, 1792)
Black-and-White Ruffed Lemur

NOTES. The pelage is predominately white to the south, with black often restricted to shoulders and flanks; it is more black to the north, with a white band around the body and white forearms and flanks (Mittermeier et al. 1994). White is often washed with yellow or even brownish.

Petter et al. (1977) and Tattersall (1982, 1986a), the latter following I. Geoffroy (1851), described several pattern morphs within this species. I have seen examples of three of the four major variants, and I am satisfied that they show sufficient geographic separation to rank as subspecies. They are not distinct species, because they grade into one another, with intermediate patterns occurring even within otherwise "typical" populations.

Varecia variegata variegata (Kerr, 1792)

1792 *Lemur macaco variegatus* Kerr. Madagascar.
1819 *M[aki] vari* Muirhead. Madagascar.
1851 *Lemur varius* I. Geoffroy. Madagascar.

DIAGNOSIS. Black (apart from crown and face, extremities, underside, and tail) limited to large square patches on shoulders and upper part of arms, separated by a longitudinal white band. In some individuals, the band reaches the crown, in others it does not, so that the black of the crown may or may not be joined to the shoulder patches. The white band varies in width, and the white ground tone is often slightly or distinctly yellowed or even brownish. This is type a of Tattersall (1986a); the typical black-and-white pattern is type 1 of Petter et al. (1977), and the variant with the brownish or yellowish posterior dorsum is their type 6.

DISTRIBUTION. I have seen specimens from Mahambo and Vidontra (these are both about 17°00'S, 49°20'E), Mananara (16°10'S, 49°46'E), and also the River Faraony, on the east coast at about 21°30'S. This last locality is well south of the others and within the distribution of *V. v. editorum*. This suggests that the differences between the patterns are very simple genetically; presumptively, the *editorum* pattern is a Mendelian dominant, the *variegata* pattern recessive, and the latter occurs at lower frequencies within the *V. v. editorum* distribution and only occasionally (as at Faraony) is found as a homozygote.

Varecia variegata editorum (O. Hill, 1953)

1953 *Lemur variegatus editorum* O. Hill. Ambatondrazaka, south of Lake Alaotra.

DIAGNOSIS. Shoulder patches extend back some way along the flanks, and there is either no longitudinal white line between them or else a short, white indentation running forward from the median dorsal region. Tattersall (1986a) called the ones without and with the indentation b_1 and b_2, respectively; in Petter et al. (1977) they are types 3 and 5.

DISTRIBUTION. Specimens without the white indentation are from Ambatondrazaka (17°50'S, 48°25'E), Manakara (22°13'S, 47°59'E), and Ampoazanambe, which is presumably in the Manakara District. Specimens with a short indentation, varying in width, are from Manakara, Tamatave (18°10'S,

49°23′E), and Sianaka, north of Tamatave. MNP 1880.2516 (no locality; collected by Humblot) has the white indentation long and skewed to the right, cutting off a discrete shoulder patch on the right, so that it has an *editorum* pattern on the left and a *variegata* pattern on the right. Two Paris (MNP) skins, unfortunately without locality, are of this pattern, but have chocolate brown tones on the posterior back and loins. A figure in Evans et al. (1995), said to illustrate the rather uniform pattern occurring in Ambatovaky Reserve (16°51′S), is of this type, not "type C" (pace Evans et al.), but with this brown saddle.

This has the largest range of any of the subspecies, occupying a zone down the coast from north of 18°S to beyond 22°S.

Varecia variegata subcincta (A. Smith, 1833)

1833 *Prosimia subcincta* A. Smith. Madagascar.

DIAGNOSIS. Dorsum black (except, usually, at the base of the tail), with just a white band around body. A few specimens have the anterior two-thirds gray. This is type c of Tattersall (1986a), type 4 of Petter et al. (1977). Another (in the Paris [MNP] collection) has the entire dorsum brown, including a longitudinal line running up the withers.

DISTRIBUTION. All specimens are from the Maroansetra region, west of the Antanambalana River, which separates *V. variegata* from *V. rubra*. Localities are Maroansetra (BM[NH], RML), 40 km NW of Maroansetra (AMNH), and Savari and Maisine (RML). These last two localities cannot be found in any gazetteer, but judging by the dates when Audebert, the collector, was at these places, they cannot be far from the same region.

Varecia variegata subsp.(?)

NOTES. Two morphs described by Petter et al. (1977) cannot be fitted into any of these three subspecies.

Their type 7 has a dark brown ground color, with two black lateral spots. They did not illustrate it, and one cannot get much of a "feel" for the animal from just this description. Only one specimen appears to be known, from "South of Madagascar."

Their type 8 they described as almost completely white on the back; there appear to be no museum specimens, but they observed it in the remains of the coastal forests south of Farafangana (i.e., in the very south of the species's range).

The Paris (MNP) collection has two skins, unfortunately with no locality, that have black crown, shoulders, thighs, and tail and a chocolate-brown back, with just a white triangle on the anterior dorsum that continues forward as a white line, and the usual white zone on rump, posterior thighs, and shanks.

Varecia rubra (E. Geoffroy, 1812)
Red Ruffed Lemur

1812 *Lemur ruber* E. Geoffroy. Madagascar.
1840 *Prosimia erythromela* Lesson. Madagascar.

DIAGNOSIS. Black on extremities, face, crown, tail, and underside, and deep reddish on upper side, except for dorsal side of neck, which is white. Some specimens have white on the ankles or white-tipped hairs on the foreparts.

DISTRIBUTION. The Masoala Peninsula; divided from *V. variegata* by the Antainambalana River. On the upper reaches of the river there are some hybrids between the two species.

Family Megaladapidae Forsyth Major, 1893

Lepilemuridae Stephan and Bauchot, 1965, is a synonym of Megaladapidae. Like *Megaladapis*, *Lepilemur* has an adult dental formula of 2133/0133 (its deciduous formula is 113/213). The relationship between the two genera is still hypothetical, however, and contrary evidence has come recently from a recent extraction of DNA from *Megaladapis* bone (Yoder et al. 1999).

Like the Lemuridae and Indridae, members of this family have the I–II adductor pedal grasp described by Gebo (1985).

Gebo and Dagosto (1988) discussed the question

of whether the family is closer to the Lemuridae or the Indridae, finding that the evidence is equivocal. It shares some features of the postcranial skeleton with Indridae, which may either be convergent or indicate that it is an early offshoot from the indrid clade; on balance, Gebo and Dagosto (1988) favored the former hypothesis. Yoder's (1997) detailed survey indicates that it is still impossible to say for certain whether the family is closer to Lemuridae or to Indridae, or even sister to a lemurid-indrid clade.

The following genus belonging to this family (unless further evidence, such as fuller DNA sequences, refutes the association [see Yoder et al. 1999]) counts, in a way, as part of the modern fauna, but was exterminated, presumably by human action, more than three centuries ago.

Genus *Megaladapis* Forsyth Major, 1893

Effectively, therefore, the only genus of this family that is part of the living fauna is:

Genus *Lepilemur* I. Geoffroy, 1851

Sportive Lemurs

1851 *Lepilemur* I. Geoffroy. *Lepilemur mustelinus* I. Geoffroy, 1851.
1855 *Galeocebus* Wagner. *Lepilemur mustelinus* I. Geoffroy, 1851.
1859 *Lepidilemur* Giebel. Emendation.
1874 *Mixocebus* Peters. *Mixocebus caniceps* Peters, 1874.
1874 *Lepidolemur* Peters. Emendation.

Schwarz (1931c) and Hill (1953) recognized only two species in this genus: *L. mustelinus* from the east, and *L. ruficaudatus* from the west and south; the latter had two subspecies, *L. r. ruficaudatus* from the western dry forests, and *L. r. leucopus* from the southern xerophytic bush. In a careful revision, Petter and Petter-Rousseaux (1960) distinguished five populations, which they assigned to a single species, *L. mustelinus*. These five, which were vicariant, they recognized as subspecies: *L. m. mustelinus, microdon, ruficaudatus, leucopus,* and *dorsalis.*

Rumpler et al. (1989) and Dutrillaux and Rumpler (1995) depicted, on the basis of chromosomes, the following phylogeny for the six species whose karyotypes are known: *L. mustelinus* first, then *L. septentrio-*

nalis; very few differences define the other nodes, but *L. ruficaudatus* seems to have separated first, then *L. leucopus,* with *L. edwardsi* and *L. dorsalis* bifurcating at the tip.

Measurements of the seven species of *Lepilemur* are given in Table 7.

Lepilemur mustelinus I. Geoffroy, 1851

Weasel Lemur

1851 *Lepilemur mustelinus* I. Geoffroy. Madagascar; fixed as Tamatave by Schwarz (1931c).
1875 *Mixocebus caniceps* Peters. Probably between Tamatave and Masindrano (Schwarz 1931c).

DIAGNOSIS. Size large, tail short; molars large (breadth of M^2 4.2–4.7 mm, according to Jenkins [1987]); ears naked, project beyond fur; dorsally brown, head often gray, tail dark brown toward tip, ventrally gray-brown. May be a vague dorsal or crown stripe. Chromosomes: $2n = 34$. In a single living female observed, eyes brown (Petter and Petter-Rousseaux 1960; Petter et al. 1977).

DISTRIBUTION. Eastern coastal forests from about 13°45′S as far, according to Jenkins (1987), as about 20°S.

Lepilemur microdon (Forsyth Major, 1894)

Small-toothed Sportive Lemur

1894 *Lepidolemur microdon* Forsyth Major (in Forbes). Eastern Betsileo, Ankafana Forest.

DIAGNOSIS. Size, tail length, ears as in *L. mustelinus,* but cheekteeth smaller (M^2 breadth 3.4–3.9 mm); dorsally red-brown, usually with dark median dorsal stripe, yellower on flanks; sides of neck light beige, forming an incomplete collar; tail as body but becoming dark brown distally; ventrally yellow-gray. Eyes, in a single living male observed, light yellow (Petter and Petter-Rousseaux 1960). Chromosomes not known.

DISTRIBUTION. Eastern coastal forest, from 18°S to 24°50′S. Sympatric with *L. mustelinus* south of Tamatave (Jenkins 1987).

Table 7

Measurements for the Seven Species of *Lepilemur,* from Jenkins (1987) and Albrecht et al. (1990)

Species	Head + body length (mm)	Tail length (mm)	Tail length as % of head + body length	Skull length (mm)	$2n$
L. mustelinus	300–350	200–300	66–85	59.9	34
L. microdon	300–350	200–300	66–85	58.6	?
L. ruficaudatus	270–323	249–274	77–88	57.5	20
L. edwardsi	272–280	269–280	92–101	58.1	22
L. leucopus	245–261	215–249	85–99	51.0	36
L. dorsalis	252–260	260–278	100–101	54.3	26
L. septentrionalis	278	247	88.8	53.6	34–38

Note: Mean skull measurements are all from Albrecht et al. (1990) except for *L. mustelinus* and *L. microdon;* because those authors did not separate these two species, the values for these two are taken from Jenkins (1987), which were based on rather smaller samples.

NOTES. One of the features by which Petter and Petter-Rousseaux (1960) originally differentiated this taxon from *L. mustelinus* was the shorter tail on average (length 20–25 cm vs. 20–30 cm in *mustelinus*). Further data have shown that this difference does not hold, apparently even on average. Mittermeier et al. (1994) expressed considerable doubt as to the validity of the species, although the yellow collar seems to be a consistent difference; curiously they placed it to the north of *L. mustelinus,* reversing the actual distributions of the two species.

Lepilemur leucopus (Forsyth Major, 1894)
White-footed Sportive Lemur

1894 *Lepidolemur leucopus* Forsyth Major. Fort Dauphin.
1894 *Lepidolemur globiceps* Forsyth Major. Ambolisatra.

DIAGNOSIS. Size small; tail relatively long; ears prominent. Cheekteeth small, M^2 breadth 3.1–3.9 mm. Very light beige-gray, with reddish tones anteriorly; tail often with reddish tones; ventrally light gray or white. Essentially similar to *L. ruficaudatus* but considerably smaller and paler, with longer tail. Skull averaging 51.4 mm in length (Petter et al. 1977). Chromosomes: $2n = 36$.

DISTRIBUTION. The xerophytic bush zone of southern Madagascar, from south of the Onilahy River, about 23°30'S, to Tolagnaro; but the type lo-

cality of *globiceps* is slightly to the north of this, at 23°03'S just north of Tulear, and puts this species in sympatry with *L. ruficaudatus.*

NOTES. Jenkins (1987) found "minor differences" in the skull of *globiceps* Forsyth Major, 1894, and considered that it is either an aberrant specimen or a distinct subspecies. It is the most northerly specimen in her sample. Tattersall (1982) considered it a synonym of *L. ruficaudatus,* but Jenkins (1987) demurred.

Lepilemur ruficaudatus Grandidier, 1867
Red-tailed Sportive Lemur

1867 *Lepilemur ruficaudatus* Grandidier. Morondava.
1873 *Lepilemur pallidicauda* Gray. Morondava.

DIAGNOSIS. Averaging smaller and longer-tailed than *L. mustelinus* and *L. microdon,* but larger than *L. leucopus;* ears prominent; cheekteeth breadth between that of *L. mustelinus* and *L. microdon,* M^2 breadth 3.9–4.3 mm; dorsally light gray-brown (but darker than *L. leucopus*) with various reddish tones, especially anteriorly; tail red-brown above, often white-tipped; ventrally pale gray or white. Eyes yellow (Petter et al. 1977), but perhaps becoming brown with age (Petter and Petter-Rousseaux 1960). Chromosomes: $2n = 20$.

DISTRIBUTION. West coast from about 19°45'S to 23°S and east of Tulear along the Onilahy River.

Lepilemur edwardsi (Forsyth Major, 1894)

Milne-Edwards's Sportive Lemur

1894 *Lepidolemur edwardsi* Forsyth Major. Bombetoka, Betsako Bay.
1898 *L[epidolemur] mustelinus rufescens* Lorenz. Bay of Bombetoka, Ambundube.

DIAGNOSIS.   Averaging smaller than *L. ruficaudatus* and longer-tailed; ears prominent; cheekteeth relatively larger, M² breadth 4.0–4.3; dorsally beige-gray, with strong reddish tones; tail reddish, often with white tip; ventrally gray. Chromosomes: $2n = 22$.

DISTRIBUTION.   Discontinuous in the west from 15°15'S to 18°50'S; according to Tattersall (1982) may extend south to the Tsiribihina River, 19°45'S.

NOTES.   Petter and Petter-Rousseaux (1960) placed this name as a synonym of what at that time they designated *L. m. ruficaudatus,* and it is almost identical externally except for the longer tail.

Lepilemur dorsalis Gray, 1871

Back-striped Sportive Lemur

1871 *Lepilemur dorsalis* Gray. Madagascar.
1894 *Lepidolemur grandidieri* Forsyth Major. Northwestern Madagascar.

DIAGNOSIS.   Size small; tail very long, longer than head plus body length (not "same length" as stated by Petter et al. [1977]); ears small, round, almost hidden in fur. Cheekteeth large, M² breadth 4.0–4.3 mm. Dorsally brown, with indistinct median dorsal stripe; tail as body; ventrally gray-buff. Skull broad, with short rostrum. Eyes brown, according to Petter et al. (1977). Chromosomes: $2n = 26$.

DISTRIBUTION.   The Sambirano region (northwest coast from 12°55'S to 13°58'S), Nosy Bé, and Nosy Komba.

Lepilemur septentrionalis Rumpler and Albignac, 1975

Northern Sportive Lemur

DIAGNOSIS.   Externally very like *L. dorsalis* but larger and longer-tailed, with ears projecting beyond fur; pelage lighter, grayer; M² smaller, 4.0 mm broad.

Skull smaller, but interorbital breadth larger, than in *L. dorsalis* (Jungers and Rumpler 1976). Chromosomes: $2n = 34$–38.

DISTRIBUTION.   Far north: north of Ambilobé, south and east of Montagne d'Ambre to the Manambato River, and probably to the Fanambana River.

Lepilemur septentrionalis septentrionalis Rumpler and Albignac, 1975

1975 *Lepilemur septentrionalis* Rumpler and Albignac. Sahafary Forest.
1975 *Lepilemur septentrionalis sahafarensis* Rumpler and Albignac. Sahafary Forest.

DIAGNOSIS.   Chromosomes: $2n = 34$, 35, or 36. There is a centric fusion between chromosomes 7 and 15.

Lepilemur septentrionalis ankaranensis Rumpler and Albignac, 1975

1975 *Lepilemur septentrionalis ankaranensis* Rumpler and Albignac. Analamerana Forest.
1975 *Lepilemur septentrionalis andrafiamenensis* Rumpler and Albignac. Andrafiamena chain.

DIAGNOSIS.   Chromosomes: $2n = 36$, 37, or 38. The centric fusion, when present (in $2n = 36$ and 37 individuals), is between chromosomes 11 and 16.

NOTES.   I noted (Groves 1989) that, as described by Rumpler and Albignac (1975) and by Petter et al. (1977), there remains a distributional difference between pairs of the "subspecies" of *L. septentrionalis,* and in the capacity of first reviser I selected *ankaranensis* to have priority for the Ankarana/Andrafiamena subspecies.

Family Indridae Burnett, 1828

Jenkins (1987) pointed out that the original use of the family name must be preserved, hence Indridae, not Indriidae, is the correct form. Gray (1870) used the form Indrisina; he also set apart *Avahi* (which he called *Microrhynchus*) as tribe Microrhynchina.

The dental formula in the adult is 2123/2123 (or possibly /2033). The deciduous formula is 213/213.

Gebo (1985) and Gebo and Dagosto (1988) summarized or cited the evidence that Indridae and Lemuridae (including Megaladapidae) form a clade, unique among the Primates in having the highly derived I–II adductor grasp of the foot, as opposed to the I–V opposable grasp of all other Primates. Within this clade the Indridae are the most derived in their foot structure, and the extant Indrinae are more derived than the subfossil Archaeolemurinae and Palaeopropithecinae.

In agreement with this anatomical assessment, Dene et al. (1976) placed a lemurid/indrid branch after the separation of Cheirogaleidae, whereas Rumpler et al. (1989), on the basis of chromosome morphology, derived the Indridae from a common stem from which Megaladapidae and a Lemuridae plus Cheirogaleidae branch emerge separately. In their scheme, *Avahi* branches off earliest, and *Propithecus* and *Indri* share numerous derived conditions; indeed, they could not be certain that *Indri* differs more from the species-groups of *Propithecus* than these latter differ from each other.

Two subfamilies of Indridae are entirely extinct, but only just (by three centuries or so) failed to make it into the living fauna.

SUBFAMILY ARCHAEOLEMURINAE GRANDIDIER, 1905
Terrestrial Lemurs

Genus *Archaeolemur* Filhol, 1895
Genus *Hadropithecus* Lorentz, 1899

SUBFAMILY PALAEOPROPITHECINAE TATTERSALL, 1973
Sloth Lemurs

Genus *Mesopropithecus* Standing, 1905
Genus *Palaeopropithecus* G. Grandidier, 1899
Genus *Archaeoindris* Standing, 1908
Genus *Babakotia* Godfrey et al., 1990

Godfrey et al. (1990) ascribed the new genus *Babakotia* to what they termed Indroidea, suggesting that it could be a primitive member of the Palaeopropithecidae. Jungers et al. (1991) confirmed this assessment, also transferring *Mesopropithecus* from the in-

drid to the palaeopropithecid clade. I am unsure about the merit of recognizing Palaeopropithecidae and Archaeolemuridae as full families; because there is no such thing as superfamily Indroidea, yet the two clades concerned are very clearly related to the living Indridae (as recently confirmed by DNA sequencing for *Palaeopropithecus* by Yoder et al. [1999]), I provisionally include them in that family, as subfamilies. Nevertheless, the terrestrial lemurs and the sloth lemurs are functionally, especially postcranially, very different from the living genera, and their affinities need to be reassessed with respect to the Lemuridae and Megaladapidae.

SUBFAMILY INDRINAE BURNETT, 1828

Genus *Indri* E. Geoffroy and G. Cuvier, 1796
Indri

1796 *Indri* E. Geoffroy and G. Cuvier. *Lemur indri* Gmelin, 1788.
1805 *Indris* G. Cuvier. Emendation.
1811 *Lichanotus* Illiger. *Lemur indri* Gmelin, 1788.
1840 *Pithelemur* Lesson. *Lemur indri* Gmelin, 1788.

Indri indri (Gmelin, 1788)
Indri

NOTES. Darker specimens are found to the north of the range, lighter ones to the south, as described by Thalmann et al. (1993), who found that at Mananara (16°23′S), between the northern and southern regions, there was a mixed pattern: although all individuals had a (narrow) face-ring, most had gray on the outer sides of limbs, and the cap and collar were variable. Thalmann et al.'s study is a classic example of the value of fieldwork in taxonomy, and the results illustrate precisely what "subspecies" should be.

Indri indri indri (Gmelin, 1788)

1788 *Lemur indri* Gmelin. Madagascar.
1796 *Indri brevicaudatus* E. Geoffroy and G. Cuvier. Madagascar.
1799 *Indri niger* Lacépède.
1825 *Indris ater* I. Geoffroy. Error for *niger* Lacépède.
1871 *Lichanotus mitratus* Peters. Nossi Vola or Saralalan.

NOTES. Thalmann et al. (1993) found that, at latitudes 15–15°30′S, indris are predominately black all over (except the sacral region), with no light occipital patch or lateral collar, nor light tones on the outer lower arm and leg, but the face-ring is full and broad. Museum specimens I have seen from Andapa ($n =$ 5) and Maroansetra (4) are of this type, although one of the Maroansetra skins has some gray on the hindlimbs. According to the description of Evans et al. (1995), the indris of Ambatovaky, at 16°51'S, are of this general type.

The earliest indris to be described and made known to science were all of the black form, hence probably from the northern part of the species's range. The type of *brevicaudatus,* in Paris (MNP), is of this type but has a gray-white posterior dorsal "saddle." Schlegel (1876:290) said that the type series of *mitratus* was from "Sera-Lalaw," and neither this nor Saralalan or Nossi Vola can be found in the modern gazetteer. The Berlin (ZMB) and Leiden (RML) specimens are consistent with this; two skins from "Nosivola" are in Paris (MNP) and one from "Sera Lalow" is in London (BM[NH]); all are of the black form.

Indri indri variegatus (Gray, 1872)

1872 *Indris variegatus* Gray. No locality: Lalo River, east of Bay of Antongil (according to Schwarz [1931c]).

NOTES. According to Thalmann et al. (1993), at latitudes 18–19°S indris mostly lack a face-ring, but have a white occipital cap, a white collar extending up to behind the ears, and grayish or whitish outer sides to the leg and lower arm.

The three skins of this type I have seen, all in the London (BM[NH]) collection, are from Tamatave, Sianaka, and Andripirina; and the type of *variegatus* is of this form. Specimens illustrated in the literature from Perinet (= Andasibe) are all like this as well (see, for example, the photo by R. A. Mittermeier on the cover of *Primate Conservation* nos. 14–15, for 1993–1994). A BM(NH) skin from Babaconte, which should be of this form, is more like the northern subspecies, but has not much of a face-ring.

Genus Avahi Jourdan, 1834
Woolly Indris or Avahis

1834 *Avahi* Jourdan. *Lemur laniger* Gmelin, 1788.
1834 *Microrhynchus* Jourdan. Not of Megerle, 1823 (Coleoptera).

1835 *Avahis* I. Geoffroy. Emendation.
1839 *Habrocebus* Wagner. *Habrocebus lanatus* Wagner, 1839.
1840 *Semnocebus* Lesson. *Semnocebus avahi* Lesson, 1840.
1841 *Iropocus* Gloger. *Lemur laniger* Gmelin, 1788.

Rumpler et al. (1990) argued, on the basis of chromosome differences, for the specific separation of the two taxa in this genus. Both have a diploid chromosome number of 70, but they differ in the morphology of individual elements.

Avahi laniger (Gmelin, 1788)
Eastern Avahi

1788 *Lemur laniger* Gmelin. Betanimena country, Antongil Bay (Rumpler et al. 1990).
1795 *Lemur brunneus* Link. Anatala country (according to Schwarz [1931c]).
1812 *Indris longicaudatus* E. Geoffroy. Anatala country.
1840 *H[abrocebus] lanatus* Wagner. Anatala country.
1840 *Semnocebus avahi* Lesson. Anatala country.
1898 *A[vahis] laniger orientalis* Lorenz. East coast of Madagascar.

DIAGNOSIS. Larger, darker; gray-brown to reddish, becoming paler toward tail; tail rusty red. Underside gray, becoming whiter on insides of thighs. Face brownish, with lighter band or patches above eyes, and lighter on cheeks and throat. Ears small, mainly hidden in fur. Weight of four males 1,033 g, of four females, 1,316 g (Mittermeier et al. 1994). Mean skull length 54.2 mm (Albrecht et al. 1990).

NOTES. All these names, except for Gmelin's and Lorenz's, were based on Sonnerat's (1781, plate 89) "Maquis à bourres"; his specimens were obtained in Anatala country.

Avahi occidentalis (Lorenz, 1898)
Western Avahi

1898 *Avahis laniger occidentalis* Lorenz. Ambundube.

DIAGNOSIS. Lighter, smaller, weight 859 g ($n = 4$); mean skull length 51.7 mm (Albrecht et al. 1990:7 [table III]). Light to medium gray, "sometimes flecked with brown or olive," paler posteriorly. Tail usually gray but sometimes reddish. Underside creamy. Face, throat, and cheeks entirely pale, with white "eyebrows." Chromosomes 1 and 2 are quite different from those of *A. laniger.*

DISTRIBUTION. Sporadically in northwest: Anka-rafantsika and Sambirano regions, and Tsingy de Be-maraha Reserve north of Ankavandra.

NOTES. The Ankarafantsika population is lighter in color than the other populations of this species (Mittermeier et al. 1994). It is likely that more than one subspecies, or even species, exists along the west coast of Madagascar (Thalmann and Geissmann 2000).

Genus *Propithecus* Bennett, 1832
Sifakas

1832 *Propithecus* Bennett. *Propithecus diadema* Bennett, 1832.
1833 *Macromerus* A. Smith. *Macromerus typicus* A. Smith, 1833.

It has always been unclear why all sifakas (*P. tattersalli* excepted) have traditionally been placed in just two species; it is probably a simple legacy of Ernst Schwarz's (1931c) overlumping. I have divided both of the two "catchall" species, but I am still not convinced that I have gone far enough.

Propithecus diadema group

Members of this species-group are large, with long, silky pelage; tail shorter than head plus body length; small heads (skull length about 18.3% of head plus body length); and a diploid chromosome number of $2n = 42$ or 44, although it is not certain how many of the "subspecies" (here mostly regarded as species) have been karyotyped.

Propithecus diadema Bennett, 1832
**Diademed Sifaka or
Simpoon**

DIAGNOSIS. The largest species of the genus. Long, silky fur, mainly white with generally silvery or golden tints.

Propithecus diadema diadema Bennett, 1832

1832 *Propithecus diadema* Bennett. Madagascar.
1833 *Macromerus typicus* A. Smith. No locality.
1862 *Indris albus* Vinson. Alanamasoatrao, between Andevoranto and Antananarivo; perhaps = Analamazaotra according to Jenkins (1987).

DIAGNOSIS. Pelage mainly white, with black crown and nape, this tone sometimes extending down onto dorsum or shading off into silvery gray; hindquarters and hindlimbs usually light golden. Tail white or golden. Hands and feet black. Weight 6–7.25 kg. Length of head plus body 520 mm, tail 465 mm (Mittermeier et al. 1994). Mean skull length 91.2 mm (Albrecht et al. 1990). Skull narrow; interorbital region flat, hardly at all depressed between the swollen medial portions of the supraorbital rims. External auditory meatus large. Posterior margin of ascending ramus abruptly expanded backward in gonial region.

DISTRIBUTION. Rain forest from the Mangoro River to south of the Antainambalana River.

Propithecus diadema candidus Grandidier, 1871

1871 *Propithecus candidus* Grandidier. North of Bay of Antongil.
1872 *Propithecus sericeus* Milne-Edwards and Grandidier. Sambava, 14°S, northeast coast of Madagascar.

DIAGNOSIS. Pelage entirely white, with usually a brownish pygal patch. Some specimens have at least traces of black on the occiput, nape, or back, as in *P. d. diadema*.

DISTRIBUTION. Replaces *P. d. diadema* north of the latitude of the Bay of Antongil.

NOTES. Petter et al. (1977) emphasized that the variations in this subspecies approach *P. d. diadema;* the whitest examples were seen by Peyrieras on the slopes of Marojejy. Mean skull length 90.7 mm (Albrecht et al. 1990). Interorbital region more depressed than in *P. d. diadema;* gonial region less expanded; skull broader.

Propithecus edwardsi Grandidier, 1871
Milne-Edwards's Sifaka

1871 *Propithecus edwardsi* Grandidier. West of Mananzary.
1872 *Propithecus bicolor* Gray. Probably Masindrano according to Schwarz (1931c); but there is no evidence for this according to Jenkins (1987).
1875 *Propithecus holomelas* Günther. Nandihizana Forest, Fianarantsoa, central Betsileo.

DIAGNOSIS. Pelage dense, black or dark chocolate brown, becoming lighter, browner, on flanks, and generally with white flank patches, sometimes meeting across back. Weight of six females, 5.9 kg; of eight males, 5.6 kg. Length of head plus body 480 mm, tail 460 mm (female), 440 mm (male). Mean skull length 88.5 mm (Albrecht et al. 1990). Skull broad; whole interorbital region very convex; gonial region hugely expanded backward. Mandibular molars small, premolars broadened. Diploid chromosome number $2n = 44$, including a pair of tiny acrocentrics (Rumpler et al. 1988).

DISTRIBUTION. Rain forest, between the Mangoro and Mananara Rivers, and possibly occurred north of the Mangoro in the past (Tatttersall [1986b] located the old locality of Ampitambé, where *P. edwardsi* was known, some way north of the Mangoro).

NOTES. Simons (1988) represented the alarm call as "shim-poon," and its spectrographic illustration is quite different from that of *P. tattersalli*, *P. verreauxi*, or *P. coquereli*. The *holomelas* form, which lacks the white flank patches, appears to have been restricted to the type locality, where it occurred in troops of normally colored *P. edwardsi* (Tattersall 1986b). It is interesting that the mean skull length value for "*holomelas*" given by Albrecht et al. (1990) is only 85.7 mm ($n = 7$), noticeably smaller than that given by the same authors for 19 specimens of "*edwardsi*."

Propithecus perrieri Lavauden, 1931

Perrier's Sifaka

1931 *Propithecus perrieri* Lavauden. Forest of Analamera, southeast of Diego Suarez.

DIAGNOSIS. Long, silky fur, deep black with brownish underside. Ears hidden in fur, black, hairless, as is muzzle. Tail relatively short. Eyes dark brown. Mean skull length 87.0 mm, average of two specimens (Albrecht et al. 1990). I have not seen the skull of this species.

DISTRIBUTION. Dry forest from Ankarana Massif to the east coast.

Propithecus tattersalli Simons, 1988

Tattersall's Sifaka

1988 *Propithecus tattersalli* Simons. Dry forest about 6–7 km northeast of Daraina, Antseranana Province, 13°09′S, 49°41′E.

DIAGNOSIS. Very small, weight averaging 3.3 kg (Simons 1988). Pelage short, sparse, mainly white with a golden orange wash on upper chest and rump; orange or golden crown-cap, often extending to shoulders and separated from bare (black) skin of face by a white ruff. Ears small but prominent, tufted. Pelage short. Tail 98.7% of head plus body length. Length of the single available skull 83.0 mm; I have not seen the skull of this species. Eyes yellow-orange.

DISTRIBUTION. A small area around Daraina, Madirabe, and Ampandraha; a range with a diameter of about 25 km (Simons 1988). Separated from the similarly dry-forest *P. perrieri* by the Andreva and Lokia Rivers and the Andrafiamena ridge; the range of *P. diadema candidus* lies in rain forest 125 km to the south.

NOTES. Simons (1988) described the alarm call as resembling that of *P. verreauxi* ("she-fak"). On the other hand the extreme protest call (not heard in the *P. verreauxi* group at all) is a "whinney" very like that of *P. diadema*.

Propithecus verreauxi group

Smaller than most members of the *P. diadema* group, with "normal," not silky pelage; tail longer than head plus body length; ears short, 9.3% of head plus body length; head relatively large, skull length 18.9% of head plus body length (Simons 1988). Chromosomes: $2n = 48$; again, it is unclear which taxa have been karyotyped; a banded karyotype of *P. coquereli* was published by Poorman (1983). The skull is characterized by extreme interorbital inflation, very wide nasal aperture, larger bulla and auditory meatus, and reduced muscular attachments, compared with the *P. diadema* group; but the differences among taxa within each species-group tend to be as great as those between the groups. In all members of the *P. verreauxi* group, however, the mandibular condyles are transversely narrow-oval, instead of nearly rounded as in the *P. diadema* group.

Propithecus verreauxi Grandidier, 1867
Verreaux's Sifaka

1867 *Propithecus verreauxi* Grandidier. Cap Sainte-Marie.
1894 *Propithecus majori* Rothschild. Antinosy country.
1936 *Propithecus verreauxoides* Lamberton. Tsiravé, south of Beroroha, Mangoky Valley (subfossil).

DIAGNOSIS. Mean weight (n = 4) 3.4 kg. Most commonly white, often with silvery or golden tints, with black or chocolate brown crown-cap; ears slightly tufted with white. The *majori* morph, which has brownish black dorsum and inner surfaces of limbs, brown chest, and dark brown tail with white tip, occurs sporadically in troops of more normally toned individuals. An entirely white morph also occurs (Mittermeier et al. 1994). Mean skull length 81.3 mm in southern specimens, 80.2 in more western ones (Albrecht et al. 1990). The snout is very deep and comparatively narrow. The braincase is longer and rounded.

DISTRIBUTION. The dry country from the Tsiribihina River south into the xerophytic bush zone and east to Tolagnaro.

NOTES. Simons (1988) described the alarm call as "shi-fak" (from which the onomatopoeic vernacular name sifaka is derived).

Propithecus coquereli (Grandidier, 1867)
Coquerel's Sifaka

1867 *Cheirogalus [sic] coquereli* Grandidier. Morondava.
1870 *Propithecus damonis* Gray. Nomen nudum.

DIAGNOSIS. Mean weight of 8 males, 3.7 kg; of 10 females, 3.76 kg. Pelage long, mostly white, with maroon chest and inner and anterior aspects of limbs; often with brownish or silvery patches on back. Usually a white patch across muzzle. Ears naked, small, but visible. Mean skull length 83.2 mm (Albrecht et al. 1990). The snout is narrow as in *P. verreauxi*, but

low, so that there is a steep drop from the (strongly inflated) interorbital region. The braincase is low and flat. The nuchal crest is very strongly developed, especially along its lateral portions.

DISTRIBUTION. Northwestern region, from Ambato-Boéni region to Antsohihy.

NOTES. The color pattern, as remarked by Petter et al. (1977), in some respects resembles that of the *majori* form of *P. verreauxi*. According to Simons (1988), the alarm call ("unsh-'kiss'") is quite different from that of supposedly conspecific *P. verreauxi*.

Propithecus deckenii Peters, 1870
Van der Decken's Sifaka

DIAGNOSIS. Muzzle short, blunt, round, with lateral pneumatization; face black, naked. Pelage white with silvery or golden tints; with or without dark areas on head and chest. The snout is deep as in *P. verreauxi*, but broad and somewhat laterally inflated. The braincase is low and flat.

NOTES. Tattersall (1986b) found both *deckenii*-like and *coronatus*-like sifakas in the same areas and concluded that the two names are probably synonyms. Thalmann and Rakotoarison (1994) refined these distributions. Their survey showed that *P. d. coronatus* tends to be a more inland form, though it reaches the coast between the Mahavavy and Betsiboka Rivers and again south of the Manambolo River; *P. d. deckenii* is a more coastal subspecies, mainly from between the Manambolo and Mahavavy Rivers, though *deckenii*-like individuals also occur on the coast north of the Mahavavy.

Propithecus deckenii deckenii Peters, 1870

1870 *Propithecus deckenii* Peters. Kanatsy.

DIAGNOSIS. Pelage white, often with silvery or golden tints on back and limbs. Usually a white patch on muzzle. Larger; skull length averaging 83.9 mm (Albrecht et al. 1990). The snout is less inflated than in *P. d. coronatus*.

DISTRIBUTION.   A coastal subspecies, from between the Manambolo and Mahavavy Rivers.

Propithecus deckenii coronatus Milne-Edwards, 1871

1871 *P[ropithecus] coronatus* Milne-Edwards. Boueny Province.
1876 *Propithecus damanus* Schlegel. South shore of Bombetoka Bay. Nomen nudum.

DIAGNOSIS.   Crown, forehead, cheeks, and throat dark chocolate brown or black; dorsally yellowish to silvery brown; chest dark. Slightly smaller; mean skull length 82.8 mm (Albrecht et al. 1990). The snout is extremely broad, inflated at and behind the canine roots.

DISTRIBUTION.   Forms a crescent enclosing that of *P. d. deckenii*: coast south of the Manambolo River, swinging inland to the uplands at 500–1,000 m, then to the coast again north and east of the Mahavavy River.

Family Daubentoniidae Gray, 1863

Genus *Daubentonia* E. Geoffroy, 1795
   Aye-aye

1795 *Daubentonia* E. Geoffroy. *Sciurus madagascariensis* Gmelin, 1788.
1795 *Scolecophagus* E. Geoffroy. Alternative name for *Daubentonia*. Not of Swainson, 1831 (Aves).
1799 *Aye-aye* Lacépède. *Sciurus madagascariensis* Gmelin, 1788.
1803 *Cheyromys* E. Geoffroy. *Sciurus madagascariensis* Gmelin, 1788.
1811 *Chiromys* Illiger. *Sciurus madagascariensis* Gmelin, 1788.
1816 *Psilodactylus* Oken. Unavailable.

1839 *Myspithecus* de Blainville. *Sciurus madagascariensis* Gmelin, 1788.
1846 *Myslemur* Anonymous (de Blainville, according to Trouessart [1878]). New name for *Myspithecus*.

Daubentonia madagascariensis (Gmelin, 1788)
   Aye-aye

1788 *Sciurus madagascariensis* Gmelin. Western Madagascar.
1800 *Lemur psilodactylus* Schreber (in Shaw). Madagascar.
1930 *Cheiromys madagascariensis* var. *laniger* G. Grandidier. Madagascar.

DIAGNOSIS.   A large lemur with coarse black hair, brindled with gray; short, pale muzzle; large ears; extremely long, slender digits; long bushy tail. Skull short and high; a single pair of incisors in each jaw, open-rooted and with enamel cusps; all three molars present but premolars reduced to a single maxillary pair.

DISTRIBUTION.   Apparently scattered over almost the whole island.

NOTES.   There is a marked difference in size between eastern and northwestern representatives of this species. Two skulls from the northwestern region are both 82.1 mm long, whereas four eastern skulls range from 85.6 to 90.5 mm, with a mean of 89.0 (Albrecht et al. 1990).

   A second species, *Daubentonia robusta* Lamberton, 1934, which was apparently exterminated over the last few centuries, lived in the xeric region of southern and southwestern Madagascar (Simons 1994).

Loriformes

In accord with Jenkins (1987) and most other modern authors, I recognize two families in this group.

<u>Family Loridae Gray, 1821</u>

Jenkins (1987) showed that, according to Article 29(b)(ii) of the *Code,* under which the stem of a non-classical name is determined by the author who first established a family-group name for it, this family has to be called Loridae, not "Lorisidae" as has hitherto been the custom. In his next paper, Gray (1825) used the form Loridina.

The two genera traditionally recognized among the African "slow-climbers" were recently unexpectedly augmented by a third, *Pseudopotto* Schwartz, 1996. The relationships between the five genera are debatable.

In 1971 I proposed that the two then-known African genera, *Perodicticus* and *Arctocebus,* form one clade, and the two Asian genera, *Loris* and *Nycticebus,* form another (Groves 1971). This was based primarily on craniodental features, such as the more laterally extended tympanic tube and tubular orbits of the Asian genera, and the different shapes of the temporal ridges and nasal bones.

Evidently unaware of this proposal, Schwartz and Tattersall (1985) suggested that the two large forms,

Perodicticus and *Nycticebus,* form one clade and the two small ones, *Arctocebus* and *Loris,* the other. This too was based on craniodental features: the small pair have a more anteriorly prolonged premaxilla, submolariform P^4, paraconid shelf on P_4, lower molar buccal cingulids, and large M^3; the large pair have an inflated anterior rostrum, robust zygomata, more bunodont molar and premolar cusps, pit lingual to the hypocone on M^2, and more robust anterior dentition.

Some of Schwartz and Tattersall's characters seem size-dependent. In the Galagonidae, the smaller the skull size, the more marked the premaxillary prolongation. Rostral inflation (by large canine roots), large anterior dentition, and robust zygomatic arches seem very clearly dependent on size, and low rounded vs. high crystalline cusps reflect relative degrees of frugivory and insectivory, which in turn are dependent on body size. These comments can be tested by examining skulls of *Nycticebus pygmaeus,* the relatively smallest member of the putative *Perodicticus-Nycticebus* clade.

The tubular nature of the external auditory meatus in the two Asian genera is a very obvious feature differentiating them from the two African genera. Their more tubular orbits, caused by the flattened, platelike lateral orbital rims, is less consistent and approached by some *Perodicticus*. The second digit of the hand is shortened in all the Loridae, but much more drastically in the two African genera. Tail reduction, on the other hand, has gone much further in the Asian genera. The facial pattern common (but not universal) in the Galagonidae, of dark eye-rings separated by a light midfacial streak, has been enhanced into a pattern of dark coronal forks plus dorsal stripe in the Asian genera, but lost entirely in the African ones.

Dene et al. (1976) found, from their immunological studies, that the African and Asian genera not only form two separate clades, but the Asian pair join the Galagonidae, and the two African genera separate earlier. Dene et al. (1976) therefore isolated *Perodicticus* and *Arctocebus* into a third family, Perodicticidae. Such a scheme requires a good deal of parallel evolution between Asian and African slow-climbers and awaits testing by morphologists.

The chromosomes of *Arctocebus* seem not yet to have been studied, but according to Rumpler et al. (1987) those of the other three genera give equivocal results. The one favored by Rumpler at al. (1987) is that the two Asian genera form a clade separate from *Perodicticus;* but it is also possible that the three spring separately from a common trunk.

All this being so, there seems little doubt but that the correct phylogeny of the four well-known genera is a geographically based one.

The position of *Pseudopotto* is still obscure. Schwartz (1996) described it as primitive with respect to *Perodicticus* and even to *Arctocebus* and perhaps to the Asian genera; it retains, for example, an entepicondylar foramen in the humerus, like galagos. Yet before a satisfactory phylogeny of all five genera can be contemplated, a good deal of further information is needed. For example, in Schwartz's paper the wrist morphology is not described; does *Pseudopotto* have the typical lorid wrist morphology, so strikingly different from that of galagos, as described by Cartmill and Milton (1977)? Many of the distinguishing features of *Pseudopotto* are found in the postcranial skeleton, and this reminds us that full comparative descriptions of the skeletons of the Loridae have not been drawn up since Hill (1953), who may even have

used, as part of his sample of *Perodicticus,* the skeleton now made the type of *Pseudopotto martini!* We need to know the variations in numbers of vertebrae, especially coccygeal and caudal; in degree of elongation of the cervical neural spines; in manual digital reduction; and so on.

SUBFAMILY PERODICTICINAE GRAY, 1870

The first "modern" classification of the African lorids is that of Schwarz (1931a), who recognized the two genera *Arctocebus* and *Perodicticus,* with a single species in each.

Genus *Arctocebus* Gray, 1863
  Angwantibos

1863 *Arctocebus* Gray. *Perodicticus calabarensis* J. A. Smith, 1863.

Schwarz (1931b) described some of the very considerable differences between the two taxa of this genus, but retained them as subspecies of a single species.

Arctocebus calabarensis (J. A. Smith, 1863)
  Calabar Angwantibo

1863 *Perodicticus calabarensis* J. A. Smith. Nigeria: Old Calabar.

DIAGNOSIS.   According to Schwarz (1931b): Size large, skull length 56.9 (56.1–58.6) mm, nasals long, 29–30% of skull length. Biorbital width less than bizygomatic. Premaxilla moderately elongated; nasal aperture wider than high. Incisors about two-thirds of canine in height, with a narrow diastema between them. P^2 and P_3 slightly raised. Cheekteeth without buccal cingulum.

DISTRIBUTION.   Rain forest north of the Sanaga River, and as far west as the Niger.

Arctocebus aureus De Winton, 1902
  Golden Angwantibo

1902 *Arctocebus aureus* De Winton. Equatorial Guinea: 80 km from mouth of Benito River.
1913 *Arctocebus ruficeps* Thomas. Cameroon: Metet, Nyong River.

DIAGNOSIS. According to Schwarz (1931b): Size small, skull length in a single specimen 54.2 mm; nasals short, 25% of skull length. Orbital rims expanded so that biorbital width is about equal to bizygomatic. Premaxilla elongated; nasal aperture higher than wide. Incisors pin-shaped, only half canine height, with a very wide gap between them. P^2 and P_3 not raised, no higher than P^3 and P_4. Cheekteeth with buccal cingulum.

DISTRIBUTION. Rain forest south of the Sanaga River in Cameroon, south and east to the River Congo.

Genus *Perodicticus* Bennett, 1831

Potto

1831 *Perodicticus* Bennett. *Perodicticus geoffroyi* Bennett, 1831.
1840 *Potto* Lesson. *Potto bosmanii* Lesson, 1840.

Perodicticus potto (Müller, 1766)

Potto

NOTES. Schwarz (1931b) considered that the pelage characters on which many of the described "species" had been based were simply age-related. The juvenile coat is silvery gray; the head and neck change to brown first, and the silvery hairs on the back are shed last.

Perodicticus potto potto (Müller, 1766)

1766 *Lemur potto* Müller. Ghana: Elmina.
1820 *Galago guineensis* Desmarest. New name for *Lemur potto*.
1831 *Perodicticus geoffroyi* Bennett. Sierra Leone.
1840 *Potto bosmanii* Lesson. New name for *Lemur potto* Müller, 1766.
1910 *Perodicticus ju-ju* Thomas. Nigeria: southern Nigeria.

DIAGNOSIS. Schwarz (1931b) distinguished this from other subspecies by its small size and small teeth, giving greatest skull length as 61.4 (60.5–63.3) mm; he did not state how many specimens this was based on. Two fully adult skulls from Liberia, in the U.S. National Museum (USNM), measure only 57 mm in greatest length, so there may be a cline in size increasing to the east.

Jenkins (1987) gave mean upper toothrow length as 20.9 mm, height of P^2 3.3 mm, and M^2 breadth 4.0

mm ($n = 18$). P_4 is almost always higher than P_3. The tail averages longer than in other subspecies (mean 87.5 mm, $n = 20$).

DISTRIBUTION. Specimens of this subspecies in Jenkins's (1987) list are from Sierra Leone, Ghana, and Nigeria east to 5°21′E.

Perodicticus potto edwardsi Bouvier, 1879

1879 *Perodicticus edwardsi* Bouvier. Congo-Brazzaville: north bank of Congo River.
1902 *Perodicticus batesi* De Winton. Equatorial Guinea: Rio Muni, 25 km from mouth of Benito River.
1910 *Perodicticus faustus* Thomas. Congo-Zaire: Maringa River, Bompona, Irneti.

DIAGNOSIS. Schwarz (1931b) gave the following measurements for this subspecies: greatest length 65.8 (63.1–68.0) mm. He placed *juju* Thomas in this subspecies, whereas Jenkins (1987) showed that it is closer to *P. p. potto* although showing some intermediacy.

Jenkins (1987) found that this subspecies has quite considerably larger teeth, on average, than *P. p. potto*: upper toothrow length 23.1 mm, P^2 height 4.3, M^2 breadth 5.0 ($n = 19, 18, 18$, respectively), and in these respects it is more or less a scaled-up version of it. But, unlike in *P. p. potto*, P_3 is nearly always higher than P_4, and the tail averages much shorter (mean 58.9 mm, $n = 13$).

DISTRIBUTION. The boundary between this subspecies and nominotypical *potto* is probably the Niger River. Specimens listed by Jenkins (1987) are from Nigeria as far west as 7°01′E, Cameroon, Equatorial Guinea, Congo-Brazzaville, and Zaire as far east as Irneti (on the equator at 19–21°E) and as far south as Mayumbe. A very typical specimen of this subspecies (in the Amsterdam collection [ZMA]) is from near Kinshasa.

NOTES. Schwarz (1931b) recognized *P. p. faustus* as a distinct subspecies, from eastern Congo-Zaire, but implied that they might be synonymous, and Jenkins (1987) confirmed this opinion. The single available adult skull of *faustus* according to Schwarz has a greatest length of only 61.6 mm, but its dental metrics fall within the range of *P. p. edwardsi*.

A remarkable study by Schwartz and Beutel (1995) noted small but apparently consistent differences within the potto sample from Cameroon in the Anthropologisches Institut und Museum, Zürich. Coefficients of variation in the "Zürich" potto were remarkably elevated, and closer inspection showed that there were five "morphs" (simila), differing in relative tooth sizes and in the presence or absence of dental features such as cingulum, raised P^2 crown, oblique orientation of P_3, and bulbous molar cusps. It is interesting that none is similar to what they call the "textbook" potto (illustrated in the previously published literature): this latter has a longer face and longer, more V-shaped mandible, lacks the diastema between upper canine and P^2 of the "Zürich" potto, and other features of the dentition.

Because the sample sizes for at least two of these simila (morphs A and B, as Schwartz and Beutel [1995] called them) are substantial ($n = 10$ and 18, respectively), the differences between them have to be taken seriously. Unfortunately the Zürich collection is poorly localized, so we do not know whether they are sympatric or allopatric. But it is clear that further tests are in order to determine whether they are really consistent or overlap, and whether there are other differences, especially external, between them.

Perodicticus potto ibeanus Thomas, 1910

1910 *Perodictius ibeanus* Thomas. Kenya: Kakamega Forest.
1917 *Perodictius [sic] arrhenii* Lönnberg. Congo-Zaire: Masisi.
1917 *Periodicticus [sic] nebulosus* Lorenz. Congo-Zaire: Ituri Forest, Ukaika.

DIAGNOSIS. Schwarz (1931b) gave the greatest skull length for this subspecies as 63.9 (60.0–66.7) mm. This makes it a large form, but it is relatively small-toothed: Jenkins (1987) gave mean values for dental measurements that are almost identical to those of *P. p. potto*, except that P^2 height averages 4.0 mm ($n = 6$), nearly as great as that of *P. p. edwardsi*. P_3 is higher than P_4, again as in *P. p. edwardsi*. A single tail length of 68 mm falls within the ranges of both the other two subspecies.

DISTRIBUTION. Jenkins (1987) listed specimens of this subspecies from eastern Congo-Zaire (Uele and Ituri Districts, and Baraka in the Itombwe region), Uganda, and the Kakamega Forest in Kenya.

Genus *Pseudopotto* Schwartz, 1996
  False Potto

This taxon was described by Schwartz (1996) as lacking the full degree of elongation of the neural spines of C3–T2 seen in *Perodicticus,* but lacking the bifid spine on C2 seen in both *Perodicticus* and *Arctocebus;* with an entepicondylar foramen in the humerus and an ulnar styloid process that is not hooked, unlike in other Loridae; with a longer tail than any other lorid; with higher, sharper cheekteeth cusps and very reduced P^3 and M^3; and, as in *Nycticebus* alone among the Loridae, with the lacrimal fossa within the inferior orbital margin instead of in front of it. The coalescence of a number of unique features seems to establish the distinction of this genus from *Perodicticus.*

As remarked earlier, there is a great deal more to be learned not only about this genus, but about the ranges of variation in the other Loridae before it is possible to draw up a phylogeny and so a taxonomic assessment of *Pseudopotto* and of the family in general. For the moment, however, and given the general tenor of the type description, *Pseudopotto* may be provisionally allocated to the Perodicticinae.

Pseudopotto martini Schwartz, 1996
  False Potto

1996 *Pseudopotto martini* Schwartz. "Equatorial Africa."

DIAGNOSIS. As for genus.

DISTRIBUTION. No definite localities are yet known for this species. The only other specimen known, apart from the type, is from "Cameroon."

NOTES. Because the type specimen, a complete skeleton, was a zoo animal that lived out its life masquerading as a potto, the external appearance of this species is presumably not unpottolike. The measurements of the type are at the lower end of the range of those of the potto.

SUBFAMILY LORINAE GRAY, 1821

There is little problem with this subfamily, which divides clearly into two genera (although these are quite closely related).

Genus *Loris* E. Geoffroy, 1796
    Slender Lorises

1784 *Tardigradus* Boddaert (in part). Preoccupied.
1796 *Loris* E. Geoffroy. *Loris gracilis* E. Geoffroy, 1796.
1811 *Stenops* Illiger (in part).
1815 *Loridium* Rafinesque. *Lemur tardigradus* Linnaeus, 1758.
1840 *Arachnocebus* Lesson. *Lemur tardigradus* Linnaeus, 1758.
1840 *Bradylemur* Lesson. *Arachnocebus lori* Lesson, 1840.

I recently argued (Groves 1998) that there are two species in the genus, not just one. The very small, narrow-snouted form from the Sri Lankan wet zone differs sharply from its neighbors, which on the other hand overlap in characters with each other. The taxa have most recently been surveyed and mapped by Schultze and Meier (1995).

Chromosomes are variable in arm ratios, and it is not known to what extent these variations are taxon-specific (Goonan et al. 1995).

Loris tardigradus (Linnaeus, 1758)
    Red Slender Loris

1758 *Lemur tardigradus* Linnaeus. Sri Lanka.
1796 *Loris gracilis* E. Geoffroy. Sri Lanka.
1804 *Lemur ceylonicus* Fischer. Sri Lanka.
1840 *Arachnocebus lori* Lesson. No locality.
1905 *Loris gracilis zeylanicus* Lydekker. Sri Lanka: Peradeniya.

DIAGNOSIS. Size very small, skull length less than 50 mm (and most of the external measurements listed by Schultze and Meier [1995] do not overlap with those of other taxa); muzzle long, very narrow, but bicanine width broad. Zygomata expanded. Pelage dorsally short, reddish brown, with no dorsal stripe; ventrally tending to be grayer than on dorsum. Eye-rings reddish brown.

DISTRIBUTION. Wet zone of southwestern Sri Lanka. A specimen in the BM(NH), mapped by Schultze and Meier (1995), following Hill (1953), as intermediate between *tardigradus* and *grandis,* is in my opinion a typical specimen of *L. tardigradus.*

Loris lydekkerianus Cabrera, 1908
    Gray Slender Loris

DIAGNOSIS. Size larger, skull length more than 50 mm; rostrum shorter, less narrowed, but less expanded anteriorly, and zygomata less expanded. Pelage varying, usually fairly short, dorsally gray-brown to red-brown, ventrally pale buffy; dorsal stripe usually present, very occasionally absent; eye-rings dark red-brown to black.

DISTRIBUTION. Dry country and medium altitudes of Sri Lanka, and southern India.

Loris lydekkerianus lydekkerianus Cabrera, 1908
1908 *Loris lydekkerianus* Cabrera. India: Madras.

DIAGNOSIS. Rather large in size, grayish buff in color, often with white frosting, usually with distinct dorsal stripe; ventrally whitish buff; eye-rings gray.

DISTRIBUTION. This subspecies is from the dry country of southern and eastern India.

Loris lydekkerianus malabaricus Wroughton, 1917
1917 *Loris malabaricus* Wroughton. India: Huvinakadu Estate, Kutta, South Coorg.

DIAGNOSIS. Averaging smaller than *L. l. lydekkerianus,* generally redder, and usually without distinct dorsal stripe. Eye-rings red-brown.

DISTRIBUTION. From the wet forests of the Western Ghats.

NOTES. Quite contrary to the situation in Sri Lanka, the Indian wet- and dry-country subspecies shade off into each other.

Loris lydekkerianus grandis Hill and Phillips, 1932

1932 *Loris tardigradus grandis* Hill and Phillips. Sri Lanka: Central Province, Mousakande, Gammaduwa, 675 m.
1933 *Loris tardigradus nordicus* Hill. Sri Lanka: North Central Province, Talawa.

DIAGNOSIS.   The Sri Lanka dry-zone and hill form. Very similar to the Indian dry-zone subspecies, nominotypical *lydekkerianus,* but usually less frosted, less pale ventrally, and with darker eye-rings. I am quite unable to distinguish the lowland dry-zone *(nordicus)* and hill country *(grandis)* lorises.

Loris lydekkerianus nycticeboides Hill, 1942

1942 *Loris tardigradus nycticeboides* Hill. Sri Lanka: Central Province, Horton Plains, 1,830 m.

DIAGNOSIS.   As large as *L. l. lydekkerianus,* but with woolly brown dorsal pelage, giving it (as the name suggests) a superficial similarity to a slow loris; no dorsal stripe; ventrally red-buff.

DISTRIBUTION.   The high plains of Sri Lanka, at 1,800–2,000 m.

Genus *Nycticebus* E. Geoffroy, 1812
    Slow Lorises

1784 *Tardigradus* Boddaert (in part). Preoccupied.
1812 *Nycticebus* E. Geoffroy. *Tardigradus coucang* Boddaert, 1784.
1820–1824 *Bradicebus* E. Geoffroy and F. Cuvier. Nomen nudum.
1839 *Bradylemur* de Blainville. *Tardigradus coucang* Boddaert, 1784.

I recently explained (Groves 1998) why three, not two, species should be recognized in this genus. It is a question whether the Java form might not also be better recognized as a species, there being probably little (no?) overlap in external features between it and other taxa.

Genetic studies so far concern only *N. pygmaeus* and *N. bengalensis.* They show that there are fixed allelic differences in several protein loci (8 of 42 studied [Su et al. 1998]).

Nomenclature and descriptions are after Groves (1971), as modified by Jenkins (1987).

Nycticebus coucang (Boddaert, 1784)
    Sunda Slow Loris

DIAGNOSIS.   Medium size, skull length 55–62 mm; third molars strongly reduced; color light brown, not paler on neck, with slight frosting; broad, dark dorsal stripe and head forks; ear short.

Nycticebus coucang coucang (Boddaert, 1784)

1784 *Tardigradus coucang* Boddaert. No locality: probably Malacca (Chasen 1940).
1821 *Lemur tardigradus* Raffles. Sumatra. Not of Linnaeus, 1758.
1867 *Nycticebus sumatrensis* Ludeking. Nomen nudum.
1881 *Nycticebus tardigradus* var. *malaiana* Anderson. Malaya.
1902 *Nycticebus coucang hilleri* Stone and Rehn. Indonesia: Sumatra, Batu Sangkar, Tanah Datar, Padang Highlands.
1902 *Nycticebus coucang natunae* Stone and Rehn. Indonesia: Bunguran, Natuna Islands.
1917 *Nycticebus coucang insularis* Robinson. Malaysia: Sungai Nipa, south end of Pulau Tioman.
1917 *Nycticebus c. buku* Robinson. Attributed to Raffles (1921), in which *buku* is a vernacular name and in any case refers to *Macaca fascicularis.*
1949 *Nycticebus coucang brachycephalus* Sody. Indonesia: Pulau Tebingtinggi.

DIAGNOSIS.   Cranial length 54.3–61.6 mm; 20% of specimens lack I^2; head markings distinct.

DISTRIBUTION.   Malaysian Peninsula from the Isthmus of Kra southward, Sumatra, and offshore islands such as Pulau Tebingtinggi; Pulau Tioman; North Natuna Islands.

Nycticebus coucang menagensis Trouessart, 1898

1898 *Nycticebus menagensis* Trouessart. Philippines: Tawitawi.
1906 *Nycticebus borneanus* Lyon. Indonesia: western Kalimantan, Sakaiam River, Sanggau District.
1906 *Nycticebus bancanus* Lyon. Indonesia: Klabat Bay, Bangka.
1908 *Nycticebus philippinus* Cabrera. Philippines: supposedly Mindanao.

DIAGNOSIS.   Slightly smaller than other subspecies, skull length 54.5–56.5 mm; always lacking I^2; head forks less distinct than in *N. c. coucang.*

DISTRIBUTION.   Borneo, Belitung, Bangka; Tawitawi.

Nycticebus coucang javanicus E. Geoffroy, 1812

1812 *Nycticebus javanicus* E. Geoffroy. Java.
1921 *Nycticebus ornatus* Thomas. Indonesia: Batavia, West Java.

DIAGNOSIS.   Slightly larger than other subspecies; I² always absent; pelage yellow-gray, with little or no frosting; neck creamy; dorsal stripe and head forks dark to black, thin but sharply marked. Ear small. Discriminant analyses (Groves 1998) distinguish this from other *N. coucang,* but not quite completely.

DISTRIBUTION.   Restricted to Java.

Nycticebus bengalensis (Lacépède, 1800)
   Bengal Slow Loris

1800 *Lori bengalensis* Lacépède. Bengal.
1867 *Nycticebus cinereus* Milne-Edwards. Thailand: Bangkok.
1905 *Nycticebus tardigradus typicus* Lydekker. Bengal.
1913 *Nycticebus tenasserimensis* Elliot. Burma: Amherst, Tenasserim.
1921 *Nycticebus incanus* Thomas. Burma: Kyeikpadein, Pegu.

DIAGNOSIS.   Size large, skull length more than 62 mm; I² always present; color orange-buff, with very strong frosting on neck and forelimbs; creamy gray on neck; dorsal stripe thin, brown, head forks indistinct. Ear short.

DISTRIBUTION.   From northeastern India, Bangladesh, and China west from about 25°N in Yunnan and the Pearl River in the east, south to the northern part of peninsular Thailand. There is no difference between specimens east and west of the Mekong River; in southern peninsular Thailand, there appears to be a zone where hybrids with *N. c. coucang* are found.

Nycticebus pygmaeus Bonhote, 1907
   Pygmy Slow Loris

1907 *Nycticebus pygmaeus* Bonhote. Vietnam: Nhatrang.
1960 *Nycticebus intermedius* Dao Van Tien. Vietnam: Hoa Binh.

DIAGNOSIS.   Small, skull length less than 55 mm; pelage fine textured, reddish buff, with medium to dark brown, broad but indistinct dorsal stripe and head forks. Ear long. A diastema between P² and P³; third molars unreduced.

DISTRIBUTION.   East of the Mekong River, in Vietnam, eastern Cambodia, Laos, and southernmost China, where it does not extend as far north or northeast as *N. bengalensis* (Zhang et al. 1997).

Family Galagonidae Gray, 1825

Jenkins (1987) showed that, according to Article 29(b)(ii) of the *Code,* under which the stem of a non-classical name is determined by the author who first established a family-group name for it, this family has to be called Galagonidae, not "Galagidae" as has hitherto been the custom.

The first "modern" arrangement of this family was by Schwarz (1931a). He recognized two genera: *Euoticus,* with one species, and *Galago,* with four *(crassicaudatus, senegalensis, alleni, demidovii [sic]).*

The still only incompletely published study by Olson in the 1970s (summarized in Nash et al. [1989]) challenged this lumped arrangement, showing that many more species exist than Schwarz recognized and that there are probably more genera than Schwarz's two. There have been several attempts since his study to bring new order into the creeping proliferation of species.

Masters et al. (1994) studied hemoglobin and nine red cell enzymes for what they conservatively called *"Galago" alleni, crassicaudatus, garnettii, moholi,* and *senegalensis,* comparing them cladistically with *Perodicticus, Nycticebus,* and certain Lemuriformes. The most parsimonious cladogram clustered G. *alleni* not with the *senegalensis/moholi* group but with the *crassicaudatus/garnettii* group. The same clustering resulted from the analysis of highly repeated DNA sequences, although all loriform taxa were highly conservative and much alike (Crovella et al. 1994).

Some of the most wide-ranging and revolutionary proposals, however, have resulted from the study of vocalizations. Zimmermann (1990) analyzed the loud call (advertisement call) of the following species, which she conservatively assigned to *Galago:*

garnettii, crassicaudatus, demidoff, zanzibaricus, moholi, senegalensis, matschiei, and *alleni.* For her cladistic analysis, she coded the following features of the calls: frequency range, call series duration, call repetition rate, call duration, frequency modulation, fundamental frequency, intonation, and relative sound intensity; the resulting cladogram, using Loridae as outgroup, was *(demidoff(((crassic,garnettii)zanzib) (moholi(seneg(alleni,matschiei)))).* Although I suggest that the character fundamental frequency might have been better coded as two characters (peak and minimum frequencies), and some of them might be argued to be ordered rather than unordered as interpreted by Zimmermann, this was a remarkable and enlightening attempt to use an entirely new source of characters for galago taxonomy.

The most recent and most comprehensive study of vocalizations has been that of Bearder et al. (1995), who divided the galagos into six groups according to the structure of their advertisement calls, as follows:

1. Click calls. Elegant galagos (*elegantulus* group)
2. Repetitive calls. Lesser galagos (*alleni, matschiei, senegalensis, moholi* group)
3. Trailing calls. Greater galagos (*crassicaudatus, garnettii*)
4. Rolling calls. Southern dwarf galagos (species at that time undescribed, but recently described as *udzungwensis* and *rondoensis*)
5. Incremental calls. Zanzibar galagos (*zanzibaricus, granti,* and a still undescribed form from Kalwe, Nkhata Bay, in Malawi)
6. Crescendo calls. Dwarf galagos (*thomasi* and *demidoff*).

I collected information from the literature and from my own studies, and used them in a cladistic analysis of morphology. The characters are listed in Table 8 and the coded data set in Table 9. It should be admitted that *G. orinus* was a kind of "all-purpose roller-caller"; cranial and vocalization characters, for example, are known for different species in the group, but are combined here. All characters were initially run ordered except for numbers 23–29, 37, and 40, which were unordered; to see what difference this might make, a second run with all characters unordered was made.

The shortest tree (Figure 2) has a length of 124, a consistency index of 0.589, and a retention index of 0.608. As expected, *garnetti/crassicaudatus* and *senegalensis/moholi* were paired in all runs, but *demidoff/thomasi* were paired in only some runs. The *crassicaudatus/garnettii* pair form the first clade; *elegantulus* comes off next. The smaller species form two clades: *matschiei-alleni-senegalensis-moholi* and *zanzibaricus-demidoff-thomasi-orinus.* The bootstrap tree (Figure 3) (length 145, consistency index 0.503, retention index 0.446) failed to show these last two clusters: only what are in effect genera *Otolemur, Euoticus,* and *Galago* (including *Galagoides*) survive.

The unordered tree shows no difference from the mixed-polarity shortest tree. It has a length of 113, a consistency index of 0.673, and a retention index of 0.611. Further runs, combining *garnettii* with *crassicaudatus* and *moholi* with *senegalensis,* produced no essential differences either.

In view of these results, I feel it is unsafe for the present to recognize any genera beyond *Otolemur, Euoticus,* and *Galago.* The latter includes all lesser and dwarf species as well as Allen's.

Genus *Otolemur* Coquerel, 1859
Greater Galagos or Thick-tailed Bushbabies

1859 *Otolemur* Coquerel. *Otolemur agisymbanus* Coquerel, 1859.
1863 *Callotus* Gray. *Callotus monteiri* Bartlett, 1863.
1863 *Otogale* Gray. *Otolicnus garnettii* Ogilby, 1838.

Schwarz (1931a), following Matschie (1904) and Thomas (1917), divided the thick-tailed bushbabies into two groups, which he placed in one species, *Galago crassicaudatus,* mentioning the key morphological features of size, relative ear size, and forehead color.

This view was revised by Olson (1981), who had found that two species are widely sympatric in northeastern Tanzania. These he called *Otolemur crassicaudatus* and *O. garnettii;* he placed the big silvery bushbabies, *argentatus* and *monteiri,* in the former because of hybridization in the lower Zambezi region.

In a comprehensive review, Masters (1988) detailed the differences between the *O. crassicaudatus/*

Table 8

Characters for a Cladistic Analysis of the Galagonidae

Character no.	Character
1.	P² shape: caniniform, raised, not raised
2.	M² compared with M¹ size: equal, slight decrease, strong decrease
3.	Dental comb angulation: less, more
4.	P²/³ diastema: large, small, absent
5.	Postero-superior zygomatic process: absent, small, large
6.	Occiput on dorsum cranii: extensive, some, none
7.	Snout breadth relative to palate: 70–80, 65–69, 55–64, <55%
8.	Number of pairs of mammae: 3, 2, 1
9.	Relative incisor size: I¹ > I², I¹ = I²
10.	Shape of molar cusps: bunodont, medium, sharp
11.	Metaconule on M¹ and M²: no, yes
12.	Shape of premaxilla: short, elongated
13.	C′–P² diastema: large, slight, absent
14.	Upper molar styles: yes, slight / absent
15.	Palate breadth to length: less, slightly less, equal
16.	Intermembral index: >65, 61–63, <60
17.	Facial marking: none, eye-rings and white stripe
18.	Nails: convex (occasionally concave), pointed
19.	Hopping gait: no or rare, yes
20.	Gestation (days): >130, <130
21.	Twinning: usually no, usually yes
22.	Penis spines: none or simple, unidentate, often more than unidentate
23.	AP alleles
24.	AK alleles
25.	CAII alleles
26.	G6PG alleles
27.	PEP A
28.	PEP B
29.	6PGD
30.	Hair follicle groupings: 8–9, 3–6
31.	Foot lever index: >36, 33, <30
32.	Cecum: sacculated, unsacculated
33.	Highest frequency, kHz: >16, <10
34.	Series duration, ms: 250–350, 3,000–4,000, 7,000, >10,000
35.	Repetition rate, calls per second: 10–20, 2–9, 1–2
36.	Call duration, ms: 20–30, 80, 150–200, 270, >400
37.	Frequency modulation: CF (almost constant), FM (single), MFM (multiple), NCF (noisy overlapped)
38.	Minimum fundamental frequency, Hz: 600–850, 350–400
39.	Peak fundamental frequency compared with minimum: equal, higher
40.	Intonation

Note: Character states are listed from most primitive to most derived. Characters from the literature are derived from Hall-Craggs (1965) for the postcranial skeleton; Hill and Carter (1939), Lowther (1940), Schultz (1948), Ansell (1963), Sauer and Sauer (1963), Ahmed and Kanagasuntheram (1965), Vincent (1969), Masters (1988), and Masters et al. (1988, 1994) for reproductive data; Yasuda et al. (1961), Montagna and Yun (1962), and Machida et al. (1966) for the skin; and Zimmermann (1990) and Bearder et al. (1995) for vocalizations.

Table 9

Character Matrix for Galagos (Bushbabies)

dimensions ntax = 12 nchar = 40
format missing = ? symbols = "0~4"

Species	Character
crassicaudatus	120010100000 0100 000012 1011221001 12241114
garnettii	120010100000 0100 001002 1021221001 11222114
elegantulus	001201020210 0001 011001 ????????00 1?200001
matschiei	3011221?1200 100? 11??0? ?????????? 13240001
alleni	301122301100 0021 101001 1020211?21 13243011
senegalensis	211121100200 2022 101001 2130102110 13230101
moholi	2?????221???? 1?22 101111 2130102?? 13120001
zanzibaricus	3112222?1211 0011 10010{0~1} ?????????? 11121013
demidoff	301222201212 0111 1001{0~1} 1 ???????120 01010002
thomasi	2??0?2211?11 0?01 101?0? ?????????? 03010002
orinus	3012123?1211 0111 101?10 ?????????? 12232013
outgroup	100010100000 0000 000000 00000{0~1}0000 00000000

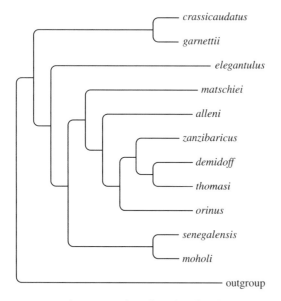

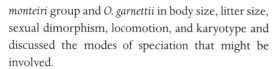

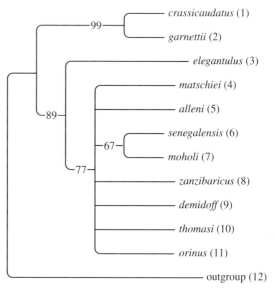

FIGURE 2. Shortest tree for galago data: length 124, consistency index 0.589, rentention index 0.608.

FIGURE 3. Bootstrap tree (50% majority rule) for galago data: length 145, consistency index 0.503, retention index 0.446. Numbers at nodes are bootstrap values.

monteiri group and *O. garnettii* in body size, litter size, sexual dimorphism, locomotion, and karyotype and discussed the modes of speciation that might be involved.

Bearder et al. (1995) illustrated oscillograms of vocalizations of the two species, which differ mainly in their advertising or loud calls.

Measurements for the two widely acknowledged species, as given by Nash et al. (1989), are given in Table 10. The ranges of these measurements seem extremely large; one suspects that many specimens that are not fully mature have been included. The means are nonetheless not dissimilar to those given

Table 10

Measurements for *Otolemur crassicaudatus* and *O. garnettii,* from Nash et al. (1989)

Species	Parameter	Head + body length (mm)	Tail length (mm)	Hindfoot length (mm)	Ear length (mm)	Weight (g)
O. crassicaudatus	Mean	313	410	93	62	1131
	Range	255–400	300–550	70–108	48–72	567–1,814
O. garnettii	Mean	266	364	91	45	767
	Range	230–338	308–440	80–103	34–55	550–1,040

Note: Sample sizes are not given.

by Masters (1988), who measured 290 *O. garnettii* from Morogoro (northeastern Tanzanian mainland), Zanzibar, and Pemba and 58 *O. monteiri* from southeastern Tanzania, western Zambia, and Malawi (listed by her as *O. crassicaudatus:* she did not recognize *O. monteiri* as a species). Specimens of true *O. crassicaudatus,* from south of the Zambezi River, are rather smaller than *O. monteiri* from north of it, and some Kenya specimens of *O. garnettii* are very large, exceeding the mean value given in Table 10 for *O. crassicaudatus.*

Kingdon (1997) recognized four species: *O. crassicaudatus* (including *monteiri*), *O. garnettii,* and two that are provisionally listed as species—*O. argentatus* and an undescribed dwarf species from Mwera District in southern Tanzania. This latter is based upon observations by Honess (1996) of a small *Otolemur,* sympatric with *O. garnettii* at Litipo and Mtopwa, but not different in vocalizations. From scaled photographs, Honess estimated head plus body length at 260 mm, tail length 300 mm, hindfoot length 77 mm, ear length 53 mm, and weight under 500 g, which makes it in fact only slightly smaller than the average *O. garnettii.*

Otolemur crassicaudatus (E. Geoffroy, 1812)
Brown Greater Galago

DIAGNOSIS. Size large, males larger than females; ear long, 18–20% of head plus body length, reaching beyond eye when laid on forehead (in a museum skin); face paler than forehead. Hindfeet shortened. Distal end of nails normal, convex. Penis with uni- or bidentate spines.

Nash et al. (1989) listed the following behavioral features as diagnostic for this species: infants carried on the back as well as, sometimes, in the mouth; bipedal hopping very rare (presumably correlated with the relatively short hindfeet), never landing on hindfeet first. They stated that gestation length is 135 days and that usually (perhaps 70% of the time [Masters 1988]) two or three young are born.

The karyotypes of individuals caught in a band from eastern Zimbabwe to Richards Bay, Kwazulu, ascribed to *O. c. loennbergi, zuluensis,* and *umbrosus* differed consistently from those of *O. garnettii,* with $2n = 62$, NF $= 76$, 12 metacentric and 48 acrocentric autosomes, and long submetacentric X and tiny acrocentric Y chromosomes (Masters 1986). The skull in this species is longer than in *O. garnettii,* with a more flattened braincase (Masters 1986).

Masters (1991) recorded vocalizations of this species in Swaziland and at several locations in Kwazulu-Natal and eastern Transvaal and described the striking differences between it and *O. garnettii.* Bearder et al. (1995) recorded the species in Natal, Transvaal, and Zimbabwe.

DISTRIBUTION. From northern Kwazulu-Natal, in mainly coastal forest and woodland, then north and west into South Africa's Northern Province and Zimbabwe, crossing the Zambezi River in Mozambique. Specimens from southernmost Malawi (at Chikwawa) are of this species.

NOTES. Olson (1980) argued that the holotype of Étienne Geoffroy-Saint-Hilaire's *Galago crassicauda-*

tus, in Paris (MNP), is very likely, but not certainly, of this species, but is in such poor condition that it should be set aside and a neotype designated. He requested that the International Commission on Zoological Nomenclature designate BM(NH) 4.12.3.6, from 15 km east of Eshowe in Ngoye Forest, Natal, as neotype. This request was acceded to by the Commission in Opinion 1329 (1985a).

Otolemur crassicaudatus crassicaudatus (E. Geoffroy, 1812)

1812 *Galago crassicaudatus* E. Geoffroy. South Africa: Kwazulu-Natal, Ngoye Forest, 15 km east of Eshowe, 28°52′S, 31°37′E (Olson [1980]; Opinion 1329 of the International Commission on Zoological Nomenclature [1985a], setting aside the designation of Mozambique: Quelimane, proposed by Thomas [1917]).
1907 *Galago zuluensis* Elliot. South Africa: Kwazulu-Natal, Zululand.

DIAGNOSIS.   Typical specimens buffy on midback and tail, grayer on flanks; creamy ventrally; tail not dark at tip; hands and feet not darkened except on digits; ear long. Mean of three skulls, 72.3 mm; of four skins (head plus body length), 303.5 mm; ear length of four skins, 62.3 mm. Tail length 120–129% of head plus body length.

DISTRIBUTION.   Known only from the Kwazulu region: localities "Durban," Eshowe (Ngoye Forest), Ubombo, Mkuzi.

Otolemur crassicaudatus kirkii (Gray, 1865)

1865 *Otogale crassicaudatus* var. *kirkii* Gray. Mozambique: Quelimane.
1905 *Otolemur badius* Matschie. Tanzania: Ugalla River.
1917 *G[alago] c[rassicaudatus] umbrosus* Thomas. South Africa: Transvaal, Zoutpansberg District, Woodbush Mountain, Tzaneen.
1930 *Galago crassicaudatus lönnbergi* Schwarz. Mozambique: Gorongoza Mountains, Tambarara.

DIAGNOSIS.   Specimens of this subspecies brown to brownish gray; tail light brown; underside creamy, slightly yellowed. Hands and feet not much darkened. Mean length of 11 skulls, 72.5 mm; of 12 skins, 297 mm. These values are not significantly different from those of nominotypical *crassicaudatus;* but ear length (*n* = 6) averages only 55.5 mm. Tail long, its length generally 131–150% of head plus body length, but two specimens (apparently complete) are only 105 and 115%.

DISTRIBUTION.   Southernmost locality in BM(NH) sample, Massangena, 26°30′S, 33°00′E. North through Mozambique to Vila Coutinho, 14°34′S, 34°21′E, and Monkey Bay at 14°S, 35°E in Malawi and northwest to Beatrice, Zimbabwe, 18°15′S, 30°55′E.

Otolemur monteiri (Bartlett, 1863)
   Silvery Greater Galago

DIAGNOSIS.   Light silvery gray-white on dorsum and tail; hands and feet distinctly dark. Creamy yellow on midline of underside. Size very large. Ear even longer than in *O. crassicaudatus,* nearly always more than 20% of head plus body length. Karyotype, known for both subspecies, is identical to that of *O. crassicaudatus.* Bearder et al. (1995) recorded calls of this species in Mtwara and Lindi regions in Tanzania, where it was sympatric with *O. garnettii:* vocalizations were similar to those of *O. crassicaudatus.*

DISTRIBUTION.   The *Brachystegia* woodland zone, from Angola in the west, north to Rwanda and to western and southeastern Kenya. In northern Tanzania (according to Olson, in Nash et al. [1989], in a band from the Usambaras to Ngorongoro) this species and *O. garnettii* coexist in the same habitats, and Bearder et al. (1995) and Honess (1996) found them coexisting at several localities in the Mtwara and Lindi regions in far southeastern Tanzania (approximately 10°00′ to 10°40′S, 39°23′ to 40°20′E).

NOTES.   I am diffident about recognizing this as a full species because of the extraordinarily wide zone where there are intermediates between it and *O. crassicaudatus.* Such a zone could represent a primary cline, or it could indicate that a simple hybrid zone has widened with uninhibited gene flow. In such a case the PSC for once breaks down and offers no guidance.

Otolemur monteiri monteiri (Bartlett, 1863)

1863 *Callotus monteiri* Bartlett (in Gray). Angola: Cuvo Bay.

DIAGNOSIS.   Very light silvery toned; forehead brown. Underside more creamy yellow. East of the Luangwa River, skins have more of a brown tinge; the tail is browner; underside is more gray-white; evidently due to gene flow from *O. c. kirkii*, although the boundary between the two in size is quite sharp. Mean length of 14 skulls from Angola and western Zambia, 75.2 mm; of 26 skins, 321 mm. Mean length of five skulls from Zambia east of the Luangwa River, 75.4 mm; of seven skins, 311 mm.

DISTRIBUTION.   Westernmost locality, Loanda; northeast (in BM[NH] sample) to Tabora, Tanzania, 5°02′S, 32°48′E, Ndola in Zambia, and Baraka in Congo- Zaire, 4°05′S, 29°05′E; east of the Luangwa River in Zambia, and in Malawi, specimens that are detectably of this subspecies are found as far south as Lusingazi, 13°21′S, 32°36′E, and Deep Bay, 10°22′S, 34°15′E. A specimen that is evidently a hybrid between this and *O. c. kirkii* is from Chikwawa, 16°02′S, 34°54′S. Olson (1981) suggested that the area of hybridization is actually much greater than this, in Zimbabwe and northern Mozambique.

Otolemur monteiri argentatus (Lönnberg, 1913)

1913 *Galago argentatus* Lönnberg. Tanzania: Ukina (east of Lake Victoria).
1953 *Galago crassicaudatus lestradei* Schouteden. Rwanda: near Rwinkwavu, northeast of Kibungu, 1,500 m.

DIAGNOSIS.   Close in external appearance to *O. c. monteiri,* but slightly darker; forehead darker. Tail often nearly white. There is a high frequency of melanism in this subspecies. Mean length of four skins from Mwanza District (Tanzania) and from Rwanda, 315 mm; a single skull from Rwanda (the type of *lestradei*) measures 75 mm long.

NOTES.   I have studied specimens from Mwanza, Ukerewe, Mau Forest, and Rwanda. Schouteden's (1953) name *lestradei* denoted an all-melanistic population from the Gissenyi District, Rwanda; because he mentioned a similarly entirely melanistic popula-

tion at Sotik, in Kenya, it is difficult to see how he intended his new subspecies to be regarded.

Kingdon (1997) listed this form as a distinct species, distinguishing it by its short ears, generally less than 6 cm long, its long face and broad muzzle, its pale tail, and its "different" penile structure. This latter detail does not appear to have been published yet; one would like to see a comparison of the penis between *argentatus* and *monteiri*, which externally can barely be distinguished, and between these and *O. crassicaudatus*. The ear character does not hold; the single available specimen from Mwanza has an ear length of 60 mm, which is virtually the same as the means for specimens from Angola (59.8 mm) and east of the Luangwa River in Zambia (59.3 mm) and only slightly less than those from western Zambia (65.8 mm).

Otolemur garnettii (Ogilby, 1838)
Northern Greater Galago

DIAGNOSIS.   Much smaller in size than *O. crassicaudatus,* males only slightly larger than females; ear short; face not paler than forehead. Distal ends of nails concave, with lateral points. Penis spines usually tridentate, some bidentate. Infants carried by mouth only; bipedal hopping is part of the locomotor repertoire; may land on hindfeet when leaping (Nash et al. 1989).

DISTRIBUTION.   From the Juba River down the East African coast to the Ruvuma River, Tanzania; inland to the Kenya highlands; on Zanzibar, Pemba, and Mafia Islands. Perhaps extending into Mozambique (Nash et al. 1989; Honess 1996). Honess (1996) found this species at Milo, on the northern end of Lake Malawi, its southwesternmost locality.

NOTES.   Nash et al. (1989) gave gestation length as 130 days and litter size as usually one or two; according to Masters (1988), some 90% of births are likely to be singletons.

The karyotype is $2n = 62$, NF $= 90$, with 26 biarmed and 34 acrocentric autosomes, and long submetacentric X and tiny acrocentric Y chromosomes.

In the skull, occipital and mastoid areas are relatively larger than in *G. crassicaudatus* (Masters 1986).

Loud calls recorded at Diani Beach, near Mombasa, were highly distinct from those of *O. crassicaudatus* (Masters 1991).

Otolemur garnettii garnettii (Ogilby, 1838)

1838 *Otolicnus garnettii* Ogilby. Tanzania: Zanzibar.
1859 *Otolemur agisymbanus* Coquerel. Tanzania: Zanzibar,
      Agisymbana Island.

DIAGNOSIS.   Tends to be rich, slightly greenish-toned reddish brown; terminal half or so of tail nearly black; underparts yellow. Mean length of 36 skulls from Zanzibar and Pemba, 65.0 mm; of 17 skins, mainly from Zanzibar, 266.0 mm.

DISTRIBUTION.   Appears to be Zanzibar and Pemba only.

Otolemur garnettii panganiensis Matschie, 1905

1905 *Otolemur panganiensis* Matschie. Tanzania: Arusha.

DIAGNOSIS.   Lacking the greenish tones common in nominotypical *garnettii*, and only last quarter of tail tends to be dark. Mean length of five skulls, 67.5 mm; of six skins, 287.7 mm.

DISTRIBUTION.   Tanzanian mainland, from the southern limits of the species (apparently the Mozambique border [see under *O. garnettii*]) north to Tanga, Mount Kilimanjaro, and Lake Manyara. Two of Schwarz's specimens assigned to this subspecies are from the same localities as some that he ascribed to *lasiotis*: Tanga (ZMB) and Taveta (BM[NH]).

NOTES.   Masters (1986) found skulls that she ascribed to this subspecies to be distinguishable on multivariate craniometry from those of other subspecies.

Otolemur garnettii lasiotis (Peters, 1876)

1876 *Galago lasiotis* Peters. Kenya: Mombasa.
1907 *Galago hindei* Elliot. Kenya: Athi River: Kitui.
1913 *Galago hindsi* Elliot. Lapsus for *hindei*.

DIAGNOSIS.   Rather lighter, more grayish than in nominotypical *O. g. garnettii*; end of tail generally only slightly darker and often sharply white at tip itself; underside gray-white rather than yellowish. Mean length of 10 skulls, 64.3 mm; of 10 skins, 268.4 mm; a skull from Bura measures 68 mm, its skin 336.6 mm.

DISTRIBUTION.   Specimens ascribed to this subspecies by Schwarz (1931a) were from the Kenya coast, north to the Juba River in Somalia, south to Tanga in Tanzania, and inland to the Taita Hills and Kibwezi (north of Mount Kilimanjaro). He assigned a specimen from Ngong to the subspecies, but this is in the highlands not far from Kikuyu country, where the type of *kikuyuensis* came from. Specimens I studied of this subspecies came from Mombasa, Tsavo (Voi and Taveta), Malindi, Kitui, and Bura.

Otolemur garnettii kikuyuensis (Lönnberg, 1912)

1912 *Galago (Otolemur) kikuyuensis* Lönnberg. Kenya: Kikuyu,
      Escarpment Station.

DIAGNOSIS.   Averages grayer than *O. g. lasiotis*, often iron gray, with a tinge of green; tail light brown, becoming nearly black on its terminal quarter; underside yellow-white. Occasional specimens are darker, more ochery. Mean length of seven skulls from Ngong, 63.7 mm; of two skulls from Mount Kenya, 66.0 mm; of seven skins from Ngong, 302.5 mm; of two skins from Mount Kenya, 302.5 mm.

DISTRIBUTION.   Kenya highlands, east of the Rift Valley: Embakasi, Langata, Ngong, Tigoni, and north to Mount Kenya.

Genus *Euoticus* Gray, 1863
Needle-clawed Bushbabies

1863 *Euoticus* Gray. *Otogale pallida* Gray.

Bearder et al. (1994) characterized the loud (advertisement) calls of this genus as "click calls," consisting of simple brief, irregular, high-pitched clicks, unlike those of any other group of galagos. Those that they recorded at Korup and Mount Kupe *(E. pallidus)* and in Gabon *(E. elegantulus)* were very similar.

Euoticus elegantulus (Leconte, 1857)
Southern Needle-clawed Bushbaby

1857 *Microcebus elegantulus* Leconte. Gabon: Ogooue River, Njola.
1860 *Otolicnus apicalis* du Chaillu. Gabon: Ogooue River, Njola.
1910 *Galago elegantulus tonsor* Dollman. Equatorial Guinea: 25 km from mouth of River Benito.

DIAGNOSIS.   Dorsally bright foxy red, with hardly any trace of dorsal stripe in most cases; foreparts little if any grayer; a very strong contrast with gray-white underside. Posterior width of nasals less than anterior width. Gap between upper central incisors less than 3.5 mm. P^2 strongly raised, caniniform; M^1 and M^2 without a mesostyle or with a very tiny one; P_2 without a talonid or with a very small one; M_3 with five cusps.

DISTRIBUTION.   Specimens from Ndjole, Como River, and Belinga (Gabon), Benito River (Equatorial Guinea), Barondo (Congo-Brazzaville), and Lelo, Obala, Bitye, and Efulen (Cameroon).

NOTES.   Most, but not all, have a white tail tip. Size large: length of head plus body 200–232 mm, tail length 279–307 mm, hindfoot length 60–67 mm, ear length 29–33 mm ($n = 5$). Two specimens from Belinga, northeastern Gabon, are smaller: length of head plus body 187–196 mm, tail length 256–257 mm, hindfoot length 53–60 mm, ear length 28.5–30.5 mm. Eight from Etoumbi and Zalabouti, Congo-Brazzaville, average noticeably darker than those from closer to the coast.

Euoticus pallidus (Gray, 1863)
Northern Needle-clawed Bushbaby

DIAGNOSIS.   Pale reddish to buffy brown or reddish gray all over dorsum; noticeably grayer on arms, shoulders, neck, and tail; a dark red-brown dorsal stripe runs from behind withers to lumbar region. Face light buff. Underside yellow-white to gray-white. Size small: head plus body length 164–198 mm, tail length 262–318 mm, hindfoot length 50–66 mm, ear length 27–35 mm ($n = 25$). Nasals wider posteriorly than anteriorly. Gap between upper central incisors more than 3.5 mm. P^2 less strongly raised; a prominent mesostyle on M^1 and M^2; P_2 with a strong talonid; M_3 with only four cusps.

Euoticus pallidus pallidus (Gray, 1863)

1863 *Otogale pallida* Gray. Equatorial Guinea: Bioko.

DIAGNOSIS.   Darker and grayer than *E. p. talboti;* dorsal stripe less clearly marked; underparts whitish gray. The black of the eye-rings continues onto the nose, so that only the interorbital region is whitish. Measurements of Bioko specimens are within the range for those from the mainland, except that ear length is 35 mm ($n = 1$).

DISTRIBUTION.   Bioko Island.

Euoticus pallidus talboti (Dollman, 1910)

1910 *Galago talboti* Dollman. Nigeria: Nkami.

DIAGNOSIS.   Lighter, less gray-toned than *E. p. pallidus;* dorsal stripe more clearly expressed; underside yellow-white; interorbital white strip tends to extend up to between the eye-rings. Ear length 27–34 mm ($n = 24$).

DISTRIBUTION.   Oban, Mamfe, Kumba, Korup.

Genus *Galago* E. Geoffroy, 1796
Lesser Bushbabies

Galago senegalensis group:

1796 *Galago* E. Geoffroy. *Galago senegalensis* E. Geoffroy, 1796.
1811 *Otolicnus* Illiger. *Lemur galago* "Schreber" (ex G. Cuvier, 1798).
1811 *Macropus* Fischer. New name for *Galago* E. Geoffroy, 1796. Not of Shaw, 1790 (= kangaroos!).
1836 *Chirosciurus* Gervais. "Khoyak" of G. Cuvier and E. Geoffroy, 1795 (= *Galago senegalensis* E. Geoffroy, 1796).

Galago demidoff group:

1833 *Galagoides* A. Smith. *Galago demidovii* Fischer, 1806.
1857 *Hemigalago* Dahlbom. *Galago demidovii* Fischer, 1806.

Galago alleni group:

1872 *Sciurocheirus* Gray. *Galago alleni* Waterhouse, 1838.

Galago senegalensis group

I appear to have been the first in modern times (Groves 1974) to suggest that the southerly taxon *moholi* is specifically distinct from *G. senegalensis*. This was verified by Zimmermann et al. (1988) on the basis of vocalizations, many of which are basically similar, whereas contact, defensive, and alarm calls are strongly different.

Nash et al. (1989) gave measurements for the three species of this group; these are tabulated in Table 11. Taken at face value, they indicate that *G. gallarum* is similar in size to *G. senegalensis* but with less-elongated hindlimbs; whereas *G. moholi* is smaller with very long ears. Although both these descriptions are certainly correct, the ranges given by Nash et al. are so wide that one suspects that some immature specimens have been inadvertently included or that there have been some misidentifications of *G. senegalensis* as *G. moholi* or *G. gallarum*. My hindlimb mean measurement for *G. gallarum,* for example, is 56.8 mm.

Bearder at al. (1995) recorded *G. senegalensis* in Tsavo and *G. moholi* in Malawi and South Africa, finding differences that seem, however, not much less than those between either of them and *G. matschiei* or the *G. alleni* group.

Galago senegalensis E. Geoffroy, 1796
Senegal Bushbaby

DIAGNOSIS. Dorsally gray or brown-gray, with yellow-toned outer surfaces of limbs; ventrally pale; dark eye-rings complete, with white interocular strip; ears relatively short. Gestation 142 days; usually a single young at birth.

DISTRIBUTION. In woodland savanna and open woodlands from Senegal to northern Somalia, south to southernmost Tanzania.

Galago senegalensis senegalensis E. Geoffroy, 1796

1796 *Galago senegalensis* E. Geoffroy. Senegal.
1798 *Lemur galago* G. Cuvier. Senegal.
1806 *Galago geoffroyi* Fischer. Substitute for *senegalensis*.
1840 *Galago acaciarium* Lesson. New name for *senegalensis*.
1843 *Otolicnus teng* Sundevall. Sudan: Bahr-al-Abiad.
1863 *Galago sennariensis* Gray. Sudan: Sennaar.
1909 *Galago braccatus albipes* Dollman. Kenya: Mount Elgon, Kirui.
1910 *Galago pupulus* Elliot. Nigeria: Yola.
1951 *Galago senegalensis camerounensis* Monard. Cameroon: Bindjal, Faro River.

DIAGNOSIS. Color tends toward pure gray; underside yellowish white; lower part of limbs creamy yellow; tail gray-brown. Mean length of 21 skulls, 43.6 mm; of 55 skins, 163.5 mm head plus body length, 40.5 mm ear length. Five skins from Senegal average larger than those from Ghana and farther east: 172.6 mm head plus body length.

Table 11

Measurements for the Three Species of the *G. senegalensis* Group (after Nash et al. [1989])

Species	Parameter	Head + body length (mm)	Tail length (mm)	Hindfoot length (mm)	Ear length (mm)	Weight (g)
G. senegalensis	Mean	165	261	69	37	206
	Range	132–210	195–303	52–78	21–57	112–300
G. gallarum	Mean	167	252	62	35	—
	Range	130–200	205–293	57–75	30–40	—
G. moholi	Mean	150	228	59	39	158
	Range	88–205	113–279	37–78	23–50	95–244

DISTRIBUTION. Extends from Senegal east across the White Nile to Mount Elgon and Karamojong.

Galago senegalensis braccatus Elliot, 1907

1907 *Galago braccatus* Elliot. Kenya: Tsavo.

DIAGNOSIS. Gray to gray-brown, with yellow underside, legs, and rump; tail brown. Medium size, but rather short ears: mean length of nine skulls, 44.4 mm; of 10 skins, head plus body length 164.1 mm, ear length 37.9 mm.

DISTRIBUTION. From Tsavo region.

Galago senegalensis sotikae Hollister, 1920

1920 *Galago sotikae* Hollister. Kenya: Sotik, Telek River.

DIAGNOSIS. Color more brownish gray dorsally than other subspecies. Size large: mean length of four skulls, 48.3 mm; a single skin has head plus body length 180 mm, ear length 43 mm.

DISTRIBUTION. Southern shores of Lake Victoria, from Mwanza to Ankole.

Galago senegalensis dunni Dollman, 1910

1910 *Galago dunni* Dollman. Somalia: Fafan River headwaters, 55 km east of Harar.

DIAGNOSIS. Similar to *G. s. sotikae* and *braccatus,* but limbs less bright yellow. Size large, with short ears: mean length of four skulls, 45.7 mm; of five skins, 188 mm head plus body length, 35 mm ear length.

DISTRIBUTION. Somalia and Ogaden, exact range unknown.

Galago moholi A. Smith, 1836

Moholi Bushbaby

1836 *Galago moholi* A. Smith. Botswana: banks of the Marikwa and Limpopo Rivers.

1851 *Galago conspicillatus* I. Geoffroy.

1876 *O[tolicnus] mossambicus* Peters. Mozambique: Tete.

1931 *Galago moholi bradfieldi* Roberts. Namibia: Waterberg (according to Shortridge).

1931 *Galago tumbolensis* Monard. Angola: 40 km east of Vila da Ponte, Tumbole River.

1931 *Galago moholi* var. *intontoi* Monard. Angola: no type locality, but Mbale River, Caquindo, Chimporo, and Vila da Ponte mentioned.

DIAGNOSIS. Relatively small with large ears; dorsally buffy or gray-brown with yellow lateral surfaces of limbs; ventrally paler; broad, dark eye-rings and prominent interocular strip. Gestation 125 days; nearly always two in a litter.

DISTRIBUTION. Woodland and woodland savanna, from northern Namibia and northern Botswana through Angola north to Lake Victoria, northeast to central Tanzania, east to northern Mozambique, southeast to Kwazulu-Natal.

NOTES. Zimmermann et al. (1988) showed that some of the vocalizations of this species, especially those associated with affiliative, contact-seeking, or contact behavior, differ in fundamental frequency, dominant frequency, number of units per call, duration of units, and intercall interval from those of *G. senegalensis;* agonistic calls differ much less. Zimmermann et al. predicted that corresponding differences will be found in the larynx and supralaryngeal tract.

There appear to be average differences between regional populations in some body measurements, but these overlap widely and there seems little evidence on which to separate any subspecies. Specimens from the western and eastern sides of the species's distribution are small, whereas those from the central part are larger:

—in South Africa, Zimbabwe, southern Mozambique, Botswana, Zambia west of the Luangwa River, and the Tanzanian portion of the range ($n = 21$), head plus body length is 142–168 (mean 149.9) mm, and ear length is 39–46 (mean 42.8) mm;

—in Namibia and Angola ($n = 33$), head plus body length is 134–157 (mean 145.0) mm, and ear length is 33–40 (mean 37.3) mm;

—of six skins from Tete, central Mozambique, the size is very small indeed: head plus body length 124–138 (mean 132.2) mm, and ear length is 34–36 (mean 35.5) mm;

—the mean head plus body length of 15 skins from eastern Zambia (east of the Luangwa River) is 157.5 mm, but ear length is only 38.9 mm. These skins are also, like the six skins from Tete, redder than those from other regions.

Galago gallarum Thomas, 1901

Somali Bushbaby

1901 *Galago gallarum* Thomas. Ethiopia: Boran country, Webi Dau.

DIAGNOSIS. Shorter hindlimbs than other members of this species-group; medium-sized ears; narrow brown eye-rings incomplete laterally, face otherwise almost white; dorsally brownish buff, with yellowish outer surfaces of limbs; ventrally off-white, with a yellowish boundary between the two zones; tail mostly black. Size like that of *G. senegalensis senegalensis*, but tail slightly shorter, hindfoot and ear markedly so.

DISTRIBUTION. Broadly, between Tana and Webi Shebeyli Rivers, west to Lake Turkana and rift lakes of Ethiopia, in semiarid woodland country.

Galago matschiei group

Galago matschiei Lorenz, 1917

Dusky Bushbaby

1917 *Galago matschiei* Lorenz. Congo-Zaire: Ituri Forest, Moera.
1931 *Galago senegalensis inustus* Schwarz. Congo-Zaire: Ituri District, Djugu.

DIAGNOSIS. Kingdon (1997), calling it the spectacled galago, described this species as dark brown with very large, amber eyes surrounded by almost black patches and a well-defined ridge bordering them above. Nails keeled, pointed.

Nash et al. (1989) gave the following measurements for this species: head plus body length 166 (147–184) mm, tail length 255 (240–279) mm, hindfoot length 68 (63–70) mm, ear length 39 (37–42) mm. They gave the weight as 210 (196–225) g. These figures are almost exactly the same as those for *G. senegalensis,* and one suspects a mix-up. I have measurements for only two specimens: head plus body length 193.3 mm, tail length 225.0 mm, hindfoot length 68 mm, ear length 39 mm; this makes the hindfeet markedly shorter (relatively) than in *G. senegalensis* and the ear somewhat shorter, which agrees with the general impression of the skins. Rahm (1966) gave (very similar) measurements of two specimens as head plus body length 190–190 mm, tail length 195–250 mm, ear length 40 mm ($n = 1$), weight 220 g ($n = 1$).

DISTRIBUTION. Rain forests of eastern Congo-Zaire.

NOTES. Schwarz (1931a), unaware of Lorenz's *G. matschiei,* described *inustus* as a subspecies of *G. senegalensis* and failed to notice features such as the differently shaped nails. He also ascribed to it at least one specimen that is actually an example of real *G. senegalensis,* from Ankole (BM[NH]).

The form of the nails was first pointed out by Hayman (1937), who specified that although pointed they are not as strongly keeled as in *Euoticus.* He noted also the dark brownish body color and the broad, deeply black eye-patches with the white line between them as differentiating the species from *G. senegalensis* and correctly reallocated Schwarz's Ankole specimen.

According to Bearder et al. (1995), the advertisement vocalizations of this species are "repetitive calls" like those of the *G. senegalensis* and *alleni* groups. Zimmermann (1990) placed all these in the same category with respect to call series duration (all long, above 10,000 ms) and intonation (sound energy uniformly distributed from start to end).

Galago alleni group

The advertising calls of this group fall into the same category as those of *G. matschiei* and the *G. senegalensis* group (Zimmermann 1990; Bearder at al. 1995). Bearder et al. (1995) recorded many vocalizations of Allen's galagos at Makande and Makokou,

Gabon; on Bioko Island; and in Korup and at Mount Kupe, northern Cameroon. These fell into two groups: Makande calls (the Gabon type) and Korup calls (Cameroon and Bioko). The advertisement calls are "broadly similar," but the other vocalization types are "very different from each other"; Bearder et al. pointed out that this contrasts with the usual situation in which it is the advertisement calls that vary taxonomically.

Although the division based on vocalizations is clear (and there are morphological differences as well), with the greatest division evidently between (1) animals from Bioko and the Cameroon-Nigeria border region and (2) those from Cameroon south of the River Sanaga, Río Muni, and Gabon, the strikingly small size and different color of the mainland representatives of the first group makes this a diagnosable entity as well. There are thus three species in the group. In addition Kingdon (1997) mentioned a very dark form with orange underparts from the Forêt des Abeilles, south of the Ogoué in Gabon, but I have seen no specimens from that area.

Galago alleni Waterhouse, 1838
   Bioko Allen's Bushbaby

1838 *Galago alleni* Waterhouse. Equatorial Guinea: Bioko Island (formerly Fernando Po).

DIAGNOSIS.   Color very dark gray, with often a heavy tinge of red, and with reddish limbs; tail black, never with white tip; underparts whitish or gray-white with little yellow tone. Face tends to be lighter, grayer than body. Size larger than other species of this group and ears relatively short: head plus body length 200–280 mm ($n = 11$), tail length 225–295 mm ($n = 11$), hindfoot length 70–76 mm ($n = 6$), ear length 32–45 mm ($n = 11$), weight 300–410 g ($n = 8$).

DISTRIBUTION.   Bioko.

Galago cameronensis (Peters, 1876)
   Cross River Allen's Bushbaby

1876 *Otolicnus alleni* var. *cameronensis* Peters. Cameroon: Duala.

DIAGNOSIS.   Smaller than *G. alleni* with relatively larger ears: head plus body length 165–192 mm, tail

length 222–282 mm; hindfoot length 58–75 mm; ear length 31–38.5 mm ($n = 12$). Color averages darker than other species of this group, head wholly dark, with usually just a gray nasal streak; tail nearly black.

DISTRIBUTION.   Cameroon northwest of the lower Sanaga River, into southeastern Nigeria. Specimens seen are from Oban, Mamfe, Obrura, and Mkpani.

Galago gabonensis Gray, 1863
   Gabon Allen's Bushbaby

1863 *Galago alleni* var. *gabonensis* Gray. Gabon.
1907 *Galago gabonensis batesi* Elliot. Gabon: 120 km from mouth of Como River.

DIAGNOSIS.   Size as in *G. alleni*: head plus body length 200–230 mm, tail length 233–260 mm, hindfoot length 65–75 mm, ear length 35–41 mm ($n = 10$). Color gingery brown, much less gray than other species of this group, with more yellow-toned underparts, limbs bright orange; crown more gray; tail dark brown or silvery gray, sometimes with a light tip. The generally light color is distinctive, but the light-colored tail is the most instantly recognizable feature.

DISTRIBUTION.   Presumably the whole forested region of Cameroon, south of the Sanaga, Río Muni, Gabon, and Congo, unless Kingdon's (1997) orange-bellied Forêt des Abeilles form is a distinct species. Specimens seen are from Como River, Benito River, Belinga, Efulen, Lomie, Batouri, Meyoss, and Obala.

Galago zanzibaricus group

For Schwarz (1931a), these galagos were subspecies of *G. senegalensis,* and it thereafter became standard to refer to them as *G. senegalensis zanzibaricus* and *G. s. granti,* until Kingdon (1971) proposed to revive *G. zanzibaricus* as a full species. I followed him in this (Groves 1974) and added *granti* as a southerly subspecies of it.

All members of the group have thick black eye-rings, continuous with the black on the sides of the muzzle, and a thick white stripe from forehead to rhinarium. They have a slight yellow tinge on the

Table 12

Measurements for Samples of the *G. zanzibaricus* Group (from My Museum Samples and from Courtenay and Bearder [1989])

Species (n)	Head + body length (mm)	Tail length (mm)	Hindfoot length (mm)	Ear length (mm)	Skull length (n) (mm)
Zanzibar (11)	140–150	205–227	51–59	31–35	41.9 (8)
NE Tanzania, SE Kenya (13)	150–170	185–230	53–60	35–37	41.7 (13)
G. granti (22)	140–160	220–239	58–61	35–41	41.8 (17)
Cholo (Malawi) (2)	170–190	210–215	60–61	38	43.4 (1)
Kisuku (Malawi) (1)	145	215	54	33	41.8

legs, though this is far from the contrasting color seen in the *G. senegalensis* group.

The skull and dentition are readily distinguished from that of *G. moholi*, a similar-sized and widely sympatric or parapatric species of the *G. senegalensis* group, as follows:

—the snout is lower, its height from palate to nasion 9.7–10.0 mm, compared with 10.4–11.8 mm;

—the dorsal outline of the rostrum is concave, not straight;

—the rostrum is narrow, the bicanine width 7.6–7.7 mm, compared with 8.1–8.5 mm;

—there is a wide diastema between upper canine and P^2;

—the premaxillae protrude, in lateral view, anterior to the nasal tips;

—the jaw angles are strongly produced downward;

—the hypocones of the upper molars are very large and protrude lingually, distorting the trapezium shape of the molars;

—P^4 has only a small lingual cusp;

—M_3 is distally narrowed and even somewhat pointed.

The vocalizations of this group are in a category all their own ("incremental calls"). Bearder et al. (1995) recorded *G. zanzibaricus* at Diani Beach, Mombasa, in Kenya, and *G. granti* in Mtopwa and Lindi Districts in the extreme southeastern corner of Tanzania andat Nambiga (8°35'S, 36°30'E). Courtenay and Bearder (1989) recorded vocalizations of an un-described form (*Galago,* undescribed species 1 later in this chapter) at Kalwe, Nkhata Bay (11°34'S, 34°13'E), and confirmed the presence of galagos of this species-group in several other parts of Malawi.

Table 12 gives measurements for localized samples of this group. Courtenay and Bearder (1989) extended the size ranges of *G. zanzibaricus* and *G. granti* somewhat: in *G. zanzibaricus,* hindfoot length up to 61 mm, and in *G. granti* head plus body length up to 165 mm, hindfoot length up to 64 mm.

Galago zanzibaricus Matschie, 1893
Zanzibar Bushbaby

DIAGNOSIS. Color buffy grayish; tail not much darkened, a darker zone occupying the terminal one-fourth at most; underside creamy gray; cheeks and throat yellowish. Tail and ears relatively shorter than in *G. granti.* Skin of ears not deeply pigmented.

DISTRIBUTION. Coastal forest from southern Somalia to central Tanzania; a few of the lower montane forests (I have seen a specimen, in the Frankfurt [SMF] collection, from Amani in the eastern Usambaras); Zanzibar Island.

Galago zanzibaricus zanzibaricus Matschie, 1893

1893 *Galago zanzibaricus* Matschie. Tanzania: Zanzibar, Yambiani.

DIAGNOSIS. Size slightly smaller than mainland subspecies.

DISTRIBUTION. Zanzibar Island.

Galago zanzibaricus cocos Heller, 1912

1912 *Galago moholi cocos* Heller. Kenya: Mazeras.

DIAGNOSIS. Specimens from the mainland are slightly larger than those from Zanzibar (see Table 12).

DISTRIBUTION. Mainland part of the range.

Galago granti Thomas and Wroughton, 1907

Grant's Bushbaby

1907 *Galago granti* Thomas and Wroughton. Mozambique: Inhambane, Coguno.
1924 *Galago mertensi* Frade. Mozambique: Inhambane, Shibuto.

DIAGNOSIS. More reddish or red-olive–toned; head darker, reddish gray; tail darker, browner, with a darkened or even blackish terminal one-third; underside more creamy; throat less yellowish; dark eyerings thick, prominent. Skin of ears black.

DISTRIBUTION. Southern Tanzania to southern Mozambique, north at least to the Ulugurus.

NOTES. In analyzing external and cranial characters by principal components analysis, I kept six skulls from Newala (in the London [BM(NH)] collection) and three from the Uluguris separate from *G. granti* from Mozambique. Both were slightly separated from typical *G. granti*, but there are probably no definitive grounds for recognizing them as separate subspecies.

Galago, undescribed species 1

Two specimens (one adult, one juvenile) from Mugesse in the Misuku Hills (9°19′S, 33°33′E) were associated by Courtenay and Bearder (1989) with the *zanzibaricus*-like but somewhat distinct vocalizations recorded by them at several localities in northern Malawi. The adult skin, as reported by them, is between cinnamon brown and Dresden brown (notations from Ridgeway), which is more like *G. zanzibaricus* than *G. granti;* and, like the former, only the tip of the tail is dark brown. This is the "Kalwe small" of Bearder et al. (1995).

The body and skull measurements fall within the range of *G. zanzibaricus zanzibaricus*. It is shorter-tailed, longer-footed, and noticeably shorter-eared than *G. granti*.

Galago nyasae Elliot, 1907

Malawi Bushbaby

1907 *Galago nyasae* Elliot. Mozambique: mountains south of Lake Malawi.

DIAGNOSIS. Not different in size or proportions from other *G. granti*, but browner, less reddish; tail less blackened; underside less yellowish. Facial pattern very poorly expressed, with hardly any eyerings, and median snout strip yellowed.

DISTRIBUTION. Recorded from the type locality and from Chikwawa, Malawi.

NOTES. This species, known from the type and five specimens in the London (BM[NH]) collection from Chikwawa, is not unlike the Cholo form (*Galago,* undescribed species 2) externally, but very different in body proportions.

Galago, undescribed species 2

I base this solely on two specimens (MCZ 44134, adult, and BM[NH] 22.12.17.250, juvenile) from Mount Cholo, Ruo, Malawi, 16°03′S, 35°08′E, 915 m. Larger, lighter, browner than *G. granti;* tail not much darker than dorsum, its tip hardly darkened at all; black eye-rings; underside creamy gray, cheeks and throat yellowish as in *G. zanzibaricus;* ears not heavily pigmented. Size large, but tail relatively shorter than in *G. granti*, though ears are proportionately large.

It is not likely that this is the same taxon as *Galago,* undescribed species 1: it is considerably larger in head and body size and in skull length, with a relatively shorter tail and much longer ears.

The skins (BM[NH] 64.982–6) from Chikwawa, Malawi, resemble the two Cholo skins in color but are more creamy below, have darker ears, and are larger. The four adults measure 150–160 mm in head plus body length and 220–235 mm in tail length and so have the proportions of G. granti.

Galago orinus group

Bearder et al. (1995) put the advertisement calls of this group in a separate category, which they dubbed "rolling calls," from those of other bushbabies. They illustrated oscillograms for what they called Matundu dwarf and Rondo dwarf galagos (now G. udzungwensis and G. rondoensis, respectively). A third species, which they called the Amani galago (G. orinus), said to be different from the other two, could not be illustrated because of background noise on the recordings; Bearder et al. (1995) classed this latter as a "roller-caller," but Honess (1996) thought it might be more similar to G. granti and G. zanzibaricus.

As far as morphological characters are concerned, members of this species-group appear to be defined by their lack of obvious dark eye-rings despite the well-defined midfacial strip, the lack of yellow on the limbs, the yellow underparts, and perhaps the absence of penis spines, though Honess (1996) cautioned that he examined only subadult males. Compared with the G. zanzibaricus group, they seem to have relatively long tails (the Ukinga species [Galago, undescribed species 3] excepted) and, especially, short ears, but in these features they are more like the G. demidoff group.

The skulls of G. orinus and G. rondoensis show the following characters:

—there is a narrow snout, with the nasals occupying most of the dorsal surface of the rostrum, and the toothrows converge strongly, more like some of the G. demidoff group;

—there are large diastemata both between upper canine and P² and between P² and P³, whereas in the G. zanzibaricus group only the former is conspicuous;

—there is a long, upturned inferior nasal spine;

—they have an even more concave facial profile than in the G. zanzibaricus group;

—the jaw angles are small, instead of being produced downward;

—the upper canines are very slender;

—P_2 is strongly bent forward;

—the molar and premolar cusps are higher and more pointed;

—the upper molar hypocones are small and do not distort the lingual surface;

—P³ is more triangular, with a well-developed but narrow lingual cusp, instead of ovoid with a mesiodistally expanded but less lingually extended one.

The main craniodental features separating this group from the G. demidoff group are as follows:

—the snout is less upturned at the tip;

—there is a diastema between P² and P³;

—the crown of P² is not raised and does not project beyond the level of P³;

—M³ is triangular, lacking much trace of a hypocone;

—P_2 is bent forward along its length.

Measurements of samples of this group are given in Table 13. The mean values given by Honess (1996) for G. rondoensis are mostly below the minimum for the four specimens given here (all from the Rondo

Table 13

Measurements for Samples of the G. orinus Group (from Museum Specimens [London (BM[NH]), Harvard (MCZ)] and from Honess [1996])

Species (n)	Head + body length (mm)	Tail length (mm)	Hindfoot length (mm)	Ear length (mm)	Weight (g)	Skull length (n) (mm)
G. rondoensis (4)	100–120	190–220	44–58	27–31	60 (5)	34.7 (3)
G. udzungwensis (3)	139–180	225–235	61.5–62	25–29	136 (3)	41.9 (1)
G. orinus (1)	125	170	40	25	—	39.2
Madehani, Ukinga (4)	140–150	165–190	50–55	35–35	—	40.0 (4)

Forest). I take this to be because, as he himself cautiously noted, at least some of the specimens he studied were subadult, but one of the BM(NH) specimens (64.1979) is nearly as small as Honess's minima, and a specimen in Frankfurt (SMF) from a different locality is actually smaller.

Galago orinus Lawrence and Washburn, 1936
Uluguru Bushbaby

1936 *Galago demidovii orinus* Lawrence and Washburn. Tanzania: Uluguru Mountains, Bagilo.

DIAGNOSIS. Dark, reddish-toned; face strip creamy yellow, not white; eye-rings dark brown; ears not deeply pigmented. Relatively short tail, very short hindfoot. Tail short-haired, of uniform thickness, reddish at base, darker at tip. I base this description both on Honess (1996) and on my examination of the type (MCZ 22453). The type skull is less modified (from a presumed *zanzibaricus*-like ancestor) than are those of *G. rondoensis* and the Madehani species (*Galago,* undescribed species 3), with less basicranial flexion, a fully four-cusped M^3, and a lingual cusp on P^3 that is less distally skewed; on the other hand the snout narrows more smoothly, and the basicranium is more narrowed at the basilar suture.

DISTRIBUTION. Honess (1996) recorded what seemed to be this species at Amani, in the eastern Usambaras. It replaces *G. udzungwensis* at high elevations in the Ulugurus and Usambaras, but apparently does not extend to the Udzungwas.

Galago rondoensis (Honess, 1997)
Rondo Bushbaby

1997 "*Galagoides* sp. nov. (named *rondoensis* by P. Honess 1996)" Kingdon, p. 106. Tanzania: Lindi District, Rondo Plateau, Rondo Forest Reserve, 10°07′S, 39°23′E.

DIAGNOSIS. The smallest species of this group, indeed possibly the smallest living galago; but long-tailed and with noticeably longer ears than the other two members of this group. Mid-brown dorsally, becoming pale yellow ventrally, darkening to deeper yellow under neck and chin. A whitish face strip but not very dark eye-rings. Tail with a reddish tinge; aptly described as "bottle-brush," with the final one-

third being much bushier than the rest (and often held rolled up when at rest). The ears seem, from Honess (1996), to be black; certainly those of the three BM(NH) skins seem to be dark (but see under Notes). In skulls of this species, in contrast to the type of *G. orinus,* M^3 is triangular rather than square.

DISTRIBUTION. Litipo, Rondo, and Ziwani, all lowland forests in extreme southeastern Tanzania.

NOTES. A specimen in the Senckenberg Museum, Frankfurt (SMF 83.201, from Pugu Forest Reserve, Kisarawe District, coast region, Tanzania [not far south of Dar-es-Salaam]), is almost certainly referable to this species. It is a nearly mature female; it was collected by Neil Baker in 1987, having died in a mistnet set for birds. It has the characteristic reddish "bottle-brush" tail; eye-rings are absent, and median facial strip is yellowish. Dorsal tone is sienna brown, with gray bases; whitish yellow ventrally. On the other hand the ears are unpigmented. It is smaller than any other specimen, with an especially short tail: head plus body length 90 mm, tail length 163 mm, hindfoot length 35 (?) mm, ear length 28 mm; skull length 32.5; the smallest of the three adult BM(NH) skulls is 34.5 mm long.

Galago udzungwensis (Honess, 1997)
Uzungwa Bushbaby

1997 "*Galagoides* sp. nov. (named *udzungwensis* by P. Honess 1996)" Kingdon, p. 106. Tanzania: Morogoro region, Kilombero District, Ichima (Honess 1996).

DIAGNOSIS. Much the largest of the three species of this group. Dorsally gray-brown grading ventrally to yellowish buff, this latter color also on cheeks; throat and interramal region orangey buff. A broad white midfacial stripe but no distinct eye-rings. Tail evenly and sparsely furred, gray-brown, darkening on distal third. Honess (1996) referred to its "remarkably large hands and feet"; this is not detectable in the hindfoot length. Ears seem not to be black.

DISTRIBUTION. Honess (1996) found this species at Matundu, West Kilombero, Kihansi, and Kiamboza; these are lowland forests southeast of the Udzungwas and east of the Ulugurus. He recorded similar vocalizations in the lowland forests of the eastern Usambaras and suggested that the species might extend south to the Ulanga District in Mahenge Swamp Forest.

NOTES.   Simon Bearder (personal communication) informed me that this species is probably actually a synonym of *G. zanzibaricus,* and that vocalizations hitherto attributed to *G. zanzibaricus* were incorrectly allocated.

Galago, undescribed species 3

Represented by four specimens (MCZ 22446 and 22449–51), from Madehani, Ukinga, in Tanzania just north of Lake Malawi. Darker, but less red than *G. orinus;* tail dark brown, with hairs black-tipped toward the end; hands and feet light, fawn. Skin of ears black. Size within the range of that of *G. udzungwensis,* but with a much shorter tail, shorter hindfeet, and much larger ears. In P^3 the lingual cusp is more skewed distally than in *G. orinus,* and M^3 is more triangular; the rostrum is more downpointing than in *G. orinus* and the basicranium is more flexed, so that a line drawn along the molar occlusal surfaces cuts the auditory bulla. In all of these features this species resembles *G. rondoensis* whereas *G. orinus* is like the *G. zanzibaricus* group.

Galago demidoff group

Zimmermann (1990) and Bearder et al. (1995) agreed that the advertisement calls of this group are quite different from those of any other galago. They share primitive features with the Loridae in frequency range (extremely wide), call repetition rate (more than 10 per second), frequency modulation (uniform from start to end), fundamental frequency (high throughout), and call duration (less than 100 ms) and are unique in their intonation (gradually increasing through the call, hence presumably Bearder et al.'s characterization of them as "crescendo calls").

Traditionally, only one species, *G. demidovii* (correctly *G. demidoff*), has been recognized in this species-group, but Olson (1979) suggested that *G. thomasi,* previously recognized as the Ituri/Kivu/Uganda subspecies, may actually be specifically distinct and may be sympatric with true *G. demidoff* as far west as Cameroon or even in West Africa. Indeed, in anybody's species concept, there have to be several species in the group, because of chromosome diversity. Thus Stanyon et al. (1992) reported three karyotypes: at Makokou, northeastern Gabon, $2n = 44$; in Nigeria, $2n = 58$, NF $= 66$; in Togo, $2n = 58$, NF $= 80$. It remains, however, to tie these down to morphological differences, document the true number of species whether allopatric or sympatric, and assign them names.

From my own all-too-brief survey of the specimens in the London (BM[NH]), Powell-Cotton (PCM), Bonn (MAKB), New York (AMNH), and Harvard (MCZ) collections, I offer the following observations:

1. In West Africa there is a fairly clear division between foxy red and more reddish gray skins. Both color morphs occur in Sierra Leone, Ghana, Nigeria, and the Cameroon plateau. Farther south and east these phases disappear, with merely a redder and a grayer end of the range, but lighter and darker morphs seem to take their place.

2. Size, as measured by both head plus body length and skull length, averages larger in the grayer animals, but there are wide overlaps. The Uganda/Kivu animals and the two from Lake Oku are noticeably large, 140–160 mm long; Gabon specimens measure 142–148 mm long; others are more usually in the 120–140 mm range. In some samples there is a clear separation between large and small specimens: in the Uele Valley, 141–154 mm and 112–123 mm, respectively. It should be added that as far as possible these are all guaranteed measurements of mature individuals, as checked by examination of skulls, and that skulls, when measured, divide similarly. (Fewer undamaged skulls are available than measured skins.)

3. Underparts vary greatly according to locality, but are homogeneous within localities. Usu-

Table 14

Measurements from Nash et al. (1989) for the Two Species They Recognized in the *G. demidoff* Group

Species	Head + body length (mm)	Tail length (mm)	Hindfoot length (mm)	Ear length (mm)	Weight (g)
G. demidoff	129	179	46	24	70
	73–155	110–215	35–60	14–35	44–97
G. thomasi	146	261	69	37	99
	123–166	150–233	39–58	23–33	55–149

ally the ventral surface is whitish gray with yellow tones, but in the far west (Ashanti, Bibianha, Sierra Leone) it is dark yellow, except for Mampong where it is more creamy; the creamy white venter recurs in specimens from Lake Mai-Ndombe, and the Uele Valley and eastward from there.

Measurements as given by Nash et al. (1989) for the two species they recognized in this group are tabulated in Table 14. There are clearly some errors in this compilation: the ranges, especially that of head plus body length in *G. demidoff,* are much too wide and suggest either a vast geographic variation or that individuals have been included that are far from mature. I suspect the latter, because in museum samples I have studied the very smallest skin has a head plus body length of 108 mm. In *G. thomasi,* the mean values for both hindfoot and ear are outside the quoted ranges; I suspect (looking at museum samples) that mean hindfoot length should read 49 mm and mean ear length 27 mm.

I ran principal components analyses on both cranial and external measurements. If there are consistently different shape morphs (candidates for distinct species), the same specimens should cluster together in the two different analyses, but they did not do so; nor were there many cases of discrete patterning within or between geographically restricted samples. Nonetheless there was some discrete patterning, but because this concerns mainly size it is as readily appreciable on raw metrical data as using multivariate statistics and is referred to in appropriate places in the following treatments.

Galago demidoff Fischer, 1806
Prince Demidoff's Bushbaby

Because the limits of this species, and its differentiation from *G. thomasi,* are so unclear (except perhaps in vocalizations!), I can offer no diagnosis at this time.

Hershkovitz (1970) called this species *G. minutus* Cuvier, 1798 (objective synonym *G. cuvieri* Fischer, 1806), but this name seems unidentifiable (J. A. Allen 1925).

Bearder et al. (1995) recorded what they called Demidoff's galagos in Korup and at Mount Kupe, in Cameroon; at Makande, Makokou, and Franceville in Gabon; and in the Uganda sector of the Semliki Forest. In all these localities, they recorded what they called Thomas's galagos as well.

In what follows, the sample has been divided up into manageable geographic chunks, with the available names listed in each case.

West Africa:

1806 *Galago demidoff* Fischer. Senegal.
1808 *Galago demidovii* Fischer. Proposed as an emendation for *demidoff,* but unnecessary under the rules.
1853 *Octolicnus [sic] peli* Temminck. Ghana: Dabocrom.

West African specimens of the *demidoff* group include some fairly large specimens, as well as some of the smallest of all galagos. Largest is the single available skin from Sierra Leone, with head plus body length of 150 mm; two Liberian skins are slightly smaller (130–136 mm). All three of these skins are reddish gray above, reddish yellow below, with brown eye-rings and a yellow-toned midfacial strip, and have fluffy tail hairs, not restricted to a tuft. A third Liberian skin is more pinkish-toned above, creamier below, and rather smaller (126 mm), though perhaps not fully mature. Note that those with redder coats are in effect retaining their juvenile pelage.

The 21 skins from Ghana divide into four phenotypic groups: (1) reddish gray, large (head plus body length 130–136 mm, $n = 4$); these are just like the larger two Liberian specimens; (2) very red, with no gray tones; large (130–142 mm, $n = 7$); (3) same color; smaller (115–125 mm, $n = 8$); (4) bright orange-yellow, very small (108–111 mm, $n = 2$). In these two the underside is more creamy white than in any of the others, and the skin of the ears is not black pigmented. There are a minimum of four, and a maximum of six, simila here. At the moment, there seems to be no way of discovering how many species these represent.

Nigeria-Cameroon Border Region:

1859 *Galago murinus* Murray. Nigeria: Old Calabar.

Skins from this region are all rather different in color from those farther west, but there is little correlation with size. Two from southeastern Nigeria are deep maroon, head plus body length 121 and 132.5 mm; four other skins are more reddish brown and vary from 114 to 136 mm long. The tail is thin, with a tuft beginning halfway along; eye-rings are black, with a whitish midfacial strip.

A probable undescribed taxon in this region is represented by four skins from Nyasoso (northeast of Mount Cameroon), Buea (on the southeastern slopes of Mount Cameroon), and the Rumpi Hills (30 km north of Mount Cameroon). They are gray with contrastingly redder middorsal region, underparts are creamy yellow; head plus body length in the sole fully mature specimen is 140 mm. Four skins said to be from Sakbayeme (on the south bank of the lower Sanaga River) have a similar pattern and have exceptionally broad, black eye-rings.

Bioko Island:

1904 *Galago demidoffi poensis* Thomas. Equatorial Guinea: Bioko, Buntabiri.

All Bioko skins ($n = 4$) are strongly red in color, though overlapping in tone with West African and Central African specimens. Head plus body length is 113–129 mm.

Between the Sanaga and Congo Rivers:

1876 *Otolicnus pusillus* Peters. Gabon: Donguila.
1893 *Galago (Hemigalago) anomurus* de Pousargues. Congo-Brazzaville: upper Kemo River, right tributary of Oubangui River.

There are three simila in the Cameroon sample (all from south of the Sanaga River), recurring in Gabon; all are large: (1) dark yellow-gray, 135–145 mm ($n = 3$); (2) dark maroon, 137–145 mm ($n = 8$ in all, but only 5 have measurements recorded); (3) light maroon-red, 145–148 mm ($n = 2$).

Two from the Dzanga-Sangha region (in the southern tip of the Central African Republic) and another from Banalia on the Sangha River, are dark red-brown above, yellow-creamy below; the two mature specimens measure 132–137 mm.

Vincent (1969) previously drew attention to differences between specimens from different regions. He described those from Brazzaville as being maroon with gray-white to golden yellow underparts, whereas those from La Maboké in the Central African Republic were agouti brown with whitish to gray underparts and a small tuft of whitish hairs in front of the ears. They also differ in the shape of the baculum, the distribution of spines on the penis, and the color of the scrotum.

Vincent (1969) reported that the gestation length of Brazzaville animals is 129 days, whereas in Nigeria it is only 109 days.

South of the Congo River:

1926 *Galagoides demidoffi phasma* Cabrera and Ruxton. Congo-Zaire: Luluabourg (now Kananga).

Specimens from Kananga (formerly Luluabourg) are nearly all dark and reddish brown in color, but head plus body length ranges from 119 to 155 mm (n = 23), a variation that seems impossibly wide for one species. Three skins from Inkongo are of the smaller type. It is interesting that, whereas the other body measurements are in proportion between large and small specimens in the other geographic batches, in this case the larger specimens seem actually to have smaller ears on average: in the larger ones, head plus body length is 132–155 mm, ear length is 21–25 mm (n = 15); in the smaller, with head plus body length 116–128 mm, ear length is 21–26 mm (n = 6).

One large Kananga specimen is much paler than the rest, and one small specimen is unusually dark and reddish. These two forms occur at nearby Lukolela, and the common Kananga type does not; from that locality three specimens are of the large pale type and one is of the small deep red-brown type.

At Lake Mai-Ndombe (formerly Lake Leopold II) there are two quite different simila. Two skins (head plus body length 137–144 mm) are very light foxy red; two (head plus body length 121–128 mm) are dark gray-maroon.

Uele-Ituri Region:

1915 *Hemigalago demidovii medius* Thomas. Congo-Zaire: upper Uele River, Poko.

The difference between large and small simila is greater in this region (Monbutu, Niangara, Medje, Avakubi) than anywhere else (head plus body length 141–154 mm, n = 6; 112–123 mm, n = 5). The type of *medius* is of the large type.

The large type is gingery red-brown on the back, forming a contrast with the grayer tone on the neck, where a nuchal whorl reveals the gray hair bases; it is yellow-creamy below, with this tone extending somewhat down the inner side of the limbs but not to the tarsus. The tail is deeper red-brown.

The small similum is a deeper red-brown color with no nuchal whorl, and so lacks the gray contrast; underside is deeper yellow, with this tone extending down the inner side of the limbs to the tarsus. The tail is a deeper red in tone.

Lake Oku:

Two specimens from Lake Oku in the Bamenda Highlands, Cameroon, stand out in many respects. The skins are large, grayish red, with head plus body length 141–159 mm. The skulls differ in several respects from others of the group: much shorter premaxillae; snout less anteriorly narrowed; wide posterior nasals (posterior breadth is 87.5–100% of anterior, whereas in others of the *G. demidoff* and *G. zanzibaricus* groups it is rarely more than 75–80%); braincase is lower; M^3 much smaller than M^2, and fully four-cusped P^4. Evidently there is an undescribed species confined to this region.

Galago thomasi Elliot, 1907
Thomas's Bushbaby

1907 *Galago (Hemigalago) thomasi* Elliot. Congo-Zaire: Beni.

Kingdon (1997) distinguished this from *G. demidoff* by its being ashy brown instead of simply brown, with a generally paler face, larger ears, and prominent eye patches. Nash et al. (1989) described it as blackish brown with venter yellowish and prominent eye-rings and interocular strip.

The distribution given by Nash et al. (1989) is Mount Cameroon; the Loanda Highlands, Angola; between Dilolo and Kolwezi in southern Congo-Zaire; in the Kivu and Ituri regions of Congo-Zaire and southwestern Uganda; and possibly Mount Marsabit, northern Kenya.

In my survey, I could identify only specimens from Kivu and Uganda with this species. All that I have seen are very large (139–160 mm, n = 5), with large ears (27–35 mm) and a prominent deep gray-black overlay to the generally reddish fur. The skull is, conveniently, always over 40 mm long (maximum 39.5 in the Uele sample and the big Sierra Leone specimen of *G.* cf. *demidoff,* and generally below 37 mm in other samples). Most are from Uganda, but I have seen a specimen from Idjwi Island, in Lake Kivu.

There are also three specimens supposed to be from Mount Marsabit in northwestern Kenya, but these are all marked with a "?," and there is no definitive evidence that the species actually does occur there. Two of them are represented by skulls only, 40 and 41 mm long; like other *G. thomasi,* they are similar in size to *G. zanzibaricus* although they are not of that species according to PCA, and their skins resemble Ugandan *G. thomasi.*

The problem with this concept of the species is that no detailed argument for it, with listing of specimens, has ever been published. Bearder et al. (1995) recorded what they called Thomas's galago in Korup and on Mount Kupe, Cameroon; on Bioko Island; at Makokou, Makande, and Franceville, Gabon; and in the Semliki, Kibale, and Bwindi Forests, Uganda. In all these places except Bwindi, Kibale, and, unexpectedly, on Bioko, they recorded Demidoff's as well. The prime differences were that Thomas's has an additional trill vocalization and a multiple crescendo call, and lacks the typical buzz call of Demidoff's. They specified, however, that whereas the Demidoff's calls were uniform throughout its range, they could not be sure of this for Thomas's.

All specimens from Uganda (and "Marsabit?") I have seen are noticeably larger and grayer (buffy gray) than any others of the *demidoff* group, even those from Uele. Their facial pattern is very conspicuous, more so (especially the white median strip) than most others.

Conclusions on the *Galago demidoff* Group

It does seem evident that this group can no longer be maintained in a single species, and that there are probably even more than two. Color and size variation, at the very least, are great even within restricted localities. Moreover, it is difficult to find similarities across wide areas, suggesting that whatever species may exist have rather narrow distributions.

Another observation worth making is that, in some regions at least, the large and small simila, if they are actually different species, appear to be sister taxa. This is quite obviously the case, for example, in the Uele-Ituri region, where a large and a small form coexist but share color features that set them apart from other members of the group. It is as if both the niches are available (I refer to the two niches assigned by Nash et al. [1989] to *G. demidoff* and *G. thomasi,* respectively) and have been filled not by vicariant species spreading their ranges but by a single indigenous taxon diversifying.

Bearder et al.'s (1995) observation, that the Demidoff's calls are spread throughout the forest belt, but that they could not be sure that the Thomas's calls really are the same everywhere, is interesting. One of the few definite conclusions to emerge from my survey is that no museum specimens can be identified with *G. thomasi* outside the Uganda-Kivu area: the diversity that clearly exists elsewhere does not include *G. thomasi.*

Tarsiiformes

The Tarsiiformes are sister group to the Simiiformes in the Haplorrhini.

Family Tarsiidae Gray, 1825

Genus *Tarsius* Storr, 1780

Tarsiers

The arrangement by Hill (1953) has formed the basis of all subsequent classifications of tarsiers. He recognized three species: *T. spectrum, T. bancanus,* and *T. syrichta,* each with subspecies. Niemitz (1984) followed this arrangement, but queried the subspecies. Musser and Dagosto (1987) refined the characterizations of the three species and added a fourth, *T. pumilus* (actually revived from more than 60 years earlier). They likewise noted that *T. bancanus* and *T. syrichta* form a group with respect to *T. spectrum* and *T. pumilus.* I concurred (Groves 1998) with Musser and Dagosto, but suggested that there were actually several species within what had been designated *T. spectrum.* One problem with revising the species is the lack of specimens from the type locality, Ujung Pandang.

Tarsius syrichta group:

1780 *Tarsius* Storr. *Simia syrichta* Linnaeus, 1758.
1794 *Macrotarsius* Link. *Simia syrichta* Linnaeus, 1758.

1835 *Cephalopachus* Swainson. *Tarsius bancanus* Horsfield, 1821.
1840 *Hypsicebus* Lesson. *Tarsius bancanus* Horsfield, 1821.

Tarsius spectrum group:

1821 *Rabienus* Gray. *Lemur spectrum* Pallas, 1779.

Tarsius syrichta group

(Definition after Musser and Dagosto [1987]; Groves [1998]): Lower incisors spaced apart at their bases and slope medially to make contact toward their tips; a prominent cingulum around maxillary incisors and canines; no diastema between I^2 and canine. Bulla very large, both anterior and posterior to carotid foramen; orbits extremely large, making skull broader than long in adult; palate very wide, V-shaped, evenly widening from front to back; snout high, pyriform aperture narrow; inion very prominent; occipital condyles form thickened rims to foramen magnum, not knoblike; jaw angle produced backward; teeth large, especially third molars; molar talonids narrow. Tail smooth on its ventral surface, its hair much reduced. No black paranasal spot; no light spot behind ear. Ears relatively small. Digital nails extremely reduced, triangular; digits very elongated. Chromosomes: $2n = 80$.

Tarsius syrichta (Linnaeus, 1758)

Philippine Tarsier

1758 *Simia syrichta* Linnaeus. "Luzon."
1777 *Lemur tarsier* Erxleben. Philippine Islands.
1894 *Tarsius philippinensis* Meyer. Philippines: Samar.
1898 *Tarsius carbonarius* Heude. Philippines: Mindanao.
1911 *Tarsius fraterculus* Miller. Philippines: Bohol.

DIAGNOSIS. Differs from *T. bancanus* in having a deeper sigmoid notch, longer toothrow, and higher-crowned incisors; P^2 reduced in size. Orbits not quite as expanded as in *T. bancanus*. Tarsus virtually naked. Tail appearing naked (actually very sparsely haired above) except at very base, poorly pigmented, smooth except for poorly developed papillary ridges. Two pairs of mammae (one pectoral, one abdominal), unlike all other tarsiers.

NOTES. Hill (1953) accepted that the description by Camel, the basis of Linnaeus's *Simia syrichta,* was that of a tarsier ("large round eyes in order to be nocturnal" sounds decisive), despite the locality being given as Luzon.

Tarsius bancanus Horsfield, 1821

**Horsfield's Tarsier or
Western Tarsier**

DIAGNOSIS. Sigmoid notch very shallow; toothrows shorter than in *T. syrichta;* incisors lower-crowned; P^2 not reduced. Orbits hugely expanded, making interorbital space very narrow and lateral edge emarginated. Nasals short; nasal aperture narrow. Tarsus haired. Tail skin dark red-brown, with light brown terminal tuft, the rest sparsely haired above, naked below; ventrally with better-developed ridges, forming V-shaped sulci with longitudinal ridges between; deeply pigmented above. Body creamy buff, contrasting somewhat with brown head; a dark spot on knee.

Tarsius bancanus bancanus Horsfield, 1821

1821 *Tarsius bancanus* Horsfield. Indonesia: Bangka.

DIAGNOSIS. Color gray-buff, occasionally rich reddish yellow; tail pencil thick.

DISTRIBUTION. From Bangka and the opposite part of the mainland of Sumatra.

Tarsius bancanus saltator Elliot, 1910

1910 *Tarsius saltator* Elliot. Indonesia: Belitung.

DIAGNOSIS. Hair tends to be less woolly than other subspecies; dorsal tone more iron gray.

DISTRIBUTION. Belitung (= Billiton).

Tarsius bancanus borneanus Elliot, 1910

1910 *Tarsius borneanus* Elliot. Indonesia: Landak River, western Borneo.
1940 *Tarsius tarsier natunensis* Chasen. Indonesia: Sirhassen, South Natuna Islands.

DIAGNOSIS. Body yellowish toned; tail pencil very sparse; size large.

DISTRIBUTION. Borneo and South Natuna Islands.

Tarsius spectrum group

(Definition after Musser and Dagosto [1987]; Groves [1998]): Bulla relatively short and narrow, especially shortened anterior to carotid foramen; orbits relatively small; jaw angle not prominent; palate narrow; sides of rostrum pinched in behind premolar region; snout low, pyriform aperture wide, nasals projecting above nasal aperture; occiput smoothly rounded with little or no inion prominence; occipital condyles knob-shaped; cranial vault flattened. Toothrows short, third molars especially reduced; molar talonids broad; lower incisors closely apposed along most of their length; poor development of cingulum on maxillary incisors and canine; a diastema between I^2 and canine. Snout with black spot on either side; a postauricular light spot. Ears long, more pointed than in *syrichta* group. Digital nails well developed. Tail appearing scaly on undersurface, this impression being derived from sparse hairs in groups of three emerging from the papillary ridges (in most individuals); tail with well-developed hair on lateral and upper surface. Hand shorter than in *syrichta* group. Male and female have a vocal duet (*T. bancanus* and

T. *syrichta* do not duet). If the chromosomes of the type of *T. dianae* are typical of the species-group, then the chromosome number is much lower than in the *syrichta* group.

Hill (1953) argued that Pallas's *Lemur spectrum*, based on specimens in a private collection in Amsterdam, very likely came from Macassar and accordingly fixed the type locality. Musser and Dagosto (1987:11) accepted this determination, but considered that the basis for it is actually "rather slim."

Tarsius spectrum (Pallas, 1779)

Spectral Tarsier

1779 *Lemur spectrum* Pallas. Macassar (= Ujung Pandang), according to Hill (1953).
1804 *Tarsius fuscus fuscomanus* Fischer. Indonesia: Macassar.
1804 *Tarsius fischeri* Desmarest. "Madagascar."
1921 *Tarsius fuscus dentatus* Miller and Hollister. Indonesia: Labua Sore, north of Parigi, Sulawesi.

DIAGNOSIS. Coronoid process low, sloping; tarsus fully haired. Tail hairs very thick, black-brown, occupying entire distal half of tail. Color gray-buff. Legs yellower than the fawn of the dorsum; feet are pale. Underside creamy.

NOTES. My discriminant analyses (Groves 1998) showed that this species almost certainly consists of two or perhaps three different species. Those from central Sulawesi differ absolutely in most analyses from those from the northern arm. Until fuller series, including specimens from southwestern Sulawesi where the type locality is supposed to be, can be studied, intraspecific taxonomy cannot be taken much further.

Shekelle et al. (1997) recorded tarsiers at various localities in North Sulawesi and by playback experiments demonstrated the acoustic uniqueness of tarsiers at the following sites: Manado region (Tangkoko-Batunagus, Ratatotok-Basaan, Molibagu, Suwawa); Libuo; Sejoli; Tinombo; Kamarora and Marantale. In each of these regions, tarsiers replied to the duet vocalizations of other tarsiers from the same region, but not to those of other regions even though they all seemed morphologically indistinguishable.

There were inconsistent results from the tarsiers of the Togian Islands (Malenge and Batudaka); tarsiers there vocalized in response to playbacks from all the other regions, but those from other regions never responded to those of Togian tarsiers. Shekelle et al. (1997) cited a Ph.D. thesis by Nietsch that proposed that the Togian tarsiers exhibit the primitive form, which is present, but differentially elaborated, in all other taxa.

Musser and Dagosto (1987) noted that central Sulawesi tarsiers tend to be longer-tailed than others from Sulawesi and slightly larger. As shown in their Table 4 (p. 28), central tarsiers and those from the northern arm west of Gorontalo average smaller than those from elsewhere in head plus body length, but slightly larger in skull length. The sample from the southwestern arm (i.e., from nearest the type locality, Ujung Pandang, of *spectrum*) average largest in head plus body length but have the shortest toothrows, and a specimen from the southeastern arm seems similar.

In my sample, two from Palopo and Malili, on the southern arm of Sulawesi, have only a very slightly marked postauricular spot. A specimen from Bumbulan has a paler friction surface on the tail than is usual and sparse hair on the tarsus. Specimens from central Sulawesi (Rorokan, Labua Sore, Parigi) sometimes lack the black paranasal spot and may have only a very poorly marked postauricular spot, but are quite variable in these characters; several of them have a pale ventral tail surface like the Bumbulan skin.

Tarsius dianae Niemitz et al., 1991

Dian's Tarsier

1991 *Tarsius dianae* Niemitz, Nietsch, Warter, and Rumpler. Kamarora, central Sulawesi.

NOTES. It is unclear from the paper of Niemitz et al. (1991) whether *T. dianae* was conceived as a species restricted to the type locality or more widely distributed; and whether it was taken to be a local or regional replacement for *T. spectrum* or sympatric with it. I searched for specimens exhibiting the features described for *T. dianae*, and, as I briefly reported

(Groves 1998), I found a specimen in the USNM whose skin seemed to show many (not all) of them. I kept this specimen separate and entered it as an unknown in my canonical analyses. It never actually fell within the dispersion of other central Sulawesi specimens, but in the general vicinity, closer to central than to north. This does seem to suggest that *dianae*-like features recur within the general central Sulawesi population, and in turn predicts that the type specimen will turn out to be part of the general central species. Shekelle et al. (1997) recorded tarsiers at both Kamarora, the type locality of *T. dianae*, and Labua Sore, the type locality of *T. s. dentatus*, and found them to be identical. They found the diagnostic features of *T. dianae* to be, in the main, found in other Sulawesi tarsiers as well.

Tarsius pelengensis Sody, 1949
   Peleng Tarsier

1949 *Tarsius fuscus pelengensis* Sody. Indonesia: Peleng Island.

DIAGNOSIS.   Relatively large in size; M$_3$ shortened. Pelage somewhat redder (red-buff) than in *T. spectrum* and *T. sangirensis;* more contrast with the creamy tipped thighs; black nose spot poorly marked.

DISTRIBUTION.   Peleng.

NOTES.   I remarked (Groves 1998) that this taxon might rank as a full species. Because it separates cleanly from all other taxa in discriminant analyses, I recommend this action here.

Tarsius sangirensis Meyer, 1897
   Sangihe Tarsier

1897 *Tarsius sangirensis* Meyer. Indonesia: Sanghir (= Sangihe) Islands.

DIAGNOSIS.   Averaging larger, broader-skulled than *T. spectrum*. Toothrows elongated; lateral incisors and canines shortened. Coronoid process high, less sloping; tarsus rather sparsely furred; tail less furred; fur rather finer, less woolly. Dorsum less mottled than *T. spectrum* or *T. pelengensis* because hair tips are darker brown; tail not as dark; postauricular spot poorly marked.

DISTRIBUTION.   Restricted to Great Sangihe, according to Shekelle et al. (1997).

NOTES.   Shekelle et al. (1997) affirmed the morphological distinctiveness of this species and reported its distinctiveness in duetting vocalizations. They also reported behavioral differences and observed that the Sangihe tarsiers are, when age and sex are taken into account, consistently heavier than those from Sulawesi.

Tarsius sp.

DIAGNOSIS.   Differs from *T. spectrum, T. pelengensis,* and *T. sangirensis* in its long toothrow, very high-crowned I^2 (like that of *T. pumilus*), high-crowned upper canine, and very elongated fingers. Has low coronoid process like that of *T. spectrum;* deep sigmoid notch, like that of *T. pumilus.* Tail tuft very short, occupying only the distal third of the tail. Tarsus tends to be sparsely haired. Head, including face, very gray; underside pale yellowed creamy fawn.

DISTRIBUTION.   Pulau Salayar, off the southwestern tip of Sulawesi. Known from three specimens: two in Leiden (RML), one in London (BM[NH]).

Tarsius pumilus Miller and Hollister, 1921
   Pygmy Tarsier

1921 *Tarsius pumilus* Miller and Hollister. Indonesia: Rano Rano, Sulawesi.

DIAGNOSIS.   Differs from others of this group in its very small size; small teeth; paraconids more mesial in position, not level with protoconids; distolingual corners of upper molars rounded off; P^2 very reduced; molars become heavily worn compared with premolars and other teeth; low-crowned incisors; interorbital space very narrow; bulla even more shortened anterior to carotid foramen than in other species of the group; no bare subauricular patch. Like the Salayar species, it has high-crowned I^2 and deep sigmoid notch. Fur is silky, contrasting with the more woolly fur typical of tarsiers (though that of *T. sangirensis* somewhat resembles it). Nails uniquely long, clawlike, extending well beyond digital pads. The detailed description of the species by Musser and Dagosto (1987) further characterizes this species.

DISTRIBUTION. Known only from Rano Rano (1°30′S, 120°28′E), at 1,800 m, and the Latimojong Mountains (3°30′S, 120°05′E), at 2,200 m, in the upper montane rain forest. Shekelle et al. (1997) reported a few unsuccessful attempts to find the species. *Tarsius spectrum* (or whatever other lowland Sulawesi tarsiers are to be called) occur up to at least 1,500 m in places.

Platyrrhini

Rylands et al. (1995) produced a fully annotated state-of-the-art listing of platyrrhine species. Most unfortunately, among platyrrhines the bulk of the family-group names that have priority were based on now superseded generic names. Thus the marmosets must be called not Callitrichinae but Hapalinae; the squirrel monkeys not Saimirini but Chrysotrichini, the howlers not Alouattinae but Mycetinae, and the night monkeys not Aotidae but Nyctipithecidae.

It is above all the influence and example of the Brazilian school of primatology that has stimulated the progression toward the adoption of the phylogenetic species concept: in particular, the insistence of Adelmar Coimbra-Filho, Mario de Vivo, and others that the distinct taxa in the genus *Callithrix* must be recognized at species level. It is probable that future revisions of *Saguinus* will follow those of *Callithrix* and result in the splitting of *S. fuscicollis* into numerous species. The rationale for this was described earlier, in the section on Resolution? in Chapter 3.

Family Cebidae Bonaparte, 1831

The original form of this name was Cebina, but because it was clearly intended to be a family-group suffix based on a generic name, it stands as the original source of Cebidae and antedates Cebidae Swainson, 1835, usually cited as author and date of the family name.

Schneider et al. (1996) placed *Aotus* in this family and could not resolve the trifurcation between *Aotus,* the Cebinae, and the marmosets with any certainty, although the IRBP evidence by itself suggested that *Aotus* might be the sister taxon to the Cebidae + Hapalinae. Horovitz et al. (1998) very clearly placed *Aotus* as sister clade to the others, so it is here separated in a family by itself.

SUBFAMILY HAPALINAE GRAY, 1821

Callitrichinae Thomas, 1903, is a synonym. There are two reasons why the name Callitrichinae (and its coordinate Callitrichidae, for those wishing to retain family-level status for the marmosets and tamarins) cannot be used.

First: Priority in the family group is accorded not to the type genus, but to the family-group name itself. The earliest family-group name given to marmosets is Harpalidae *[sic]* by Gray (1821), who misread *Hapale* Illiger, 1811, as *Harpale.* The current (fourth) edition of the *International Code of Zoological Nomenclature* (2000) states in Article 40:

> (a) **After 1960.** When, after 1960, the generic name on which a valid family-group name is based is rejected as

126

a junior synonym, that family-group name is not to be replaced unless the conditions of Subsection (i) apply.

(i) If the senior generic synonym is itself the basis of a family-group name, or if a reclassification also involves other family-group names, the Principle of Priority applies to all the family-group names concerned.

(b) **Before 1961.** If a family-group name has been replaced before 1961 because of such synonymy, and the replacement name has won general acceptance, it is to be maintained.

In this case, Thomas (1903) discovered that the generic name *Callithrix* Erxleben, 1777, referred to marmosets, not to titis as had been previously assumed, and took three actions: (1) he replaced the commonly used name *Hapale* Illiger, 1811, with *Callithrix*, (2) he replaced the family name Hapalidae with Callithricidae (recte Callitrichidae), and (3) he gave the titis a new generic name, *Callicebus*. The first and third of these actions were justified; the second, admittedly retroactively, was not. But the *Code* must be followed. The provisions of Article 40(b) apply: the family-group name was replaced before 1961, but the replacement name cannot be said to have "won general acceptance," by virtue especially of the continued use of Hapalidae in W. C. O. Hill's influential monograph series *Primates: Comparative Anatomy and Taxonomy*.

Second: Because the name *Callithrix* was long used, incorrectly, for the titis (as just discussed), family-group names for the marmoset/tamarin group were understandably based on what was thought to be the correct name. Gray (1821) misread the name as *Callitrix* and based the family name Callitricidae on it. This might, at a pinch, be taken as effectively a different name, but the same cannot be said of Callithricina Gray, 1825 (during the intervening four years, he had corrected his misspelling). This means that Callitrichinae/-idae Thomas, 1903, for the marmosets and tamarins is preoccupied by the same name of Gray, 1825, for the titis.

The first point is difficult, but arguable. The second point is fundamental, not arguable. The correct name for the subfamily containing marmosets is

therefore not Callitrichinae Thomas, 1903, as listed by Simpson (1945, as Callithricidae), or Napier and Napier (1967), or Hershkovitz (1977).

On morphological grounds, *Callimico* is the genus that clearly stands out in this subfamily; Hershkovitz (1977) did not consider it part of the group at all, but Rosenberger (1983b) was adamant that it is, although he allocated it to a tribe, Callimiconini, separate from the other three genera (Callitrichini). *Callimico* stands apart in its retention of third molars; it is apparently unique in bearing single offspring rather than twins; and its vocalizations mark it as sister group to other genera (Snowdon 1993).

The molecular data are in strong disagreement with these morphological, physiological, and ethological data. Thus Cronin and Sarich (1975) found that *Callithrix* and *Callimico* form a clade, which with *Leontopithecus* and *Saguinus* forms a trifurcation. Barroso (1995) and Schneider et al. (1996) found the order of clade separation to be *(Saguinus(Leontopithecus(Callimico, Callithrix)))*, whereas Horovitz et al. (1998) reversed the order of separation between *Leontopithecus* and *Saguinus*.

In common between two of the three sets of findings, however, is the conclusion that *Leontopithecus* forms a clade with *Callithrix*, separate from *Saguinus*. Rosenberger (1983b) mentioned, as synapomorphies, the enlarged upper central incisors, tall lower incisors, somewhat vertical lower canine, tricuspid upper molars, compressed and sloping symphysis, anteriorly convergent maxillary arcade, and absence of entepicondylar foramen (in distal humerus). The vocalizations also set *Saguinus* apart from the *Callithrix-Leontopithecus* clade (Snowdon 1993). For those wishing to set the tamarins apart as a separate tribe, the name Saguinini Gray, 1825, is available.

Genus *Callithrix* Erxleben, 1777
Marmosets

Hershkovitz (1977) recognized only three species in the genus: *C. jacchus, C. argentata,* and *C. humeralifer*. Vivo (1991) recognized 12 species, which he divided into two species-groups, the *jacchus* group from coastal Brazil and the *argentata* group (including

C. humeralifer, correctly *humeralifera*) from the Amazon forests. Natori (1986) demonstrated the monophyly of the two species-groups on dental characters, and Vivo (1991:25) likewise noted dental distinctions; Souza Barros et al. (1990) compared the chromosomes of *C. emiliae* and *C. jacchus* in detail, representing the two species-groups.

Mittermeier et al. (1988) resurrected *C. kuhlii,* Ferrari and Lopes (1992) described *C. nigriceps* as new, Mittermeier et al. (1992) described a further new species, *C. mauesi,* and finally Alperin (1993) described *C. argentata marcai* (here treated as a full species).

SUBGENUS *CALLITHRIX* ERXLEBEN, 1777

Atlantic Marmosets

1777 *Callithrix* Erxleben. *Simia jacchus* Linnaeus, 1758.
1784 *Callitrix* Boddaert. Lapsus (?) for *Callithrix*.
1792 *Sagoinus* Kerr. *Simia jacchus* Linnaeus, 1758.
1799 *Sagouin* Lacépède. *Simia jacchus* Linnaeus, 1758.
1803 *Saguin* Fischer. *Simia jacchus* Linnaeus, 1758.
1804 *Sagoin* Desmarest. *Simia jacchus* Linnaeus, 1758.
1811 *Hapale* Illiger. *Simia jacchus* Linnaeus, 1758.
1812 *Jacchus* E. Geoffroy. *Simia jacchus* Linnaeus, 1758 (by tautonymy).
1819 *Arctopithecus* G. Cuvier. *Simia jacchus* Linnaeus, 1758.
1821 *Harpale* Gray. Lapsus for *Hapale*.
1823 *Iacchus* Spix. Emendation of *Jacchus*.
1826 *Ouistitis* Burnett. *Simia jacchus* Linnaeus, 1758.
1828 *Midas* E. Geoffroy (in part). Not of Latreille, 1796 (Diptera).
1829 *Hapales* F. Cuvier. Lapsus for *Hapale*.
1829 *Anthropithecus* F. Cuvier. Lapsus for *Arctopithecus*.

All species of this subgenus have mottled or speckled pelage, black with white, yellow, buff, or reddish; tail strongly ringed gray and black; and white, black, or yellow ear tufts springing from the region surrounding or in front of the ear or from inner aspect of ear.

A cladistic analysis of dental characters (Natori 1986) found a basal trifurcation between *C. flaviceps, C. aurita,* and the rest; of the others, *C. geoffroyi* separated first, leaving *C. penicillata* and *C. jacchus* as a terminal pair. In that analysis, *C. kuhlii* was not included. Later, in a dental morphometric analysis, Natori (1990) included *C. kuhlii* and affirmed its distinctiveness.

Rosenberger (1983b) reported that captive-bred hybrids between different taxa of this group did not resemble specimens that Hershkovitz (1977) had proposed as wild hybrids. The evidence for widespread natural interbreeding between taxa is therefore much less clear-cut than Hershkovitz (1977) supposed, although there are some (very narrow) contact zones where hybridization does occur (Mittermeier et al. 1988; Rylands et al. 1993). Marroig (1995) in fact considered that there are very few localities where hybridization occurs, but this was disputed by Mendes (1997), who maintained that on the contrary all species pairs interbreed where their ranges meet, with the possible exception of *C. aurita* and *C. penicillata* in São Paulo.

Callithrix (Callithrix) jacchus (Linnaeus, 1758)

Common Marmoset

1758 *Simia jacchus* Linnaeus. America; restricted to Pernambuco by Thomas (1911).
1792 *Simia (Sagoinus) jacchus moschatus* Kerr. Status fide Hershkovitz (1977).
1812 *Jacchus vulgaris* Humboldt. "French Guiana and Brazil" (mentioned only in synonymy of *Simia jacchus*).
1823 *Jacchus albicollis* Spix. Brazil: Salvador.
1829 *Jacchus vulgaris rufus* Fischer. Based on a variety of *Jacchus vulgaris* E. Geoffroy (ex Humboldt), 1812.
1840 *Hapale leucotis* Lesson. New name for *jacchus* Linnaeus, 1758.
1845 *Hapale communis* South. Status fide Hershkovitz (1977).

DIAGNOSIS.   Body mottled black, gray, and yellow, becoming arranged into transverse stripes on posterior half of dorsum; a white blaze on forehead; white ear tufts, about 30 mm long, forming a semicircular arc above and in front of ears, and hiding ears.

DISTRIBUTION.   Vivo (1991) mapped this in a wide area in easternmost Brazil: almost the whole of the states of Ceará, Paraíba, and Alagoas, and inland to Valença (Piauí); coastal Pernambuco, south to Ilhéus (Bahia) and southwest as far as Ibipetuba, where, according to Vivo (1991), it hybridizes with *C. penicillata.* But at Senhor do Bonfim and at Ilhéus itself the ranges of this species and *C. penicillata* apparently meet without any evidence of hybridization.

NOTES.   De Vivo (1991) described some minor geographic variation in this species: paler in drier regions, more chestnut in more mesic regions.

Callithrix (Callithrix) penicillata (E. Geoffroy, 1812)
Black-tufted Marmoset

1812 *Jacchus penicillatus* E. Geoffroy. Brazil, restricted to Lamarão, near Salvador, by Thomas (1911).
1840 *Hapale melanotis* Lesson. New name for *penicillatus*.
1862 *Hapale trigonifer* Reichenbach. Brazil.
1904 *Callithrix penicillata jordani* Thomas. Brazil: near Araguary, Río Jordão.

DIAGNOSIS.   Pelage as in *C. jacchus* and *C. geoffroyi*, but face colored more as in *C. jacchus;* ear tufts long, about 45 mm, black, springing only from just in front of ears.

DISTRIBUTION.   According to Vivo (1991), from the region of Salvador, where it hybridizes with *C. jacchus,* along the coast to Rio Jucurucu in southernmost Bahia, inland occupying most of Goias, and north (inland of the range of *C. jacchus*) to Humberto de Campos on the coast of Maranhão. Rylands et al. (1993) ascribed the coastal range essentially to a hybrid zone with *C. jacchus,* but mapped it over an enormous inland region, in the hinterland of all the other species of the *C. jacchus* group.

NOTES.   This species hybridizes with *C. geoffroyi* in the Caeté region of Minas Gerais and in the Guilman-Amorim Private Reserve in that state (Passamani et al. 1997), and with *C. jacchus* near Salvador, in Bahia (Mittermeier et al. 1988), although these latter are ascribed by Coimbra-Filho et al. (1993) to casual introductions of *C. jacchus* into the range of *C. penicillata*. The records from the Bahia–Minas Gerais border areas are actually, according to Mittermeier et al. (1988), *C. kuhlii*.

The whereabouts of Lamarão, the restricted type locality of *penicillata,* is quite uncertain (A. Rylands, in litt., 19 September 1998).

Callithrix (Callithrix) kuhlii Coimbra-Filho, 1985
Wied's Marmoset

1826 *H[apale] penicillata Kuhlii* Wied-Neuwied. Brazil: Rio Belmonte, Bahia.
1985 *Callithrix kuhlii* Coimbra-Filho. Ascribed to "Wied, 1926 [sic]."

DIAGNOSIS.   Ear tufts black as in *C. penicillata* but posterior crown buffy, cheeks and forehead and anterior crown whitish buffy; a median nasal white stripe.

DISTRIBUTION.   Between the Rio de Contas and Rio Jequitinhonha in southern Bahia and in extreme northwestern Minas Gerais.

NOTES.   Hershkovitz (1977) considered that this supposed species actually represents wild hybrids between *geoffroyi* and *penicillata,* but Rosenberger (1983b) reported that it resembles *C. geoffroyi* more and is not intermediate as one would expect of a hybrid. Natori (1990) showed that in dental characters *C. kuhlii* is different from both *C. geoffroyi* and *C. penicillata* and not intermediate between them. Between the Rio de Contas and Salvador a population occurs resembling *C. kuhlii* but lighter in color, intermediate between this species and *C. penicillata* (Mittermeier et al. 1988).

The original reference is generally cited as *Hapale penicillata kuhlii* Wide-Neuwied. Vivo (1991), however, argued that Wied-Neuwied in fact habitually placed the author of a name in italics after the name itself and in the Latinized genitive; he was, therefore, intending to refer to *H. penicillata* Kuhl. Vivo considered that the earliest person to use the name in a Linnaean fashion was actually Hershkovitz (1977); but all that Hershkovitz did was to treat *kuhlii* as if Wied-Neuwied's reference intended an available scientific name—he decidedly did not treat it as valid, a requirement for availability under the *Code*. Although the name was mentioned several times over the next few years, often enough as the name of a valid taxon, there was never a description nor a bibliographic reference to one until a special article by Coimbra-Filho in a Brazilian conservation journal in 1985.

Callithrix (Callithrix) geoffroyi (Humboldt, 1912)
White-headed Marmoset

1812 *Simia geoffroyi* Humboldt. Brazil; restricted to between Rios Espírito Santo and Jucú, near Victoria, 20–21°S, by Wied-Neuwied (1826, cited by Napier [1976]).

1812 *Jacchus leucocephalus* Humboldt. Name mentioned in
synonymy of *geoffroyi*.
1819 *Simia albifrons* Thunberg. Brazil.
1840 *Hapale melanotis* Lesson. New name for *leucocephalus*.
1862 *Jacchus maximiliani* Reichenbach. Brazil: Rio Jucú, near
Vitória, Espírito Santo (restricted by Hershkovitz [1977]).
1870 *Jacchus leucogenys* Gray. Nomen nudum.

DIAGNOSIS. Pelage as *C. jacchus*, except head, which is black but with white face; ear tufts longer, about 35 mm, and black.

DISTRIBUTION. States of Minas Gerais and Espírito Santo, along the coast from the Rio Jucú north to the Rio Itaunas (Coimbra-Filho et al. 1993) and inland to Mato Dentro in the west and near Caratinga in the north; it has been sporadically introduced outside this range.

NOTES. This species hybridizes with *C. penicillata* where their ranges meet in the Caeté region, near Belo Horizonte, in Minas Gerais (Mittermeier et al. 1988); Coimbra-Filho et al. (1993) discussed claimed and probably hybrid zones, concluding that some, at least, may result from destruction of the Atlantic forest and consequent penetration of *C. penicillata* into the range of *C. geoffroyi* and also possibly from introductions of the species into areas outside its former range.

Callithrix (Callithrix) flaviceps (Thomas, 1903)
Buffy-headed Marmoset

1903 *Hapale flaviceps* Thomas. Brazil: Engenheiro Reeve, Espírito
Santo (20°46'S, 41°28'W).
1924 *H[apale] flaviscente* Miranda Ribeiro. Lapsus for *flaviceps*.

DIAGNOSIS. Pelage black, speckled with reddish and buffy; forehead and crown yellow; ear tufts short, about 25 mm, yellow, springing from inner aspect of pinna.

DISTRIBUTION. Restricted to a small region above 400 m altitude in Rio de Janeiro, Espírito Santo, and Minas Gerais.

NOTES. This species hybridizes with *C. geoffroyi* at three sites in Espírito Santo and with *C. aurita* at three sites in Minas Gerais (Mendes 1997). At two of these sites, apparently unmixed groups were observed as well as groups containing hybrids. As Mendes (1997) indicated, from these observations it seems that gene exchange is limited; gene exchange by *C. flaviceps* with *C. aurita* seems less limited than with *C. geoffroyi*, because only in the first instance were there groups consisting entirely of hybrids.

Callithrix (Callithrix) aurita (E. Geoffroy, 1812)
Buffy-tufted Marmoset

1812 *Jacchus auritus* E. Geoffroy. Brazil, restricted to Silveira
Lobo, Minas Gerais, by Moojen (1950, cited by Napier
[1976] and Vivo [1991]).
1924 *Hapale coelestis* Miranda Ribeiro. Brazil: Teresópolis, Serra
dos Orgãos, Rio de Janeiro.
1924 *Hapale petronius* Miranda Ribeiro. Brazil: Silveira Lobo,
Minas Gerais.
1959 *Hapale caelestis [sic] itatiayae* Avila-Pires. Brazil: Serra do
Itatiaia, Rio de Janeiro.

DIAGNOSIS. Pelage black with buffy speckling; forehead with white blaze; crown with red patch; ear tufts short, about 25 mm, springing from inner surface of pinna.

DISTRIBUTION. Minas Gerais, Rio de Janeiro, and São Paulo, along the coast between Serra de Macaé and Ubatuba, and inland to Anhembi, Pádua Sales, and Alfenas (Vivo 1991).

NOTES. There is some geographic variation in this species, as described by Vivo (1991): specimens from São Paulo average lighter in color than those from Rio de Janeiro, farther northeast, whereas some from localities such as Volta Grande and Além Paraíba, in northeastern Minas Gerais, are black with little or no light tipping to the hairs. Because of the considerable overlaps between these variants, Vivo (1991) recommended that no subspecies be recognized.

SUBGENUS *MICO* LESSON, 1840
Amazonian Marmosets

1840 *Mico* Lesson. *Simia argentata* Linnaeus, 1766.
1840 *Liocephalus* Wagner. *Simia argentata* Linnaeus, 1766.
1870 *Micoella* Gray. *Mico sericeus* Gray, 1868.

Members of this subgenus have pelage that is either black with white flecks, or fully gray, or white; the tail is without true rings, but often with the lie of the hair giving a ringlike effect; the ears are very large, posteriorly notched, either bare or white with a fringe from both aspects of the ear. The face is hairless, usually pinkish red. The diploid chromosome number is 44 (in the *C. jacchus* group it is 46), and there are large amounts of distal constitutive heterochromatin (Souza Barros 1990; Pieczarka et al. 1996).

The chromosome studies of Nagamachi et al. (1996) and Canavez et al. (1996) showed that *C. emiliae* is the most distinctive species of those they studied, and *C. argentata* is sister to a clade containing *C. humeralifera, mauesi,* and *chrysoleuca*. The tufted and untufted sections of the species-group are therefore not monophyletic. Note, however, that all genetic studies on what purports to be *C. emiliae* have been on a captive breeding group of the Rondônia form, which may not actually represent that species at all (A. Rylands, personal communication).

According to Goodman et al. (1998) this subgenus together with subgenus *Cebuella* separated from *Callithrix (Callithrix)* 5 million years ago. This is longer ago than some Old World monkey genera have been separated from their sister genera.

This group has undergone an astounding taxonomic proliferation in recent years. In addition to the species listed here, Rylands et al. (1995) and Roosmalen et al. (1998) listed one called *"Callithrix saterei* Silva e Sousa Jr and Noronha 1995" and "in press," respectively. Description of this apparently very distinctive taxon has still not appeared in print, but it is depicted in Stephen Nash's illustration (fig. 5 in Roosmalen et al. [1998]) of all the species of this group.

Callithrix (Mico) argentata (Linnaeus, 1766)
Silvery Marmoset

1766 *Simia argentata* Linnaeus. Brazil: Pará, restricted to Cametá, left bank of lower Rio Tocantins, by Avila-Pires (1969, cited by Napier [1976]).
1840 *"Cebus canus* de Blainville" Lesson. De Blainville's figure is probably *Lagothrix cana* according to Hershkovitz (1977).

DIAGNOSIS. Pale silvery gray; tail entirely black, unringed; ears hairless, pinkish red like face. The pre-dominantly white body hairs have medium gray-brown tips. Hands and feet are slightly darker than body. Underparts creamy yellow, sharply demarcated from upper parts. Head creamy white.

DISTRIBUTION. Limited by the Rio Tapajós and Rio Tocantins, down the right bank of the Tapajós and left bank of the Rio Tocantins to about 4°S or a little more (Alperin 1993). The range appears to cross the Rio Xingu with no noticeable differentiation on either side.

NOTES. Some specimens have just a tinge of brown, and a white thigh stripe is sometimes distinguishable. Details of these variations, all of which can be found at a single locality on the lower Tapajós, are given by Vivo (1991).

Callithrix (Mico) leucippe (Thomas, 1922)
White Marmoset

1922 *Mico leucippe* Thomas. Brazil: Pimental, near mouth of Rio Jamanxim, right bank of Rio Tapajós.

DIAGNOSIS. Both body and tail creamy; feet and shanks and (to a lesser degree) hands have pale orange tones. Ears hairless, pinkish red like face.

DISTRIBUTION. Right bank of the Rio Tapajós, slightly upstream of the range of *C. argentata,* between the Rios Jamanxim and Cupari, according to Alperin (1993), who regarded it as a subspecies of *C. argentata.*

Callithrix (Mico) emiliae (Thomas, 1920)
Emilia's Marmoset

1920 *Hapale emiliae* Thomas. Brazil: Rio Curuá, Maloca.

DIAGNOSIS. Ears hairless except for a few very sparse blackish hairs on outer surface. Dorsum gradually darkening, from silvery fawn anteriorly to chestnut-fawn posteriorly, with no hip stripe; underside dark creamy; a dark chestnut cap or black on vertex, with a small white spot behind it; cheeks and forehead white; no thigh stripe (occasionally very poorly marked); tail black, unringed. Hands and feet slightly darker. Facial skin light fawn. The predominantly gray-white hairs have very dark brown tips. Stated to be a relatively heavy, robust, and short-bodied species.

DISTRIBUTION.   Found in two distinct areas, apparently separated by a northern extension of the range of *C. melanura* between the Rio Tapajós and upper Rio Aripuanã. The western area, mainly in Rondônia, was mapped by Ferrari and Lopes (1992) from the west bank of the Rio Aripuanã, as far north as Rio Castanho (7°33′S, 60°20′W), south to Nova Brasilia (11°09′S, 61°34′W) and Jiparaná (10°52′S, 61°57′W), and thence northwest to between the Rios Jamari and Jiparaná. (East of the Rio Jiparaná, *C. nigriceps* is found.) The eastern area is between the upper courses of the Rio Tapajós and Rio Xingu.

NOTES.   This species had not been recognized, even at subspecific level, since its original description until Avila-Pires (1985) resurrected it on the basis of specimens from Rio Peixoto de Azevedo, in the eastern area to the south of the type locality. According to Vivo (1991), specimens from Rondônia, and from the mouth of the Rio Castenho, Amazonas, are slightly less dark than those from Maloca and (presumably) those from Rio Peixoto de Azevedo. Rylands et al. (1993) considered that the Rondônia form is likely to be specifically distinct from the typical Pará form; Alperin (1993), however, united them and placed *emiliae* as a subspecies of *C. argentata*.

There is a skin (USNM 239460) from Maicá, above the mouth of the Rio Tapajós on its right bank, a locality from which *C. argentata* is also known. For Hershkovitz (1977), this was a reason to synonymize the two: *C. emiliae* is simply a color morph of *C. argentata*. In my opinion, the difference is so sharp, and the Maicá skin so typical of *C. emiliae,* that it may rather be taken as evidence for a range extension of *C. emiliae,* and that the two are sympatric in places.

Chromosomes in the Rondônia form: $2n = 44$ (Souza Barros et al. 1990). There are slight differences between this species and *C. argentata* and *C. humeralifera,* which have a similar diploid number (Pieczarka et al. 1996).

Callithrix (Mico) nigriceps Ferrari and Lopes, 1992
Black-headed Marmoset

1992 *Callithrix nigriceps* Ferrari and Lopes. Brazil: Lago dos Reis, 7°31′S, 62°52′W, 17 km east of Humaitá, Amazonas.

DIAGNOSIS.   No ear tufts. Face and ears pigmented; dorsum brownish gray, ventral zone and arms yellow to orange; hindlegs orange-red; thigh stripe present; mantle indistinct, grayish brown. A robust, heavily built, short-bodied species.

DISTRIBUTION.   A very small area east of the Rios Jiparaná and Madeira, perhaps as far as the Rio Aripuanã, but probably only as far east as the Rio dos Marmelos (Ferrari and Lopes 1992).

NOTES.   This is undoubtedly very closely related to *C. argentata,* of which both Alperin (1993) and Rylands et al. (1993) regarded it as a subspecies.

A juvenile skin (AMNH 255861) from Beni, 6 km west of Casarabe (14°48′S, 64°14′W), Bolivia, seems intermediate between this species and *C. melanura.* It is slightly lighter brown than *C. melanura,* with a white thigh stripe; legs are much more orange on their inner surfaces, and there are some orange tones even on the dorsum; underside is yellow-buff, becoming yellower on chest and inner side of legs, and pale buff on throat and inner side of arms; tail is black; crown is brown, with a whitish forehead band; face and ears are black. Against this identification is that the hair on the backs of the ears is slightly less sparse than in *C. melanura* and *C. argentata,* and the insides of the ears have long, dark brown hairs, whereas the ears themselves are partially hidden by long, light gray-brown hairs sweeping back from the temples.

Callithrix (Mico) marcai Alperin, 1993
Marca's Marmoset

1993 *Callithrix argentata marcai* Alperin. Mouth of Rio Castanho, a tributary of Rio Roosevelt, a left-hand tributary of the Rio Aripuanã, Amazonas.

DIAGNOSIS   (after Alperin [1993]). Mantle long-haired, light ochre; rest of dorsum brown washed with chestnut, and brindled; arms paler than dorsum; legs more chestnut; tail dark chestnut; no white thigh stripe or white mark on vertex; head dark chestnut. Skin of face dark, depigmented around muzzle.

DISTRIBUTION. Known only from the area of the type locality, which is only a short distance east of the northernmost extension of the range of Rondônia *C. emiliae.*

NOTES. Alperin (1993), who alone of recent commentators recognized the bare-eared marmosets as subspecies of *C. argentata* rather than as full species, also described this new form as only a subspecies.

From photos of two skins kindly forwarded to me by Anthony Rylands in 1996, I describe the dorsum as brindled ochre and gray, tending more to gray anteriorly and chestnut posteriorly; the arms are, as suggested by the type description, light gray-ochre, whereas the legs are darker, more chestnut (very dark, almost blackened, in one of the two skins). The underparts and flanks are ochery like the arms, with this tone being sharply defined from the grayer dorsum. The tail is extremely dark.

Callithrix (Mico) melanura (E. Geoffroy, 1812)
Black-tailed Marmoset

1812 *Jacchus melanurus* E. Geoffroy. Brazil: restricted to Cuiabá, Mato Grosso, by J. A. Allen (1916).
1846 *Jacchus leucomerus* Gray. Bolivia.
1883 *Simia leukeurin* Pelzeln. Brazil: Cuiabá, Mato Grosso (restricted by Vivo [1991]).
1977 *Callithrix argentata melanius* Hershkovitz. Lapsus for *melanura.*

DIAGNOSIS. Brown, with a striking white or yellow-white hip stripe extending down front and inner thigh; crown, hind part of back, and hindlegs dark brown; a dark line bounds crown-cap in front, this tone extending along brows in a V; tail black; underside fawn-white; inner side of shank tends to be orange; ears hairless, pigmented like face.

DISTRIBUTION. Central Brazil, northern Bolivia, and the Paraguayan chaco, between Rios Ariguaia and Mamoré, as far as Rio Beni in Bolivia (Alperin

1993). The northernmost extension is at about 10°S, on the right bank of the Rio Aripuanã.

NOTES. There is some variation in this species. Series from Cáceres, Palmeiras, Rio Aricá, Santo Antonio do Leverger, Chapada dos Guimarães, Aripuanã, Agua Dolce (Chaco), and Urucum have chestnut on the back and proximal part of arms; but some from Corumbá, southeast of Aripuanã and Agua Dolce, are much less light colored, with the mantle and arms yellowish. These were further described by Vivo (1991).

Callithrix (Mico) humeralifera (Humboldt, 1812)
Santarem Marmoset

1812 *Simia humeralifera* Humboldt. Brazil: restricted to Paricatuba, left bank of Rio Tapajós near its mouth by Hershkovitz (1958).
1812 *Jacchus humeralifer* E. Geoffroy. Lapsus for *humeraliferus,* the masculine of *humeralifera.*
1893 *Hapale santaremensis* Matschie. Brazil: Paricatuba.

DIAGNOSIS. Dorsum black, with white flecks, giving marbled impression; hairs of dorsum are black, white, black again, in equal thirds. Shoulders, arms, and occiput gray-brown/white, with the hair bases on shoulders dark brown, then banded white and dark brown, with white tips; the white and dark bands drop out on crown and arms, and the white tips predominate. Hindlegs brownish black; a white thigh stripe, formed by black-based hairs with long, white tips. Tail black, with white-tipped hairs arranged to give the impression of narrow white rings. Ears with long white hairs growing from both aspects and on rims; these tufts are continuous with the mantle hairs. Face pigmented (yellow-brown). Front of crown, whence hairs extend in midline onto forehead, entirely black, with this zone extending down cheeks, expanding across nose and on jaw angles; sides of crown with short, silvery hairs. Underside with sparse orange hair.

DISTRIBUTION. A small area along the left bank of the Rio Tapajós from about 5°S to the Amazon, and on the south bank of the Amazon west to the Rio Maués.

NOTES. Humboldt based his description on an unpublished manuscript by Étienne Geoffroy-Saint-Hilaire; according to the *Code*, this gives Humboldt credit for the authorship (Vivo 1991:40).

Callithrix (Mico) mauesi Mittermeier et al., 1992
    Maués Marmoset

1992 *Callithrix mauesi* Mittermeier, Schwarz, and Ayres, 1992. Brazil: west bank of Rio Maués-Açú, opposite Maués town, Amazonas.

DIAGNOSIS. With full ear tufts surrounding ears and ringed tail like *C. humeralifera* and *C. chrysoleuca*, but darker; ear tufts erect, "neatly trimmed"; dorsum all brindled, lacking the light mantle of *C. humeralifera;* thigh stripe silvery; underside with light orange tint; black on face restricted to a dark circumbuccal zone, whereas in *C. humeralifera* the whole face, including cheeks, is dark.

DISTRIBUTION. Apparently along the Rio Maués-Açú south of the type locality and west to the Rio Urariá and Rio Abacaxis. The range forms a small wedge between those of *C. humeralifera* and *C. chrysoleuca*.

Callithrix (Mico) chrysoleuca (Wagner, 1842)
    Gold-and-white Marmoset

1842 *Hapale chrysoleucos* Wagner. Brazil: Borba, lower Rio Madeira, Amazonas.
1868 *Mico sericeus* Gray. Brazil: Pará.

DIAGNOSIS. White, with golden- or orange-toned limbs and tail, but head always white; tail yellow, with hairs arranged in rings, giving spurious impression of pigment rings; underside yellow-creamy to more orange; no crown patch; ear tufts as in *C. humeralifera*, springing from both surfaces and from rims, fanning outward to a fanlike tuft; tufts white. Face and ears unpigmented.

DISTRIBUTION. Along the right bank of the Rio Madeira and its tributary the Aripuanã, south to about 8°S, and north to the Amazon.

NOTES. There is variation in the intensity of the yellow on the tail and the posterior part of the back. Vivo (1991) said that there is no indication of intergradation with its neighbors, even though it is found at Prainha, on the right bank of the Rio Aripuanã, very close to *C. intermedia*, which occurs at Foz do Rio Guariba nearby.

Callithrix (Mico) intermedia Hershkovitz, 1977
    Hershkovitz's Marmoset

1977 *Callithrix humeralifer intermedia* Hershkovitz. Brazil: near mouth of Rio Guariba, left bank of Rio Aripuanã, southeastern Amazonas.

DIAGNOSIS. Face and ears not or poorly pigmented. Rudimentary white ear tufts, originating from external surface of ears only. Anterior half of body whitish silvery, whitish or yellowish white; posterior half irregularly spotted with chestnut; crown dark. Thigh stripe golden yellow ("orange" [Hershkovitz 1977]). Tail not ringed.

DISTRIBUTION. A short section of the right bank of the Rio Aripuanã, between the ranges of *C. chrysoleuca* and *C. melanura*. It is possibly sympatric with both these species where their ranges meet (Vivo 1991).

NOTES. On Hershkovitz's illustration, the tail is represented as ringed like that of *C. humeralifera* and *C. chrysoleuca*, but actually both the type and a second specimen examined by Vivo (1991) have smooth tails with no rings, either of color or of hair tracts.

Although Hershkovitz described this form as a subspecies of *C. humeralifer*, intermediate between what he considered its other subspecies, *humeralifer* and *chrysoleuca*, all subsequent commentators have considered it closer to *C. argentata* (Avila-Pires 1985; Rylands et al. 1993) and either a subspecies of the latter or a distinct species. Thus the tail is not ringed, and although it has ear tufts, these grow only from the external surface of the ear and do not surround it like those of *C. humeralifera* and *C. chrysoleuca* (Vivo 1991). If it is intermediate between any other taxa, it is between the tufted and untufted sections of the *C. argentata* species-group.

Callithrix (Mico) humilis Roosmalen et al., 1998
Roosmalen's Dwarf Marmoset

1998 *Callithrix humilis* Roosmalen, Roosmalen, Mittermeier, and
da Fonseca. Brazil: west bank of the lower Rio Aripuanã, 1
km south of Nova Olinda, 5°30′63S, 60°24′61W, 45 m.

DIAGNOSIS. Size very small (head plus body only
15 mm long, weight 120–200 g); evenly olive-brown
above, orange yellow to golden below including inner sides of limbs; crown black; white "eyebrows,"
extending back to temples; face unpigmented, except for blackish circumocular zone; tail at most ob-
scurely ringed.

DISTRIBUTION. Between the lower Rio Aripuanã
and the Rio Madeira, probably south to beyond 6°S,
and west to the Rio Atininga or the Rio Manicoré.

NOTES. This species was discovered unexpectedly
by Marc van Roosmalen and Russ Mittermeier in a
small triangle of land south of the confluence of the
Rio Aripuanã with the Rio Madeira (Quammen
1997). Though nearly as small as *C. pygmaea,* it is
clearly a member of the *C. argentata* group, with
prominent "notched" ears; no cape of hair; a black,
virtually unringed tail; brown (unbrindled) body;
and contrasting black crown. There is no thigh
stripe. The face is depigmented as in all of the group,
with an arc of white hairs around the upper half. The
color changes with age are complex; a juvenile described by Roosmalen et al. (1998) was lighter, with
a clearly ringed tail and the ears concealed by long,
backswept hairs as in *C. pygmaea,* and only a broad,
black longitudinal stripe on crown, with the rest of
the head white.

In an addendum (Roosmalen et al. 1998:13, foot-
note 1), the discovery of the Rio Atininga population
was announced; these are said to retain the white lateral crown streaks into maturity and have a more
orange-ochraceous underside.

According to the describers (1998:20, table 3)
C. humilis resembles the larger *Callithrix* in its behavioral characters, thus differentiating it from *C. pygmaea;* but it does not exhibit anogenital or urine
scent marking or territoriality, thereby differing from
all other known *Callithrix.*

SUBGENUS *CEBUELLA* GRAY, 1866
Pygmy Marmoset

1866 *Cebuella* Gray. *Iacchus pygmaeus* Spix, 1823.

Members of this subgenus have no ear tufts, but the
ears are hidden by the long, backswept hair of the
cheeks. The pelage is agouti gray-brown, and the tail
is ringed.

Canavez et al. (1996) found that, on the basis of
chromosome studies and number ($2n = 44$), these
pygmy marmosets are closer to the *C. argentata*
group (subgenus *Mico*) than to the *C. jacchus* group
(subgenus *Callithrix*). Goodman et al. (1998) came to
the same conclusion on the basis of their molecular
studies. The three subgenera of *Callithrix* are, therefore, of rather unequal value.

Callithrix (Cebuella) pygmaea (Spix, 1823)
Pygmy Marmoset

DIAGNOSIS. Size very small, weighing only 110–
132 g; pelage tawny agouti; tail clearly ringed; long
cheek hairs, swept back concealing ears and continued as a conspicuous mane and shoulder mantle.
Skull with very small facial skeleton and large,
rounded braincase. There are many behavioral
differences from large species of *Callithrix* (Roosmalen et al. 1998), some at least of which may be attrib-
utable to the obviously pedomorphic nature of this
species.

DISTRIBUTION. Between the Rio Madeira and Rio
Caquetá, as far west as the slopes of the Andes.
Formerly the Rio Purus seemed to be a reasonable
southern boundary, but Roosmalen and Roosmalen
(1997) recorded a population from between the
Purus and the Madeira.

Callithrix (Cebuella) pygmaea pygmaea (Spix, 1823)

1823 *Iacchus pygmaeus* Spix. Brazil: near Tabatinga, Rio
Solimões.
?1844 *Hapale nigra* Schinz. Peru: Rio Ucayali and Rio Marañon.
Probable attribution to this species, fide Hershkovitz
(1977).
?1864 *Midas leoninus* Bates. Brazil: "upper Amazons." Not of
E. Geoffroy, 1812. Probable attribution to this species, fide
Hershkovitz (1977).

DIAGNOSIS. The nominotypical subspecies has buffy underparts, not sharply distinct from the color of the upper side.

DISTRIBUTION. Roosmalen and Roosmalen (1997) mapped it as occurring between the Rio Solimões and Rio Caquetá. According to specimens in the BM(NH), USNM, and AMNH collections, and as described by Hershkovitz (1977), the picture is more complicated. All specimens from north of the Rio Solimões between the Rio Napo and Rio Caquetá have buffy underparts, but so do some from south of the Solimões (Tabatinga, the type locality; and João Pessõa, on the middle Rio Jurúa).

Callithrix (Cebuella) pygmaea niveiventris (Lönnberg, 1940)

1940 *Cebuella pygmaea niveiventris* Lönnberg, 1940. Brazil: Lago Ipixuna, 3°52′S, 63°52′W, Rio Solimões.

DIAGNOSIS. Hershkovitz (1977) considered the color of the underside in *C. pygmaea* too variable to permit the recognition of this subspecies, but its validity as a subspecies from south of the Rio Solimões has been affirmed by Roosmalen and Roosmalen (1997) and Roosmalen et al. (1998), who discovered a further population between the Rio Purus and Rio Madeira.

DISTRIBUTION. Mapping the distributions of ventral color as given by Hershkovitz (1977), it seems evident that white underparts (and inner limb surfaces) occur to the south of the Solimões, including its southern bank (but with more variability as one approaches the mouth of the Rio Napo, and with a gray-bellied population pocket farther south on the middle Rio Jurúa), and also in the upper reaches of the river systems between the Rio Napo and at least the Rio Ucayali.

Genus *Leontopithecus* Lesson, 1840

Lion Tamarins

1840 *Leontopithecus* Lesson. *Leontopithecus marikina* Lesson, 1840.
1913 *Leontocebus* Elliot. Not of Wagner, 1840 (= *Saguinus*).
1956 *Leontideus* Cabrera. New name for *Leontocebus* Elliot, 1913, preoccupied.

The three long-known species, *L. rosalia, L. chrysomelas,* and *L. chrysopygus,* were generally regarded as subspecies of a single species, but Rosenberger and Coimbra-Filho (1984) found that the differences, including those in skulls and teeth, are quite marked and more clear-cut than previously thought. They proposed to recognize all three as distinct species, which indeed is mandatory under a phylogenetic species concept.

Leontopithecus rosalia (Linnaeus, 1766)

Golden Lion Tamarin

1766 *Simia rosalia* Linnaeus. Brazil: restricted to right bank of Rio São João, north of Cabo Frio, Rio de Janeiro, by Carvalho (1965).
1829 [*Jacchus rosalia*] *guyannensis* Fischer. Supposedly from Cayenne.
1829 [*Jacchus rosalia*] *brasiliensis* Fischer. Brazil.
1840 *Leontopithecus marikina* Lesson. Substitute for *rosalia.*
1913 *L[eontopithecus] aurora* Elliot. Lapsus for *marikina.*
1914 *Leontocebus leoninus* Pocock. Lapsus?

DIAGNOSIS. Color golden yellow to old gold, often variegated with black patches.

DISTRIBUTION. Formerly throughout the central and southern parts of Rio de Janeiro state, now restricted to a few isolated forest patches.

Leontopithecus chrysomelas (Kuhl, 1820)

Golden-headed Lion Tamarin

1820 *Midas chrysomelas* Kuhl. Brazil: forests of the Rio Ilhéos, Pardo, and Belmonte; restricted to Ribeirão das Minhocas, left bank of upper Rio dos Ilhéus, upper Bahia (Hershkovitz 1977).
1840 *Leontopithecus ater,* var. *A* and *B* Lesson.

DIAGNOSIS. Color black, with golden mane and ruff, forearms, hands and feet, and proximal half of tail.

DISTRIBUTION. Formerly the southern part of Bahia, north as far as the Rio de Contas, and into Minas Gerais as far as the Rio Jequitinhonha (Rylands et al. 1993); now greatly reduced.

Leontopithecus chrysopygus (Mikan, 1823)

Black Lion Tamarin

1823 *Jacchus chrysopygus* Mikan. Brazil: near Ipanema, São Paulo.
1840 *Leontopithecus ater* Lesson. Replacement for *chrysopygus*.

DIAGNOSIS. Black, with golden reddish forehead, rump, legs and root of tail; a few red-gold hairs in mane.

DISTRIBUTION. In the state of São Paulo, formerly between the Rio Paranapanema and Rio Tieté, but now apparently reduced to two small remnant forest patches.

Leontopithecus caissara Lorini and Persson, 1990

Superagui Lion Tamarin

1990 *Leontopithecus caissara* Lorini and Persson. Brazil: Paraná, Guaraqueçaba Municipality, Superagui Island, Barra do Ararapira (ca. 25°18′S, 48°11′W).

DIAGNOSIS. Head, circumfacial ruff and beard, nape, withers, arms and tail black; forehead, preauricular tuft, rest of body, and hindlegs golden; feet black with golden hairs intermixed. The transitional region behind the shoulders consists of long golden hairs that are dark brown at base.

DISTRIBUTION. Superagui Island, and on the opposite mainland on the Rio Sebuí and Rio dos Patos and in the Cananéia region in São Paulo (Rylands et al. 1993).

NOTES. As recorded by Lorini and Persson (1990), the limited evidence suggests that this is a large species, perhaps exceeding the size of *L. chrysopygus*, hitherto considered the largest of the genus; from their fig. 3, the skull seems to resemble those of *L. chrysomelas* and *L. chrysopygus* in having narrow biorbital width, relatively broad braincase, and low ascending ramus compared with *L. rosalia*. The premaxillae are said to be exceptionally reduced.

Coimbra-Filho (1990) considered this a subspecies of *L. chrysopygus*, which sometimes approaches this species in color, especially in the southern part of its range.

Genus *Saguinus* Hoffmannsegg, 1807

Tamarins

Hershkovitz (1977) divided this genus into three sections: hairy-faced (with *S. nigricollis*, *mystax*, and *midas* groups), mottle-faced (only the *S. inustus* group), and bare-faced (with *S. bicolor* and *oedipus* groups). They were distinguished by whether the face and temples are fully haired, or naked, or only sparsely haired, and by the blackish pelage of the *S. inustus* group.

A craniometric analysis, examining size by the Penrose statistic and shape by Rohlf and Sokal Q-mode correlation coefficients, succeeded in replicating Hershkovitz's hairy-faced and bare-faced groups (*S. inustus* and *S. bicolor* were not examined), and the *midas*, *nigricollis*, and *mystax* subgroups within the former (Natori and Hanihara 1988). Cladistic analysis of cranial and dental characters, this time including *S. inustus* and *S. bicolor*, again emphasized the separateness of the bare-faced *S. oedipus* group, but associated *S. inustus* and *S. bicolor* with the hairy-faced group: the latter, indeed, clustered with *S. midas* (Natori 1988). It should be noted, however, that two of the three derived conditions shared between the *bicolor* and *midas* groups (no. 8, reduction of entoconid on M_1, and no. 11, position of inferior petrosal sinus foramen) are shared also with the *oedipus* group, and that the only derived state shared between them and the hairy- and mottle-faced tamarins, the structure of the dental enamel, is actually of uncertain polarity, being subject to homoplasy both among species of *Saguinus* and between them and other genera. Instead, these characters could be taken as uniting the bare-faced species, and as evidence that the *mystax* group is related to the bare-faced clade.

Analysis of vocalizations (Snowdon 1993) found the *S. oedipus* group to be distinct from the *S. nigricollis* and *mystax* groups, with *S. bicolor* apparently somewhat intermediate.

In contrast to all this, a mtDNA (cytochrome *b* and D-loop) study recovered a small-bodied and a large-bodied clade, the former being the *S. nigricollis* group and the latter consisting of all others, whether hairy- or bare-faced (Cropp et al. 1999). Within the large-bodied clade, the *bicolor* and *midas* groups were associated, as in Natori's (1988) analysis, but there were no other stable associations except for the species-groups themselves.

Saguinus midas group:

1807 *Saguinus* Hoffmannsegg. *Saguinus ursula* Hoffmannsegg, 1807.
1812 *Midas* E. Geoffroy. *Simia midas* Linnaeus, 1758. Preoccupied.
1870 *Tamarin* Gray. *Saguinus ursulus* Hoffmannsegg, 1807.

Saguinus nigricollis group:

1840 *Leontocebus* Wagner. *Simia leonina* Humboldt, 1812.

Saguinus bicolor group:

1840 *Marikina* Lesson. *Midas bicolor* Spix, 1823.
1870 *Seniocebus* Gray. *Midas bicolor* Spix, 1823.

Saguinus oedipus group:

1840 *Oedipus* Lesson. *Simia oedipus* Linnaeus, 1758. Not of Tschudi, 1838 (Amphibia).
1862 *Oedipomidas* Reichenbach. *Simia oedipus* Linnaeus, 1758.
1870 *Hapanella* Gray. *Hapale geoffroyi* Pucheran, 1845.

Saguinus mystax group:

1870 *Mystax* Gray. *Simia mystax* Spix, 1823.
1899 *Tamarinus* Trouessart. *Simia mystax* Spix, 1823.

Saguinus midas group

Facial skin and hair entirely black; head and shoulders black, rest of body and legs mottled black and buffy or reddish. In the skull, the group is distinguished by the downward shift of the rhinion (Natori and Hanihara 1988).

They share with the bare-faced tamarins, the *S. bicolor* and *oedipus* groups (or, rather, with two of the three species of the latter), a reduction of the entoconid on M_1 and the loss of the inferior petrosal sinus foramen (or, more likely, its coalescence with the jugular foramen). With the *S. bicolor* group alone is shared a derived form of the premetacristid on P_2 (Natori 1988).

Further cranial and dental morphometric analyses by Hanihara and Natori (1989) reinforced the distinc-

tiveness of this species-group among the hairy-faced tamarins.

Saguinus midas (Linnaeus, 1758)

Red-handed Tamarin

1758 *Simia midas* Linnaeus. "America." Restricted to Suriname by Schreber in 1775 (Hershkovitz 1958).
1795 *C[ebus] tamarin* Link. French Guiana: Cayenne (according to Hershkovitz 1977).
1806 *Simia lacepedii* Fischer. "America."
1812 *Midas rufimanus* E. Geoffroy. Substitute for *midas*.
1912 *Leontocebus midas egens* Thomas. Brazil: Obidos, Colônia de Veado, Pará.

DIAGNOSIS. Hands and feet orange, contrasting with blackish body and limbs.

DISTRIBUTION. North of the Amazon, east of the Rio Negro, extending north and east to the coast in the Guyanas and Brazil.

Saguinus niger (E. Geoffroy, 1803)

Black Tamarin

1803 *Sagouin niger* E. Geoffroy. Supposedly from Cayenne, redetermined by Hershkovitz (1977) as Belém.
1807 *Saguinus ursula* Hoffmannsegg. Brazil: near Pará (= Belém).
1922 *Mystax ursulus umbratus* Thomas. Brazil: Cametá, Rio Tocantins, Pará.

DIAGNOSIS. Hands and feet black like body and limbs.

DISTRIBUTION. South of the Amazon, east of the Rio Xingu as far as the coast and probably the Rio Gurupi, and on Marajó Island.

NOTES. Natori and Hanihara (1988) found that the dentition of this species is distinct from that of *S. midas*, which more closely resembles that of *S. bicolor*.

Hershkovitz (1977) showed that the specific name *tamarin* Link, 1795, often used for this species, was actually based on a red-handed tamarin *(S. midas)* described by Buffon.

Saguinus nigricollis group

Facial skin and hair black, except for white hair on muzzle.

Saguinus nigricollis (Spix, 1823)
Black-mantled Tamarin

DIAGNOSIS. Foreparts of back black, grading into red or brownish red hindparts; tail with reddish tones at base, then black for remaining four-fifths. Underside brown.

DISTRIBUTION. North of the Rio Solimões, between the Rio Napo and the Rio Caquetá, in Brazil, Peru, and Colombia (Hernández-Camacho and Cooper 1976; Hershkovitz 1977).

Saguinus nigricollis nigricollis (Spix, 1823)

1823 *Midas nigricollis* Spix. Brazil: near São Paulo de Olivença, north bank of Rio Solimões.
1848 *Midas rufoniger* I. Geoffroy and Deville. Peru: Rio Marañon.

DIAGNOSIS. Rich reddish on hindparts with no yellowish tones in the saddle.

DISTRIBUTION. From between the lower Rio Putumayo and the Rio Napo and Rio Solimões.

Saguinus nigricollis hernandezi (Hershkovitz, 1982)

1982 *Saguinus nigricollis hernandezi* Hershkovitz. Colombia: Rio Peneya, a left-bank tributary of the Rio Caquetá, 15 km above the mouth of the Rio Caguan.

DIAGNOSIS. This is the form described by Hernández-Camacho and Cooper (1976), from the upper Putumayo and Caguan, as having "a dull and brownish cast to the lower back and hind limbs, as well as some grizzled yellow and black in the saddle."

Saguinus graellsi (Jiménez de la Espada, 1870)
Graells's Tamarin

1870 *Midas graellsi* Jiménez de la Espada. Peru: Tarapoto, opposite mouth of Rio Curaray, Rio Napo, Loreto (Hershkovitz 1977).

DIAGNOSIS. Foreparts of back black with buff ticking, grading into buffy brown hindparts; proximal third of tail buffy brown, remainder black.

DISTRIBUTION. West of the Rio Napo, from the Rio Putumayo south to the Rio Marañon, and west to the Rio Santiago, in Ecuador, Peru, and across the Colombian border (Hershkovitz 1977).

NOTES. This was placed as a subspecies of *S. nigricollis* by Hershkovitz (1977). According to Hernández-Camacho and Cooper (1976) it is sympatric with the latter in the region of Puerto Leguízamo, in southernmost Colombia along the Ecuadorian border, but later it was found that the claim was due to a mislabeled specimen from the north bank of the Rio Caquetá, at Curiplaya (Hershkovitz 1977).

Saguinus fuscicollis (Spix, 1823)
Brown-mantled Tamarin

DIAGNOSIS. Head, arms, legs, and both forepart and hindpart of back solid or ticked black, brown, or red, middle zone of back forming a mottled saddle; tail reddish or brownish at base, at least on underside, remainder black.

NOTES. Hershkovitz (1977) recognized 14 subspecies of this species. One of these, *S. f. tripartitus,* was awarded full species status by Thorington (1988), and another, *S. f. acrensis,* was reduced to synonymy, under *S. m. melanoleucus,* by Peres (1993). The remainder divide into three groups in Hershkovitz's (1977) scheme:
  —northern races: *fuscus, avilapiresi;*
  —central races: *fuscicollis, nigrifrons, illigeri, leucogenys, lagonotus;*
  —southern races: *weddelli, cruzlimai, crandalli, melanoleucus;* by extrapolation, *primitivus* also belongs to this group.
The northern and central forms overlap in their characters: blackish or agouti forehead and crown, (usually) agouti nape and mantle, and agouti or blackish tail. The southern forms all have a white frontal blaze (gray in *S. f. primitivus*), and the crown, nape, mantle, and tail are never agouti.

Craniometric analysis (Cheverud and Moore 1990), in a study of eight of these subspecies, showed a remarkably similar matrix of relationships to that proposed by Hershkovitz (1977).

Thorington (1988), in documenting apparent sympatry between *S. f. lagonotus* and *S. tripartitus,* noted that some of the subspecies of *S. fuscicollis* are at least as distinct from each other as are these two. Rylands et al. (1993) suggested that the white *S. f. melanoleucus* in particular should stand as a full species. It does seem likely that, in fact, *S. fuscicollis* ought to be broken up into several species, though what they might be is unclear. Napier (1976) was impressed with the consistency of the differences among them, remarking that only at the headwaters of some rivers were there any intermediates between them (see also Peres et al. 1996).

Saguinus fuscicollis fuscicollis (Spix, 1823)

1823 *Midas fuscicollis* Spix. Brazil: near São Paulo de Olivença, between Rio Solimões and Rio Içá, Amazonas (restricted by Hershkovitz [1977]).
1848 *Midas flavifrons* I. Geoffroy and Deville. "Upper Amazon," Peru.

DIAGNOSIS. Crown and circumfacial hair speckled distinctively yellowish; mantle and arms agouti brown; legs reddish brown; base of tail reddish brown.

DISTRIBUTION. From the Rio Javari to the Rio Jutaí and upper Rio Juruá, south of the Amazon.

Saguinus fuscicollis nigrifrons (I. Geoffroy, 1851)

1851 *Hapale nigrifrons* I. Geoffroy. Locality unknown.
1928 *Mystax nigrifrons pebilis* Thomas. Peru: south bank, Pebas.

DIAGNOSIS. Crown speckled red-brown, circumfacial hair black; mantle and arms agouti red-brown; legs reddish brown; base of tail reddish brown.

DISTRIBUTION. Between the Rio Ucayali and Rio Javari, south of the Rio Solimões as far as the Rio Blanco.

Saguinus fuscicollis illigeri (Pucheran, 1845)

1845 *Hapale illigeri* Pucheran. Colombia: restricted to Colombian bank of Rio Solimões by Hershkovitz (1949), but altered to left bank of Rio Ucayali near its mouth, Peru, by Hershkovitz (1966).

1850 *Hapale deville* I. Geoffroy. Peru: Sarayacu, Loreto.
1915 *Mystax bluntschlii* Matschie. Peru: Rio Samaria.
1920 *Leontocebus mounseyi* Thomas. Peru: Lower Rio Ucayali, Rio Pacaya, opposite Sapote.

DIAGNOSIS. Crown and circumfacial hair black; mantle and arms agouti reddish maroon; legs reddish; base of tail reddish.

DISTRIBUTION. From south of the Rio Marañon and east of the Rio Huallaga, to somewhat east of the Rio Ucayali.

NOTES. The main difference between this form and *S. f. nigrifrons* is that in *S. f. nigrifrons* the entire crown is black, whereas in *S. f. illigeri* the black is restricted to the front of the crown.

Saguinus fuscicollis leucogenys (Gray, 1866)

1866 *Midas leucogenys* Gray. Peru: Sarayacu (fixed by Napier 1976).
1914 *Leontocebus pacator* Thomas. Peru: Rio Pachitea.
1928 *Mystax devillei micans* Thomas. Peru: Yurac Yacu, 760 m, San Martín.

DIAGNOSIS. Crown and circumfacial hair black; mantle and arms black or with brown agouti bands, and longer haired than in previous subspecies; legs reddish; base of tail reddish.

DISTRIBUTION. From north-central Peru, west of the Rio Huallaga and the upper Rio Ucayali, and south of the Rio Pisqui. This is a very distinctive form.

Saguinus fuscicollis lagonotus (Jiménez de la Espada, 1870)

1870 *Midas lagonotus* Jiménez de la Espada. Peru: Destacamento (= Francisco de Orellana), confluence of Rio Napo and Rio Amazonas.
1904 *Midas apiculatus* Thomas. Ecuador: Rio Copataza.

DIAGNOSIS. Crown and circumfacial hair black; mantle, arms, and legs red; base of tail reddish.

DISTRIBUTION. From Ecuador and Peru, between the Rio Napo and Rio Marañon, west to the Andes.

NOTES. This subspecies is rather like *S. f. illigeri,* but has the long hairs of *S. f. leucogenys* and the black of the crown grades more into the deep maroon of the mantle, instead of being rather sharply demarcated.

Saguinus fuscicollis fuscus (Lesson, 1840)

1805 *Simia leonina* Humboldt. Not of Shaw, 1800 (= *Trachypithecus johnii,* according to Hershkovitz [1977]).
1840 *Leontocebus fuscus* Lesson. Colombia: Mocoa, Putumayo.

DIAGNOSIS. Crown and circumfacial hair black; mantle, arms, and legs agouti buff and black; basal one- to two-thirds of tail brown on ventral surface only.

DISTRIBUTION. North of the Rio Solimões, between the Rios Japurá-Caquetá and Içá-Putumayo. Hernández-Camacho and Cooper (1976) referred to a specimen from the right bank of the Rio Guayabero in Angostura in the Macarena Mountains.

NOTES. Overall this is extremely like *S. f. illigeri* but is more agouti on shoulders, and the mantle begins farther back, so that the red of the foreback occupies a larger area.

Saguinus fuscicollis avilapiresi Hershkovitz, 1966

1966 *Saguinus fuscicollis avilapiresi* Hershkovitz. Brazil: mouth of the Lago de Tefé, Rio Solimões, Amazonas.

DIAGNOSIS. A very dull, uncontrastingly colored form. Crown speckled buff, forehead black; mantle, arms, and legs black and buff; base of tail agouti buff.

DISTRIBUTION. Known for certain only from the type locality and from Ayapuá, left bank of Rio Purus (Napier 1976).

Saguinus fuscicollis weddelli (Deville, 1849)

1849 *Midas weddelli* Deville. Bolivia: Apolobamba (= Caupolicán) Province, La Paz.
1914 *Leontocebus purillus* Thomas. Brazil: Rio Xapuri, upper Rio Acre, upper Rio Purus.
1940 *Mystax imberbis* Lönnberg. Bolivia: Victoria, Rios Beni and Madre de Dios.

DIAGNOSIS. Crown and circumfacial hair black, with white eyebrows; mantle and arms dark brown; legs red; base of tail red.

DISTRIBUTION. A very wide distribution in Bolivia, southern Peru, and Brazil, south and east of the Rio Purus to east of the Rio Madeira.

Saguinus fuscicollis cruzlimai Hershkovitz, 1966

1966 *Saguinus fuscicollis cruzlimai* Hershkovitz. Brazil: said to be from upper Rio Purus, Amazonas.

DIAGNOSIS. Crown, mantle, and limbs reddish orange; saddle agouti; with a white transverse eyebrow band.

DISTRIBUTION. Unknown. Assuming the type locality to be correct, Hershkovitz (1977) extrapolated a distribution on the west bank of the upper Rio Purus.

Saguinus fuscicollis primitivus Hershkovitz, 1977

1977 *Saguinus fuscicollis primitivus* Hershkovitz. Brazil: Rio Juruá, Amazonas.

DIAGNOSIS. Overall almost unvarying agouti, and so corresponding to the plesiomorphic state according to Hershkovitz's principle of metachromism; but with a well-defined grayish forehead band; tail mainly or entirely black, except for basal 2–7 cm.

DISTRIBUTION. Known only from the upper Rio Purus and Rio Pauiní.

Saguinus melanoleucus (Miranda Ribeiro, 1912)
White-mantled Tamarin

DIAGNOSIS. Color white or with localized buffy tones and orange rump and thighs. There is no agouti banding. Skin is black.

DISTRIBUTION. As far as is known, between Rios Juruá and Tarauacá.

NOTES. Rylands et al. (1993) suggested that this constitutes a distinct species. Certainly it is dramatically distinct from unbleached saddled tamarins (here, *S. fuscicollis*), and the chain of intermediates that links the subspecies of the latter does not extend to these white forms.

This species hybridizes with *S. fuscicollis fuscicollis* at the headwaters of the Rio Juruá; intermediate phenotypes and genotypes (cytochrome *b*) occur on the left bank of the river's headwaters at 8°36–40'S, 72°47–51'W, where they occur along with pure *S. f. fuscicollis* in about a 2:1 ratio; in this region, the presence of oxbow lakes on the left bank indicates that the course of the river has shifted slightly to the east in very recent times (Peres et al. 1996).

Saguinus melanoleucus melanoleucus (Miranda Ribeiro, 1912)

1912 *Mico melanoleucus* Miranda Ribeiro. Brazil: Pará. Restricted by Hershkovitz (1977) to Santo Antonio, Rio Eirú, upper Rio Juruá, Amazonas.
1937 *Leontocebus hololeucus* Pinto. Brazil: Santo Antonio, Rio Eirú.
1957 *Leontocebus melanoleucus acrensis* Carvalho. Brazil: Pedra Preta, about 15 km below Vila Taumaturgo, right bank of upper Rio Juruá, Acre.

DIAGNOSIS.   Head with white hair on pigmented (blackish) skin; body and tail wholly whitish or yellowish white (*melanoleucus* morph) or with drab and buffy mixed saddle and orange rump and thighs (*acrensis* morph).

DISTRIBUTION.   Occurs between the Rio Tarauacá and Rio Juruá.

NOTES.   Peres (1993) recorded hybridization between this form and *S. f. fuscicollis* at the headwaters of the Rio Juruá. He found that the *acrensis* pelage is not geographically distinct, but is actually a morph occurring in various frequencies within populations of *S. m. melanoleucus*.

Saguinus melanoleucus crandalli Hershkovitz, 1966

1966 *Saguinus fuscicollis crandalli* Hershkovitz. Locality unknown.

DIAGNOSIS.   Crown, mantle, and arms drab (grayish); rump and thighs orange; saddle mixed buffy and black; tail pale brown; a broad, white frontal band; underparts buffy or nearly white.

NOTES.   Rylands et al. (1993) regarded this as forming part of a species *S. melanoleucus,* separate from *S. fuscicollis.* It, like *acrensis,* could likewise be a simple color morph of a monotypic *S. melanoleucus* (A. B. Rylands, personal communication; Peres et al. 1996).

Saguinus tripartitus (Milne-Edwards, 1878)

Golden-mantled Tamarin

1878 *Midas tripartitus* Milne-Edwards. Ecuador: Rio Napo, Oriente.

DIAGNOSIS.   Body and legs speckled gray-buff; forepart of back golden yellow; arms and underside orange-red; basal third of tail red on ventral surface only; head sharply defined black, with a prominent gray frontal chevron.

NOTES.   Thorington (1988) found evidence that this is probably sympatric with *S. fuscicollis lagonotus* on the right bank of the Rio Napo and so removed it from *S. fuscicollis,* of which Hershkovitz (1977) had ranked it as a subspecies. Hernández-Camacho and Cooper (1976) recorded it from Puerto Narino on the Colombian bank of the Amazon.

Saguinus mystax group

Facial skin black, except under long, white mustache hairs, where it is not pigmented. I_1 is only a little wider than long, instead of being much wider as in other species-groups (Natori 1988).

According to Natori and Hanihara (1988), this group is close, craniometrically, to the *S. nigricollis* group. Among the species, *S. imperator* stands out as being only distantly related to the rest (Hanihara and Natori 1989).

Subspecies in *S. mystax* and *S. imperator* are distinctive, and further study may show them to be specifically distinct. On the other hand the three subspecies of *S. labiatus* are barely different, despite their quite isolated ranges.

Saguinus mystax (Spix, 1823)
Mustached Tamarin

DIAGNOSIS. Forehead, crown, nape, and mantle black ticked with buff; back mottled black and reddish or buff; tail entirely black; limbs agouti; white mustache forms clover-leaf pattern.

DISTRIBUTION. South of the Amazon-Solimões-Marañon, approximately between the lower Rio Huallaga on the west and the Rio Madeira on the east. On the upper Rio Juruá, upriver of Porto Walter, this species is sympatric with *S. imperator subgrisescens* of the same species-group (Peres 1993). Rylands et al. (1993) drew attention to the records of populations ascribed to this species from Bolivia, where they are sympatric with *S. labiatus*. It may well be parapatric or marginally sympatric with *S. pileatus* (see Notes under that species).

Saguinus mystax mystax (Spix, 1823)

1823 *Midas mystax* Spix. Brazil: between Rio Solimões and Rio Iça, Amazonas; fixed at São Paulo de Olivença by Hershkovitz (1977).

DIAGNOSIS. Back (saddle) mottled rufous, the white bases often showing through; crown black; tail black; legs mottled rufous; mustache extensive; genitalia largely unpigmented, with very sparse white hairs, contrasting with blackish tone of rest of underside.

DISTRIBUTION. From Peru, south of the Amazon-Marañon, west to a little beyond the Rio Ucayali and Rio Tapiche (probably to the lower Rio Huallaga), east to the Rio Tefé, and south to the mouth of the Rio Urubamba.

Saguinus mystax pluto (Lönnberg, 1926)

1926 *Mystax pluto* Lönnberg. Brazil: Lago Ayapuá, west of Rio Purus, Amazonas; revised by Hershkovitz (in Napier 1976) to east bank of Rio Purus, but this seems dubious according to Rylands et al. (1993).

DIAGNOSIS. Back (saddle) mottled with pale, whitish buff; legs agouti like mantle, which has drab, not white, hair bases; crown black; tail black; mustache extensive; whole genital area white, including region from posterior belly to circumanal region and underside of tail base. This and *S. m. mystax* are very alike; the color of the posterior belly is the best distinguishing characteristic.

NOTES. Hershkovitz (1977) thought that this subspecies is likely to be separated from *S. pileatus* by the Rio Purus and so maintained that, of the three known localities, Lago Ayapuá and Jaburú, both on the left bank of the Rio Purus, should actually be interpreted as "opposite" these locales; only the third locality, Lago de Mapixí, actually is on the eastern side of the river. Rylands et al. (1993) took the opposite view and surmised that it probably occurs along the left bank of the Rio Purus, below the Rio Tapauá, and west to the Rio Coarí, approaching the range of *S. pileatus*, which has been recorded at São Luis de Marmoriá. No new records seem to have been obtained in the meantime, so there is no evidence either way.

Saguinus pileatus (I. Geoffroy and Deville, 1848)
Red-capped Tamarin

1848 *Midas pileatus* I. Geoffroy and Deville. Brazil: Rio Javari; restricted by Hershkovitz (1977) to Lago de Tefé, and should almost certainly be further restricted to its eastern margin (Rylands et al. 1993).
1904 *Midas pileatus juruanus* Ihering. Brazil: Rio Juruá about 7°S.

DIAGNOSIS. Limbs blackish brown, becoming black on hands and feet, and not contrasting much with brownish black body; tail black; mustache somewhat less extensive; forehead and crown deep maroon red; genitalia largely unpigmented, with sparse white hairs, contrasting with blackish tone of rest of underparts.

DISTRIBUTION. East of the Rio Tefé, west of the Rio Purus, and south as far as the Rio Pauiní or Rio Marmoriá.

NOTES. This taxon, placed as a subspecies of *S. mystax* by Hershkovitz (1977), is here given species rank because it is diagnosably different from *S. m. mystax* and *S. m. pluto,* which, in contrast, are only marginally distinct from each other. The red crown of *S. pileatus,* in particular, is strikingly distinctive. Its range inserts within that of *S. mystax* and it seems not unlikely to be at least parapatric with *S. m. pluto* along the left bank of the Rio Purus (see Notes for *S. m. pluto*).

Saguinus labiatus (Humboldt, 1812)
White-lipped Tamarin

DIAGNOSIS.   Nape and mantle agouti, hindparts with paler wash; tail with red or orange on ventral surface of base; white mustache outlines lips; white crown spot, sometimes with red mark as well; underside reddish.

DISTRIBUTION.   Occupies two widely separated areas, one south of the Amazon between the Rio Purus and Rio Madeira, the other north of it, between the Amazon itself and the Rio Japurá.

NOTES.   Rylands et al. (1993) discussed the implications of a report by Izawa and Bejarano (1981) of *S. labiatus*–like tamarins on the left bank of the Rio Acre in Bolivia, described as having a red underside like *S. l. labiatus,* but gray instead of black above, and with a large, almost *mystax*-like mustache; they might be either hybrids between *S. l. labiatus* and *S. imperator* or a new taxon related to both.

Saguinus labiatus labiatus (Humboldt, 1812)

1812 *Simia labiata* Humboldt. Brazil: restricted to environs of Lago Joanacan (= Janauacá), Amazonas, by Cabrera (1957), but more probably from between the Rio Purus and Rio Madeira, south of the Rio Ipixuna, according to Hershkovitz (1977).
?1862 *Midas erythrogaster* Reichenbach. Brazil.
?1862 *Midas elegantulus* Slack. Brazil: Amazon region.
1907 *Midas griseovertex* Goeldi. Brazil: Bom Lugaz, Rio Purus, Amazonas.
1913 *Midas griseoventris* Elliot. Lapsus for *griseovertex*.

DIAGNOSIS.   Overall black, with white marbling especially on hindparts; underside (except throat) and basal one-third or one-fourth of ventral surface of tail

sharply red; an indistinct red line on front of crown, with a conspicuous white spot behind.

DISTRIBUTION.   From between the Rio Purus and Rio Madeira-Abuná system, south of the Rio Ipixuna, crossing the headwaters of the Rio Abuná in Bolivia; south as far as the Rio Acre in the Rio Tahuamantú region (Rylands et al. 1993).

NOTES.   Hershkovitz (1977) noted variation in the degree to which the black of the throat invades the anterior part of the chest (but it never goes beyond the top of the sternum) and pointed out the variation in the expression of the red and whitish crown marks: the most southerly specimen lacks a red patch altogether.

Saguinus labiatus rufiventer (Gray, 1843)

1843 *Jacchus rufiventer* Gray. "Mexico" (!).

DIAGNOSIS.   Like *S. l. labiatus,* but has a red, Y-shaped mark on front of crown, with a weakly expressed silvery area behind.

DISTRIBUTION.   From the lower reaches of the region south of the Rio Solimões, between the Rios Purus and Madeira, probably below the Rio Ipixuna.

NOTES.   Hershkovitz (1977) noted that the northerly specimens have the red crown mark well expressed, with only a vague silvery tone behind, whereas the southerly ones have a very insignificant red mark but a sharply marked white spot. In the London (BM[NH]) material, this differentiation is quite noticeable (the northern and southern populations are at least as distinct as either is from *S. l. thomasi*), and I distinguish them here as subspecies. The types of *labiata* and *griseovertex* are the white-spot form, and that of *rufiventer* is red-marked; I have not seen the types of *erythrogaster* and *elegantulus* and provisionally place them in the synonymy of *S. l. labiatus.*

Saguinus labiatus thomasi (Goeldi, 1907)

1907 *Midas thomasi* Goeldi. Brazil: Tonantins, Rio Tonantins, left bank of Rio Solimões below mouth of Rio Içá.

DIAGNOSIS. Underside and basal fifth or sixth of ventral surface of tail orange; black of throat extending onto upper sternum; very poorly developed red line on front of crown, or none at all, with a small pale silvery line behind.

DISTRIBUTION. Known to Hershkovitz (1977) only from the type locality, this was recorded by malaria researchers 200 km farther west, at Barreirinha, on the Rio Auatí-Paraná, suggesting that it may inhabit a wide area between the Rio Solimões and Rio Japurá (Rylands et al. 1993).

NOTES. This subspecies is barely differentiated from nominotypical *S. l. labiatus,* as Hershkovitz (1977) remarked. Its differential features lie at the edge of the range of variation in *S. l. labiatus.* It is less close to *S. l. rufiventer.*

Saguinus imperator (Goeldi, 1907)
Emperor Tamarin

DIAGNOSIS. White mustache greatly elongated; body agouti gray-buff; tail usually reddish with dark tip; hands and feet black.

NOTES. At one site, on the Rio Juruá upriver of Porto Walter, this species is sympatric with *S. mystax* (Peres 1993).

The facial pattern of the young resembles that of *S. mystax* (Hershkovitz 1979).

The two subspecies recognized by Hershkovitz (1979) seem strikingly different, but I do not have evidence to assess whether they are really only subspecies or should rank as full species.

Saguinus imperator imperator (Goeldi, 1907)

1907 *Midas imperator* Goeldi. Brazil: Rio Purus, Amazonas; either Bom Lugar or, perhaps, Monte Verde (Carvalho [1959], cited by Hershkovitz [1979]).

DIAGNOSIS. White chin hairs short, not hiding the black chin patch; underside and inner side of arms reddish or orange with white hairs; upper parts buffy or grayish agouti with dark brown hair bases (Hershkovitz 1979).

DISTRIBUTION. Hershkovitz (1979) mapped it in the triangle between the lower Rio Purus and the Rio Acre, but noted that the probable type locality, Bom Lugar, is actually on the right bank of the Rio Purus below its confluence with the Rio Acre, whereas Monte Verde, a possible alternative as type locality, is on the left bank opposite the mouth of the Rio Acre, neither being within the "triangle." Hershkovitz expressed caution about the accuracy of these localities, but by implication did not rule them out. There are isolated populations on both banks of the Rio Acre, 20 km west of Inapari, in Brazil (Rylands et al. 1993).

Saguinus imperator subgrisescens (Lönnberg, 1940)

1940 *Mystax imperator subgrisescens* Lönnberg. Brazil: Santo Antonio, Rio Eirú, Amazonas.

DIAGNOSIS. White chin whiskers elongated into long tufts, largely hiding blackish chin patch; underside and inner side of arms brown mixed with white hairs; upper parts brown agouti with light (brown or white) hair bases (Hershkovitz 1979).

DISTRIBUTION. According to Hershkovitz (1979), this has a wide range, between the Rio Eirú and upper Rio Juruá, southwest into Peru to the Rio Urubamba and then southeast to the upper Rio Madre de Dios. According to Peres (1993), it also occurs, but irregularly, on the left bank of the Rio Juruá.

Saguinus bicolor group

Face and forepart of crown almost naked; tail bicolored, dark brown above and orange below. They share the derived states of entoconid reduction on M_1 and of the loss of a separate inferior petrosal sinus foramen with some of the *S. oedipus* group and with the *S. midas* group, and of the morphology of the P_2 premetacristid with the *S. midas* group alone (Natori 1988).

Saguinus bicolor (Spix, 1823)
Pied Tamarin

1823 *Midas bicolor* Spix. Brazil: Manaus.

DIAGNOSIS.   Body bicolored: anterior part, including arms, neck, chest, and mantle, white, posterior part agouti brown above, orange below including inner surfaces of hindlimbs.

DISTRIBUTION.   North of the Amazon, east of the Rio Negro, extending east for only about 100 km, as far as Itacoatiara, and north of Manaus for just 30–45 km.

Saguinus martinsi (Thomas, 1912)
Martins's Tamarin

DIAGNOSIS.   Body uniformly agouti brown above from crown to tail base, paler on flanks and shoulders, to buffy on limbs; orange below.

DISTRIBUTION.   A very small area on the north bank of the Amazon on either side of the Rio Nhamundá.

NOTES.   This is extremely distinct from *S. bicolor*, in which it has usually been included as a subspecies. I have not seen examples of *S. m. ochraceus*, which I here provisionally retain in *S. martinsi* as a subspecies.

Experimental hybridization between *S. bicolor* and *S. m. martinsi* produced twin offspring that were similar, but not absolutely identical, to *S. m. ochraceus* (Coimbra-Filho et al. 1997). In *S. m. ochraceus* the neck and base of the mantle are described as "greyish-ochre or yellowish-grey," whereas in the hybrids they were white in one individual and grayish white in the other. It is possible that the taxon *ochraceus* could be a stabilized hybrid between the other two; but it is also possible that it could be a largely plesiomorphic taxon, retaining the phenotype ancestral for the species-group, as appears to be the case with the gray-legged species of *Pygathrix* (see under Cercopithecoidea).

Saguinus martinsi martinsi (Thomas, 1912)

1912 *Leontocebus martinsi* Thomas. Brazil: Faro, Pará.

DIAGNOSIS.   No white ruff or nape; body somewhat darker than *S. m. ochraceus*.

DISTRIBUTION.   East of the Rio Nhamundá, as far as the Rio Erepecurú.

Saguinus martinsi ochraceus Hershkovitz, 1966

1966 *Saguinus bicolor ochraceus* Hershkovitz. Brazil: mouth of Rio Paratucú, right bank tributary of Rio Nhamundá, Amazonas.

DIAGNOSIS.   Paler and less distinctly agouti banded than *S. m. martinsi* with silvery to buffy tones on ruff and nape.

DISTRIBUTION.   From west of the Rio Nhamundá, perhaps as far as the Rio Uatumã.

Saguinus oedipus group

Face almost naked, with short white or silvery hairs locally. Natori and Hanihara (1988) found that this group could be distinguished from all hairy-faced tamarins by the narrow maxilla; they affirmed the full species status of *S. geoffroyi*. The posterior part of the malar-maxillary suture runs postero-inferiorly along the zygomatic arch instead of inferiorly as in other tamarins, and the suture crosses the inferior orbital border more medially (Natori 1988). Two members of the group (*S. oedipus* and *S. geoffroyi*) share derived conditions of P_2 and the basicranial foramina with the *S. bicolor* and, unexpectedly, the *S. midas* groups (Natori 1988).

Saguinus oedipus (Linnaeus, 1758)
**Cottontop Tamarin or
Pinché Tamarin**

1758 *Simia oedipus* Linnaeus. America (restricted to lower Rio Sinú, Bolívar, Colombia, by Hershkovitz [1949]).
1840 *Oedipus titi* Lesson. Replacement for *oedipus*.
1912 *Seniocebus meticulosus* Elliot. Colombia: Rio San Jorge.

DIAGNOSIS.   Face skin black with very sparse white hairs on cheeks and brows; back agouti gray-brown, rump and backs of thighs red; limbs and underside white; tail red for proximal half, then dark brown; a long white crest on crown.

DISTRIBUTION.   Confined to Colombia, where it occupies a very small area, west of the lower Cauca and lower Magdalena Rios, west to the Atlantic coast and perhaps to the lower Rio Atrato (Hernández-Camacho and Cooper 1976).

Saguinus geoffroyi (Pucheran, 1845)
Geoffroy's Tamarin

1845 *Hapale geoffroyi* Pucheran. Panama: restricted to Canal Zone by Hershkovitz (1949).
1862 *Jacchus spixii* Reichenbach. Supposedly from the Guyanas.
1912 *Oedipomidas salaquiensis* Elliot. Colombia: Rio Salaquí.

DIAGNOSIS. Face skin black with very sparse white hairs on cheeks and brows; back, upper arms, and legs mottled black and yellow; forearms and underparts white or yellowish; tail reddish for proximal one-third, rest black; nape dark red; white crown crest very short.

DISTRIBUTION. From the Coto region of Costa Rica, through Panama into Colombia; the eastern limit is the Rio Atrato in Colombia (Hernández-Camacho and Cooper 1976).

NOTES. Hershkovitz (1977) observed a tendency to become paler from south to north: mahogany mantle and yellowish or ochery underparts with red-brown mottling in a series from Sandó in the extreme south, and red mantle (with narrower black bands) and whiter underparts in those from the Panama Canal. The Panamanian ones were said to have more thickly haired faces.

Saguinus leucopus (Günther, 1876)
White-footed Tamarin

1876 *Hapale leucopus* Günther. Colombia: near Medellín, Antioquia.
1913 *Seniocebus pegasis* Elliot. Colombia: Puerto Berrío, Rio Magdalena.

DIAGNOSIS. Face skin black with very sparse white hairs on cheeks and crown; body brown, with long whitish hair tips; forearms, hands, and feet white; underparts red-brown; tail brown, sometimes with whitish tip; long brown hairs on back of head.

DISTRIBUTION. Restricted to a small area of Colombia, between the lower Rio Cauca and the middle Rio Magdalena, extending south along the western bank of the Rio Magdalena at least as far as Mariquita; Mompos Island; and some of the islands in the Magdalena (Hernández-Camacho and Cooper 1976).

NOTES. According to Hernández-Camacho and Cooper (1976), the southernmost populations tend to be darker, with dark stripes on the forearms and shanks.

Saguinus inustus group

The only potential shared derived condition with any other species-group is the dental enamel structure, shared with the hairy-faced and *S. bicolor* groups (Natori 1988), which is, however, subject to too much homoplasy to constitute convincing evidence of its affinities.

Saguinus inustus (Schwarz, 1951)
Mottle-faced Tamarin

1951 *Leontocebus midas inustus* Schwarz. Brazil: Tabocal, Amazonas.

DIAGNOSIS. Some white hairs on nose and lips; rest of face bare, pink with black mottle patches; color black, becoming chocolate brown on back and cinnamon on flanks; tail black.

DISTRIBUTION. In Brazil, known between the upper Rio Negro and Rio Japurá, and it was mapped by Hernández-Camacho and Cooper (1976) in southeastern Colombia, south of the Rio Guaviare in Vaupés and Guainia Departments.

Genus Callimico Miranda Ribeiro, 1911
Goeldi's Marmoset

1911 *Callimico* Miranda Ribeiro. *Callimico snethlageri* Miranda Ribeiro, 1911.
1912 *Callimidas* Miranda Ribeiro. Lapsus.

Callimico goeldii (Thomas, 1904)
Goeldi's Marmoset

1904 *Midas (Hapale) weddelli* Goeldi and Hagmann. Not of Deville, 1849.
1904 *Hapale goeldii* Thomas (in Goeldi and Hagmann). Amazonas; restricted to Rio Xapuri, Acre, by Cabrera (1957).
1911 *Callimico snethlageri* Miranda Ribeiro. Brazil: Rio Mu, 7–8°S, 72°W.

DIAGNOSIS.   A large, black marmoset with a heavy cape over head and shoulders; third molars present.

DISTRIBUTION.   Headwaters of upper Amazon tributaries, from Rio Japurá in the north to Rio Madre de Dios in the south, along the Andean foothills.

NOTES.   Hershkovitz (1977) described the individual variation in this species, and hypothesized that the karyotypic polymorphism described in the literature might prove to be geographic in nature.

SUBFAMILY CEBINAE BONAPARTE, 1831

Dutrillaux et al. (1986), studying karyology, appear to be the only authors in modern times to doubt the association of *Saimiri* and *Cebus* in a single clade.

Genus *Cebus* Erxleben, 1777
Capuchin Monkeys

1777 *Cebus* Erxleben. *Simia capucina* Linnaeus, 1758.
1792 *Sapajus* Kerr (in part).
1815 *Agipan* Rafinesque. Replacement for *Cebus*. Unavailable.
1827 *Claeetes* Billberg. Nomen nudum.
1862 *Pseudocebus* Reichenbach. *Cebus ochroleucus* Reichenbach, 1862 (here nominated).
1862 *Calyptrocebus* Reichenbach. *Cebus hypoleucus* E. Geoffroy (recte Humboldt), 1812 (here nominated).
1862 *Otocebus* Reichenbach. *Cebus frontatus* Kuhl, 1820 (here nominated).
1862 *Eucebus* Reichenbach. *Cebus fistulator* Reichenbach, 1862 (here nominated).

Hershkovitz (1949) was the first to clearly distinguish the "untufted" and "tufted" groups of capuchins. Such a division appears to be supported by chromosome studies (Dutrillaux et al. 1986).

The following names were regarded by Hershkovitz (1949) as probably referring to capuchins but not identifiable as to species:

1774 *Simia flavia* Schreber. No locality.
1777 *Cebus lugubris* Erxleben. No locality.
1792 *Simia (Sapajus) capucinus albulus* Kerr. No locality.
1809 *Cercopithecus flavus* Goldfuss. The type, in Paris (MNP), is a mount in very poor condition; it seems to be a specimen of *C. albifrons*.
1812 *Cebus barbatus* E. Geoffroy. An untufted capuchin. Not of Humboldt, 1812 (a tufted capuchin). The type, in Paris (MNP), is a badly preserved mount, perhaps *C. albifrons albifrons*.

1812 *Cebus albus* E. Geoffroy. No locality.
1856 *Cebus pucheranii* Dahlbom. South America. This, from the fairly well-preserved type (in Paris [MNP]), is almost certainly the light form of *C. olivaceus*.
1862 *Cebus paraguayanus* Reichenbach. A composite of a tufted and an untufted capuchin.

Cebus capucinus group

Hairs on crown smooth and run backward. May be a superciliary brush in females. Crown-cap often pointed anteriorly and narrow, but may be broader and rounded anteriorly. No dark sideburns; circumfacial pelage and throat light, white to pale brown.

(After Hershkovitz [1949]): Skull with weakly developed temporal ridges, never united to form a sagittal crest; well-developed, nearly horizontal supraorbital ridges; forehead flat or only slightly convex; braincase long and low; ascending ramus low; vertical plate of vomer generally visible in basal view behind posterior margin of palate; legs of vomer heavy, close together, hiding presphenoid in basal view; pterygoid fossa and posterior nares wide; medial pterygoid plate and hamuli tending to converge. Hershkovitz (1949) stated that these characters are inconstant, but taken together should serve to allocate all specimens.

Apparently there are 22 or 23 thoracolumbar vertebrae (Hershkovitz 1949).

Cebus capucinus (Linnaeus, 1758)
White-headed Capuchin

1758 *Simia capucina* Linnaeus. Fixed as northern Colombia by Goldman (1914).
1812 *Cebus hypoleucus* E. Geoffroy. Colombia: Rio Sinú, Bolívar.
1903 *Cebus imitator* Thomas. Panama: Chiriquí, Boquete, 1,350 m.
1905 *Cebus curtus* Bangs. Colombia: Gorgona Island.
1909 *Cebus capucinus nigripectus* Elliot. Colombia: Las Pubas, Cauca Valley; corrected by Hershkovitz (1949) to Pavas, western slope of Cordillera Occidental between Cali and Buenaventura, Department of Valle del Cauca, 1,350 m.
1914 *Cebus capucinus limitaneus* Hollister. Honduras: Segovia River; corrected by Hershkovitz (1949) to Cabo Gracias a Dios, mouth of Segovia River, Honduras-Nicaragua border.
1942 *Cebus capucinus albulus* Pusch. Colombia: Cartagena. Not of Kerr, 1792.

DIAGNOSIS.   Color black, on crown, body, limbs, and tail; chest white, this color extending forward to face and front of crown, and upward to shoulders and upper arms. White area on forehead tending to become thinly haired or even somewhat bald with age.

DISTRIBUTION. From southeastern Honduras through Central America into northern and western Colombia (west of the Rio Magdalena-Cauca), including Gorgona Island, and northern Venezuela (see Hernández-Camacho and Cooper 1976).

NOTES. Hershkovitz (1949) retained a number of subspecies "pending a thorough study of ample material," though admitting that none of the characters on which any was described appears valid. Hernández-Camacho and Cooper (1976) found pelage characters too variable to permit recognition of any subspecies. My limited experience with this species (from material in the USNM, AMNH, and MCZ) likewise inclines me to consider the species monotypic.

Cebus albifrons (Humboldt, 1812)
White-fronted Capuchin

DIAGNOSIS. Color is some shade of brown where *C. capucinus* is black, but hands, feet, and distal part of tail are lighter; areas that are white in *C. capucinus* are more creamy. Crown-cap extends farther forward, rounded in front, well demarcated from light forehead.

DISTRIBUTION. According to Hershkovitz (1949), this species has a broken distribution: (1) the Amazon region, west of the Rio Negro / Rio Branco and Rio Tapajós, in Brazil, Colombia, Venezuela, Ecuador, and Peru; (2) the upper Orinoco and Lake Maracaibo drainage in northern Colombia and Venezuela, and the lower Sinú, Ranchería, and Cauca-Magdalena drainages in Colombia; (3) the Pacific coast of Ecuador; and (4) Trinidad, where it was almost certainly introduced.

NOTES. I examined skins and skulls in the USNM, MCZ, and AMNH and conclude that the subspecies recognized by Hershkovitz (1949) can be reduced in number. The main geographically varying characters are degree of darkening on dorsum, development of reddish tones especially on limbs, color of pale areas (on face, throat, shoulders, and underside), and darkness and forward extension of dark cap on crown.

Hernández-Camacho and Cooper (1976) recorded possible hybrids between this species and *C. capucinus,* supposed to have originated from the middle San Jorge Valley. The specimens differ from *C. albifrons* in being uniformly very dark brown on the body, with more extensive white area on the forehead extending well back onto the crown and tending to baldness in that region, and much lighter shoulders and upper arms. Other putative hybrids, presumed to be from the lower Cauca, were similar but not darker than *C. albifrons.* Hernández-Camacho and Cooper (1976) pointed out at the same time that the two species are supposed to be sympatric in northwestern Ecuador, but this proposed sympatry needs further investigation.

Cebus albifrons albifrons (Humboldt, 1812)

1812 *Simia albifrons* Humboldt. Reputedly from Venezuela: forests near Santa Barbara and the cataracts of the Rio Orinoco, Amazonas.

DIAGNOSIS. The type description noted a brownish black tail tip, the rest of the tail being pale, a feature that appears to differentiate it from any other form (Hershkovitz 1949).

DISTRIBUTION. It lives in the cataracts region of the Rio Orinoco, on the Venezuela-Colombia border.

NOTES. Hershkovitz (1949) knew this only from the original description; no topotypes were recorded.

Cebus albifrons unicolor Spix, 1823

1823 *Cebus unicolor* Spix. Brazil: Rio Tefé, near its mouth, Amazonas.
1823 *Cebus gracilis* Spix. Brazil: Rio Tefé, near its mouth, Amazonas.
1827 *Cebus chrysopus* Lesson. "North America" [sic].
1865 *Cebus flavescens* Gray. Brazil: probably Rio Negro (Thomas 1911).

DIAGNOSIS. Uniformly colored; crown nearly black, ends just behind brows; bright ochery or grayish brown on body; tail, legs, and arms red-yellow or reddish; flanks grayer than midback, which is dark brown; no white on front of shoulder; pale areas tend to be yellowish or creamy fawn.

DISTRIBUTION. The distribution suggested by Hershkovitz (1949) includes all the Amazonian range of the species except that part that is assigned to *C. a. yuracus* and *cuscinus*. Bodini and Pérez-Hernàndez (1987) reported its occurrence in southernmost Venezuela.

Cebus albifrons cuscinus Thomas, 1901

1901 *Cebus flavescens cuscinus* Thomas. Peru: Callanga, 1,500 m, Cuzco.
1949 *Cebus albifrons yuracus* Hershkovitz. Colombia: Montalvo, north bank of Rio Bobonaza, 45 km from its confluence with the Rio Pastaza.

DIAGNOSIS. Medium to rather dark brown (sometimes with red tones) and limbs less contrasted with back, but still with some reddish tones, more on legs than on arms. Tail with a dark stripe along dorsal side. Crown dark brown or black, contrasting with body; ends bluntly, just behind brows. Temples and forehead grayish. Often has long fur.

DISTRIBUTION. From the eastern slopes of Ecuador and Peru, at high altitudes.

Cebus albifrons trinitatis Pusch, 1942

1942 *Cebus capucinus trinitatis* Pusch. Trinidad.

NOTES. Hershkovitz (1949) found that this form differs from other subspecies in its pale color, but could not decide whether it represents a relict distribution (the opposite mainland is occupied by *C. olivaceus* today) or has been introduced to Trinidad. Specimens I have seen are distinctively pale, yellowish, with a poorly distinguished crown patch.

Cebus albifrons aequatorialis Allen, 1914

1914 *Cebus aequatorialis* Allen. Ecuador: Río del Oro, Manaví Province.

DIAGNOSIS. A brightly colored, medium brown subspecies (individuals from montane regions are darker). It may have red tones like *C. a. versicolor,* but is darker, and temples are darkened. Underside is creamy yellow. As in the northern Colombian and Venezuelan subspecies, the cap ends well behind the brows.

DISTRIBUTION. Western Ecuador. It is unknown whether this form is truly isolated, or perhaps extends north along the coast to meet *C. capucinus*.

Cebus albifrons versicolor Pucheran, 1845

1845 *Cebus versicolor* Pucheran. Colombia: Santa Fé de Bogotá.
1865 *Cebus leucocephalus* Gray. Colombia. Restricted by Hershkovitz (1949) to El Tambor, 25 km NW of Bucaramanga, Santander.
1909 *Cebus malitiosus* Elliot. Colombia: Bonda, northwest corner of base of Sierra Nevada de Santa Marta, Magdalena.
1949 *Cebus albifrons adustus* Hershkovitz. Venezuela: Rio Cogollo, 5 km NW of Machiques, Zulia.
1949 *Cebus albifrons cesarae* Hershkovitz. Colombia: Rio Guaimaral, Department of Magdalena.
1949 *Cebus albifrons pleei* Hershkovitz. Colombia: swamps around Norosí, base of northern extremity of Cordillera Central, Mompós, Bolívar.

DIAGNOSIS. Tending to be rather pale, with red tones in middorsal region and more especially on forearm and foreleg, generally contrasting with rest of body; extension of pale areas variable; dark (but not black) crown, contrasting strongly with creamy yellow temples and forehead.

DISTRIBUTION. The Middle Rio Magdalena region, from the Rio Cesar west into the Sierra Nevada de Santa Marta to about 500 m altitude, to about the Rio Cauca; east into Zulia State, Venezuela.

NOTES. *Cebus a. malitiosus* was regarded by Hershkovitz (1949) and Hernández-Camacho and Cooper (1976) as a distinct subspecies inhabiting a very small area on the northwestern base of the Sierra Nevada de Santa Marta, on the northern coast of Colombia. *Cebus a. adustus* (described as paler than *leucocephalus;* limbs reddish, well contrasted with body; pale area extending to upper part of shoulder) was recorded only from the type locality; Hershkovitz (1949) stated that it "differs in a comparatively small degree" from *C. a. malitiosus,* the other coastal lowland form, but it was recognized as distinct by Bodini and Pérez-Hernàndez (1987), who recorded it in Venezuela. Hernández-Camacho and Cooper (1976) described individuals typical of three described forms *(versicolor, leucocephalus,* and *pleei)* as being found within the same troops along the eastern bank of the middle Rio Magdalena. They regarded *C. a. cesarae* as distinct, but there seems to me to be no more than an average difference between specimens ascribed to *pleei* and typical *cesarae.*

Cebus olivaceus Schomburgk, 1848
Weeper Capuchin

1848 *Cebus olivaceus* Schomburgk. Venezuela: southern base of Mount Roraima, 4°57′N, 61°01′W, 930 m, Bolívar State.

1848 *Cebus nigrivittatus* Wagner. Brazil: San Joaquim, upper Rio Branco. Rejected by Pusch (1941) as a homonym of *Cebus nigrivittatus* (Wagner, 1846), a squirrel monkey, and so permanently rejected (*Code,* Article 59[c]: see Notes).

1851 *Cebus castaneus* I. Geoffroy. French Guiana: Cayenne.

1865 *Cebus annellatus* Gray. "Brazil."

1907 *Cebus apiculatus* Elliot. Venezuela: La Union, lower Rio Orinoco.

1914 *Cebus apella brunneus* Allen. Venezuela: Aroa, Yaracuy State.

1941 *Cebus capucinus leporinus* Pusch. Brazil: Cachoeira do Sucuriju, upper Rio Catrimani, Amazonas. To replace *nigrivittatus* Wagner, which (see Notes) Pusch considered preoccupied by another *nigrivittatus* of Wagner, a squirrel monkey.

DIAGNOSIS.   Color brown, with black agouti banding on flanks, tail, and limbs, and hands, feet, and tail tip tend to be darkened; lighter areas are light gray-brown, restricted to face and forehead, setting off a dark brown (not agouti) wedge-shaped crown patch. Underside tends to be dark. Pelage relative coarse. Skull large, braincase elongated.

DISTRIBUTION.   Northeastern Venezuela east to Guyana, south to the Rio Amazon, east of the Rio Negro. In southernmost Venezuela, it is sympatric with *C. albifrons,* according to Bodini and Pérez-Hernàndez (1987).

NOTES.   The nomenclature of this species is tortuous. Pusch (1941) united *Cebus* and *Saimiri* into one genus, and so in his system *Cebus nigrivittatus* Wagner, 1848, became a secondary homonym of

Chrysothrix nigrivittatus Wagner, 1848 (for a form of *Saimiri,* see under *Saimiri sciureus sciureus*). Husson (1957) considered that this action entailed the permanent rejection of *Cebus nigrivittatus,* but Hershkovitz (1958) strenuously disputed this, pointing out that there are good reasons why a transitory secondary homonymy simply cannot be used as a criterion for permanent rejection of a name; moreover the name *Cebus olivaceus* Schomburgk, 1848, used for the species by Husson to replace *nigrivittatus,* was equally a secondary homonym because Fischer in 1829 included a woolly monkey called *olivaceus* in his own concept of the genus *Cebus.* Husson (1978) responded that under Article 59(c) of the *International Code of Zoological Nomenclature,* a junior homonym created before 1960 is not to be revived, so *Cebus nigrivittatus* Wagner, 1848, must be permanently rejected; and as for Fischer's placement of *olivaceus* Spix, 1823 (a woolly monkey), in *Cebus,* this would entail its permanent rejection *only if* some author before 1960 had rejected *Cebus olivaceus* Schomburgk, 1848, on that account, but no such action appears to have been taken. So the name of this species is *Cebus olivaceus* Schomburgk, 1848.

Hershkovitz (1949) and Bodini and Pérez-Hernàndez (1987) recognized some subspecies provisionally; I note a differentiation between paler, more buffy colors in Venezuela and the Demerara region of Guyana, and dark brown in the hinterland of Guyana, but because I have seen too few specimens from Brazil I could not allocate subspecific names to them. The type of *apiculatus* is of the pale type, that of *castaneus* is the dark type, and unfortunately the type of *leporinus* is neither, but a medium fawn skin with light arms and temples.

Queiroz (1992) showed that, according to cranial measurements, this is the largest capuchin (or at least, according to my data, the largest untufted species), though with a relatively and absolutely narrow biorbital width.

Cebus kaapori Queiroz, 1992
Kaapori Capuchin

1992 *Cebus kaapori* Queiroz. Brazil: Quadrant 7, 10 km southwest of the Chaga-Tudo Prospection (0°30′S, 47°30′W) (Carutapera, Maranhão), near right bank of Rio Gurupi, Maranhão.

DIAGNOSIS. Long-bodied compared with other untufted species; face, shoulders, mantle, and tail tip silvery gray; arms and legs agouti; hands and feet black and dark brown. Crown with a triangular black cap extended to a dark stripe down nose.

DISTRIBUTION. Possibly restricted to region between Rios Gurupi and Pindaré, south of the Amazon, although it may extend a few kilometers west of the Gurupi, according to Queiroz (1992).

NOTES. This recently described species, the only one of the untufted group to be found south of the lower Amazon, is undoubtedly related to *C. olivaceus*, from which it is separated by more than 400 km. Queiroz (1992) warned that the conclusion that it is longer-bodied and less robust than other untufted species rests on the measurements of a single specimen, the paratype.

A multivariate morphometric study (Materson 1995) placed the single available skull just outside the range of *C. olivaceus*, though in some respects, especially bizygomatic width, closer to *C. albifrons*.

Cebus apella group

Hairs on crown are erect and commonly elongated at sides of front of crown to form ridges or "horns" or as a single tuft in middle. Pelage relatively coarse. Cap very broad, never pointed anteriorly, and covering almost whole crown. Dark sideburns, often meeting below chin.

(After Hershkovitz [1949]): Temporal ridges well developed, often uniting to form a sagittal crest; supraorbital ridges generally weakly developed; forehead strongly convex; braincase relatively short, high; ascending ramus high (its height in males generally greater than C–M³ length); vomer not or hardly visible in basal view behind palate; legs of vomer more slender than in untufted group, wider apart, making presphenoid visible in basal view; pterygoid fossa and choanae relatively narrow; medial pterygoid plates parallel-sided, and hamuli divergent. These features, according to Hershkovitz, are not entirely constant but probably serve to differentiate all specimens when taken together.

Limited evidence (Hershkovitz 1949) suggests only 21 thoracolumbar vertebrae.

This group certainly consists of several species, not just the one *(Cebus apella)* traditionally recognized. Thus chromosome no. 11 of the distinctive *C. xanthosternos* ($n = 6$) has intercalar heterochromatin, apparently uniquely in the genus. Mittermeier et al. (1988) cited a thesis written in 1983 by the late M. Torres de Assumpçao in which she placed them all in one species but divided it into six "core areas," where there are stable characters, with great variability in between; her divisions seem to correspond in many respects to those arrived at independently by me in my study of specimens in the U.S. collections (USNM and AMNH). The main divisions are so impressive that, despite the undeniable evidence of some natural interbreeding in intervening areas, I propose to recognize them as full species (four in number; the characters of *nigritus* and *robustus* seem to overlap, and it is interesting that one of Torres de Assumpçao's "core areas" was said by Mittermeier et al. [1988:48] to "lack a typical phenotype").

The following names seem impossible to allocate subspecifically:

1812 *Cebus niger* E. Geoffroy. "Brazil."
1812 *Cebus cirrifer* E. Geoffroy in Humboldt. "Brazil." The type, in Paris (MNP), is almost certainly a senior synonym of *C. nigritus cucullatus*.
1812 *Simia variegata* Humboldt. "Brazil." The type is a very poorly preserved mount in Paris (MNP); from its dark color, it probably represents some form of *C. nigritus*.
1820 *Cebus monachus* F. Cuvier. Locality unknown.
1820 *Cebus frontatus* Kuhl. Locality unknown. The type, like that of *variegata*, is probably a (poorly preserved) specimen of one of the subspecies of *C. nigritus*.
1820 *Cebus lunulatus* Kuhl. Locality unknown.
1829 *Cebus cristatus* G. Cuvier. Locality unknown.
1857 *Cebus crassiceps* Pucheran. "South American." The type, in Paris (MNP), is quite well preserved and could represent a senior synonym of *C. apella peruanus*.
1865 *Cebus capillatus* Gray. "Brazil."
1865 *Cebus subcristatus* Gray. "South America."
1866 *Cebus leucogenys* Gray. Locality unknown. The type, in London (BM[NH]), is a very dark, all-over blackish skin with small "horns," black below, and with slight whitish spots on upper cheeks behind eyes; it probably belongs to one of the dark Amazonian forms of *C. apella*.

Cebus apella (Linnaeus, 1758)
Tufted Capuchin

DIAGNOSIS.   Crown black, contrasting with body; gray-fawn to dark brown; face light gray-brown; sideburns thick, black; lower limbs and tail black, humerus fawn; underside yellowish or red; variably well-developed dorsal stripe; tufts on crown tend to be relatively small at all ages. Size large, but externally not sexual dimorphic; skull length 100 mm in a single adult male, 87–92 mm in females ($n = 4$); head plus body length 380–430 mm in both sexes. Tail about equal in length to head plus body.

DISTRIBUTION.   Northern and western South America: from Guyana, Venezuela (Bolívar State), and Colombia south across the Amazon to about 5°S in the east and in the west to the headwaters of the upper tributaries, nearly 10°S.

Cebus apella apella (Linnaeus, 1758)

1758 *Simia apella* Linnaeus. "America"; restricted to French Guiana by Humboldt in 1812 (see Cabrera [1957] and Hershkovitz [1958]).
1766 *Simia trepida* Linnaeus. Suriname.
1792 *Simia (Sapajus) trepidus fulvus* Kerr. For status of this name, see Hershkovitz (1949).
1812 *Cebus barbatus* Humboldt. Hershkovitz (1949) argued that this is a synonym of *Cebus apella*.
1820 *Cebus griseus* Desmarest. Based on the same source as *fulvus* Kerr, 1792, and *barbatus* Humboldt, 1812.

DIAGNOSIS.   Dull brown in color; face and temples light gray-brown; underside often slightly more reddish toned; crown and sideburns black, strongly distinct from the body color; crown tufts small; facial surround pale, often whitish; lower limbs and tail black; dorsal stripe a vague, broad, fuzzy zone.

DISTRIBUTION.   Specimens of this description come from Guyana, French Guiana, and Suriname.

Cebus apella fatuellus (Linnaeus, 1766)

1766 *Simia fatuellus* Linnaeus. Colombia: upper Magdalena Valley, according to Tate.

DIAGNOSIS.   Bright brown, fiery red below; dorsal stripe prominent.

DISTRIBUTION.   I have seen specimens from Villavicencio and other localities on Rio Guaviare, Colombia, and on the Rio Negro.

Cebus apella ?margaritae Hollister, 1914

1914 *Cebus margaritae* Hollister. Venezuela: Margarita Island.

NOTES.   The status of the Margarita Island capuchin is unclear. It is separated by 800 km from the nearest mainland *C. apella* in southern Venezuela and was presumably introduced in pre-Columbian times. Although the geographically closest form is *C. a. apella,* its affinities seem to be with *C. a. fatuellus.* It is in danger of extinction and is in desperate need of close study (Romari A. Martinez, personal communication).

Cebus apella macrocephalus Spix, 1823

1823 *Cebus macrocephalus* Spix. Brazil: Lago Cactuá, Rio Solimões, Amazonas.

DIAGNOSIS.   Very well-developed crown tufts ("horns") in the adult; often a gray-white eye-ear stripe; color gray-brown or gray ochery to dark brown with dark dorsal stripe; in darkest specimens crown contrasts very little with body; sides of neck lighter; upper arms pale yellowish; legs black; yellow-fawn or red-gold below; occasionally blackish. The shape of the crown-cap in front, whether angular or broad, and the presence of white margins, is variable. The tufts develop in adults, becoming very high and pointed, then become reduced and flattened backward with age. The differences between *macrocephalus* and *fatuellus,* as Lönnberg (1939) pointed out, are grading and very much a matter of emphasis.

DISTRIBUTION.   This subspecies ranges widely through the middle Amazon: Lönnberg (1939) recorded it from Codajáz, Itacoatiara, and Lago Cuitena north of the Amazon, the Rio Purus, and Prainha and other localities east of the Rio Tapajós, and I have seen specimens from a similar range.

Cebus apella peruanus Thomas, 1901

1901 *Cebus fatuellus peruanus* Thomas. Peru: Huaynapata, Marcapata, 760 m, Cuzco.
1941 *Cebus apella magnus* Pusch. Peru: Rio Putumayo, 1°N, 76°W.
1941 *Cebus apella maranonis* Pusch. Peru: Hamburgo, 5°S, 75°W.

DIAGNOSIS.   Uniform dark chestnut brown, becoming more chestnut rufous toward the flanks, and has forearms generally colored like the back, but sometimes (as in the type) black; legs and tail are black; deep yellow-brown below; trace of dorsal stripe. Cap distinctly black; temples and sides of crown often white. No crown tufts. There is, because of the dark body color, less contrast than usual between the black crown and brown body.

DISTRIBUTION.   Specimens I have seen come from a range between Iquitos and the Rio Napo, Ecuador.

Cebus apella tocantinus Lönnberg, 1939

1939 *Cebus fatuellus tocantinus* Lönnberg. Brazil: Cametá, Rio Tocantins.

DIAGNOSIS.   Dark gray-fawn to dark chestnut brown, the younger individuals being darker; more rufous toward the flanks; trace of dorsal stripe; head black with little trace of white on the head; "horns" well developed. Arms as back; below elbows, black. Thighs black inside and on front; tail, shanks, and feet black; bright reddish below.

DISTRIBUTION.   I have seen specimens from the south bank of the lower Amazon, as far west as the lower Rio Madeira.

Cebus libidinosus Spix, 1823
Black-striped Capuchin

DIAGNOSIS.   Head yellow-white, with black sideburns; body light so that black of crown and sideburns contrasts with body; a prominent dark dorsal stripe; limbs mainly dark to blackish, upper arms not lighter than body; underside yellowish or reddish, often overlain with black. Size small, no sexual dimorphism in external measurements: skull length, male 93–98 mm ($n = 6$), female 83–89 mm ($n = 8$); length

of head plus body in both sexes 340–440 mm. Tail much longer than head plus body.

DISTRIBUTION.   Southern Brazil (highland region) into Bolivia and Paraguay.

Cebus libidinosus libidinosus Spix, 1823

1823 *Cebus libidinosus* Spix. Brazil: Rio Carinhanha, north of Minas Gerais.
1850 *Cebus elegans* I. Geoffroy. Brazil: near Rio Los Piloes, Goias.

DIAGNOSIS.   Color lighter than in most other subspecies: dull dark fawny or slightly grayish brown with at least a trace of dorsal stripe; no white on face; crown-cap contrasting, with white margin in front; very little development of crown tufts; limbs dark but not black; dull red or gold below; a dark stripe down tail.

DISTRIBUTION.   A wide area on the Brazilian uplands behind the coast: Carinhanha, Carolina, Caldas Novas, Ponso Alto, Macoca, Pedro Alfonso, Presidente Venceslau.

Cebus libidinosus pallidus Gray, 1866

1866 *Cebus pallidus* Gray. Bolivia: restricted by Cabrera (1957) to Rio Beni.
1941 *Cebus apella sagitta* Pusch. Bolivia: Chimate, 15°S, 68°W, 700 m.

DIAGNOSIS.   A very contrasting black cap, with hair partially erect, often forming a low tuft; a yellowish white forehead band of varying extent; temples also yellowish white; upper parts vary from pale (yellowish) brown, with a dark dorsal stripe that is often not very conspicuous, to reddish brown; throat and upper breast buffy reddish in the pale morph, golden brown in the darker morph; upper arms pale brown to yellowish white; forearms blackish; thighs brown with a black anterior stripe; shanks black; fingers have pale gray hairs mixed in; tail dark brown to black, but proximally brown.

DISTRIBUTION.   According to Avila (in press), it occurs in Bolivia with an isolated population in the Yungas cloud forests of northwestern Argentina.

NOTES. I have not seen the red-brown (darker) morph, and I thank Ignacio Avila (personal communication) for allowing me to cite his description.

Cebus libidinosus paraguayanus Fischer, 1829

1829 *Cebus apella paraguayanus* Fischer. Paraguay: north bank of the Pilcomayo (restricted by Cabrera [1957]).
1830 *Cebus azarae* Rengger. Paraguay.
1910 *Cebus versuta* Elliot. Brazil: Araguary, Río Jordão, 700–900 m, Minas Gerais.
1941 *Cebus apella morrulus* Pusch. Paraguay: Santa Barbara.
1941 *Cebus apella chacoensis* Pusch. Paraguay: the Chaco.

NOTES. Lönnberg (1939) considered *paraguayanus* (synonym *azarae*) a subspecies of *C. libidinosus* and stated that a Paraguay specimen differs from the Bolivian ones in its darker color, lacking the reddish tones. But Paraguay specimens in the BM(NH) are very pale, often with no dark cap at all; they are also very obviously short-limbed.

Avila (in press) is working on this subspecies and on *C. l. pallidus,* and has kindly permitted me to summarize some of his findings (Ignacio Avila, personal communication). *Cebus l. paraguayanus* is found in eastern Paraguay, south of the Mato Grosso region of Brazil; field observations confirm that there is color variability. Craniometrically the two subspecies separate clearly.

Cebus libidinosus juruanus Lönnberg, 1939

1939 *Cebus libidinosus juruanus* Lönnberg. Brazil: Frente a João Pessôa, Rio Juruá.

DIAGNOSIS. Darker (rufescent brown), becoming rufous at the border toward the belly; there is a very pronounced blackish dorsal stripe; head as usual in *C. libidinosus,* except that whitish cheek band is very poorly developed. Throat and upper breast blackish or pale reddish buff. Upper arms brownish; forearms black; thighs dark rufous brown, black in front and inside; shanks black; tail black, except at base.

DISTRIBUTION. Range, from the BM(NH) material, extends to Mato Grosso.

Cebus nigritus (Goldfuss, 1809)
Black Capuchin

DIAGNOSIS. Very dark brown or gray, even blackish; no dorsal stripe, or very vague; face contrastingly white; sideburns poorly marked; limbs darker than body, usually blackish; underside deep reddish with black overlay; crown tufts always present, pointed, in adult but wearing away with age. Size large: skull length, male 96–102 mm ($n = 16$; in fully mature males, a large sagittal crest), female 88–94 mm ($n = 13$); length of head plus body, male 380–440 mm, female 340–400 mm. Tail somewhat longer than head plus body.

DISTRIBUTION. The Atlantic forests of the Brazilian coast, 16–30°S.

Cebus nigritus nigritus (Goldfuss, 1809)

1809 *Cercopithecus nigritus* Goldfuss. Brazil: Rio de Janeiro, Sierra dos Orgãos.
1823 *Cebus xanthocephalus* Spix. Brazil: Rio de Janeiro.

DIAGNOSIS. Crown contrasting; very dark brown or gray, often with white sprinkling; face light, with white hairs around it; sideburns poorly marked; limbs colored as body or somewhat contrastingly darker; underside reddish or yellow-fawn, generally with a strong black overlay; dorsal stripe inconspicuous; crown tufts well developed in adult; tail black.

DISTRIBUTION. A coastal strip in Brazil from about 20°S to the Rio de Janeiro District.

Cebus nigritus robustus Kuhl, 1820

1820 *Cebus robustus* Kuhl. Brazil: Rio Doce (restricted by Cabrera [1957]).

DIAGNOSIS. Very dark wood brown or blackish; trace of dorsal stripe; hands and feet black, rest of limbs as body; face usually dark with light surround; crown-cap outlined with white on forehead and temples; deep maroon-red or yellow-brown below; tall conical crown tufts.

DISTRIBUTION. A small coastal region north of the range of *C. n. nigritus,* about 16–20°S.

Cebus nigritus cucullatus Spix, 1823

1823 *Cebus cucullatus* Spix. Brazil: São Paulo Province.
1851 *Cebus vellerosus* I. Geoffroy. Brazil: São Paulo, further restricted by Cabrera (1957) to Porto Cabral.
1910 *Cebus caliginosus* Elliot. Brazil: Santa Catarina, Colonia Hansa.

DIAGNOSIS. Crown not contrasting, cap only slightly darker than body; body very dark wood brown or (more usually) blackish above; face usually light; limbs contrasting with body; maroon-red to yellowish below, generally with strong black overlay; dorsal stripe inconspicuous; tufts variable, usually small; tail black.

DISTRIBUTION. The most southerly capuchin, from about 24 to 30°S.

NOTES. The application of the name *cucullatus* is uncertain and the name *vellerosus* may be applicable under the *Code*, Article 23(b) (Ignacio Avila, personal communication).

Cebus xanthosternos Wied-Neuwied, 1826

Golden-bellied Capuchin

1826 *Cebus xanthosternos* Wied-Neuwied. Brazil: Rio Belmonte.

DIAGNOSIS. Crown not contrasting with body; body brindled reddish above; face and temples fawn; cap all black; limbs black; underside sharply marked golden red; tail black. Size relatively small: skull length within range of *C. libidinosus* (male 94 mm, female 89 mm [*n* = 1 in both cases]), but flesh measurements larger (head plus body length in male 410–420 mm, in female 360–390 mm). Tail not much longer than head plus body.

DISTRIBUTION. The original distribution was between the Rio São Francisco and the Rio Jequitinhonha, possibly even south of the latter in the region occupied by *C. n. robustus* (Coimbra-Filho, Rylands, et al. 1995); today the range is severely reduced, and the species ranks as one of the most endangered South American primates. I have seen specimens from Ilheus, Vitoria, Belmonte, Passúi, and Carinhanha; as Coimbra-Filho, Rylands, et al. (1995)

pointed out, Carinhanha is to the west of Rio São Francisco, suggesting a further range extension west into the distributional area of *C. l. libidinosus*, of which I have also seen specimens from the same locality.

SUBFAMILY CHRYSOTRICHINAE CABRERA, 1900

Saimirinae Miller, 1924, is a synonym.

Genus *Saimiri* Voigt, 1831

Squirrel Monkeys

1831 *Saimiri* Voigt. *Simia sciurea* Linnaeus, 1758.
1835 *Chrysothrix* Kaup. *Simia sciurea* Linnaeus, 1758.
1840 *Pithesciurus* Lesson. *Sapajou saimiri* Lacépède, 1803.
1843 *Saimiris* I. Geoffroy. *Simia sciurea* Linnaeus, 1758.

Traditionally, squirrel monkeys have been divided into a South American species, *S. sciureus,* and a Central American one, *S. oerstedti.* Two studies, published almost simultaneously and based on different material, challenged this but in different ways, whereas a more recent one (Costello et al. 1993) argued for a reversion to the South vs. Central division.

Hershkovitz (1984) divided the genus into two species groups: the *S. boliviensis* group ("Roman arch" type, in which the dark color of the crown extends down onto the forehead, leaving only a restricted circle of white above each eye) and the *S. sciureus* group ("Gothic arch" type, in which the white above each eye extends well up in a high arch). The latter has three species: *S. sciureus, S. oerstedti,* and *S. ustus.*

Thorington (1985) independently produced a taxonomic revision; unlike Hershkovitz (1984), he made his underlying philosophy explicit (the BSC, and a multistate concordance criterion for subspecies) and used craniometry as an important part of his study. He recognized only two species, *S. sciureus* and *S. madeirae;* the latter was what Hershkovitz (1984) had called *S. ustus.*

Costello et al. (1993) summarized genetic and behavioral data, and produced new genetic data of their own. They deduced evidence for hybridization between *S. sciureus, S. madeirae,* and *S. boliviensis* and so, using a combination of BSC and Paterson's Recognition concept, proposed to unite them, leaving only *S. oerstedti* as a species separate from *S. sciureus*.

The gender of the name *Saimiri* is masculine, as argued by Cabrera (1957) and confirmed by Thorington (1985). The specific name of the type species therefore takes the form *sciureus*.

Now is the time to reveal that the type specimen of *Simia sciurea* of Linnaeus, a complete specimen (male) preserved in spirits in the Stockholm (NRS) collection, is of "Roman" type and has a black crown: an example of *Saimiri boliviensis*. Because of the long-standing usage of the name *S. sciureus* for the Guyanan form, and because the next available name for the Guyanan species would be *ustus* (used by Hershkovitz [1984] for quite a different species!), it is best to avoid wholesale confusion by some nomenclatural fiat.

Saimiri sciureus group

A high white arch above each eye ("Gothic" type); blackish supraorbital tufts usually conspicuous; tail pencil bushy; dorsum "tending toward pheomelanin dominance." According to Hershkovitz (1984), the species exhibit the following behavioral differences from the *S. boliviensis* group: sexes more integrated outside mating season; male dominance; penile display to mirror image; "stiff leg" scratch not used; males do not "kick wash"; less aggressive.

Saimiri oerstedti (Reinhardt, 1872)
Central American Squirrel Monkey

DIAGNOSIS. Pelage predominantly orange to reddish orange; crown blackish in female, blackish or agouti in male; ears tufted. Palate without a posterior median spine. Chromosomes: $2n = 44$ as in all *Saimiri,* with 16 biarmed and 5 acrocentric autosome pairs.

NOTES. Thorington (1985), who did not recognize *citrinellus* as a distinct taxon, made *S. oerstedti* a subspecies of *S. sciureus,* although agreeing that it is very distinctive: the orange band on the dorsal hairs is

some 20 mm long, whereas in *S. sciureus* from the Colombian llanos it is only 5–7 mm, and the length of the black tip is correspondingly much shorter in *oerstedti*.

A single specimen of this species had some protein alleles not found in any of the other species tested, and its teeth differed more from the South American taxa than those differed from each other (Costello et al. 1993).

Saimiri oerstedti oerstedti (Reinhardt, 1872)

1872 *Chrysothrix oerstedti* Reinhardt. Panama: vicinity of David, Chiriquí.

DIAGNOSIS. Crown blackish in male as well as in female; preauricular patch blackish; outer side of legs orange like arms.

DISTRIBUTION. Found on the Pacific coast of Puntarenas Province, Costa Rica, and Chiriquí and Veraguas Provinces, Panama.

Saimiri oerstedti citrinellus Thomas, 1904

1904 *Saimiri oerstedti citrinellus* Thomas. Costa Rica: Pozo Azul, Pirris.

DIAGNOSIS. Crown agouti in male, so sexually dichromatic; preauricular patch agouti or buffy like cheek; outer side of legs buffy or grayish agouti.

DISTRIBUTION. Known only at the type locality; geographically isolated from *S. o. oerstedti.*

Saimiri sciureus (Linnaeus, 1758)
Common Squirrel Monkey

DIAGNOSIS. Crown mainly agouti, never predominately black; body grayish or greenish to reddish agouti; ears tufted. Chromosomes: 14–15 biarmed, 6–7 acrocentric autosome pairs. Chromosome 5 has a terminal C-band.

DISTRIBUTION. South American rain forest mainly north of the Amazon-Juruá system, extending south of the Amazon mouth to about the Rio Pindari and west to the Rio Xingu or the Rio Iriri.

NOTES. Thorington (1985) noted the same geographic variations as did Hershkovitz (1984), except for those in chromosomes, but did not consider them of subspecific distinction, except for *cassiquiarensis*.

Saimiri sciureus sciureus (Linnaeus, 1758)

1758 *Simia sciurea* Linnaeus. "India"; Guyana: Kartabo (restricted by Tate [1939]).
1758 *Simia apedia* Linnaeus. On the identification of this name, see Fooden (1966).
1758 *Simia morta* Linnaeus. "America."
1803 *Sapajou saimiri* Lacépède. French Guiana.
1804 *Lemur leucopsis* Hermann. "Egypt" (!).
1848 *Chrysothrix nigrivittata* Wagner. Brazil: Tefé, Rio Solimões, Amazonas.
1916 *Saimiri sciurea collinsi* Osgood. Brazil: Fazenda Teso, near Soure, Marajó Island.

DIAGNOSIS. Outer side of thighs buffy agouti like body; hands, feet, and forearms orange or yellowish; crown gray-agouti in male, intermixed or bordered with black in female. No defined nuchal band. Chromosomes: 14 biarmed, 7 acrocentric autosome pairs.

DISTRIBUTION. Hershkovitz (1984) gave the range of this subspecies as Guyanas and Brazil east of the Rio Demeni-Negro and Rio Xingu, on both sides of the lower Amazon. Thorington (1985) described those from south of the Amazon as being slightly yellower on the back and with more yellowish suffusion on the crown.

NOTES. For Hershkovitz (1984) Wagner's *nigrivittata* is a synonym of *S. s. cassiquiarensis*. He reported that one of the three syntypes (chosen by Hershkovitz as lectotype) shows blackish preauricular patches, and the "detailed description . . . leaves no doubt that *nigrivittatus* is Gothic and probably the same as *cassiquiarensis*." He considered the type locality, Ega (now Tefé), on the south bank of the Amazon in *S. boliviensis* country, to be in error: probably the specimens originated on the north bank, opposite Tefé. Thorington (1985), on the other hand, accepted the type locality as correct and placed the name in the synonymy of *boliviensis*.

Ayres (1985) confirmed that the type locality is indeed correct: *S. sciureus* really does occur south of the Amazon, on the left bank of Lago Tefé. Females, but not males, from that region have the *nigrivittatus* preauricular band; *S. s. cassiquiarensis*, from the Rio Japurá, never possesses this band. Because Hershkovitz (1984) identified *nigrivittatus* with confidence only to species level and placed it with *cassiquiarensis* apparently only on locality grounds, Ayres's new information seems to confirm it as a synonym of *S. s. sciureus*, as indeed he suggested.

Saimiri sciureus albigena Pusch, 1942

1942 *Cebus (Saimiri) sciureus albigena* Pusch. Colombia: Medina, eastern slope of Cordillera Oriental, 50 km northeast of Villavicencio, Cundinamarca, 576 m.

DIAGNOSIS. Forearms and hands mostly grayish agouti, with at most a pale orange tinge; back slightly more orange-toned; crown and nape mainly grayish or buffy agouti. Karyotype unknown.

DISTRIBUTION. A small range on the eastern slopes of the Cordillera Oriental, an unknown distance eastward into the llanos gallery forests.

Saimiri sciureus cassiquiarensis (Lesson, 1840)

1840 *Simia sciurea cassiquiarensis* Lesson. Venezuela: Rio Casiquiare.
1843 *Saimiris lunulatus* I. Geoffroy. Same locality.
1940 *Saimiri sciurea codajazensis* Lönnberg. Brazil: Codajáz, north bank of Rio Solimões, Amazonas.

DIAGNOSIS. According to Hershkovitz (1984), this is not unlike *S. s. sciureus* and *S. s. macrodon*, but has a (fairly weakly) contrasting light nuchal collar. For Thorington (1985:18), the most striking features were the "tawny orange" body tone, extending to the forearm, and especially "the suggestion of tawniness in the crown." The skull is different from that of other subspecies, apparently mainly correlated with absolute size (Thorington 1985). Karyotype is unknown.

DISTRIBUTION. This distinctive subspecies inserts between the ranges of the externally indistinguishable *S. s. sciureus* and *S. s. macrodon*, a curiosity emphasized by Thorington (1985).

Saimiri sciureus macrodon Elliot, 1907

1907 *Saimiri macrodon* Elliot. Ecuador: Copataza, upper Rio Pastaza.
1916 *Saimiri caquetensis* Allen. Colombia: Florencia, Caquetá.
?1927 *Saimiri sciurea petrina* Thomas. Peru: Yurac Yacu, 750 m, San Martín.
?1940 *Saimiri boliviensis juruana* Lönnberg. Brazil: João Pessôa and Igarapé do Gordão, Amazonas. NHMS 2196 was selected as lectotype by Thorington (1985).

DIAGNOSIS. Externally identical to *S. s. sciureus*. Chromosomes: 15 biarmed, 6 acrocentric autosome pairs.

DISTRIBUTION. Range is west of Rio Apoporis-Japurá and Rio Juruá, on both sides of the upper Amazon (Solimões); it is separated from the phenotypically identical *S. s. sciureus* by the externally rather more distinctive *S. s. cassiquiarensis*.

Saimiri ustus (I. Geoffroy, 1844)

Bare-eared Squirrel Monkey

1844 *Saimiris ustus* I. Geoffroy. Brazil: Humaitá, Rio Madeira, Amazonas (as restricted by Cabrera [1957] and further restricted by Hershkovitz [1984]); but more probably the Rio Tapajós according to Thorington (1985), although the type was a captive specimen of unknown origin.
1908 *Saimiri madeirae* Thomas. Brazil: Humaitá, Rio Madeira.

DIAGNOSIS. Ears untufted; outer side of thighs grayish agouti, contrasting with body; forearms and hands and feet orange or yellowish; crown agouti, but with black intermixture or border in females. Size is large compared with *S. sciureus* (Thorington 1985). Chromosomes: 16 biarmed autosome pairs, 5 acrocentric (as in *S. oerstedti*).

DISTRIBUTION. Brazil south of the Amazon; considered by Hershkovitz (1984) and Thorington (1985) to be between the Rio Purus and probably the Rio Xingu, but Ayres (1985) reported specimens from well to the west of the Rio Purus, on the right bank of Lago Tefé, at approximately 61°30′W.

NOTES. There may be geographic variation in this species. Thorington (1985) described southerly specimens as having orange forearms, whereas over most of the range this color does not extend above the wrists; and skulls from Borba are much larger than those from Villa Bella.

Hershkovitz (1984) called this species *Saimiri ustus*, with *madeirae* as a junior synonym, but Costello et al. (1993), like Thorington (1985), argued that I. Geoffroy's name *ustus* has nothing to do with *S. madeirae*, but very likely refers to a specimen of *S. s. sciureus*. For both Thorington (1985) and Costello et al. (1993), the type of *ustus*, an animal in captivity of an ultimately unknown origin, was *S. s. sciureus* because of its olive crown and reddish back. I disagree; the type (in Paris [MNP]) has quite naked, untufted ears, grayish agouti outer thighs, and the crown is in fact gray-agouti. There is a red wash on the back, but it is very slight.

Hershkovitz (1984) stated that there is a small area of sympatry between this species and *S. s. sciureus* in the lower Rio Madeira basin. Thorington (1985) recorded sympatry between *S. ustus* and subspecies of *S. sciureus* at Lago Berury, on the Rio Purus and at Borba; and close approach between them on the south bank of the Amazon, where *S. ustus* occurs at Villa Bella do Imperatriz and *S. sciureus* at Parintins nearby. But Thorington (1985) recorded that they do appear to hybridize in four localities on the east side of the Rio Tapajós (Fordlândia, Caxiricatuba, Piquiatuba, and Tauri), although skulls from this region are like *S. sciureus;* between the Rio Madeira and the Rio Tapajós, according to the interpretation of Costello et al. (1993), there are both *S. sciureus* and *S. ustus* and hybrids between them.

The discovery of the species on the eastern shore of Lago Tefé (Ayres 1985) raises further interesting questions. Because *S. boliviensis* is found on the west bank of the Rio Purus at Jaburú (type locality of *jaburuensis* Lönnberg), the two must be parapatric or marginally sympatric on the south bank of the Amazon (= Solimões) between Rio Purus and Lago Tefé. Costello et al. (1993) considered that in fact *S. ustus* east of Lago Tefé is not "pure-bred," yet, curiously, it is with *S. sciureus*, not *S. boliviensis*, that it is hybridized. It may be that *S. ustus* occurs (occurred?) even farther west, because two specimens from the west bank of Lago Tefé were interpreted by Costello et al. (1993) as being hybrids of *S. ustus* with *S. sciureus* in one case and with *S. boliviensis* or *S. vanzolinii* in the other.

It is quite clear that *S. ustus* and *S. sciureus* are different according to the PSC and deserve separate species status. Although evidently *S. ustus* hybridizes in the wild with *S. sciureus,* and arguably with *S. boliviensis* as well, it does coexist with *S. sciureus,* even in some places where hybrids are also found, and the two probably rank as separate species according to the BSC as well.

Saimiri boliviensis group

White regions above eyes restricted to narrow line on brows ("Roman" type); blackish supraorbital vibrissal tufts inconspicuous; tail pencil thin; dorsal color "tending toward eumelanin dominance." Hershkovitz (1984) listed the following behavioral differences from the *S. sciureus* group: adult males socially segregated outside mating season; female dominance; no penile display to mirror image; "stiff leg" scratch; males "kick wash"; more aggressive. There are also characteristic vocalization differences.

Saimiri boliviensis (d'Orbigny, 1834)
Black-capped Squirrel Monkey

DIAGNOSIS.   As for species-group. Chromosomes: Apparently 16 biarmed, 5 acrocentric autosome pairs. According to Thorington (1985), the skull is shorter than in *S. sciureus,* with much shorter occipital bones and wider zygomatic arches.

NOTES.   Thorington (1985) placed *boliviensis* not as a distinct species at all, but as a subspecies of *S. sciureus,* on the grounds that, as he interpreted the situation, there is clinal intergradation between them, as follows:

1. At 12–14°S, typical *boliviensis* is found, with Roman arches and black crowns in both sexes.
2. At about 9–10°S, 74°W, only the females have black crowns, and males have gray crowns edged and sprinkled with black.
3. At Yarinacocha, 8°15'S, 74°34'W, males have gray crowns with various amounts of black.

4. At Contamana and Sarayacu, which are at 6°44'S and 7°15'S, respectively, most males are similar, but one has a Gothic arch; among the females from Sarayacu three have the usual black crown, two have no black, and two have black-edged, black-sprinkled crowns (Thorington did not say whether they have Gothic or Roman arches).

It is obvious that specimens from areas 2 and 3 are what Hershkovitz (1984) described as *S. boliviensis peruviensis,* which from all the evidence really is a stable taxon over a reasonably wide distribution. It is the specimens from area 4 that are in question. Without seeing the specimens it is impossible to be absolutely sure whether the Gothic-arched male is an intermediate as Thorington suggested or a specimen of *S. sciureus macrodon* as Hershkovitz evidently interpreted it; and whether the two intermediate females described by Thorington are actually intermediate or fall within the range of *S. b. peruviensis* as Hershkovitz presumably implied by his description of the normal variation of the female crown as "blackish agouti to dominantly blackish." My only contribution is to note that in zoo colonies of *S. b. peruviensis* I have seen the females all have jet black crowns.

Thorington (1985) found that skulls from the region in question (essentially, along the Ucayali) were intermediate between *sciureus* and *boliviensis,* but, as he noted, his material was inadequate for finer analysis. A sample could be craniometrically intermediate either because it is mixed (the Hershkovitz model) or because it represents a homogeneous population that is genuinely transitional.

Because of the very considerable differences between *S. sciureus* and *S. boliviensis,* the questionable existence of wild hybrids between them (pace Costello et al. 1993), and their diagnosability over wide areas, I follow Hershkovitz's interpretation and recognize them as distinct species and, indeed, as belonging to different species-groups.

Saimiri boliviensis boliviensis (d'Orbigny, 1834)

1834 *Callithrix boliviensis* d'Orbigny. Bolivia: Guarayos Mission, Rio San Miguel, Santa Cruz.
1835 *Calltrix [sic] entomophagus* d'Orbigny. Bolivia: Santa Cruz.

1902 *Saimiri boliviensis nigriceps* Thomas. Peru: Cosñipata, upper Rio Marcapata, Cuzco.
1940 *Saimiri boliviensis pluvialis* Lönnberg. Brazil: Lago Grande, Santo Antonio (Rio Eirú) and Rio Juruá opposite João Pessõa. Thorington (1985) selected NHMS 2446 as lectotype.
1940 *Saimiri boliviensis jaburuensis* Lönnberg. Brazil: Jaburú, Rio Purus, Amazonas.

DIAGNOSIS.   Crown and preauricular patch blackish in both sexes; tail grayish or buffy agouti to blackish above, with black tip. Chromosomes: 6 acrocentric autosomal pairs.

DISTRIBUTION.   Hershkovitz (1984) gave the distribution as south of Rio Juruá-Amazonas and west of the Rio Purus-Guaporé, west to the Andes, in Brazil, Peru, and Bolivia. Peres (1993) confirmed, on the basis of fieldwork, that this taxon and *S. sciureus macrodon* are separated by the Rio Juruá along its entire length.

NOTES.   Lönnberg's (1940) subspecies *pluvialis* was distinguished by its very dark coloration. Hershkovitz (1984:192) synonymized it with *S. b. boliviensis,* noting first that a specimen from Buenavista (southern Bolivia) was just as "saturate," and second that of two specimens from Lago Grande that he had studied one was very dark but the other was perfectly normal for *S. b. boliviensis.* After seeing the type material in the NHMS, however, Hershkovitz (1987a) revived the subspecies, describing it as "highly melanotic." Thorington (1985) stated that the darkest specimens, sometimes with even a black median dorsal zone, do indeed come from the highest-rainfall areas, but in his material (which included the NHMS specimens) the darkest of all was from Tefé. It appears from all this that dark phenotypes recur sporadically throughout the range of *S. b. boliviensis* and are not concentrated in a contiguous region as they would have to be if a special dark subspecies were to be recognized.

Hershkovitz (1987a) likewise revived Lönnberg's subspecies *jaburuensis,* stating that it is differentiated by the crown being agouti in males as in *S. b. peruviensis* and blackish in females as in both sexes of *S. b. boliviensis.* Because males answering to this description are not stated to occur beyond the type locality, it is unlikely to be more than a simple polymorphism of restricted distribution.

Saimiri boliviensis peruviensis Hershkovitz, 1984

1984 *Saimiri boliviensis peruviensis* Hershkovitz. Peru: right bank of Rio Samiria, opposite Biological Station Pithecia, Loreto.

DIAGNOSIS.   Crown and preauricular patch agouti in male (but with black hairs along the edge of the crown and scattered through it), mainly black or blackish agouti in female; tail grayish to blackish agouti above, with black tip. Chromosomes: 5 acrocentric autosome pairs.

DISTRIBUTION.   Said by Hershkovitz (1984) to be Amazonian Peru south of the Rio Marañon from Rio Tapiche west to Rio Huallaga, and south at least to Rio Abujao. He quoted field data that the Rio Tapiche separates this taxon on the west from *S. s. macrodon* on the east, but noted that both appear to have been collected at Puerto Punga on the west bank of the Rio Tapiche and, farther west, at Sarayacu on the Ucayali. On this evidence he proposed that there is a small area of sympatry between them between the two rivers.

Saimiri vanzolinii Ayres, 1985
Black Squirrel Monkey

1985 *Saimiri vanzolinii* Ayres. Brazil: left (north) bank of Lago Mamirauá, at the mouth of the Rio Japurá, Amazonas (2°59'S, 64°55'W).

DIAGNOSIS   (after Ayres [1985]). Dorsum very dark, the black hairs predominating over agouti, forming a broad continuous black band from crown to tail tip; hand, forearm, and foot light "burnt yellow" rather than orange as in *S. boliviensis;* pelage short and dense. Crown black in both sexes. Smaller than *S. boliviensis.* Chromosomes more like those of *S. sciureus macrodon* than *S. boliviensis,* but unique in the presence of heterochromatin blocks in the long arm of chromosomes 13, 15, 17, and 19 and the absence of an interstitial C-band in the short arm of chromosome 6.

DISTRIBUTION.   Known only from a small triangle of land between the Rios Japurá and Solimões, including two islands, Capucho (= Uanacá) and Tarará (= Pananim). Ayres (1985) considered it "very probable" that the western limit is the Paraná do Jarauá.

Family Nyctipithecidae Gray, 1870

Aotidae Poche, 1908, is a synonym. This is yet another instance where a family-group name based on a junior synonym has priority over one based on the senior synonym.

According to almost (but not quite) all recent molecular studies, *Aotus* is part of the cebine/marmoset clade; different studies disagree whether the evidence suggests that it is (Schneider et al. 1997; Horovitz et al. 1998) or is not (Goodman et al. 1998) more different from these other two subfamilies than they are from each other. Most of the evidence supports the second suggestion, even putting it about as far from the cebine/marmoset clade as are the Pitheciidae from the Atelidae.

Genus *Aotus* Illiger, 1811
Night Monkeys or
Owl Monkeys or
Douroucoulis

1811 *Aotus* Illiger. *Simia trivirgata* Humboldt, 1812.
1823 *Nyctipithecus* Spix. *Simia trivirgata* Humboldt, 1812.
1824 *Nocthora* F. Cuvier. *Simia trivirgata* Humboldt, 1812.
1829 *Cheirogaleus* Vigors and Horsfield. *Simia trivirgata* Humboldt, 1812.

The genus traditionally has been regarded as monotypic, with the earliest name for the putative sole species being *Aotus trivirgatus*. In the 1970s the discovery that some of the phenotypically different forms had different karyotypes led to increasing unease with this "traditional" arrangement.

Hershkovitz (1983) divided the genus into two groups, a "Gray-neck" and a "Red-neck" group, which differ also chromosomally and are largely found north and south of the Amazon, respectively. Nine species were recognized: *A. trivirgatus*, *A. lemurinus* (with two subspecies), *A. brumbacki,* and *A. vociferans* in the Gray-neck group, and *A. miconax, A. nancymai* (correctly *nancymaae*), *A. nigriceps,*

A. azarae (with two subspecies), and *A. infulatus* in the Red-neck group.

Ford (1994b) used multivariate analysis on both craniodental metrical and pelage coded data to try to discriminate interrelationships among the 11 taxa (nine species, two of them with two subspecies) recognized by Hershkovitz (1983). Principal components analyses differentiated samples only slightly, but canonical variates analyses produced some strong separations. *Aotus azarae boliviensis* was distinct in both analyses, especially craniodental, from *A. a. azarae,* whereas *A. infulatus* and *A. nancymaae* fell within the range of *A. a. azarae* in the craniodental but not in the pelage analysis. North of the Amazon, *A. trivirgatus* stood out sharply on both criteria. There was some separation in the pelage analysis between the two subspecies of *A. lemurinus,* which were considered different only in karyotype by Hershkovitz (1983).

Ford proposed to reduce the number of species to seven, or perhaps five, as follows: *A. trivirgatus; A. vociferans,* including *A. lemurinus* and *A. brumbacki; A. miconax; A. nancymaae* (perhaps conspecific with *A. miconax); A. nigriceps; A. azarae* (but without *A. a. boliviensis);* and, provisionally, *A. infulatus,* including *A. azarae boliviensis.*

I examined the specimens in the BM(NH), leading me to agree with Ford (1994b) in most cases. I agree that *A. brumbacki* should be included in *A. lemurinus,* though not that the whole complex should be included in *A. vociferans. Aotus infulatus-boliviensis-azarae* form a complex with overlapping characters, and I combine them; *A. nigriceps* is close to them, though clearly specifically distinct. So the specimens listed by Napier (1976:48–51) allocate as follows:

Bolivia: *A. azarae boliviensis*
Brazil: *A. vociferans:* Oyapock; *A. nigriceps:* Canabouca, Manacaparu, Ayapuá, Fonte Boa, Rio Juruá; *A. azarae infulatus:* Serra do Chapada
Colombia: *A. lemurinus* (the two subspecies are morphologically indistinguishable): Chili, Tolima, Jericho, Muzo, San Juan de Rio Seco, Onaca, Santa Marta, Bogota, Concordia, Quipama, Santa Fé de Bogotá; *A. lemurinus brumbacki:* Vilavicencio

Ecuador: *A. vociferans*

Nicaragua (?) and Panama: *A. lemurinus*

Paraguay: *A. azarae azarae*

Peru: *A. miconax:* San Nicolas, Tingo Maria; *A. nigriceps:* La Merced, Chanchamayo, Chicosa, Contamana, Pozuzo; *A. nancymaae:* Pebas, Rio Pacaya, Puinuahua, Iquitos

Venezuela: *A. lemurinus* subsp.: Mérida; *A. trivirgatus:* San Carlos

Gray-neck group

Aotus lemurinus I. Geoffroy, 1843
Gray-bellied Night Monkey

DIAGNOSIS.   Upper parts grayish to buffy agouti, with at most a poorly defined brownish median dorsal band; under tail red nearly to tip, which is reddish black. Hands and feet dark. Underside yellowish to pale orange, this tone extending to inner surface of limbs, usually only as far as knee and elbow. Rather shaggy, long-haired. Chromosomes: $2n = 50$–56, with at least 38 acrocentric and only 8 to 12 biarmed autosomes.

DISTRIBUTION.   Panama and northern Colombia, into northwestern Venezuela. The distributions of the first two subspecies were mapped in detail by Hershkovitz (1949).

NOTES.   I agree with Ford (1994b) that the color of *A. brumbacki* overlaps with that of *A. lemurinus* subspp. as recognized by Hershkovitz and that this is likely for the karyotypic features. Given also her findings that the whorl/crest difference is inconstant, it seems entirely probable that *brumbacki* is a third subspecies of *A. lemurinus*. Ford (1994b) suggested in addition that *A. lemurinus* is itself conspecific with *A. vociferans,* but in this case the differentiation remains well marked in both chromosome number and pelage, and I keep them separate here.

A. *l. brumbacki* apart, Hershkovitz (1949, 1983) recognized two subspecies, *A. l. lemurinus* and *A. l. griseimembra,* but Hernández-Camacho and Cooper (1976) recognized a third in addition, *A. l. zonalis.*

Aotus lemurinus lemurinus I. Geoffroy, 1843

1843 *Aotus lemurinus* I. Geoffroy. Colombia: restricted by Hershkovitz (1949) to Quindio forests, Caldas Department.
1847 *Nyctipithecus villosus* Gray. Colombia: Santa Fé de Bogotá.
1870 *Nyctipithecus hirsutus* Gray. Error for *villosus.*
1909 *Aotus lanius* Dollman. Colombia: Tolima Mountains, 1,830 m.
1913 *Aotus aversus* Elliot. Colombia: Fusagasugá, Cundinamarca.
1913 *Aotus pervigilis* Elliot. Colombia: La Candela, Huila.

DIAGNOSIS.   A montane form, with long shaggy coat, rather variable in color (Hernández-Camacho and Cooper 1976). According to Ford (1994b), head stripes and distal tail tend to be much darker than in *A. l. griseimembra;* note however that Hernández-Camacho and Cooper (1976) emphasized the variability in the subspecies, and indeed the syntypes (in the BM[NH]) have hardly any black on the head. Chromosomes: $2n = 55$ or 56, with 44 to 46 acrocentric and 8 or 9 biarmed autosomes.

DISTRIBUTION.   Hershkovitz (1949) mapped this subspecies in northern Colombia, east of the Rio Cauca, and into Venezuela, but Hernández-Camacho and Cooper (1976) restricted it to the northern Andes of Colombia at about 1,000 to 3,200 m. In the north the range is divided by the lowlands of the middle Magdalena Valley, where *A. l. griseimembra* is found, and in the south it may extend somewhat over the border into Ecuador.

Aotus lemurinus griseimembra Elliot, 1912

1912 *Aotus griseimembra* Elliot. Colombia: Hacienda Cincinnati, Santa Marta, and Rio Sinú Cereté, Bolívar.
1937 *Aotus bipunstatus* Bole. Panama: Paracolé, Azuero Peninsula.

DIAGNOSIS.   No interscapular whorl or crest; upper parts grayish to buffy, with at most a poorly defined brownish median dorsal band; hands and feet light brownish. Facial pattern inconspicuous. Underside yellowish to pale orange, this tone extending to inner surface of limbs, usually only as far as knee and elbow, and not extending forward to throat. Chromosomes: $2n = 52$, 53, or 54, with 38 to 42 acrocentric and 10 to 12 biarmed autosomes; Y chromosome metacentric.

DISTRIBUTION. Hershkovitz (1949, 1983) gave the range as Panama, probably to the Rio Cauca in Colombia; but Hernández-Camacho and Cooper (1976) restricted it in Colombia, between about the Rio Sinú and Rio Cauca, in the lowlands from the upper Magdalena Valley to the Caribbean coast and extending east into Venezuela to about 70°W (Bodini and Pérez-Hernàndez 1987).

NOTES. According to the key in Hershkovitz (1983), this subspecies is rather shaggy and long-haired, but the concept of the subspecies in Hernández-Camacho and Cooper (1976) was rather different: a lowland form with short, close pelage.

Aotus lemurinus zonalis Goldman, 1914

1914 *Aotus zonalis* Goldman. Panama: Gatún, Canal Zone.

DIAGNOSIS. According to Hernández-Camacho and Cooper (1976), this is the Panamanian subspecies, extending into the Pacific lowlands of Colombia east to the Sinú Valley and north into the Department of Cordoba. It resembles *A. l. griseimembra* except for the dark brown or blackish hands and feet.

Aotus lemurinus brumbacki Hershkovitz, 1983

1983 *Aotus brumbacki* Hershkovitz. Supposedly "Rio Paraguay area," but redetermined by Hershkovitz (1983) as Colombia: Villavicencio region, Department of Meta.

DIAGNOSIS. Generally a short longitudinal interscapular crest and a longitudinal gular gland with hairs parted on either side. Upper side grayish buffy agouti, dark brown in middorsal zone but without distinct middorsal stripe. Well-marked, thin, brownish black temporal stripes, continuous with dark malar stripes; white above eyes yellowed, not forming triangles; white on face extends to chin. Underside pale orange, this tone extending to elbow and knee and (unlike other subspecies) to posterior part of throat. Chromosomes: $2n = 50$, with 38 acrocentric and 10 biarmed autosomes; Y chromosome acrocentric.

DISTRIBUTION. Known only from western Meta (Agua Dulce and Villavicencio) and Boyacá (Valle de Tenza) Departments, at 467–1,543 m a.s.l. It is unclear whether it occupies an enclave within the range of *A. l. lemurinus* or replaces it in the northeastern part of its range as mapped by Hernández-Camacho and Cooper (1976).

NOTES. Ford (1994b) found this taxon rather more variable than was indicated by Hershkovitz (1983) (or indicated in the BM[NH] sample), even in the presence of the supposedly diagnostic interscapular crest. In pelage characters, she found it to span the gap between the two subspecies of *A. lemurinus*, which are otherwise more distinctive than hitherto recognized. In this light, it does seem likely that, as she indicated, it should be placed as a subspecies of *A. lemurinus*, and that the characters of the (now three) subspecies are somewhat mosaic.

Aotus hershkovitzi Ramirez-Cerquera, 1983
Hershkovitz's Night Monkey

1983 *Aotus hershkovitzi* Ramirez-Cerquera. Colombia: eastern side of Cordillera Oriental, Department of Boyacá, 1,750 m.

DIAGNOSIS. Described as being close to *A. lemurinus* and *A. vociferans*, but with diploid chromosome number 58, the highest in the genus.

NOTES. This species was described on the evidence of four specimens from a single locality. It is presumably not impossible that it may turn out to be a localized pocket of *A. lemurinus* homozygous for a high acrocentric total that in nearby regions is heterozygous; but for the moment the diploid number seems diagnostic.

Aotus trivirgatus (Humboldt, 1812)
Three-striped Night Monkey

1812 *Simia trivirgatus* Humboldt. Venezuela: forests of the Rio Casiquiare, near the foot of Mount Duida, Amazonas.
1823 *Nyctipithecus felinus* Spix. Brazil: Belém do Pará.
1829 *Cheirogaleus commersoni* Vigors and Horsfield. No locality.
1840 *Nyctipithecus duruculi* Lesson. Brazil: Pará.

DIAGNOSIS. No interscapular whorl or crest. Upper parts grayish to buffy agouti, with contrasting orange median dorsal band. Hands and feet dark brown. Face pattern very inconspicuous; the stripes brown, the lateral stripes only just reaching eyes; the face a dirty gray rather than white. Underside brighter orange than *A. vociferans* or *A. lemurinus,* this tone sometimes extending to wrist and ankle but not at all to throat. Relative to other gray-necked species, this species has large teeth and high bicanine width (Ford 1994b).

DISTRIBUTION.   Eastern Colombia (east of the Andes), southern Venezuela (north to about the middle Orinoco), and part of northern Brazil, north of the Amazon; recently discovered in Guyana, where it may represent a new subspecies (Bodini and Pérez-Hernàndez 1987). The boundary with *A. vociferans* is, according to Hershkovitz (1983), the Rio Negro.

NOTES.   It seems particularly ironic that the chromosomes of this species, the type species of the genus and formerly thought to be the only valid species, are unknown. Ford (1994b) found it to be perhaps the most distinct of all species in the genus, in both craniometric and pelage analyses.

Aotus vociferans (Spix, 1823)
Spix's Night Monkey

1823 *Nyctipithecus vociferans* Spix. Peru: Tabatinga, upper Rio Marañon.
1848 *Nyctipithecus oseryi* I. Geoffroy and Deville. Peru: upper Amazon River.
1857 *Nyctipithecus spixi* Pucheran. "South America."
1872 *Nyctipithecus rufipes* Sclater. Nicaragua? Probably Brazil, according to Hershkovitz (1949).
1909 *Aotus gularis* Dollman. Ecuador: mouth of Rio Coca, upper Rio Napo.
1909 *Aotus microdon* Dollman. Ecuador: Macas, Chimborazo forest.

DIAGNOSIS.   An interscapular whorl; circular throat gland, the hairs radiating from it. Brown-toned above, either all over or concentrated in the median dorsal zone. Hands and feet black. Face white except for chin; white above eyes restricted to two small patches (not full triangles), quickly grading into the agouti of crown. Crown stripes thick but brownish, temporal stripes united behind. Tail below is reddish or gray-red for its proximal third or half, the rest be-

ing black. Underside off-white with the merest trace of orange, this tone extending vaguely to wrists and ankles and to chin, and not at all up to sides of neck. Chromosomes: $2n = 46$ to 48; Y chromosome usually acrocentric but sometimes metacentric. Albumin slow-migrating.

DISTRIBUTION.   Occupies a wide range in Colombia, northern Brazil, northern Peru, and eastern Ecuador. Hershkovitz (1983) proposed that the border with *A. trivirgatus* is the Rio Negro, with *A. brumbacki* the Rio Meta and Rio Tomo, with *A. miconax* the Rio Marañon, with *A. nancymaae* and *A. nigriceps* the Amazon, except that there is marginal sympatry with *A. nancymaae* and also a small enclave south of the Amazon along the Rio Purus, within *A. nigriceps* country.

Red-neck group

Aotus miconax Thomas, 1927
Peruvian Night Monkey

1927 *Aotus miconax* Thomas. Peru: San Nicolas, 1,350 m, Amazonas.

DIAGNOSIS.   No interscapular whorl; upper side light gray with a brownish tint to quite infused with red-brown; tail bushy, underside infused with reddish toward tip. Facial pattern inconspicuous. Underside pale orange, this color extending halfway up sides of neck, forward as far as chin, and on inner side of limbs usually not quite reaching ankle.

DISTRIBUTION.   A very small region of northwestern Peru, bounded in the north by the Rio Marañon; its boundaries in the east, with *A. nancymaae* and *A. nigriceps,* are unknown.

NOTES.   Chromosomes are unknown.

Aotus nancymaae Hershkovitz, 1983
Nancy Ma's Night Monkey

1983 *Aotus nancymai* Hershkovitz. Peru: right bank of Rio Samiria above Estación Pithecia, Reserva Nacional Pacaya-Samiria, Loreto.

DIAGNOSIS.   No interscapular whorl. Pelage short; upper parts grayish agouti, less brownish than *A. mi-*

conax; a darkened middorsal zone; face more gray than white, but the pale zone extensive; crown stripes dark brown, narrow, not extending forward past eyes. Proximal part of tail orange, with blackish stripe or wash above, mostly blackish below; underside pale orange, this tone extending up sides of neck and on inner aspect of limbs to wrist and ankle, or nearly so; sides of throat and jaw colored like body. Most cranial and upper facial measurements are very large, whereas lower facial dental measurements are rather small (Ford 1994b). Chromosomes: $2n = 54$, with 32 acrocentric and 20 biarmed autosomes; Y chromosome either acrocentric or metacentric.

DISTRIBUTION. A small region across the Peru-Brazil border, mostly south of the Amazon/Marañon; southern boundary, with *A. nigriceps,* is somewhere north of the Rio Juruá. In a tiny region along the lower Rio Tigre, it is found north of the Amazon, sympatric in one area with *A. vociferans,* and in another area forming an enclave within the distribution of the latter.

NOTES. The *Code* stipulates that a scientific name honoring a woman, in this case Nancy Shui Fong Ma, should end in *-ae,* not *-i.* I (Groves 1989) therefore emended it to *nancymaae.* Ford (1994b) also recognized this, but spelled it *nancymae.*

Hershkovitz (1983) noted that the karyotype of *A. miconax* is unknown and that, should it prove to be the same as that of *A. nancymaae,* the two might well be combined into a single species. Ford (1994b) agreed that the pelage characters are rather similar, but found that her (single available) specimen of *A. nancymaae* is cranially distinctive.

Equally, Hershkovitz (1983) noted that there appear to be intermediates between *A. nancymaae* and *A. nigriceps* along the Rio Ucayali south of 7°S. There is no doubt that it is very like *A. nigriceps,* but other factors, notably their chromosomal distinctiveness, keep them separate.

Aotus nigriceps Dollman, 1909
Black-headed Night Monkey

1909 *Aotus nigriceps* Dollman. Peru: Chanchamayo, 1,000 m.
1909 *Aotus senex* Dollman. Peru: Pozuzo, 800 m.

DIAGNOSIS. No whorl. Iron gray above with brownish wash on dorsum, giving a hint of a dorsal stripe. Face stripes broad; white on face conspicuous. Underside orange, with white tones, this color extending well up sides of neck between ear and shoulder, and to throat, chin, and sides of jaw and, on inner aspect of limbs, to wrist and ankle or nearly so. Chromosomes: $2n = 51$ in male, 52 in female; 34 acrocentric and 16 biarmed autosomes. A single individual had the slow-migrating albumin allele like *A. azarae boliviensis.*

DISTRIBUTION. Brazil, south of the Amazon, west of the Rio Tapajós-Juruena into Peru, south about to the Rio Madre de Dios and Rio Guaporé. Peres (1993) found that populations of what had been supposed to be this species differed on the two sides of the Rio Juruá; those on the right bank are graybacked, those on the left bank have a reddish brown back and perhaps resemble *A. nancymaae.*

Aotus azarae (Humboldt, 1812)
Azara's Night Monkey

DIAGNOSIS. Interscapular whorl present. Upper parts grayish to buffy or olive agouti; proximal part of tail orange to straw color, often with blackish stripe or wash above, orange or deep red below for proximal half, then blackened; digits dark, often black. Face stripes usually narrow, tending to unite on crown, and median one tending to extend forward down nose and lateral ones past corners of eyes; rest of face white, including chin and triangles above eyes, but this contrast is less conspicuous or extensive than in *A. nigriceps.* Underside pale whitish orange, this tone not extending far up sides of neck (not as high as level of ear) and only halfway up throat (not to interramal region), and on inner aspect of limbs not beyond elbow or knee. Chromosomes: Karyotype characterized by Y-autosome fusion; $2n = 51$ in male, 52 in female; 35 or 36 acrocentric and 12 or 13 biarmed autosomes.

DISTRIBUTION. Brazil, south of the Amazon, between the Rio Tocantins and Tapajós-Juruena, southwest into Bolivia and south into Paraguay.

NOTES. Ford (1994b) provisionally distinguished *A. infulatus* as a species and transferred *boliviensis* from *A. azarae* to *A. infulatus*. She noted that in fact *A. azarae,* thus restricted, much more closely resembles *A. nigriceps* than does geographically intervening *A. infulatus;* but added an addendum from karyologist J. Pieczarka to the effect that the structure of the Y-autosome combination in *A. infulatus* is closer to that of *A. nigriceps* than to that of restricted *A. azarae.*

The problem is difficult, but the evidence does seem to me to indicate a mosaic distribution of characters, and I continue to include *boliviensis* and *infulatus* in *A. azarae.* More chromosome studies in particular are necessary.

Aotus azarae azarae (Humboldt, 1812)

1812 *Simia azarae* Humboldt. "Paraguay": restricted by Elliot (1913) to right bank of Rio Paraguay, in northeastern Argentina.
1812 *Pithecia miriquouina* E. Geoffroy. Argentina: right bank of Rio Paraguay.
1898 *Aotus azarai* Lahille.

DIAGNOSIS. Pelage thick, shaggy, grayish to pale buffy agouti. Chromosomes: $2n = 49$ in male, 50 in female; apparently 35 or 36 acrocentric and 12 or 13 biarmed autosomes. Y chromosome fused to an acrocentric autosome, instead of to the short arm of a biarmed chromosome as in the other two subspecies (Pieczarka and Nagamachi 1988).

DISTRIBUTION. Paraguay and neighboring region of southern Bolivia.

Aotus azarae boliviensis Elliot, 1907

1907 *Aotus boliviensis* Elliot. Bolivia: Sara Province, 700 m.
1941 *Aotus bidentatus* Lönnberg. Bolivia: Sud Yungas, Chulumani, Department of La Paz.

DIAGNOSIS. Pelage relatively short. Whorl on withers conspicuous. More olive above, but the limbs grayer, often contrastingly so. Face stripes very narrow, except where middle one expands on crown; white on face less striking than in *A. nigriceps,* triangles above eyes fade out rapidly. Black temporal stripe often poorly defined; black malar stripe faint or absent; usually a whitish band between eye and temporal stripe. Cranial size is large in most metrics

(Ford 1994b). Karyotype similar to that of *A. a. azarae* except that the Y chromosome is submetacentric. Albumin always with the slow-migrating fraction.

DISTRIBUTION. Bolivia, except the Paraguay border region.

Aotus azarae infulatus (Kuhl, 1820)

1820 *Callithrix infulatus* Kuhl. Brazil: Pará.
1909 *Aotus roberti* Dollman. Brazil: Serra do Chapada, 700–900 m, Mato Grosso.

DIAGNOSIS. Pelage generally short. Whorl on withers. Very like *A. a. boliviensis* in general, but white on face more prominent, even chin, though restricted above eyes. Black temporal stripe usually well defined, continuous with malar stripe; generally no whitish band between eye and temporal stripe. Tail tends to be reddish throughout, becoming extremely black only toward tip. Karyotype identical to that of *A. a. boliviensis,* except that there are four G-bands on chromosome B12, instead of only three.

DISTRIBUTION. Brazilian portion of the species's range. From the right bank of the Rio Jamari, south of the Rio Madeira at 9°S, 63°30'W, specimens occurred that are polymorphic for the difference between this subspecies and *A. a. infulatus* or with a narrow interstitial band (Pieczarka et al. 1993).

Family Pitheciidae Mivart, 1865

It is clear from many studies (most lately Horovitz et al. 1998) that the pitheciines are associated in a clade with *Callicebus.* They are therefore united into a single family but separated at subfamily level, given the evidence that *Callicebus* is the sister group to the others.

SUBFAMILY PITHECIINAE MIVART, 1865

Fortunately Mivart's name escapes, by the simple inclusion of the second "i," being preoccupied by the name Pithecidae proposed by Gray (1821) for the Great Apes. The synonym Brachyurina Gray, 1870, includes *Chiropotes* and *Ouakaria* (i.e., *Cacajao*), which Gray separated at tribal level from *Pithecia* (tribe Pithecina).

Molecular studies (Schneider et al. 1996) concluded that *Chiropotes* and *Cacajao* are closest to each other, with *Pithecia* having split off earlier; Goodman et al. (1998) agreed and even combined *Chiropotes* and *Cacajao* in a single genus.

I have not studied original specimens of this subfamily and so have no option but to accept the arrangement of Hershkovitz (1985, 1987a,b).

Genus *Pithecia* Desmarest, 1804

Saki Monkeys

1804 *Pithecia* Desmarest. *Simia pithecia* Linnaeus, 1766.
1822 *Calletrix* Fleming. Error for *Callithrix* Erxleben: *Simia pithecia* Linnaeus, 1766.
1840 *Yarkea* Lesson. *Simia leucocephala* Audebert, 1797.

Pithecia pithecia (Linnaeus, 1766)

White-faced Saki

DIAGNOSIS.   Characters as per Hershkovitz (1985, 1987a,b).

Pithecia pithecia pithecia (Linnaeus, 1766)

1766 *Simia pithecia* Linnaeus. "Guiana"; Hershkovitz (1987b) argued that the origin was probably Cayenne.
1797 *Simia leucocephala* Audebert. Locality unknown.
1812 *Simia rufiventer* E. Geoffroy. French Guiana.
1818 *Pithecia adusta* Olfers. Replacement for *Simia pithecia* Linnaeus, 1766.
1818 *Simia nocturna* Olfers. No locality.
1819 *Pithecia saki* Muirhead. Guyana.
1820 *Pithecia rufibarbata* Kuhl. Suriname.
1820 *Pithecia ochrocephala* Kuhl. French Guiana: Cayenne.
1823 *Pithecia capillamentosa* Spix. "South America."
1842 *Pithecia pogonias* Gray. "Brazil."

Pithecia pithecia chrysocephala I. Geoffroy, 1850

1850 *Pithecia chrysocephala* I. Geoffroy. Restricted by Hershkovitz (1987b) to Brazil: Archipelago das Anavilhanas Ecological Station, lower Rio Negro above Manaus, Amazonas.
1925 *Pithecia monachus lotichiusi* Mertens. Brazil: Manacapurú, Amazonas.

DIAGNOSIS.   Hershkovitz (1987b:416) keyed this subspecies as having facial and circumfacial hair "dominantly or usually" darker, more orange than in *P. p. pithecia,* and remarks on p. 422 suggest that his arrangement was provisional, especially with regard to distribution. The implication is that the two sub-

species of *P. pithecia* grade into one another and are genuinely conspecific.

Pithecia monachus (E. Geoffroy, 1812)

Monk Saki

Pithecia monachus monachus (E. Geoffroy, 1812)

1812 *Simia monacha* E. Geoffroy. Brazil; restricted by Hershkovitz (1987b) to left bank of Rio Solimões between Tabatinga and the Rio Tonantins.
1823 *Pithecia hirsuta* Spix. Brazil: north bank of Rio Solimões below Tabatinga, Amazonas.
1823 *Pithecia inusta* Spix. Brazil: Rio Tonantins, Amazonas.
1844 *Pithecia guapo* Schinz. Peru: Rio Marañon, Loreto.
1938 *Pithecia monachus napensis* Lönnberg. Ecuador: La Coca, Rio Napo.

Pithecia monachus milleri J. Allen, 1914

1914 *Pithecia milleri* J. Allen. Colombia: La Muralla, Caquetá.

DIAGNOSIS.   Hershkovitz (1987b:425) considered this subspecies "weakly defined."

Pithecia irrorata Gray, 1842

Rio Tapajós Saki

DIAGNOSIS.   Hershkovitz (1987b:426) remarked that this taxon is close to *P. m. monachus* but is nonetheless diagnosably distinct.

Pithecia irrorata irrorata Gray, 1842

1842 *Pithecia irrorata* Gray. "Brazil." Restricted by Hershkovitz (1987b) to Rio Tapajós, Amazônia National Park, Pará.

Pithecia irrorata vanzolinii Hershkovitz, 1987

1987 *Pithecia irrorata vanzolinii* Hershkovitz. Brazil: Santa Cruz, Rio Eirú, a right-bank (southern) tributary of the Rio Juruá, Amazonas.

DISTRIBUTION.   This has a very small range at the western edge of the far larger range of *P. i. irrorata.*

Pithecia aequatorialis Hershkovitz, 1987

Equatorial Saki

1987 *Pithecia aequatorialis* Hershkovitz. Peru: Santa Luisa, lower Rio Nanay, Loreto.

DISTRIBUTION.   This is the only example of sympatry in the genus. The species shares its (very small) range with *P. monachus.*

Pithecia albicans Gray, 1860

White-footed Saki

1860 *Pithecia albicans* Gray. Brazil: Tefé, south bank of Rio Solimões.

Genus *Chiropotes* Lesson, 1840

Bearded Sakis

1840 *Chiropotes* Lesson. *Pithecia (Chiropotes) couxio* Lesson, 1840.
1862 *Cheiropotes* Reichenbach. Emendation of *Chiropotes*.
1876 *Saki* Schlegel. *Pithecia satanas* I. Geoffroy, 1851.

This genus and *Cacajao* are sister groups; Goodman et al. (1998) combined them, selecting *Chiropotes* to have priority. A final decision on the acceptability of this proposal must await more complete evidence and a consensus on rank/time association.

Chiropotes satanas (Hoffmannsegg, 1807)

Black Bearded Saki

NOTES.   In contrast to the subspecies within the two species of *Cacajao*, the subspecies of *Chiropotes satanas*, as described, are sharply distinguished (in external and cranial characters) and seem not to intergrade. *Chiroptes s. utahickae* has 20 biarmed and 32 acrocentric autosomes ($n = 23$), whereas two specimens of *C. s. chiropotes* have 22 biarmed and 30 acrocentric (Seuánez et al. 1992). More observations are needed to confirm whether they are truly only subspecies or should be raised to species rank.

Chiropotes satanas satanas (Hoffmannsegg, 1807)

1807 *Cebus satanas* Hoffmannsegg. Brazil: Pará (= Belém).
1870 *Chiropotes ater* Gray. Brazil?
1897 *[Pithecia] nigra* Trouessart. No locality.

Chiropotes satanas chiropotes (Humboldt, 1811)

1811 *Simia chiropotes* Humboldt. Venezuela: upper Rio Orinoco south of the cataracts, Amazonas.
1821 *Simia sagulata* Traill. "Demerara" (= Guyana).
1823 *Brachyurus israelita* Spix. Brazil: Rio Negro or Rio Japurá, Amazonas.
1840 *Pithecia (Chiropotes) couxio* Lesson. New name for *chiropotes* (Var. A), *sagulata* (Var. B), and *israelita* (Var. C).
1897 *[Pithecia] satanas* var. *fulvo-fusca* Trouessart. No locality.

Chiropotes satanas utahickae Hershkovitz, 1985

1985 *Chiropotes satanas utahicki [sic]* Hershkovitz. Brazil: Tapará, right (east) bank of Rio Xingu, near mouth, Pará.

Chiropotes albinasus (I. Geoffroy and Deville, 1848)

White-nosed Saki

1848 *Pithecia albinasa* I. Geoffroy and Deville. Brazil: Santarém, lower Rio Tapajós (restricted by Cabrera [1957]).
1914 *Cacajao roosevelti* J. Allen. Brazil: Barão Melgaço, upper Rio Gy-Paraná (= Rio Jiparaná), Rondônia.

Genus *Cacajao* Lesson, 1840

Uakaris

1823 *Brachyurus* Spix. *Brachyurus ouakary* Spix, 1823. Not of Fischer, 1813 (Rodentia).
1840 *Cacajao* Lesson. *Brachyurus ouakary* Spix, 1823.
1842 *Cercoptochus* Gloger. *Brachyurus ouakary* Spix, 1823.
1849 *Ouakaria* Gray. *Ouakaria spixii* Gray, 1849.
1891 *Uakaria* Flower and Lydekker. Emendation of *Ouakaria*.
1899 *Cothurus* Palmer. New name for *Brachyurus* Spix, 1823. Not of Champion, 1891 (Coleoptera).
1903 *Neocothurus* Palmer. Replacement for *Cothurus* Palmer, 1899.

I have no personal experience with members of this genus and rely upon the taxonomic schema of Hershkovitz (1987a), which seems well justified on the face of it. More distributional data on the described taxa are necessary to substantiate their status more fully as subspecies rather than, in some cases, as potential species.

Cacajao melanocephalus (Humboldt, 1811)

Black-headed Uakari

Cacajao melanocephalus melanocephalus (Humboldt, 1811)

1811 *Simia melanocephala* Humboldt. Venezuela: a pet at San Francisco Solano Mission, Rio Casiquiare.

Cacajao melanocephalus ouakary (Spix, 1823)

1823 *Brachyurus ouakary* Spix. Brazil: Rio Içá at confluence with Rio Solimões, but "from known distribution, more likely the Rio Japurá on the north bank of its confluence with the Solimões" (Hershkovitz 1987a).
1849 *Ouakaria spixii* Gray. No locality.

Cacajao calvus (I. Geoffroy, 1847)

Bald Uakari

Cacajao calvus calvus (I. Geoffroy, 1847)

1847 *Brachyurus calvus* I. Geoffroy. "Para," but fixed by Hershkovitz (1987a) as "left bank of the Solimões opposite Fonte Boa."
1876 *Pithecia alba* Schlegel. Brazil: near mouth of Rio Japurá.

Cacajao calvus ucayalii Thomas, 1928

1928 *Cacajao rubicundus ucayalii* Thomas. Peru: Cerro Azul, Contamana, east bank of Rio Ucayali, 625 m.

Cacajao calvus rubicundus (I. Geoffroy, 1848)

1848 *Brachyurus rubicundus* I. Geoffroy. Brazil: north bank of Rio Solimões, opposite São Paulo de Olivença.

Cacajao calvus novaesi Hershkovitz, 1987

1987 *Cacajao calvus novaesi* Hershkovitz (1987a). Brazil: Santo Antônio, Rio Eirú, near mouth of Rio Tarauacá, a south-bank tributary of upper Rio Juruá, Amazonas.

SUBFAMILY CALLICEBINAE POCOCK, 1925

The synonym Callitricidae Gray, 1821, is based on *Callithrix* E. Geoffroy, which Gray misspelled *Callitrix*. Callitrichina Gray, 1870, is derived from it.

One of the rather few consistent "simplifications" to emerge from recent studies of platyrrhine taxonomy has been the association of *Callicebus* with the Pitheciinae. It is supported by the recent "total evidence" approach of Horovitz et al. (1998).

Genus *Callicebus* Thomas, 1903
   Titis

1812 *Callithrix* E. Geoffroy. *Callithrix personatus* E. Geoffroy, 1812. Not of Erxleben, 1777.
1903 *Callicebus* Thomas. Replacement for *Callithrix* E. Geoffroy, 1812.
1998 *Torquatus* Goodman, Porter, Czelusniak, Page, Schneider, Shoshani, Gunnell, and Groves. *Callicebus torquatus* (Hoffmannsegg, 1807).

Hershkovitz (1963) and Napier (1976) showed how E. Geoffroy's use of *Callithrix* for this genus was perpetuated throughout the nineteenth century, until Thomas (1903) pointed out that this name, as originally used by Erxleben (1777), belongs to the true marmosets. Until 1903, therefore, this well-known and speciose genus was without a valid generic name! Hershkovitz (1963) recognized only three species: *Callicebus moloch, torquatus,* and *personatus,* and that revision was universally adopted until 1990, when Hershkovitz returned to the subject and this time divided up the genus into 13 species in four species-groups (two of them represented by species that he had not considered valid in his 1963 paper). It should, of course, be borne in mind that his 1963 paper was based entirely on specimens in the FMNH, whereas by 1990 he had studied collections throughout North and South America and Europe; but there is a very clear difference in the underlying taxonomic philosophy, even though it is not entirely clear exactly where it lies.

In his 1963 paper, Hershkovitz remarked (1963:12) that two of the species he retained, *C. moloch* and *C. torquatus,* might well have been regarded as no more than subspecifically distinct but for the fact that they are sympatric in some regions. This places him, at that time (Hershkovitz Mark 1, in "lumper" mode), well in the Ellerman and Morrison-Scott school of taxonomy. It is unfortunate that Hershkovitz Mark 2 (in "splitter" mode) never, in his 1990 *Callicebus* monograph or anywhere else, stated his criteria for recognition of species or subspecies, except obliquely when he stated (1987b:426) that *Pithecia irrorata irrorata* is close to *P. m. monachus* but is sharply distinct and does not grade into it. This seems to place him near the PSC school, but I find (below) that some of his subspecies are as sharply distinct as his species, and his detailed criteria remain a little mysterious.

In what follows, I generally follow Hershkovitz (1990) as far as alpha taxonomy is concerned, having no personal experience to query his decisions; but in a few cases, where I made a special effort to examine some of his propositions, I have made changes. I asked the following questions (Groves 1992):

1. Are *C. moloch* and *C. cinerascens* really the same, or at least only subspecifically distinct?
2. Are there really three species in the allopatric series *C. brunneus, C. hoffmannsi hoffmannsi, C. hoffmannsi baptista, C. moloch*?
3. Are *C. caligatus* and *C. cupreus cupreus* really the same?
4. Is *C. dubius* a geographic variant of *C. caligatus* (if that itself is distinct)?

In a visit to the AMNH in 1997, I investigated these four questions. The answers appear to be as follows:

1. No, they are not the same.
2. No, but instead of finding a single species with four subspecies as I had expected, they are four full species (i.e., *baptista* should be raised to species rank).
3. Yes.

4. Yes and no. *Callicebus dubius* is not in fact distinct at all.

I also comment, where appropriate, on changes between Hershkovitz's 1963 and 1990 arrangements.

Hershkovitz (1990) arranged his species into four species-groups, as follows:

1. *C. modestus* group: *C. modestus* only.
2. *C. donacophilus* group: *C. donacophilus, C. olallae, C. oenanthe.*
3. *C. moloch* group: *C. cinerascens, C. moloch, C. hoffmannsi, C. brunneus, C. cupreus, C. caligatus, C. dubius, C. ornatus, C. personatus.*
4. *C. torquatus* group: *C. torquatus* only.

Kobayashi (1995) examined and measured skulls and slightly altered this arrangement. The only Hershkovitzean species unavailable to him was *C. oenanthe.* He divided the genus into five groups:

1. *C. torquatus* group: *C. torquatus* only.
2. *C. personatus* group: *C. personatus* only.
3. *C. moloch* group: including *C. cinerascens, C. hoffmannsi,* and *C. brunneus.*
4. *C. cupreus* group: including *C. cupreus* and *C. caligatus.*
5. *C. donacophilus* group: including *C. donacophilus, C. olallae,* and *C. modestus.*

The first two of these groups were quite distinct and slightly closer to each other than to the other groups. The other three were more closely related to each other and formed a single "super-cluster."

The single available specimen of *C. dubius* was, oddly, affined to the *C. personatus* group, and Kobayashi was clearly at a loss to understand why (and so am I), attributing the anomaly to the lack of "an adequate number of specimens."

Kobayashi (1995) also surveyed the other sources of evidence (chromosomes, geographic distribution, pelage characters), concluding that they were at any rate consistent with his five groups. In pelage, he characterized the *donacophilus* and *personatus* groups as having "a contrastless design," buffy in the first case and blackish to yellowish in the second; the *cupreus* group "weakly contrasted"; and the *moloch* and *torquatus* groups "contrasted," with contrasting underside in the first case and white "muffler" in the second.

There seems little dispute that the *torquatus* group

is very distinct from all others. The subgenus *Torquatus* of Goodman et al. (1998) is probably justified, but a full arrangement into subgenera, or even full genera, will have to await a detailed assessment of all the evidence based on a complete survey of all species.

Callicebus modestus group

Externally resembles the *C. moloch* group, but cranially primitive according to Hershkovitz (1990), with an elongate, low-slung cranium, very small cranial capacity, only 20% of greatest skull length, and short occiput, condylobasal length averaging 86% of greatest skull length. Median pterygoids very large; mandibular angle large. Postcranial skeleton unknown; chromosomes unknown.

Callicebus modestus Lönnberg, 1939
Rio Beni Titi

1939 *Callicebus modestus* Lönnberg. Bolivia: El Consuelo, Rio Beni.

DIAGNOSIS (after Hershkovitz [1990]). Brown- or red-agouti, with whitish ear tufts; forehead and crown red-brown agouti, with a thin black brow fringe; toes with black hairs; sideburns not contrasting; tail blackish agouti.

DISTRIBUTION. Known only from the type region.

NOTES. Cabrera (1957) made this a synonym of *donacophilus,* and Hershkovitz (1963) placed it in the synonymy of what he called *Callicebus moloch brunneus,* stating that in fact it "points to complete intergradation between *brunneus* and *donacophilus,*" between which it intervenes geographically; he noted that the black hands and feet and predominantly black tail resemble *brunneus,* whereas the white-tufted ears resemble *donacophilus.*

Callicebus donacophilus group

Cranial capacity 21–25% of greatest skull length, condylobasal length 81–84% of greatest skull length. Arm (radius plus humerus) 52–58% of trunk length, leg (tibia plus femur) 71–78%. Chromosomes: $2n = 50$.

Callicebus donacophilus (d'Orbigny, 1836)
White-eared Titi

1836 *Callithrix donacophilus* d'Orbigny. Bolivia: Rio Marmoré basin, Beni Province.

DIAGNOSIS (after Hershkovitz [1990]). Small in size, females larger than males: head plus body length 278–330 mm in male, 305–420 mm in female. Pelage thick, long; buffy to orange agouti above; ear tufts white; little or no trace of black brow fringe; a buffy malar stripe; underside orange; limbs orange agouti, hands buffy; tail buffy to blackish.

DISTRIBUTION. Both banks of the Rio Marmoré, in West-Central Bolivia, at 100–500 m.

NOTES. In 1963, Hershkovitz recognized this as a subspecies of *C. moloch,* with *pallescens* and *geoffroyi* as synonyms. At the time he had not seen the type of *donacophilus,* but only d'Orbigny's colored figure, on which basis he placed *pallescens* in synonymy with it, and regarded the whole taxon as showing a cline of increasing paleness to the south and southeast.

Callicebus pallescens Thomas, 1907
White-coated Titi

1907 *Callicebus pallescens* Thomas. Paraguay: 48 km north of Concepcion, Chaco.

DIAGNOSIS (after Hershkovitz [1990]). Pelage extremely long, shaggy, around face nearly concealing skin; very pale: dorsum very pale buffy or grayish agouti; limbs buffy agouti.

DISTRIBUTION. West of the Rio Paraguay in the Gran Chaco, and the Pantanal of Mato Grosso do Sul, Brazil.

NOTES. This is an extraordinarily distinctive species. Why Hershkovitz (1990) ranked it as a subspecies of *C. donacophilus* is quite unexplained.

Callicebus olallae Lönnberg, 1939
Olalla Brothers's Titi

1939 *Callicebus olallae* Lönnberg. Bolivia: La Laguna Station, Rosa, Beni.

DIAGNOSIS (after Hershkovitz [1990]). The most primitive of the species-group, with cranial capacity only 21% of greatest skull length; condylobasal length 84% of greatest skull length. Orange-toned, hairs with a single, very broad orange band; tail dark agouti; face framed with blackish; forehead red-brown-agouti; weak white ear tufts; limbs red-brown, hands and feet blackish; conspicuous black supraorbital vibrissae.

DISTRIBUTION. Known only from the type region.

NOTES. Cabrera (1957) regarded this as a synonym of *C. brunneus,* and in this he was followed by Hershkovitz (1963). Neither had seen the type specimen, but Hershkovitz (1963:33) stated that it "has all the important diagnostic characters of *Callicebus moloch brunneus* and others which suggested intergradation with *C. m. donacophilus."*

Callicebus oenanthe Thomas, 1924
Rio Mayo Titi

1924 *Callicebus oenanthe* Thomas. Peru: Moyobamba, San Martin.

DIAGNOSIS (after Hershkovitz [1990]). Pelage thick, resembling that of *C. olallae* in general; frontal blaze usually present, continuous with long circumfacial crest; a pale malar stripe; limbs agouti; underside and inner surfaces of limbs orange; tail dark brown agouti.

DISTRIBUTION. Upper Mayo Valley, 750–950 m, northern Peru.

NOTES. For Hershkovitz (1963), this was a synonym of what he called *C. moloch discolor;* although it is pale in color, individuals of *C. moloch ornatus* "may be as pale or paler," but the color pattern is "unmistakably" that of *discolor.*

Callicebus moloch group

Cranial capacity 26–29% of greatest skull length; condylobasal length 78–82%. Forelimb (known only for *C. cupreus*) 53–61% of trunk length, hindlimb 72–81%. Chromosomes: $2n = 48$ *(C. moloch, C. brunneus)* or 46 *(C. cupreus, C. ornatus).*

Callicebus cinerascens (Spix, 1823)
Ashy Black Titi

1823 *Callithrix cinerascens* Spix. Probably Peru/Brazil border, Rio Putumayo or Rio Içá.

DIAGNOSIS   (after Hershkovitz [1990]). Body mainly grayish to blackish agouti, except middorsal region, which is contrastingly tawny agouti. Sideburns and throat grayish or yellowish agouti.

DISTRIBUTION.   Southeastern Amazonas, Rondônia, and Mato Grosso, upper Rio Madeira.

NOTES.   Hershkovitz (1990) doubted the correctness of the type locality, because otherwise only *C. cupreus* and *C. torquatus* are known from there. He noted that the frequently yellowish color of the sideburns and throat approaches that of *C. hoffmannsi* and the contrasting dorsum resembles *C. moloch;* only the gray-agouti underparts are really distinctive.

Callicebus hoffmannsi Thomas, 1908
Hoffmanns's Titi

1908 *Callicebus hoffmannsi* Thomas. Pará: Urucuritiba, Rio Tapajós.

DIAGNOSIS.   Body medium to dark brown; hairs with short, light gray-brown base, three pairs of alternating very pale tawny and black bands, and tip tawny. Crown gray, the hairs with a white base, black shaft, white tip. Limbs as body, or a bit grayer, becoming grayer toward hands and feet. Tail black, with tendency to have a light tip. Underside orange-red, this tone broadly extending to inner aspects of limbs (including hands and feet), cheeks, and chin.

DISTRIBUTION.   South of Rio Amazon, from Rio Tapajós-Jurena to Rio Canumã.

NOTES.   A subspecies of *C. moloch* in Hershkovitz (1963), but a valid species in Hershkovitz (1990).

Callicebus baptista Lönnberg, 1939
Baptista Lake Titi

1939 *Callicebus baptista* Lönnberg. Brazil: Lago do Baptista, Amazonas.

DIAGNOSIS.   Body very dark brown; hairs with short gray-black base, shaft with two pairs of bands, pale tawny and black, and tip tawny. Crown gray; the hairs with very short white base, black shaft, white tip. Limbs as body, or a bit grayer, darkening toward hands and feet. Tail black, with tendency to have a light tip. Underside maroon-red, this tone broadly extending to inner aspects of limbs (including hands and feet), cheeks, and chin. Relatively large but short-tailed.

DISTRIBUTION.   Very restricted: Lago do Baptista and Lago do Tapaiuna, on Tupinambaranas Island, lower Rio Madeira.

NOTES.   Regarded by Hershkovitz (1963) as a synonym of what he called *C. moloch hoffmannsi,* but as a valid subspecies of *C. hoffmannsi* by Hershkovitz (1990). The differences between this form and *C. hoffmannsi* are in fact greater than described by Hershkovitz (1990). As I have opined (Groves 1992), the relationships between *hoffmannsi, baptista,* and *moloch* are very even. I expected that when I was able to examine skins for myself, I would reduce all three to subspecies under *C. moloch;* instead, study of the specimens in the AMNH demonstrated to me that each is specifically distinct.

Callicebus moloch (Hoffmannsegg, 1807)
Red-bellied Titi

1807 *Cebus moloch* Hoffmannsegg. Brazil: near Pará (= Belém). Redetermined by Hershkovitz (1963) as right bank of lower Rio Tapajós, Santarém District.
1815 *Callithrix hypoxantha* Illiger. Nomen nudum.
1818 *Callithrix hypokantha* Olfers. New name for *moloch* Hoffmannsegg.
1855 *Simia sakir* Giebel. No locality.
1908 *Callicebus remulus* Thomas. Brazil: Santarém, Rio Tapajós.
1911 *Callicebus emiliae* Thomas. Brazil: Lower Amazonas.
1914 *Callicebus geoffroyi* Miranda Ribeiro. Brazil: Urupá, upper Rio Jiparaná.

DIAGNOSIS. Body lighter, redder than in *C. hoff-mannsi* and *C. baptista;* the hairs with short, light gray base, usually four alternating pairs of bands, light red and black, tip usually black. Crown light gray, the hairs with very long white base, black shaft, white tip. Limbs much grayer than body, becoming light tawny toward hands and feet, which are buffy white. Tail black, with tendency to have a light tip. Underside fiery red, this tone broadly extending to inner aspects of limbs (including hands and feet), cheeks, and chin.

DISTRIBUTION. South of the Amazon, from Rio Tocantins-Araguaia to Rio Tapajós.

Callicebus brunneus (Wagner, 1842)
Brown Titi

1842 *Callithrix brunea [sic]* Wagner. Brazil: Cachoeira da Bananeira, Rio Guaporé, Rondônia. Corrected to *brunnea* by Wagner in 1848.

DIAGNOSIS. Body wood brown, with very fluffy pelage, the hairs with long black-brown base, usually three pairs of alternating bands, light brown and black, usually a light tip. Crown black, the hairs with very short, light gray bases. Limbs black. Tail black, with tendency to have a light tip. Underside black-brown, not sharply set off from body; inner aspect of limbs, chin, and cheeks much blacker.

DISTRIBUTION. Middle and upper Rio Madeira basin, west to upper Rio Purus basin.

NOTES. Regarded by Hershkovitz (1963) as only a subspecies of *C. moloch,* this is in fact the most distinctive of the southern Middle-Amazon ($2n = 48$) subgroup.

Callicebus cupreus (Spix, 1823)
Coppery Titi

1823 *Callithrix cuprea* Spix. Brazil: Rio Solimões, Amazonas; restricted by Hershkovitz (1963) to Tabatinga, but amended to Rio Solimões opposite Tabatinga by Hershkovitz (1990).
1842 *Callithrix caligata* Wagner. Brazil: Borba, lower Rio Madeira and Rio Solimões; restricted to Borba by Thomas (1908),

but more probably on the west side of the Rio Madeira opposite Borba according to Hershkovitz (1963).
1848 *Callithrix discolor* I. Geoffroy and Deville. Peru: Sarayacu.
1866 *Callithrix castaneoventris* Gray. Brazil.
1900 *Callithrix cuprea leucometopa* Cabrera. Peru: junction of Rio Napo and Rio Aguarico.
1907 *Callicebus subrufus* Elliot. Peru: Pachitéa, Rio Ucayali, 120–150 m.
1907 *Callicebus usto-fuscus* Elliot. Brazil.
1908 *Callicebus egeria* Thomas. Brazil: Teffé (= Tefé), Rio Solimões, Amazonas.
1909 *Callicebus paenulatus* Elliot. Ecuador: Andoas, Rio Pastaza.
1914 *Callicebus toppini* Thomas. Peru: Rio Tahuamanu, about 12°20′S, 68°45′W.
1922 *Callicebus cupreus napoleon* Lönnberg. Ecuador: Rio Napo, 750 m.
1923 *Callicebus rutteri* Thomas. Peru: Puerto Legula, Rio Pachitéa, upper Ucayali, 455 m.
1952 *Callicebus cupreus acreanus* Vieira. Brazil: Iquiri, upper Rio Purus.
1988 *Callicebus dubius* Hershkovitz. Brazil: Lago do Aiapuá, west bank of lower Rio Purus, but more probably the east bank opposite the lago according to Hershkovitz (1990).

DIAGNOSIS. Hairs have a long maroon-brown base, a straw-colored band, a black band, another straw band, and sometimes a black tip. Hands and feet red. Tail brindled (hairs have straw-colored base, long blackish shaft, straw tip). Crown agouti, becoming black anteriorly for various distances, but including forehead; sideburns reddish or orange; a white brow band variably present. Underside sharply marked reddish or orange, this tone extending to sides of neck and inner surface of limbs; hands and feet reddish to whitish.

DISTRIBUTION. South of the Napo-Solimões, from the Rio Purus-Ituxi to the Andes; Hernández-Camacho and Cooper (1976) mapped it as far north as the southern bank of the Rio Guamués, in extreme southern Colombia. A population on the Rio Sucusari, a lower left-bank tributary of Rio Napo, was reported by Brooks and Pando-Vasquez (1997).

NOTES. Hershkovitz (1990) recognized four different taxa here: *C. cupreus cupreus, C. cupreus discolor, C. caligatus,* and *C. dubius.* The first two were said to be predominately phaeomelanic, the second two more eumelanic; in *C. c. cupreus* and *C. caligatus* there is no frontal band or just a small, agouti median tuft;

in *C. c. discolor* and *C. dubius* there is a variably developed white frontal blaze. The distributions of the four, taken as a group, are coterminous, but the two easterly forms have the frontal blaze whereas the two westerly ones are without it, so that at the very least there is a remarkable parallelism between phaeomelanic and eumelanic species in their geographic variation. I suggested (Groves 1992) that this odd situation could be resolved if the eumelanic and phaeomelanic "species" are actually morphs of a single species.

This hypothesis was tested by my examination of skins in the AMNH in early June 1997. It is corroborated; not only are eu- and phaeomelanic skins most readily interpreted as morphs of a single species, there is in fact very little difference between them at all. In skins identified as *caligatus,* the hair bases are redder, less brown than those labeled *cupreus;* the underside is more maroon, less foxy red; the red of the hands and feet is darker; the maroon of the cheeks is blackish rather than red.

Close reading of Hershkovitz (1990), indeed, suggests that even the geographic variation in presence or absence of the frontal blaze may not be so clearcut. Thus, two series ascribed to *C. c. discolor,* collected by the Olallas at the mouth of the Rio Inuya (Rio Urubamba) and Lagarto (upper Ucayali), were regarded as misplaced by Hershkovitz because they are on the "wrong" side of the Ucayali. He similarly regarded the localities of Olalla specimens of *C. caligatus* (Sarayacu and mouth of Rio Inuya). If the localities of these specimens are indeed incorrect, then an easterly trend in development of the frontal blaze is very marked; if not (if the Olallas correctly reported the localities), then blaze development ceases to have much geographic significance.

Finally, Hershkovitz (1990) also queried the locality of the type specimen of his *C. dubius.* If it originated from the east bank of the Rio Purus, as he hypothesized, then *C. dubius* is preserved as a distinct taxon with a small, river-bounded range; if from the west bank, as the label records, then the taxon has no identifiably separate distribution. From my findings and conclusions given here, I see no reason at all to query the locality as given.

Callicebus ornatus (Gray, 1866)
Ornate Titi

1866 *Callithrix ornata* Gray. Colombia: Villavicencia, Rio Meta.

DIAGNOSIS (after Hershkovitz [1990]). Body and limbs buffy agouti; forearms reddish; hands and feet contrastingly buffy or whitish. Forehead with blaze or a simple median tuft. Crown red-brown to nearly black, the hairs with buffy or whitish bases. Sideburns, underparts, and inner surfaces of limbs reddish.

DISTRIBUTION. A small region in the headwaters of the Rio Meta and Rio Guiviare, Colombia.

NOTES. This taxon has always been regarded as distinctive, although Hershkovitz (1963), calling it *C. moloch ornatus,* claimed that its intergradation with *C. m. discolor* (here, *C. cupreus*) is "complete."

Callicebus personatus (E. Geoffroy, 1812)
Atlantic Titi

DIAGNOSIS (after Hershkovitz [1990]). Largest species; coat coarse, shaggy, with brownish underwool; hands and feet blackish; facial hairs long; forehead, sideburns, and ear tufts blackish.

DISTRIBUTION. Atlantic coast of Brazil, from (according to Hershkovitz [1990]) the Rio Tietze north to the Rio Itipicuru or the Rio São Francisco.

NOTES. Kobayashi and Langguth (1999) divided this into several species, and almost certainly they are right; but I have no personal experience of the taxa and so will not take this step here.

Callicebus personatus personatus (E. Geoffroy, 1812)

1812 *Simia personata* E. Geoffroy. Brazil: restricted by Hershkovitz (1990) to lower Rio Doce, Espírito Santo.

DIAGNOSIS (after Hershkovitz [1990]). Forehead blackish, as is crown as far as the level of the ears, then sharply buffy or orange like the nape; throat blackish.

DISTRIBUTION. Probably the whole of Espírito Santo Province.

Callicebus personatus nigrifrons (Spix, 1823)

1823 *Callithrix nigrifrons* Spix. Brazil: restricted by Hershkovitz (1990) to Rio Onças, Campos municipality, Rio de Janeiro.
1829 *Pithecia melanops* Vigors. "Mexico."
1840 *Callithryx [sic] chlorocnemius* Lund. Brazil: Magoa Santa (Pleistocene).
1841 *Callithrix crinicaudus* Lund. Brazil: Lagoa Santa.
1841 *Jacchus grandis* Lund. Brazil: Lagoa Santa (Pleistocene).
1913 *Callicebus personatus brunello* Thomas. Brazil: Piquete, São Paulo.

DIAGNOSIS (after Hershkovitz [1990]). Forehead blackish, as is crown to about halfway back; rest of crown grading into the multibanded buffy or orange of the nape; throat is pale like chest.

DISTRIBUTION. Probably between Rio Tiete and Rio São Francisco and upper Rio Jequitinhonha.

Callicebus personatus melanochir (Wied-Neuwied, 1820)

1820 *Callithrix melanochir* Wied-Neuwied. Brazil: Morro d'Arara or Fazenda Arara, Bahia.
1820 *Callithrix canescens* Kuhl. Based on same specimen as *melanochir*.
1823 *Callithrix gigot* Spix. Brazil: Bahia.

DIAGNOSIS (after Hershkovitz [1990]). Pelage multibanded; forehead, crown, and throat gray- to buffy- or pale-brown agouti.

DISTRIBUTION. Probably between the Rio Paraguaçu and lower Rio Jequitinhonha.

NOTES. Hershkovitz (1990) described the confusion in the literature over the use and significance of the name *gigot* Spix, 1823; the holotype is an example of *C. p. melanochir*.

Callicebus personatus barbarabrownae Hershkovitz, 1991

1991 *Callicebus personatus barbarabrownae* Hershkovitz. Brazil: Lamārao, Bahia, 300 m.

DIAGNOSIS (after Hershkovitz [1990]). Body pale, buffy to silvery. Forehead, crown, and throat buffy, the hairs black-tipped.

DISTRIBUTION. Probably between the Rio Itapicuru and Rio Paraguaçu; coastal highlands of north-central Bahia.

Callicebus coimbrai Kobayashi and Langguth, 1999

1999 *Callicebus coimbrai* Kobayashi and Langguth. Brazil: Aragão, Santana dos Frades region, Sergipe State.

DIAGNOSIS. Kobayashi and Langguth, whose description was published just in time to be included in this book, distinguished their new species by its black forehead, crown, and ears and buffy body; pale cheek whiskers, this color reaching to nape; hands and feet blackish, tail orange, and zebra stripes on forepart of back.

DISTRIBUTION. It seems to be restricted to the coastal region between the Rio São Francisco and Rio Real; in the drier hinterland it is replaced by *C. barbarabrownae,* which, like the other supposed subspecies of *C. personatus,* they regarded, doubtless correctly, as full species.

NOTES. The species is named after the eminent Brazilian biologist Adelmar F. Coimbra-Filho, a living legend in primatology.

Callicebus torquatus group

This group (all one species, according to Hershkovitz [1990]) is distinguished by several dental, postcranial, and cranial characters. Mesostyle and distostyle on the upper premolars are well defined (in other groups they are absent on P^2 and weak or absent on P^{3-4}). An entepicondylar foramen is present (it is lacking in all other species). Limbs are very long: arm 67–73% of trunk length, leg 90%. First ethmoturbinal very large. Cranial capacity 29% of greatest skull length; condylobasal length 81% of greatest skull length. Chromosomes distinctive: $2n = 20$. Forehead, forearms, sideburns, feet, and tail are blackish; the crown is reddish to blackish; sideburns little developed. There is generally a whitish or buffy throat collar. Hershkovitz's statement that these are unusually large animals is not borne out by the available measurements.

There may well be several species in the group but, lacking personal knowledge, I here follow Hershkovitz (1990), except in recognizing *medemi* as a species.

Callicebus medemi Hershkovitz, 1963

Black-handed Titi

1963 *Callicebus torquatus medemi* Hershkovitz. Colombia: Rio Mecaya, near mouth at right bank of Rio Caquetá, approximately 180 m.

DIAGNOSIS (after Hershkovitz [1990]). Hands and feet, tail, sideburns, and underside (except for throat) blackish. Rather small in size, head plus body only 232–360 mm.

DISTRIBUTION. A very small range between the Rio Caquetá and Rio Putumayo, Colombia.

NOTES. This is the only instance where I propose to modify the revisions of Hershkovitz (1963, 1990). The black hands and feet of this taxon are diagnostic. According to Hershkovitz (1990) its range is isolated well to the north of other members of the group, but Hernández-Camacho and Cooper (1976) mapped *C. torquatus lugens* to its immediate north in Colombia, separated by the Rio Caquetá.

Callicebus torquatus (Hoffmannsegg, 1807)

Collared Titi

DIAGNOSIS. Larger than *C. medemi*, with light-colored, not blackish, hands; underparts usually not black.

DISTRIBUTION. A large area from about the Orinoco, west of the Rio Negro, to south of the Amazon and west of the Rio Purus.

NOTES. As in the case of *C. personatus* (treated earlier), almost certainly there are several species among what Hershkovitz (1990) designated as subspecies. I have little experience of this group and so leave them as subspecies of the one species, except for the strikingly distinct black-handed *medemi*.

Callicebus torquatus lugens (Humboldt, 1811)

1811 *Simia lugens* Humboldt. Venezuela: San Fernando de Atabapo.
1840 *Saguinus vidua* Lesson. New name for *lugens*.
1914 *Callicebus lugenus duida* J. Allen. Venezuela: Rio Cunucunumá, western base of Mount Duida.

DIAGNOSIS (after Hershkovitz [1990]). Color dark, even blackish, uniform or faintly banded; hair around ears black; hands and feet golden or yellowish; ventral side brown or blackish.

DISTRIBUTION. North of the Rio Juruá (except where *C. t. torquatus* is found), into Colombia and Venezuela. According to Hernández-Camacho and Cooper (1976) this occurs throughout the Amazonian region of Colombia, except south of the Rio Caquetá where *C. medemi* occurs; and in Venezuela north to the middle Orinoco (Bodini and Pérez-Hernàndez 1987).

Callicebus torquatus lucifer Thomas, 1914

1914 *Callicebus lucifer* Thomas. Peru: Yahuas territory, near Pebas, Loreto.

DIAGNOSIS (after Hershkovitz [1990]). Upper parts brownish or red-brown, hairs agouti-banded but often only weakly so; underside brown or blackish; hair around ears black; hands and feet orange.

DISTRIBUTION. From between the Rio Solimões and the Rio Caquetá.

NOTES. For Hershkovitz (1963), this was a synonym of *C. t. torquatus;* but in his 1990 monograph he resurrected it as a subspecies.

Callicebus torquatus regulus Thomas, 1927

1927 *Callicebus regulus* Thomas. Brazil: Fonte Boa, upper Rio Solimões.

DIAGNOSIS (after Hershkovitz [1990]). Hairs around ears banded; underside brown or blackish; hands and feet orange.

DISTRIBUTION. From between the Rio Solimões and the Rio Juruá.

NOTES. For Hershkovitz (1963), this was a synonym of *C. t. torquatus;* he revived it in his 1990 monograph.

Callicebus torquatus torquatus (Hoffmannsegg, 1807)

1807 *Callitrix torquatus* Hoffmannsegg. Brazil: Codajáz, north bank of Rio Solimões above the Rio Negro.
1812 *Simia amicta* E. Geoffroy. "Probably Brazil."
1927 *Callicebus torquatus ignitus* Thomas. Brazil: Rio Tonantins, upper Rio Solimões.

DIAGNOSIS (after Hershkovitz [1990]). Underside reddish; hair around ears not black; hands and feet yellow, golden, or orange. Body weakly banded to uniform red-brown above; throat collar weak or absent.

DISTRIBUTION. A small strip on the north bank of the Rio Japurá-Solimões, within the southern part of the range of *C. t. lugens.*

Callicebus torquatus purinus Thomas, 1914

1914 *Callicebus lugens* Thomas. Brazil: Aiapuá, lower Rio Purus, Amazonas.

DIAGNOSIS (after Hershkovitz [1990]). Underside reddish; hair around ears not black; hands and feet yellow, golden, or orange. Upper parts buffy, more clearly banded.

DISTRIBUTION. South of the Amazon, between the Rios Juruá and Purus.

NOTES. For Hershkovitz (1963), this was a synonym of *C. t. torquatus;* he revived it in his 1990 monograph.

Family Atelidae Gray, 1825

The form of the name as proposed by Gray was Atelina.

There seems little doubt that *Alouatta* belongs to the *Ateles/Lagothrix* clade, on both morphological grounds (for example, the modified prehensile tail) and molecular evidence: see, for example, Horovitz et al. (1998).

SUBFAMILY MYCETINAE GRAY, 1825

Alouatinae Trouessart, 1897, and Alouattinae Elliot, 1904, are synonyms. This is the last of the cases in the Platyrrhini where the senior family-group synonym is not based on the senior generic synonym. The form of the name as used by Gray was Mycetina; it was included (with Atelina, Callithricina, Saguinina, and Harpalina) in a family Sariguidae, which appears not to be based on any generic name.

Distinguished from other atelids in the extremely small braincase compared with the face, and (from all other platyrrhines) in the posterior position of the foramen magnum and occipital condyles. According to Goodman et al. (1998), the separation between Alouattinae and Atelinae occurred about 16 million years ago.

Genus *Alouatta* Lacépède, 1799
Howler Monkeys

1799 *Alouatta* Lacépède. *Simia belzebul* Linnaeus, 1766.
1811 *Mycetes* Illiger. Substitute for *Alouatta.*
1812 *Stentor* E. Geoffroy in Humboldt.

The forehead is very flat; supraorbital ridges are prominent, with no ophryonic groove behind; jaw angle is enlarged; anterior and posterior margins of ascending ramus are parallel; coronoid process is small, upright; zygomatic arch is very deep; teeth are large, toothrows are parallel. Alveolar margins are very swung up anteriorly (airorhynch).

The chromosomes, as in most platyrrhine genera, differ strongly between species and sometimes within what are supposed to be the same species. Many species, but apparently not all, have various numbers of microchromosomes. Most or perhaps all species have an $X_1X_2Y_1Y_2$ sex chromosome system (Mudry et al. 1998).

Alouatta palliata group

Skull, according to Hershkovitz (1949), wider and less prognathous than in *A. seniculus* group; nasals angulated in middle; mesopterygoid fossa narrower, the walls convergent behind, with short hamuli that are bent sharply downward and end in broad, blunt tips; foramen magnum oval, wider than long.

Hyoid (after Hershkovitz [1949], recorded only for *A. palliata*): in male, smaller than in female of *A. seniculus* group; opening fills entire posterior surface; no tentorium. Cornu for thyreohyal bone broad; corniculum for stylohyoid attachment well developed. In female smaller but similar in structure.

Alouatta pigra Lawrence, 1933
Guatemalan Black Howler

1845 *Mycetes villosus* Gray. "Brazils"; Sclater (1872) suggested Vera Paz, Guatemala.

1933 *Alouatta palliata pigra* Lawrence. Guatemala: Uaxactún, Petén.

1933 *Alouatta palliata luctuosa* Lawrence. Belize: Mountain Cow, Cayo District.

DIAGNOSIS. Pelage soft, dense, black with brown bases on cheeks, shoulders, and anterior part of dorsum. Hair comes forward from a whorl on nape, but suddenly becomes very short, upright on crown, with a slight parting going back to the whorl. The palate is distinctively narrow, deep, and V-shaped toward the back.

DISTRIBUTION. Yucatan and southern Chiapas, Mexico; Guatemala; Belize.

NOTES. The name *villosa* was used for this species by Napier (1976), who reviewed the different views held on the status of this name. Sclater (1872) compared the type skin closely with a Guatemalan specimen, but there are several other taxa of completely black howlers of which he was unaware. Because the type skin is now lost and the associated skull is a subadult female (and so without diagnostic characters), the name does seem most appropriately rejected as indeterminable, as argued by Lawrence (1933) and others.

Alouatta palliata (Gray, 1849)
Mantled Howler

1849 *Mycetes palliata* Gray. "Caracas, Venezuela," but actually Nicaragua: Lago Nicaragua (see Sclater [1872]).

1880 *Mycetes niger* Thomas. Not of E. Geoffroy, 1812 (= *A. caraya*).

1902 *Alouatta palliata mexicana* Merriam. Mexico: Minatitlan, Veracruz.

1903 *Alouatta aequatorialis* Festa. Ecuador: Vinces, Guayas Province.

1908 *Alouatta palliata matagalpae* Allen.

1913 *Alouatta inclamax* Thomas. Ecuador: Intac, northwest of Quito.

1913 *Alouatta palliata quichua* Thomas. Ecuador: 32 km west of Mindo, Rio Blanco, 750 m.

1913 *Alouatta palliata inconsonans* Goldman. Panama: Cerro Azul.

DIAGNOSIS. Pelage smooth, silky, black with long brown or golden flank hairs. Hyoid small with wide aperture, even in male. Hair very short, upright; on crown, meeting backward-directed forehead hair in a straight transverse crest. Palate deep and narrow, as in *A. nigerrima*. Nasal bones short and wide, constricted in the middle (Watanabe 1982).

DISTRIBUTION. From northern Chiapas, Vera Cruz, and Tabasco, in Mexico, through Central America to western Colombia and Ecuador. In Colombia, found through the Pacific lowlands into Ecuador, and north to the Sinú Valley, where it is sympatric with *A. seniculus* (Hernández-Camacho and Cooper 1976).

NOTES. I cannot comment extensively on the subspecies ascribed to this species. The northern *A. p. mexicana* is said to have the hair bases silvery; in those from Central America, Ecuador, and Colombia they are "walnut." Both northerly and southerly populations are said to have the light flank veil extending up onto the dorsum, whereas in Central American ones it is restricted to the flanks. In the rather sparse BM(NH) material, Ecuadorian skins tend to have the mantle dark; in Panamanian specimens it is lighter, more golden; and in Costa Rican ones it is lighter, more grading. A dermatoglyphic study (Froehlich and Froehlich 1987) found some distinction between samples from eastern Panama (Darien, Canal Zone, and Cerro Azul) and western Panama (Chiriquí, Bocas del Toro), Costa Rica, and Nicaragua.

Alouatta coibensis Thomas, 1902

Coiba Island Howler

1902 *Alouatta palliata coibensis* Thomas. Panama: Coiba Island.
1933 *Alouatta palliata trabeata* Lawrence. Panama: Capina, Herrera Province.

DIAGNOSIS. High frequencies of complex dermatoglyphic patterns, compared with *A. palliata;* more strongly expressed sexual dimorphism in size. Veil is very dark, restricted in area, infused with red-brown; the hairs of the back have long, light tips, so differentiation between back and veil is poor.

DISTRIBUTION. Panama: Coiba Island and the Azuero Peninsula on the opposite mainland.

NOTES. The separation of this species is based on studies, largely relating to dermatoglyphics, by Froehlich and Froehlich (1987). The patterns of Coiba Island and Azuero Peninsula howlers differed very little from each other in a multivariate analysis, but greatly from those of *A. palliata;* the differences between these two species are comparable with those between them and South American howler species. It must be admitted, however, that these genetic distances are the summation of differences in pattern frequencies: there is no one pattern that is unique to, let alone fixed in, either of the two species.

The mainland and insular populations of this species differ considerably and are presumably (at least?) subspecifically distinct. In the mainland form the whole dorsum is brownish, and the flank hairs are golden; in the insular form the veil is more restricted to the flanks. The insular form is smaller in size.

Alouatta seniculus group

Hershkovitz (1949) detailed the following cranial and hyoid differences from the *A. palliata* group:

Skull is longer and narrower, more prognathous; nasals evenly concave; sphenomaxillary fissure reduced or absent; mesopterygoid fossa wide, its walls parallel-sided, with hamuli elongate, backwardly directed with long, tapering tips; foramen magnum subtriangular in outline.

Hyoid in male huge (>38 by 55 mm), globular, its opening restricted; above the opening is an inflated tentorium, demarcated internally from main chamber by lateral bony plates (in *A. seniculus* and *A. macconnelli,* not in *A. nigerrima* and *A. guariba*). Thyreohyal bones insert on lateral corners of tentorium by a wide depression (in *A. seniculus*) or a blunt cornu (in *A. macconnelli*); a rudimentary corniculum, for stylohyoid ligament, on lateral side of opening on each side. In female only one-fourth to one-fifth as large in volume, less inflated; no walls between tentorium and main chamber.

Alouatta seniculus (Linnaeus, 1766)

Venezuelan Red Howler

DIAGNOSIS. Color golden-toned on body, contrasting with maroon head, shoulders, limbs, and proximal part of tail. Crown hair runs completely forward, as it does in *A. caraya,* meeting the backward-directed forehead hair in a forwardly concave V. Palate very shallow, with some posterior deepening, as in *A. belzebul.* Nasal bones long and narrow. Hyoid lacking cornua; its mouth constricted at rim. Chromosomes: $2n = 44$ (male), 45 (female), including 4 microchromosomes.

DISTRIBUTION. The species is marginally sympatric with *A. palliata.* As reported by Hershkovitz (1949) and Hernández-Camacho and Cooper (1976), both occur on the lower Rio Truando and Rio Atrato, from the Cartagena region, the mouth of the Rio Sinú, and from the Chocó. The distribution extends south and east into Venezuela and northwestern Brazil.

NOTES. The young have (very slightly) less contrast than the adults. The sexes are similar.

Evidence has been accumulating from chromosome studies since 1985 that what had hitherto been considered a single species actually consists of several. First, Minezawa et al. (1985) showed that the karyotype of red howlers from Bolivia was unlike that of those from Colombia and proposed to resurrect the species *A. sara.* My examination of museum material confirms that this separation is plausible. Banding studies showed that A. sara is distinct also from specimens from Venezuela (Stanyon et al. 1995), and this was reinforced by chromosome painting studies (Consigliere et al. 1996); but they are similar in possessing microchromosomes, in contradistinction to those from the Jari River (Brazil) and Guyana. Chromosome number in Colombian and Venezuelan howlers differs from that of Guyana specimens (Stanyon et al. 1995). Those from French Guiana resemble those from the Jari River, Brazil (Vassart et al. 1996).

According to Stanyon et al. (1995), for A. s. arctoidea, there is a complex $X_1X_2Y_1Y_2$ sex chromosome system, with both Y chromosomes translocated onto autosomes, accounting for the difference in number between the sexes. This is the same as in "A. stramineus," A. palliata, A. guariba, and A. belzebul, but differs from A. sara, in which only one Y is translocated.

Bodini and Pérez-Hernàndez (1987) indicated the existence of an undescribed subspecies in the Venezuelan llanos north of the Orinoco.

Alouatta seniculus seniculus (Linnaeus, 1766)

1766 Simia seniculus Linnaeus. Colombia: Cartagena, Bolívar.
1812 Simia ursina Humboldt. Text only (plate is A. guariba: see Napier [1976]). Venezuela.
1829 Stentor chrysurus I. Geoffroy. "Spanish Guiana or Colombia."
?1845 Mycetes auratus Gray. Brazil: Orinoco.
?1845 Mycetes laniger Gray. Supposedly from Colombia.
1904 Alouatta seniculus rubicunda Allen. Colombia: Bonda, near Santa Marta.
1904 Alouatta seniculus caucensis Allen. Colombia: Charingo, upper Cauca Valley.
1914 Alouatta seniculus bogotensis Allen. Colombia: Subia, Cundinamarca.
1914 Alouatta seniculus caquetensis Allen. Colombia: La Muralla, Caquetá.

DIAGNOSIS. Color very contrasty: body old gold to deep golden-red; head (including beard), shoulders, limbs, and proximal part of tail deep maroon. Lower flanks with a maroon line, due to lengthened hairs. Tail becomes paler along its length, tip ending up nearly as light as body.

According to Watanabe (1982), this is a relatively small form; in males from northern Colombia,

prosthion-inion length is 114.5 mm, SD 4.39; facial length 99.3 mm, SD 5.82 (both, $n = 5$); basihyal length, 64.72 mm, SD 9.16 ($n = 7$).

DISTRIBUTION. This is represented by specimens from northern Colombia and northwestern Venezuela (overall about 7–11°30′N, 71–75°53′W).

NOTES. The types of auratus and laniger, in the BM(NH) collection, seem to resemble A. macconnelli rather than A. seniculus; it is probably best to keep the names incertae sedis.

Alouatta seniculus arctoidea Cabrera, 1940

1805 Simia ursinus Humboldt, 1805. (Text.) Not of Kerr, 1792 (a baboon).
1940 Alouatta seniculus arctoidea Cabrera. To replace ursinus Humboldt, 1805. Venezuela: Caracas (designated by Allen [1916]); further restricted by Cabrera (1957) to valley of Aragua.

DIAGNOSIS. Color is blackish maroon, with no saddle.

DISTRIBUTION. A coastal form, mapped along the Venezuelan coast from about 66 to 71°W, in the Rupununi savanna and Paramaraibo.

Alouatta seniculus juara Elliot, 1910

1910 Alouatta juara Elliot. Brazil: Rio Juruá, Amazonas.
1941 Alouatta seniculus puruensis Lönnberg. Brazil: district around Rio Purus.
1941 Alouatta seniculus juruana Lönnberg. Brazil: upper Rio Juruá.
1941 Alouatta seniculus amazonica Lönnberg. Brazil: Codajáz, north of Rio Solimões.

DIAGNOSIS. Somewhat less contrasty: body is deeper, golden brown, with sometimes a golden splotch in middle of back; rest is brown-maroon; limbs, head, and tail can be nearly black.

Larger than A. s. seniculus. In males, according to Watanabe (1982), prosthion-inion length 122.0 mm, SD 6.43, facial length 105.5 mm, SD 5.00 (both, $n = 22$); basihyal length 68.91 mm, SD 4.99 ($n = 6$). Bonvicino et al. (1995) gave skull length for males (number unstated) as 119.5 mm, SD 10.6. These two figures are very similar. Watanabe's sample was from southern Colombia, southern Venezuela, and northern Peru; Bonvicino et al.'s (mainly or entirely?) from Acre and Amazonas States, Brazil.

DISTRIBUTION. The type localities of the synonyms assigned to this subspecies are all in Brazil; I have seen specimens from Auaty Paraná on the Rio Solimões, and specimens I have seen from Ecuador, northern Peru, Colombia, and Amazonas Province of Venezuela, about 8°17′N to 1°44′S, 67°07′ to 78°19′W, most likely belong to it as well. Although the difference between this subspecies and *A. s. seniculus* is quite clear overall, there is much overlap; for example, one from Rio Tigre, Oriente Province, Ecuador, falls well within the range of northern Colombian skins, but four skins from Bolívar and one from Yacua, Venezuela, are more like southerly examples.

Alouatta macconnelli Elliot, 1910
Guyanan Red Howler

1910 *Alouatta macconnelli* Elliot. Guyana: coast region.
1910 *Alouatta insulanus* Elliot. Trinidad: Quinam, south coast.

DIAGNOSIS. Back yellow-brown, or this tone restricted to a golden "saddle," if that; anterior part of back, hindquarters, head, and limbs very dark rufous brown often with a blackish tone; a dark dorsal stripe goes through the saddle; tail rapidly becomes paler distally. Skull measurements given by Bonvicino et al. (1995) indicate a very large species; prosthion-opisthocranion (here, = inion) length of males 131.7 mm, SD 6.5 (number not stated, but apparently substantial). Hyoid of male with short cornua; its mouth smaller than in *A. seniculus,* wider than high. Diploid chromosome number 47, 48, or 49.

DISTRIBUTION. I have seen skins only from Guyana, and the type of *insulanus* from Trinidad, but Vassart et al. (1996) recorded individuals with a similar sex-chromosome arrangement from French Guiana. The taxon extends into Brazil, north of the lower and middle Amazon (Lima et al. 1990), but not apparently south of it: Lima and Seuánez (1989) noted that six howlers "with copper-brown pelage" from Tucuruí Dam, resembling this species superficially, are actually *A. belzebul.*

NOTES. There is a X_1X_2Y sex chromosome system, with only one of the Y chromosomes being translocated onto an autosome, unlike *A. seniculus.* The chromosomes are similar to those of *A. belzebul* except for an extra number of nuclear organizing regions (NORs) (Lima and Seuánez 1989).

Rylands and Brandon-Jones (1998) noted that the holotype of *macconnelli* is a rather uniform orange color, with the merest trace of a dorsal stripe, and does not match other Guyanan specimens in the BM(NH); nor indeed does it resemble those I have seen from Guyana in the USNM. Is it, they asked, representative of a very localized coastal population? As they noted, the types of *Mycetes auratus* and *M. laniger* are not unlike Guyanan howlers.

A specimen in the BM(NH) from Suriname, described by Rylands and Brandon-Jones (1998), is "uniform dark titian."

Specimens from the lower Amazon and the coast just to the north (reported as *Alouatta seniculus macconnelli*) differ from those from the middle Amazon and Guyana (designated *A. seniculus stramineus*) in a number of important chromosome rearrangements (Lima and Seuánez 1989). The latter are also larger (male skull length 127.5 mm, SD 7.0), and the two can be distinguished by discriminant analysis of skull measurements (Bonvicino et al. 1995). These populations require investigating. Sampaio et al. (1996) found that genetic distances between three populations along the north bank of the Amazon, from the Rio Uatumã east to the estuary, are very low.

Alouatta sara Elliot, 1910
Bolivian Red Howler

1910 *Alouatta sara* Elliot. Bolivia: Sara Province, 450 m.

DIAGNOSIS. Color brick red; limbs, head, and proximal part of tail only slightly darker, more rufous. Much larger than *A. seniculus;* in males, prosthion-inion length 133.7 mm, SD 1.15, facial length 117.5 mm, SD 2.65 (both, $n = 3$); but hyoid smaller, basihyal length (in a single male) only 62 mm (Watanabe 1982). Chromosomes: $2n = 50$.

DISTRIBUTION. I have seen a series from Sara; Cerro Azul, Loreto, and Rio Comberciato, Urubamba, Peru; and, in Brazil, the Lago de Arara, Solimões; Canabouca, Paraná do Jacari; Rio Negro; Codajáz; Rio Juruá; Rio Comberciato; and Cachoeiro do Bananeira, Rondônia.

Alouatta belzebul (Linnaeus, 1766)
Red-handed Howler

1766 *Simia belzebul* Linnaeus. Brazil; restricted by Cabrera (1957) to Rio Capim, Pará.
1800 *Simia beelzebub* Bechstein.
1820 *Mycetes rufimanus* Kuhl. Locality unknown.
1823 *Mycetes discolor* Spix. Brazil: Gurupá, Pará.
?1845 *Mycetes bicolor* Gray. No locality.
1863 *Mycetes flavimanus* Bates. Error for *rufimanus*.
1908 *Mycetes belzebul mexianae* Hagmann. Brazil: Mexiana Island.
1912 *Alouatta ululata* Elliot. Brazil: Miritiba, Maranhao.
1941 *Alouatta belzebul tapajozensis* Lönnberg. Brazil: Aveiros, Amazonas (restricted by Bonvicino et al. 1989).

DIAGNOSIS. Pelage smooth, thin, black, with reddish hands, feet, and tail tip, occasionally spreading to entire body. A short-haired toupee on brows meets a forward-directed stream (from a whorl between nape and withers) in a low transverse crest on front of crown; this forward stream diverges to the sides, to run laterally. Palate wide, shallow, of even depth anteriorly but suddenly deepening at M^2 level. Nasals straight. Incisive foramina large and often fused to a single heart-shaped foramen. Hyoid of male (Cruz Lima 1945; Hershkovitz 1949) large, inflated, with large mouth and less-developed tentorium than in *A. seniculus* and *A. macconnelli*, without lateral partitions; with rudimentary cornua; in female smaller than in male.

DISTRIBUTION. South of the Amazon, from somewhat west of the Rio Tapajós (Lago do Baptista, on the south bank of the Amazon at 58°11′W [Bonvicino et al. 1989]), eastward to Marajó and Mexiana

Islands, in the Amazon estuary, and along the Atlantic coast to about 40°W; there is an apparently isolated population on the eastern seaboard between 7 and 10°S.

Coimbra-Filho et al. (1993) mapped the range of the species in northeastern Brazil, using place names as well as confirmed localities as evidence for its former distribution. The former boundary with *A. guariba* was probably the Rio São Francisco. Its recent occurrence in remnant forests such as at Angico, Parnaguá, in Piauí State, suggests the presence until quite recent times of a forest continuum between the Amazonian and Atlantic forests in that general region.

NOTES. Lima and Seuánez (1989) found a karyotype of $2n = 50$ (female) and 49 (male) in red-handed howlers from Tucuruí Dam, on both sides of the Rio Tocantins. Six specimens of coppery red color, resembling *A. seniculus* and its allies, were karyotypically identical to normal black-and-red examples.

The species was revised by Bonvicino et al. (1989), who distinguished four subspecies, one of them now regarded as a distinct species, *A. nigerrima*. The others differ in the relative frequency of color variants: apart from red hands, feet, and tail tip, individuals may have red on the crown and nape, on the anterior part of the dorsum, and in patches on the limbs. In the Atlantic coast isolate, only the usual red-handed morph occurs. The red-naped form is frequent (28% according to Bonvicino et al. [1989]) south of the Amazon estuary; individuals with red backs predominate (>50%) to the east and west of that region.

In the Tucuruí Dam region, on the Rio Tocantins, Bonvicino et al. (1989) recorded great variability in tone, including wholly black and wholly red individuals. Because it now seems clear that *A. nigerrima* is specifically distinct and that it too occurs at Tucuruí Dam (Lima and Seuánez 1989), it may be that the all-black specimens recorded there by Bonvicino et al. (1989) are actually all *A. nigerrima;* this should be checked by examination of their skulls.

Wholly red individuals were recorded by Bonvicino et al. (1989) from only two localities, both in the

Tucuruí Dam region. Specimens from that general area had chromosomes identical to those of normally colored specimens and so were not *A. seniculus* group (which they resembled externally [Lima and Seuánez 1989]).

Alouatta nigerrima Lönnberg, 1941
Amazon Black Howler

1941 *Alouatta nigerrima* Lönnberg. Brazil: Patinga, Amazonas (restricted by Cabrera [1957]).

DIAGNOSIS.   Color completely black. The transverse crest, where the backward-directed toupee meets the forward-directed stream from the nape whorl, runs across middle to back of crown unlike in *A. belzebul*. Palate slopes evenly down from post-incisive region to level of M²; mesopterygoid fossa is narrow, as is back of cranium. Nasal concave. Hyoid of male (Cruz Lima 1945) with very small, semicircular opening and large convex tentorium; female (Hershkovitz 1949) resembling *A. seniculus* but with long cornicula as in *A. macconnelli*. Chromosomes: $2n = 50$ as in *A. belzebul*, but the structure was somewhat different (Lima and Seuánez 1989).

DISTRIBUTION.   From their survey of the literature, Rylands and Brandon-Jones (1998) located this species in the region east of the Rio Trombetas and in a presumably isolated population within the range of *A. macconnelli* at Oriximiná, at the mouth of the Rio Trombetas, and west of the Rio Madeira at Lago Janauacá. It extends east to the Rio Tapajós; but if the identifications of Lima and Seuánez (1989) are correct, and from their description this seems to be likely, then this species extends east as far as the Rio Tocantins and is sympatric with *A. belzebul* at Tucuruí Dam (4°S, 49°W).

NOTES.   The skull characters differentiating this from *A. belzebul* are, in specimens I have seen in the USNM, exactly as described by Cruz Lima (1945). There are usually two separate incisive foramina.

For a given toothrow length, the palate is narrower than in *A. belzebul*; it is deeper in females than

in female *A. belzebul* (except in the region of M³), but there is no difference in males. The similarity of the skull to those of *A. seniculus, macconnelli,* and *sara* supports the suggestion of Rylands and Brandon-Jones (1998) that this species is more closely related to those three species than to *A. belzebul*, and their observation that the blackened tones of *A. macconnelli* form an intermediate stage between *A. seniculus* and *A. nigerrima* (Peres et al. 1996).

Alouatta guariba (Humboldt, 1812)
Red-and-Black Howler

DIAGNOSIS.   Pelage black or dark brown to reddish, generally with light tips, giving a marked brindled effect. A triangular or transverse crest toward front of crown where forward-pointing crown hairs meet short backward-pointing forehead hairs. Palate is the flat type (as in *A. nigerrima*). Hyoid (Hershkovitz 1949) as in *A. belzebul* and *A. nigerrima*, but lacking cornua.

DISTRIBUTION.   The northern boundary of this species, separating it from *A. belzebul*, may be the Rio São Francisco (Coimbra-Filho et al. 1993).

NOTES.   This species has traditionally been divided into subspecies, a northern one *(A. g. guariba)* with the sexes the same color, and a southern one *(A. g. clamitans)* with the male lighter than the female. Even limited data indicate that the position is more complex than this. Those in the center of the distribution are dark in both sexes, but with long, pale tips giving a conspicuous brindle; to the north the males, at least, are red-fawn, and to the south both sexes, but particularly the males, are bright red. Chromosome data add a further dimension: Oliveira et al. (1995) found differences between members of this species from Espírito Santo ($2n = 52$), Rio de Janeiro (49), Santa Catarina (46), and Paraná (45); Lima et al. (1997) found both $2n = 46$ and $2n = 49/50$ karyotypes in Rio de Janeiro State. Study of karyotypes from intermediate areas is desirable to determine whether the changes are gradual or discrete.

Alouatta guariba guariba (Humboldt, 1812)

1812 *Simia guariba* Humboldt. Brazil: restricted by Cabrera
(1957) to the Rio Paraguaçu, Bahia. Hershkovitz (1963)
regarded this name as preoccupied by *guariba* E. Geoffroy,
1812 (= *A. belzebul*), but Rylands and Brandon-Jones (1998)
showed that this latter was not a binomial. The name
Simia guariba predates *Stentor fusca* by 2 months.
1812 *Stentor fusca* E. Geoffroy. Based on the same specimen as
Simia guariba.
1812 *Simia ursina* Humboldt (plate only). The text (based on
Humboldt [1805]) describes *A. seniculus arctoidea.*
1845 *Mycetes bicolor* Gray. No locality.
?1941 *Alouatta beniensis* Lönnberg. Bolivia: Puerto Salinas, Beni.

NOTES. In this subspecies, known from a small
area in northeastern Minas Gerais and neighboring
parts of Bahia and Espírito Santo, both sexes are described as being a brownish color, the female duller
than the male. I have seen no female specimens, only
two males (from Linhares and San Domingos, Espírito Santo), both of which are red-fawn. Chromosomes: $2n = 52$.

Mittermeier et al. (1988) believed that *beniensis*
may actually be *A. seniculus.* I have not seen the
type specimen.

Alouatta guariba clamitans Cabrera, 1940

1940 *Alouatta guariba clamitans* Cabrera. Brazil: Serra do
Cantareira, São Paulo (restricted by Cabrera [1957]).
1941 *Alouatta fusca iheringi* Lönnberg. Southern Brazil.

NOTES. Males tend to be lighter than females, but
the variations seem complex. Three females from
São Sebastião (São Paulo) and Roça Nova (Paraná)
are very dark brown or black, with a dark red-brown
brindle especially on the tail and extremities and
sometimes on the head; four males vary from dark
brown-red to again black with reddish brindle. Chromosomes: $2n = 45$ or 46.

Specimens from Enghiniere Reeve (Rio de Janeiro) are lighter: two females are dark red-brown
with light brindling, and four males are lighter red-
brown with much straw-colored brindling; the diploid chromosome number was given as 49 by Ol-
iveira et al. (1995), but Lima et al. (1997) found $2n = 46$ as well as $2n = 49/50$. Descriptions of the external
phenotypes of the studied individuals would be most

useful to determine whether these two karyotypes
are ends of a spectrum, in which diploid numbers 47
and 48 have by chance simply not been discovered
as yet, or whether they on the contrary associate
with phenotypic differences such as those described
above, and so perhaps of taxonomic significance.

Four males from farther south (Colonia Hansa
and Ibirama, both in Santa Catarina, and Santa Rita
in Rio Grande do Sul) are strongly red, from golden-
red to bright brick red with no brindling; a single female from Santa Rita is red-brown, much blackened
on the foreparts. Presumably these represent an undescribed subspecies, although more specimens (or,
better, field observations) are desirable before it can
be defined. The chromosomes seem to be the same
as those farther north.

Alouatta caraya group

Hyoid of male (Hershkovitz 1949) smaller than in
A. seniculus group and more angular, less inflated;
mouth large, occupying over half of posterior face;
tentorium represented only by a large bony plate and
without lateral bony walls; rudimentary cornua; cornicula reduced or absent.

Alouatta caraya (Humboldt, 1812)
Brown Howler

1812 *Simia caraya* Humboldt. Paraguay.
1812 *Simia straminea* Humboldt. "Grand Para." For the correct
identification of this name, see Rylands and Brandon-Jones
(1998).
1812 *Stentor niger* E. Geoffroy in Humboldt. "Brazil."
1823 *Mycetes barbatus* Spix. Brazil: between Rio Negro and Rio
Solimões.

DIAGNOSIS. Pelage black in male, with light hairs
at most on scrotum, throat, and midline of underside
and perhaps some yellow tips of feet and tail; yellowish or brindled tawny in female. Hair grows entirely
forward, to overhang brows, on crown (from a whorl
on nape), except that there is a small upright fringe
behind brows, which meets the forward hairs in a
small crest. The palate is flat as in *A. nigerrima* and
A. guariba. Chromosomes: $2n = 52$ (Lima et al. 1997;
Mudry et al. 1998).

DISTRIBUTION. Paraguay and northern Argentina, and probably extends into Uruguay (Villalba et al. 1995).

NOTES. In the adult female the hairs are black at base with fawn tips; the tips are longer on flanks, crown, limbs, and tail. The young resemble the adult female; in the male, the light hairs or tips disappear with age.

Rylands and Brandon-Jones (1998) discussed the identity of Humboldt's *Simia straminea* (taken from a manuscript name of E. Geoffroy-Saint-Hilaire, later in the same year published by him as *Stentor stramineus*). The type specimen is a female *A. caraya,* and the supposed type locality ("Grand Para" = Pará State, Brazil) may derive from an unwarranted identification of the new species with howlers seen and briefly described by the explorer Gumilla in the Orinoco region of Venezuela (former use of "Grand Para" meant the Amazon Basin in general).

SUBFAMILY ATELINAE GRAY, 1825

Lagotrichina Gray, 1870, and Brachytelini Goodman et al., 1998, are synonyms.

Dutrillaux et al. (1986), on chromosome banding evidence, supported a *Lagothrix/Ateles* clade, with *Brachyteles* diverging earlier. On the other hand the molecular evidence of Schneider et al. (1996), Sampaio et al. (1997), and Goodman et al. (1998) favors a sister-group relationship of *Lagothrix* and *Brachyteles,* with *Ateles* more distantly related. Bootstrap support for this is 92% (Schneider et al. 1996); the separation of the three genera took place, on molecular clock calculations, between 13 and 9 million years ago (Sampaio et al. 1997) or fully 13 million (Goodman et al. 1998).

Genus *Ateles* E. Geoffroy, 1806
   Spider Monkeys

1799 *Sapajou* Lacépède.
1806 *Ateles* E. Geoffroy. *Simia paniscus* Linnaeus, 1758.
1815 *Paniscus* Rafinesque. *Simia paniscus* Linnaeus, 1758.
1911 *Montaneia* Ameghino.
1929 *Ameranthropoides* Montandon. *Ameranthropoides loysi* Montandon, 1929.

Compared with other genera, frontals are very high and rounded; vault is arched; supraorbital ridges are very small, merely thickenings of upper orbital rims; superior temporal lines are very far apart, even in the male; bregmatic bone is very frequent; jaw angle is comparatively little enlarged; the anterior margin of ascending ramus slopes strongly back, converging upward with the posterior margin; the coronoid process is thin and curved back; the zygomatic arch is deep; teeth are very small; toothrows diverge posteriorly. The alveolar margin is parallel to the Frankfurt Plane, or slightly divergent in front.

Froehlich et al. (1991), in a study of 50 cranial and dental characters, divided the genus into three groups:

1. *A. paniscus;*
2. *A. hybridus, geoffroyi, fusciceps,* and *robustus* (= *rufiventris*); and
3. *A. belzebuth, chamek,* and *marginatus.*

These they suggested are species. Within two of the groups they detected hybrid populations: in group 2 between *fusciceps* and "*robustus*" and in group 3 between *belzebuth* and *chamek* and between *chamek* and *marginatus.* The two hybrid cases in group 3, however, are unusual: only the skulls are intermediate; the colors are typical of *belzebuth* in the one case and of *chamek* in the other.

These and other findings of their study led Froehlich et al. (1991) to conclude that rivers are effective barriers between taxa in *Ateles,* to the extent that pelage differences are often sharper at taxon boundaries than between geographically remote populations of the same taxa. The implication is that assortative mating is restricting gene flow in two of the three cases of hybrid populations.

Medeiros et al. (1997) divided the taxa in this genus into four karyomorphs:

1. *A. geoffroyi* and *hybridus;*
2. *A. rufiventris* (the chromosomes of possibly conspecific *A. fusciceps* are unknown);
3. *A. belzebuth, chamek,* and *marginatus;* and
4. *A. paniscus.*

Of these, *A. paniscus* is the most distinct; it even has a different chromosome number from other species, with $2n = 32$; the others have 34. *Ateles rufiventris* differs from its neighbors, *A. geoffroyi* and *hybridus*, in the morphology of chromosome pairs 5 and 6 and is probably reproductively isolated from them (Medeiros et al. 1997).

Given all these findings, and the sharpness of the differences between most geographically adjacent taxa, I suggest that the most reasonable solution, and the one most consistent with the phylogenetic species concept (and even, at a pinch, with the biological species concept), is to recognize all taxa from east of the Andes as distinct species. The evidence that gene flow is restricted between *chamek* and *marginatus* on the one hand and *belzebuth* on the other (meaning the offset between cranial and pelage data reported by Froehlich et al. [1991]) strongly implies that these three are specifically distinct from one another even though they interbreed.

Ateles paniscus (Linnaeus, 1758)

Red-faced Spider Monkey

1758 *Simia paniscus* Linnaeus. "South America: Brazil": restricted by E. Geoffroy in 1803 to French Guiana, as recounted by Hershkovitz (1958), who noted that the original concept was a composite of this species and a howler monkey, but Linnaeus (1766) dropped the reference to the howler and to Brazil, so restricting the name to a non-Brazilian spider monkey.
1806 *Ateles pentadactylus* E. Geoffroy. No locality.
1820 *Ateles subpentadactylus* Desmarest. No locality.
1823 *Ateles ater* F. Cuvier. French Guiana.
1829 *Cebus paniscus surinamensis* Fischer. Locality not stated, but presumably Suriname.
1829 *Cebus paniscus cayennensis* Fischer. French Guiana.

DIAGNOSIS. Long, silky, black pelage, with no white markings. Skin of face red or pink in adult with at most a very few short, white or silvery hairs on muzzle.

DISTRIBUTION. The Guyanas and northeastern Brazil.

Ateles belzebuth E. Geoffroy, 1806

White-fronted Spider Monkey

1806 *Ateles belzebuth* E. Geoffroy. Locality unknown; restricted by Kellogg and Goldman (1944) to Venezuela: Esmeralda, west of mouth of Rio Guapo, Rio Orinoco, and south of Mount Duida.
1820 *Ateles fuliginosus* Kuhl. Locality unknown.

1829 *Cebus brissonii* Fischer. Venezuela: Rio Orinoco.
1840 *Ateles variegatus* Wagner. Brazil: Serra do Cucuí, upper Rio Negro, Amazonas.
1867 *Ateles bartlettii* Gray. Peru: Jeberos, Upper Rio Amazon.
1876 *Ateles chuva* Schlegel. No locality.
1883 *Ateles braccatus* Pelzeln. No locality.

DIAGNOSIS. Pelage black, with white or golden triangular patch on forehead ("widow's peak" of Froehlich et al. [1991]); underside (including inner side of limbs, backs of thighs, and underside of tail) contrastingly white or yellowish.

DISTRIBUTION. The border between this species and *A. paniscus* is the Rio Branco / Rio Napo system (Froehlich et al. 1991).

NOTES. Skulls from the southwest of the range, between the Rio Napo and Rio Marañon, were rated by Froehlich et al. (1991) as being intermediate between this species and *A. chamek*. Yet 17 of the 18 available skins had very conspicuous forehead patches, only one as inconspicuous as in most *A. chamek;* this despite the fact that away from the border zone the patch in *A. belzebuth* rates as "conspicuous" or "inconspicuous" about equally. Froehlich et al. (1991) interpreted this as a sharpening of the species recognition characters at the border, implying in turn that gene flow between the two species is restricted.

Ateles chamek (Humboldt, 1812)

Peruvian Spider Monkey

1812 *Simia chamek* Humboldt. Peru; restricted by Kellogg and Goldman (1944) to Rio Combercinto, Cuzco.
1914 *Ateles longimembris* Allen. Brazil: Barrão de Melgaço, headwaters of Rio Gy-Paraná.
1940 *Ateles ater peruvianus* Lönnberg. Eastern Peru.

DIAGNOSIS. Pelage black, except for a silvery genital patch (from which a few gray or gray-banded hairs may extend a short distance down inner surface of thigh), and sometimes a few white hairs on muzzle, cheeks, and forehead. Skin of face black.

DISTRIBUTION. From Peru east to the Rio Tapajós, Brazil.

NOTES. Heltne and Kunkel (1975) noted the Bartlett (1871) description of two forms of spider monkey near Chamarros, Peru, implying a zone of sympatry or parapatry between this species and *A. belzebuth*. The apparently restricted hybrid zone between them north of the Rio Marañon has already been described, under *A. belzebuth*.

Between this species and *A. marginatus* there is likewise a zone of apparent interbreeding, on the upper left bank of the Rio Tapajós. Because the animals here are typical *A. marginatus* in pelage characters, the case is discussed under that species.

Ateles hybridus I. Geoffroy, 1829
Brown Spider Monkey

1829 *Ateles hybridus* I. Geoffroy. Colombia: valley of Rio Magdalena, restricted by Kellogg and Goldman (1944) to La Gloria.
1870 *Ateles albifrons* Gray. Locality unknown.
1870 *Ateles belzebuth brunneus* Gray. No locality.
1929 *Ameranthropoides loysi* Montandon.

DIAGNOSIS. Pelage wood brown above, whitish or buffy below (including inner side of limbs and ventral surface of tail), with white triangular patch on forehead.

DISTRIBUTION. Northern Colombia and northwestern Venezuela.

NOTES. Froehlich et al.'s (1991) sample of this species included specimens from the border area with *A. fusciceps rufiventris* on the lower Rio Cauca; they detected no evidence of gene flow. Collins and Dubach have analyzed both mitochondrial and nuclear DNA and confirm the status of *hybridus* as a full species (Andrew C. Collins, personal communication).

The border with *A. belzebuth* could not be defined in detail by Froehlich et al. (1991); they estimated it to be somewhere north of Serrania de la Macarena, Colombia.

Ateles marginatus E. Geoffroy, 1809
White-cheeked Spider Monkey

1809 *Ateles marginatus* E. Geoffroy. Brazil: Rio Janeiro, restricted by Kellogg and Goldman (1944) to Cametá, left bank of Rio Tocantins, Pará.
1831 *Ateles frontalis* Bennett. No locality.

DIAGNOSIS. Pelage entirely black, except for (almost always) a white triangular patch on forehead and white cheek whiskers.

DISTRIBUTION. South of the Amazon, in eastern Brazil. The boundary between this species and *A. paniscus* is the Amazon (Froehlich et al. 1991).

NOTES. Froehlich et al. (1991) described a zone of apparent interbreeding between this species and *A. chamek,* on the east bank of the upper Rio Tapajós. Here the skulls are intermediate, but the skins are not; indeed, the distinctive semilunar frontal patch of *A. marginatus* is invariably present here, whereas it is occasionally absent in the Xingu region.

Ateles fusciceps Gray, 1866
Black-headed Spider Monkey

DIAGNOSIS. Pelage coarse, black, with sparse white hairs on lips and chin. Dark brown cheek hairs meet the crown hairs in a small eye-to-ear crest.

Ateles fusciceps fusciceps Gray, 1866

1866 *Ateles fusciceps* Gray. South America; restricted by Kellogg and Goldman (1944) to Ecuador: Hacienda Chinipamba, near Peñaherrera, Imbabura Province, 1,500 m.

DIAGNOSIS. Dorsum brownish black; anterior crown yellow-brown, quickly grading through brown to black on the nape.

Ateles fusciceps rufiventris Sclater, 1872

1872 *Ateles rufiventris* Sclater. Colombia: Rio Atrato.
1914 *Ateles robustus* Allen. Colombia: Gallera, Cauca, western Andes.
1915 *Ateles dariensis* Goldman. Panama: headwaters of Rio Limón, Mount Pirre, 1,565 m.

DIAGNOSIS. Skin of face black; dorsum glossy black; whitish or golden hairs on muzzle and interramal region and a few white or golden hairs on cheeks; crown black; may be a slight brownish tinge on forehead; a genital patch of red-banded hairs extending to inner aspect of thigh.

DISTRIBUTION. The documented western limit of this taxon, according to Heltne and Kunkel (1975), is the Rio Majecito, 9°09′N, 78°49′W, about 95 km east of the Panama Canal. There is therefore an approximately 20-mile-wide (32 km) zone between its known range and that of *A. geoffroyi ornatus* (called *A. g. panamensis* by Heltne and Kunkel [1975]). The border with *A. g. grisescens* may be Cerro Pirre or the Rio Tucutí, according to the same authors.

The eastern limit is the lower Rio Cauca. Here all individuals lack a brown frontal patch, even though it occurs commonly in Panamanian specimens, as described by Froehlich et al. (1991); this, they noted, intensifies the contrast with parapatric *A. hybridus,* which always has such a patch.

The transition between this subspecies and *A. f. fusciceps* is across the upper Rio Cauca; the sample from this region was rated as fully intermediate (Froehlich et al. 1991), and this is taken here to imply a genetic continuum between them.

NOTES. The name *robustus* has been commonly used for this subspecies, but as shown by Heltne and Kunkel (1975) the name *rufiventris* is a senior synonym.

Ateles geoffroyi Kuhl, 1820
Geoffroy's Spider Monkey

DIAGNOSIS. Body yellowish, reddish to blackish brown, contrastingly black on head and at least part of limbs. The forward-directed hairs from the nape whorl form a cowl, ending in a triangular crest above the brows; the cheek hairs stand out prominently.

DISTRIBUTION. From about 24°N in Tamaulipas, Mexico, through Central America to Panama.

Ateles geoffroyi yucatanensis Kellogg and Goldman, 1944

1944 *Ateles geoffroyi yucatanensis* Kellogg and Goldman. Mexico: Puerto Morelos, northeastern coast of Quintana Roo, 30 m.

DIAGNOSIS. Body light brown above, silvery white below; blackish brown on head, arms, knees, and feet.

Ateles geoffroyi vellerosus Gray, 1866

1866 *Ateles vellerosus* Gray. ?Brazil; restricted by Kellogg and Goldman (1944) to Mexico: Mirador, 25 km northeast of Huatusco, Vera Cruz.
1873 *Ateles neglectus* Reinhardt. Mexico: Mirador, Vera Cruz.
1876 *Ateles pan* Schlegel. Guatemala: Cobán, Alto Vera Paz.
1914 *Ateles tricolor* Hollister. Mexico: Hacienda Santa Efigenía, 13 km north of Tapanatepec, southeastern Oaxaca.

DIAGNOSIS. Black or brownish black above, but posterior part of back and flanks light brown; underside and inner side of limbs yellowish buff.

Ateles geoffroyi geoffroyi Kuhl, 1820

1820 *Ateles geoffroyi* Kuhl. Locality unknown; restricted by Kellogg and Goldman (1944) to Nicaragua: San Juan del Norte.
1820 *Ateles melanochir* Desmarest. Locality unknown.
1842 *Eriodes frontatus* Gray. Costa Rica: Culebra, Guanacaste.
1862 *Ateles belzebuth trianguligera* Weinland. Locality unknown.

DIAGNOSIS. Body yellowish buff to pale reddish brown; black markings restricted to forehead, hands, and feet and patches on knees and elbows.

Ateles geoffroyi ornatus Gray, 1870

1870 *Ateles ornatus* Gray. Locality unknown.
1937 *Ateles azuerensis* Bole. Panama: Altos Negritos, 16 km east of Montijo Bay, Veraguas Province.
1944 *Ateles geoffroyi panamensis* Kellogg and Goldman. Panama: Cerro Brujo, 610 m, 24 km southeast of Portobello, Colón.

DIAGNOSIS. Red-brown to yellow-brown above and below; black markings more extensive, on head, arms, outer side of thighs, feet, and distal part of tail.

DISTRIBUTION. Heltne and Kunkel (1975) traced the eastern limit of this taxon as San Juan, Cerro Brujo, Cerro Azul, and the Rio Pequeñi region, about 50 km east of the Panama Canal on Modden Lake watershed.

NOTES. Napier (1976) showed that *A. g. panamensis* is a synonym of *A. g. ornatus.*

Ateles geoffroyi grisescens Gray, 1866

1866 *Ateles grisescens* Gray. Locality unknown; restricted by Kellogg and Goldman (1944) to Panama: Rio Tuira.
1866 *Ateles cucullatus* Gray. Locality unknown.

DIAGNOSIS. Black, with variable sprinkling of gray and yellow hairs, giving a variegated effect. Hairs are very long, mostly gold at base, brown at tips, but some are entirely gold.

Genus *Brachyteles* Spix, 1823
Woolly Spider Monkeys or Muriquis

1823 *Brachyteles* Spix. *Brachyteles macrotarsus* Spix, 1823.
1829 *Eriodes* I. Geoffroy. *Eriodes tuberifer* I. Geoffroy, 1829.

Frontals are fairly rounded; the vault is flattened; supraorbital ridges are fairly small, with (unlike *Ateles*) no ophryonic groove; superior temporal lines are very far apart, even in male; there is no bregmatic bone; the jaw angle is very enlarged; anterior and posterior margins of the ascending ramus are parallel; the coronoid process is substantial, fairly backcurved; the zygomatic arch is not deepened; teeth are large; toothrows are parallel. The alveolar margin swings up in front (airorhynch).

It seems fairly clear that there are actually two species of this genus, suggested to be subspecies by Lemos de Sá et al. (1990, 1993), but promoted to species rank by Rylands et al. (1995). The boundary may be the Rio Grande, the Rio Paraiba do Sul, or the Serra da Mantiqueira (Lemos de Sá et al. 1993).

Brachyteles arachnoides (E. Geoffroy, 1806)
Southern Muriqui

1806 *Ateles arachnoides* E. Geoffroy. Brazil; restricted by Vieira (1944) to Rio de Janeiro.
1823 *Brachyteles macrotarsus* Spix, 1823. No locality.
1829 *Eriodes tuberifer* I. Geoffroy. No locality.
1876 *Ateles eriodes* Brehm. Replacement for *arachnoides*.

DIAGNOSIS. Thumb entirely missing. Face and scrotum black. Canines of males considerably longer than those of females: crown height of upper canines in males 7.9–12.5 mm, in females 4.0–7.7 mm (Lemos de Sá et al. 1993; Leigh and Jungers 1994).

DISTRIBUTION. Southern Brazil (states of Rio de Janeiro and São Paulo).

Brachyteles hypoxanthus (Kuhl, 1820)
Northern Muriqui

1820 *Ateles hypoxanthus* Kuhl. Brazil: Bahia.
1829 *Eriodes hemidactylus* I. Geoffroy. "Brazil."

DIAGNOSIS. Thumb present, though small. Face and scrotum mottled pink and black in adult (though black at birth). Canines not markedly sexually dimorphic, upper canines being about 4.6 to 9.8 mm crown height in both sexes (Lemos de Sá et al. 1993).

DISTRIBUTION. Northern Brazil (states of Bahia, Minas Gerais, and Espírito Santo).

NOTES. The striking differences between these two species were first picked up by Lemos de Sá et al. (1990) during a comparative survey of two sites: Fazenda Esmeralda, Minas Gerais (in the northern region) and Fazenda Barreiro Rico, São Paulo (in the southern). The study was extended to the whole range of the species by Lemos de Sá et al. (1993), who in addition related the presence/absence of canine sexual dimorphism to a reported difference in mating strategies. Leigh and Jungers (1994) took issue with the conclusion of a difference in sexual dimorphism, finding that although unprotected analysis of variance tests indicated a significant difference (for upper canines only), using protected probabilities failed to record a significant difference. This may be so, but it remains true that samples for males and females of the southern population do not overlap in upper canine height.

Genus *Lagothrix* E. Geoffroy, 1812
Woolly Monkeys

1812 *Lagothrix* E. Geoffroy. *Lagothrix humboldtii* E. Geoffroy, 1812.
1823 *Gastrimargus* Spix. *Gastrimargus infumatus* Spix, 1823.

Frontals are somewhat rounded; the vault is flattened; supraorbital ridges are fairly small, with no ophryonic groove; superior temporal lines are very far apart, even in male; there is no bregmatic bone; the jaw angle is very enlarged; anterior and posterior margins of the ascending ramus are parallel; the coronoid process is large and upright; the zygomatic arch is not deep; teeth are large; toothrows are parallel. The alveolar margin is parallel to the Frankfurt Plane or slightly divergent in front.

According to the molecular clock calculations of Goodman et al. (1998), this genus separated from *Lagothrix* 11 million years ago, and the two together separated from *Ateles* 13 million years ago.

I have examined skins of this genus in the AMNH and also in the USNM and MCZ. Fooden (1963) recognized only one species (though he also included what I here call *Oreonax flavicauda* in the genus as a second species) and divided it into four subspecies. My examination confirmed the accuracy of Fooden's arrangement, but the four taxa are sharply distinct, with no intermediates between them, and they are here raised to specific rank, in apparent agreement with Mittermeier et al. (1988). In addition some of them show geographic variation, as indeed noted by Fooden.

Lagothrix lagotricha (Humboldt, 1812)
Brown Woolly Monkey

1812 *Simia lagothricha* [sic] Humboldt, 1812. Colombia: Rio Guaviare, above mouth of Rio Amanaveni, Uaupés.
1812 *Lagothrix humboldtii* E. Geoffroy. Replacement for *lagothricha* Humboldt, 1812.
1823 *Gastrimargus infumatus* Spix. Apparently based on same specimen as *Simia lagothricha* Humboldt, 1812.
1840 *Lagothrix caparro* Lesson. Replacement for *lagothricha* Humboldt, 1812.
1857 *Lagothrix geoffroyi* Pucheran. Supposedly from Cayenne, where woolly monkeys do not occur.
1883 *Macaco barrigo* Natterer. Brazil: Rio Uaupés.
1931 *Lagothrix caroarensis* Lönnberg. Brazil: lower Rio Içá, near Lago di Caroaro.

DIAGNOSIS. Brown in color with ticking only on loins and tail or none at all; head slightly paler; withers paler than shoulders; hands and feet grayer, darker; midline of underside jet black.

DISTRIBUTION. Brazil, probably north of the Rio Napo-Amazon system, and recorded by Hernández-Camacho and Cooper (1976) in southeastern Colombia (as far north as the northern bank of the Rio Gua-

viare and extreme northern Peru and northeastern Ecuador.

NOTES. Four skins from the Rio Napo, at the mouths of the Rio Curaray and Rio Lagarto Cocha, are darker, warmer brown than those from farther east, but have all the characters of the species. On the face of it *L. lagotricha* seems to be sympatric at these localities with *L. poeppigii*. Fooden (1963) mapped the two species on opposite sides of the Rio Napo, and this may be correct, but there is no precise evidence. The Rio Napo specimens are typical of their respective species.

Lagothrix cana (E. Geoffroy, 1812)
Gray Woolly Monkey

DIAGNOSIS. Grayer than *L. lagotricha*, and the pelage is ticked all over the upper side; head contrastingly dark gray; hands, feet, and tail darker, browner (on tail, especially toward tip); underside broadly blackish gray, often with a dark reddish tinge.

DISTRIBUTION. Brazil south of the Amazon, into the southern highlands of Peru, in parts of which it may be narrowly sympatric with *L. poeppigii*. Peres (1993) observed both this form and *L. poeppigii* at different sites on both sides of the Rio Juruá.

Lagothrix cana cana (E. Geoffroy, 1812)

1812 *Simia cana* E. Geoffroy. Brazil; restricted by Fooden (1963) to the south bank of Rio Solimões near mouth of Rio Tefé.
1823 *Gastrimargus olivaceus* Spix. South bank of Rio Solimões near mouth of Rio Tefé.
1909 *Lagothrix ubericola* Elliot. Brazil: Rio Juruá, Amazonas.
1940 *Lagothrix puruensis* Lönnberg. Brazil: Redemçao.

NOTES. The lowland specimens are pale fawn-gray with darker hands, feet, and tail; but I have seen almost identical specimens from Santa Rosa, on the upper Ucayali, and from Monte Alegrea, on the Rio Pachitea. Both these localities are very close to sites from which *L. poeppigii* has been collected, and Peres's (1993) observation of these two on both sides of the Juruá was quoted above.

Lagothrix cana tschudii Pucheran, 1857

1857 *Lagothrix tschudii* Pucheran. Peru or Bolivia.
1900 *Alouata nigra* J. Allen. Peru: Inca mines.
1909 *Lagothrix thomasi* Elliot. Peru: Cuzco, Rio Comberciato, Callanga, 1,500 m.

DIAGNOSIS. Members of this species from the highlands at the southwestern part of the range are much darker than others, a deep blackish gray, with a tinge of red; the head, limbs, and tail are black.

NOTES. The type locality of *tschudii* is unknown, but according to Fooden (1963:233) it is a dark individual.

Lagothrix lugens Elliot, 1907
Colombian Woolly Monkey

1907 *Lagothrix lugens* Elliot. Colombia: Tolima, upper Rio Magdalena Valley, 2°20′N, 1,500–2,100 m.

DIAGNOSIS. This is, as Fooden (1963) noted, a variable taxon, but the overwhelming majority of specimens that I have seen are ticked iron gray, with somewhat blacker underside; hands, feet, tail, and anterior part of crown vary from slightly darker to black. Some specimens are brownish black all over, with maroon tones, and there is also a lighter, buffy gray morph with black restricted to the ventral midline except that the tail becomes black toward the end.

DISTRIBUTION. This is restricted to a strip of country mainly through the headwaters of the Orinoco tributaries, in Colombia, extending into Venezuela in the Sarare River drainage (Hernández-Camacho and Cooper 1976). Those authors drew attention to two previously unknown isolated populations, perhaps referable to this taxon, in northern Colombia, in the upper San Jorge Valley and in the Serranía de San Lucas.

Lagothrix poeppigii Schinz, 1844
Silvery Woolly Monkey

1844 *Lagothrix poeppigii* Schinz. "Brazil": banks of the Rio Marañon, restricted by Cabrera (1957) to Peru: lower Rio Huallaga, north of Yurimaguas, Loreto.

1848 *Lagothrix castelnaui* I. Geoffroy and Deville. Brazil and Peru; restricted by Fooden (1963) to Peru: Rio Ucayali near Sarayacu.

DIAGNOSIS. This is also a variable species, usually with a silvery sheen on red-based hairs, the degree of silver varying with the length of the gray tips. Head, hands, and feet black; also chest, groin, and a median strip connecting them through the belly, the rest of the belly usually being deep reddish. Some specimens are dark brown, with black tones on the loins, but these too have a silvery sheen. A rare color morph is light yellow-gray, but these too have dark head and extremities.

DISTRIBUTION. Mainly the highlands of eastern Ecuador and northern Peru, but extending east to about 70°W, 5°S in Brazil. Along the Juruá, both this species and *L. cana* occur, at different sites, on both sides of the river (Peres 1993).

NOTES. The most silvery specimens come from the highest altitudes along the western edge of the species's range, including the type localities of both *poeppigii* and *castelnaui*. The more easterly (mainly lowland) populations should probably be separated subspecifically.

Genus *Oreonax* Thomas, 1927
Yellow-tailed Woolly Monkey

1927 *Oreonax* Thomas. *Lagothrix (Oreonax) hendeei* Thomas, 1927.

Among the Atelinae, the genera *Alouatta, Ateles,* and *Brachyteles* can all be satisfactorily diagnosed by apomorphic character states. The same cannot be said for the fourth generally recognized genus, *Lagothrix,* which is commonly said to contain two strongly distinctive species, *L. lagotricha* (Humboldt's woolly monkey) and *L. flavicauda* (yellow-tailed or Hendee's woolly monkey). I have here broken up the first of these two into four species, and I now focus on the second.

I examined what appear to be the only two available skulls of *L. flavicauda* in American collections (AMNH 73222, male, and 73223, female) and three

Table 15

Characters for a Cladistic Analysis of the Atelinae

No.	Character
1.	Superolateral angle of orbit: rounded; squared
2.	Lower margin of malar: evenly rounded; notched
3.	Nasal profile: slightly concave; snub-nosed
4.	Depth of zygomatic process of temporal: very shallow (4 mm or less); deeper (3.5 mm or more)
5.	Length of zygomatic process of malar: restricted (21.5 mm or less); long (21.8 mm or more)
6.	Glabella: flat; prominent
7.	Postorbital constriction, viewed from above: smoothly rounded; angular
8.	Pterionic foramen: tiny or virtually absent; large
9.	Frontomalar sutural crest (in pterionic region): absent, or virtually so; present
10.	Frontal bone at pterion: restricted; extends down to pterionic foramen
11.	Internal nares: high, wide; narrow and angular
12.	Median incisive foramen: absent; present
13.	Foramen lacerum: sphenoid; on spheno-petrous suture
14.	Glenoid fossa: mediolaterally concave; nearly flat
15.	Vomer on floor of mesopterygoid fossa: flattened; raised
16.	Medial pterygoid: free from lateral pterygoid; small, mainly fused to lateral pterygoid
17.	Jaw angle: not enlarged; somewhat enlarged; very enlarged (ordered)
18.	Cheekteeth: small; medium; enlarged (from Zingeser [1973])
19.	Alouattine dentition: no; yes (from Zingeser [1973])
20.	I1 size compared with M1: small; large (from Zingeser [1973])

Note: The primitive state, as shown by the outgroup comparison, is listed first in each case.

(two adult, one juvenile) in London (BM[NH] 27.1.1.1, -2, and -3) and compared them with skulls of *L. lagotricha* ($n = 6$) and of the other ateline genera (*Ateles* and *Alouatta*, $n = 4$ each; *Brachyteles*, $n = 3$) and of *Chiropotes* ($n = 4$) and *Pithecia* ($n = 2$), to serve as outgroups. These were picked at random from the AMNH collection. I added the characters listed by Zingeser (1973), recording their condition in *L. flavicauda* on the assumption that Zingeser's *Lagothrix* sample consisted entirely of *L. lagotricha*. To avoid ascertainment bias, I then compared the other ingroup samples with each other, adding a few more characters, and removed all those characters that showed uniquely derived states (i.e., showed the derived state in only one of the taxa concerned, and not in the outgroups either). Twenty characters remained (Table 15), and the resulting character matrix is given in Table 16. The shortest tree (Figure 4) has length 31, consistency index 0.900, retention index 0.796. It puts *Brachyteles* as sister taxon to the other

Table 16

Character Matrix for Atelinae (position of *Oreonax*)

dimensions ntax = 5 nchar = 20

format missing = ? symbols = "0~3"

Genus	Character
Oreonax	11000011110101111201
Lagothrix	00111100101110001101
Ateles	0111000111000{0,1}110002
Brachyteles	01010010000100112210
Alouatta	{0,1}{0,1}011{0,1}1000101{0,1}012310

atelines; the rest split into two clades, *Lagothrix,* and *Oreonax* plus *Ateles.* When *Lagothrix* is constrained in a clade with *Oreonax,* the length rises to 34, and the consistency index is reduced to 0.706.

But recently some new and extraordinarily complete, well-preserved material of subfossil atelines,

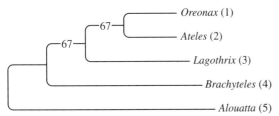

FIGURE 4. 50% Majority-rule consensus of three trees for Atelinae: length 31, consistency index 0.9, retention index 0.796. Numbers at nodes are bootstrap values.

Protopithecus and *Caipora*, has been described (Cartelle and Hartwig 1996; Hartwig and Cartelle 1996). Walter Hartwig has kindly coded the skull characters of these two genera (Table 17). The length of the shortest tree (Figure 5) is 36, consistency index 0.667, retention index 0.538; it too links *Ateles* and *Oreonax*, with *Lagothrix*, *Caipora*, *Protopithecus*, and *Brachyteles* successively more and more distant. If, in line with some molecular findings, *Lagothrix* is linked with *Brachyteles*, length goes up to 42, consistency index is reduced to 0.571, retention index to 0.308. Again, the genus *Lagothrix* as commonly recognized is nonmonophyletic.

But as well as these characters, the two skulls of *L. flavicauda* differ from *L. lagotricha* as follows:

1. *L. flavicauda* has a ridge bordering the carotid foramen on the bullar side; because none of the other studied taxa has this ridge, it can be interpreted as uniquely derived and so uninformative as far as phylogeny is concerned.

2. The median incisive foramen in *L. flavicauda* lies in the plane of, and so between, the pair of main incisive foramina; in all the other taxa it lies in front of the main pair. If the main foramina are fused, as in the type of *hendeei*, the conjoint foramen forms a kidney-shaped whole, concave anteriorly, and the median foramen nestles in the concavity.

3. The foramen at the anterior end of the base of the lateral pterygoid plates (and partly hidden by the tuber maxillaris) is enormous in *L. flavicauda*, but tiny in the other taxa.

4. The foramen magnum is anteroposteriorly

Table 17

Character Matrix for Atelinae, with *Protopithecus* and *Caipora* Included

dimensions ntax = 7 nchar = 20
format missing = ? symbols = "0~3"

Genus	Character
Oreonax	11000011110101111201
Lagothrix	00111100101110001101
Ateles	0111000111000{0,1}110002
Brachyteles	01010010000100112210
Alouatta	{0,1}{0,1}011{0,1}1000101{0,1}012310
Protopithecus	011?1110001111112??1
Caipora	011?1000001110110001

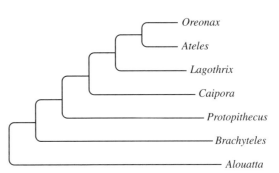

FIGURE 5. Shortest tree for Atelinae, when *Protopithecus* and *Caipora* are included: length 36, consistency index 0.667, retention index 0.538.

elongated in *L. lagotricha*, and so differs from the other taxa in which it is nearly as wide as it is long (though there is some variability in this, especially in *Ateles*).

5. The mandibular condyle is, uniquely, anteroposteriorly expanded and bilaterally reduced in most (but not all) *L. lagotricha*.

In conclusion, the yellow-tailed woolly monkey should be separated generically from *Lagothrix*, because there are no derived character states that appear to unite them; and from all other atelines, because the bootstrap value for the clade uniting it with *Ateles*, its putative closest relative, is not high. Its taxonomic position is isolated.

Oreonax flavicauda (Humboldt, 1812)

**Yellow-tailed Woolly Monkey or
Hendee's Woolly Monkey**

1812 *Simia flavicauda* Humboldt..Peru: banks of the Amazon, restricted by Fooden (1963) to Puca Tambo, 1,550 m, 80 km east of Chachapoyas, San Martín.

1927 *Lagothrix (Oreonax) hendeei* Thomas. Puca Tambo, 1,550 m, 80 km east of Chachapoyas, San Martín.

DISTRIBUTION. The northeastern montane cloud forest of Peru, between 1,500 and 2,700 m, along the Marañon-Huallaga divide (Leo Luna 1987; Butchart et al. 1995). It is highly endangered.

Old World Monkeys—Superfamily Cercopithecoidea Gray, 1821

Genetic studies (Zhang and Ryder 1998) confirmed the monophyly of the two main subfamilies. I gave them family status (Groves 1989); according to Goodman et al. (1998), however, they separated only 14 million years ago, which is well below the time level for full families.

Family Cercopithecidae Gray, 1821

The differences between the two subfamilies were discussed by Napier (1981). The position of the lacrimal fossa, whether wholly within the lacrimal bone or on the maxillolacrimal suture, has commonly been supposed to distinguish between them, but Benefit and McCrossin (1993) and Mouri (1994) have shown that there is far too much variability, particularly in the Cercopithecinae, for this to be used as a diagnostic character. Nor is there any difference in the relative length of the hallux, which instead relates to arboreality vs. terrestriality; there are some differences in the feet, though they are not completely diagnostic (Strasser 1994).

SUBFAMILY CERCOPITHECINAE GRAY, 1821

Strasser and Delson (1987) defined this subfamily by the following derived character states: cheek pouches present, lower incisors lack lingual enamel, nasal bones elongated, lacrimal fossa lacrimal in position, cranial vault low, face tends to be elongated. The lacrimal character is now known to be invalid (see above).

Cercopithecine feet are normally mesaxonic (that is, with the third ray the longest [Strasser 1994]). The foot is adapted for extreme eversion/plantarflexion and inversion/dorsiflexion: facet for medial malleolus occupies entire plantar border of astragalus, distal astragalocalcaneal facet is convex and at an acute angle to the calcaneonavicular facet, and entocuneiform is flatter, lacking much buttressing (Strasser 1988).

The structure of the nails on the feet is variable, even within the two tribes (Soligo and Müller 1999).

In Goodman et al.'s (1998) molecular clock assessment, the two tribes of Cercopithecinae separated 10 million years ago.

Tribe Cercopithecini Gray, 1821

The problems of defining this tribe, and of what to include within it, were discussed by Groves (1978) and Napier (1981). The third lower molar is invariably four-cusped, lacking a hypoconulid, and there are never any suborbital (maxillary) or mandibular fossae; but these features are variable in the Papionini. Other commonly cited differences are variable within the Cercopithecini. Diploid chromosome number varies, but is always more than 42, which is the invariable number in the Papionini: this is perhaps the only character that is absolutely diagnostic of the two tribes.

Rowell (1988) listed some behavioral differences between Cercopithecini and Papionini:

1. The Papionini organize their troops by quite complex communicative signals, whereas the Cercopithecini use what she called the "monitor-adjust" method.

2. Adult male Papionini may form dominance hierarchies based on agonistic encounters, but they coexist in the same social groups and even form coalitions with each other, whereas male Cercopithecini are simply intolerant of each other in the presence of females, and this normally enforces the existence of exclusive one-male groups.

3. Papionin troops generally avoid each other, but their home ranges overlap; those of cercopithecins occupy territories, defended by the females.

4. Male cercopithecins are only weakly integrated into the troop's social organization, unlike male papionins.

Yet in these behavioral traits, too, *Miopithecus* is intermediate in some respects: thus, they use contact calls more like papionins, although this may be somewhat illusory because their frequency declines in more open environments, such as in captivity. They also have multimale troops, but actually there is almost no interaction between the sexes outside the (restricted) breeding season; the troop is effectively an all-female troop and an all-male troop that happens to travel together. Rowell (1988) predicted that the social organization of *Allenopithecus* will turn out to have talapoinlike aspects. The multimale troops of *Chlorocebus* seem to be another exception to the above generalizations, but the structure is in other respects typically cercopithecin.

Ruvolo (1988), based on a study of protein electrophoretic variants, found that *Allenopithecus* was cer-

tainly the first branch to separate, then *Erythrocebus*, then *Miopithecus* and *Chlorocebus*, but the relative positions of these last three were uncertain (note that Ruvolo referred to all of them as species of *Cercopithecus*).

Dutrillaux et al. (1988), on the basis of chromosome banding studies, favored a basal dichotomy, with *Miopithecus*, *Erythrocebus*, *Chlorocebus*, and, unexpectedly, the *C. lhoesti* group on one branch and *Allenopithecus* with the remaining *Cercopithecus* species on the other. Gautier (1988), too, using vocalizations (but excluding warning calls, of uncertain polarity), placed the *C. lhoesti* group with *Miopithecus*, *Erythrocebus*, and *Chlorocebus*, and separated *Allenopithecus* from them. Likewise Martin and MacLarnon (1988), using craniometric and odontometric data, agreed in putting *C. lhoesti* with *Miopithecus*, *Erythrocebus*, *Chlorocebus*, and (this time) *Allenopithecus*, and separating it from the rest of *Cercopithecus*.

Genus *Allenopithecus* Lang, 1923
Swamp Monkey

1923 *Allenopithecus* Lang. *Cercopithecus nigroviridis* Pocock, 1907.

Molar flare is present; ischial callosities fuse across the midline in adult males. These two conditions are unique to *Allenopithecus* among the Cercopithecini, but are shared with some or all of the Papionini. Periodic sexual swelling occurs in females; this feature is shared with most of the Papionini and with *Miopithecus* among the Cercopithecini. Build is very robust, with short limbs and short and broad hands and feet. Ears are pointed, macaquelike. Tail is relatively short. Chromosomes: $2n = 48$.

Allenopithecus nigroviridis (Pocock, 1907)
Allen's Swamp Monkey

1907 *Cercopithecus nigroviridis* Pocock. Upper Congo.

DIAGNOSIS. Eyelids and chin pale, rest of face dark. Agouti black/yellow, giving dark olive green effect. A narrow black eyebrow band, extending back to ears. Cheek whiskers bushy. Underside gray, slightly speckled; throat light gray, unspeckled. Perineum with reddish tuft; scrotum bluish white. Neonatal coat yellow-brown, the crown almost white.

DISTRIBUTION. Both sides of the lower and middle Congo, in Congo-Zaire and Congo-Brazzaville; south of the river, extends east to the River Lomami (Colyn 1988).

Genus *Miopithecus* I. Geoffroy, 1842

Talapoins

1842 *Miopithecus* I. Geoffroy. *Simia talapoin* Schreber, 1774.

Differs from *Cercopithecus, Chlorocebus,* and *Erythrocebus* in its small size, in the occurrence of cyclical sexual swelling in the female, in the distal reduction of the third upper molar, and in the high degree of development of facial gestures. Chromosomes: $2n = 54$.

It was Machado (1969) who first pointed out that talapoins north and south of the River Congo are specifically distinct, and that all names thus far given to putative subspecies actually designate the Angolan species. He stated that a formal description, with a new name, was under way, in collaboration with P. Dandelot, but no such description has appeared in the nearly 30 years since then, and recently Kingdon (1997) at last rectified the omission.

Miopithecus talapoin (Schreber, 1774)

Angolan Talapoin

1774 *Simia talapoin* Schreber. No locality.
1792 *S[imia] (Cercopithecus) niger* Kerr. West Africa.
1812 *Cercopithecus pileatus* E. Geoffroy. No locality.
1842 *Miopithecus capillatus* I. Geoffroy. No locality.
1844 *Cercopithecus melarhinus* Schinz. "Africa."
1907 *[Cercopithecus talapoin] ansorgei* Pocock. Angola: Canhoca.
1919 *Miopithecus talapoin pilettei* Lorenz. Supposedly from Congo-Zaire: Rwenzori, Kabawaki.
1940 *Cercopithecus talapoin vleeschouwersi* Poll. Congo-Zaire, Kinseke (selected by Hill [1966]).

DIAGNOSIS (after Machado [1969]). Pelage relatively long, coarsely banded yellow and black; underparts contrastingly white or whitish; underside of tail base gray; cheek hairs directed downward and backward, covering suborbital fossae; preauricular tufts very well developed, central part lighter; ears large, black; nose black, and bordering facial skin also black; rest of facial skin pinkish ochre; upper lip lighter in middle than at sides; scrotum salmon-

colored medially, blue laterally. Tail relatively long, 360–525 mm in seven specimens (Machado 1969).

DISTRIBUTION. The east coast of Angola, south to about 13°S (Machado 1969), and into Congo-Zaire as far as the River Cuango and both sides of the River Kasai.

Miopithecus ogouensis Kingdon, 1997

Gabon Talapoin

1997 *Miopithecus ogouensis* Kingdon. No type locality: "Endemic to the equatorial coastal watersheds between Cabinda and the River Nyong" (Kingdon 1997:55).

DIAGNOSIS (after Machado [1969]). Pelage short, finely banded yellow and black; underparts gray-white or yellowish, less contrasting; underside of tail base yellowish; cheek hairs directed laterally, not covering suborbital fossae, which are outlined with black; preauricular tufts smaller; ears smaller, dark but not black; nose smudged with black or blackish, but bordering facial skin not blackened; facial skin less bright; upper lip no different in middle and at sides; scrotum blue. Size similar to *M. talapoin* but tail shorter, 375–420 mm in three specimens (Machado 1969). There are also differences in facial gestures.

DISTRIBUTION. Cameroon, south of the River Sanaga; Río Muni, Gabon.

NOTES. Kingdon (1997) noted that a description of this species had been awaited since 1969 and stated that he was using the name as "a *nomen nudum* . . . in anticipation of a formal description." But an excellent description is given by him, and the name is certainly no nomen nudum, but is available from Kingdon's work.

Genus *Erythrocebus* Trouessart, 1897

Patas Monkey

1897 *Erythrocebus* Trouessart. *Simia patas* Schreber, 1774 (Allen 1939a).

There is no molar flare, and no sexual swellings. Ischial callosities remain separate in both sexes. Facial skeleton is long, protruding. Build is slender but deep-chested. Chromosomes: $2n = 54$.

Erythrocebus patas (Schreber, 1774)

Patas Monkey

1774 *Simia patas* Schreber. Senegal.
1788 *Simia rubra* Gmelin. No locality.
1792 *Simia (Cercopithecus) ruber nigro-fasciatus* Kerr. Senegal, Congo.
1792 *Simia (Cercopithecus) ruber albo-fasciatus* Kerr. Senegal, Congo.
1829 *Cercopithecus pyrrhonotus* Hemprich and Ehrenberg. Sudan: Dongola.
1839 *Simia rufa* Wagner. No locality.
1863 *Cercopithecus poliophaeus* Heuglin in Reichenbach. Sudan: Blue Nile, Fazughli.
1863 *Cercopithecus circumcinctus* Reichenbach. West Africa.
1877 *Cercopithecus poliolophus* Heuglin. Substitute for *poliophaeus.*
1905 *Erythrocebus baumstarki* Matschie. Tanzania: Ikoma.
1905 *Erythrocebus langheldi* Matschie. Cameroon: Garoua, upper Benue.
1905 *Erythrocebus zechi* Matschie. Togo: Kete Kradji.
1906 *Cercopithecus patas sannio* Thomas. Nigeria: Yo, Lake Chad.
1906 *Cercopithecus kerstingi* Matschie. Togo: Sokode.
1909 *Erythrocebus formosus* Elliot. Uganda.
1909 *Erythrocebus albigenus* Elliot. Sudan.
1910 *Erythrocebus whitei* Hollister. Kenya: Guas Ngishu, Nzoia River.
1912 *Cercopithecus (Erythrocebus) patas albosignatus* Matschie. Congo-Zaire: Uele Valley, Mbomu River.
1912 *Cercopithecus (Erythrocebus) patas poliomystax* Matschie. Upper Congo.
1950 *Erythrocebus patas villiersi* Dekeyser. Niger: Irabellaben, Air, 1,200–1,300 m.

DIAGNOSIS. Color reddish above, white below including insides of limbs, neck and chin, and on outsides of lower segments of limbs; much more brightly contrasting in male than in female. Scrotum bright blue.

DISTRIBUTION. Semidesert, grassland and woodland savanna from Senegal to borders of Ethiopia, north to Nubia, and south to Serengeti and Athi Plains and west side of Mount Kilimanjaro; Air and Ennedi.

NOTES. Kingdon (1997), who placed the species in *Cercopithecus,* recognized four subspecies; the only one he described, however, is *baumstarki,* as "pallid Serengeti isolate." Dandelot (1971), who likewise recognized *Erythrocebus* as only a subgenus of

Cercopithecus, distinguished the same four, as follows: *C. p. baumstarki,* pale and lacking black frontal and temporal bands, otherwise like *pyrrhonotus;* West African *patas,* with pink face and black nasal spot; Central and Eastern *pyrrhonotus,* with blackish face and white nose; and *villiersi,* from Air and "north of the bend of the Niger River," small in size, otherwise resembling *patas.*

As Loy (1987) has shown, females have a black face, including the nose, black superciliary and temporal lines, and white moustache; in late pregnancy, the nose becomes white, the temporal lines disappear, and the face becomes light gray, and this continues until about 45 days postpartum. These previously unsuspected changes cover at least some of the supposed subspecific variation.

Genus *Chlorocebus* Gray, 1870

Vervet Monkeys

1862 *Callithrix* Reichenbach. *Cercopithecus callithrix* I. Geoffroy, 1851. Not of Erxleben, 1777 (Platyrrhini).
1870 *Chlorocebus* Gray. *Simia sabaea* Linnaeus, 1766 (fixed by Pocock [1907]).
1870 *Cynocebus* Gray. *Cercopithecus cynosuros* Scopoli, 1786.

There is no molar flare nor sexual swellings. The ischial callosities are well separated in the midline in both sexes.

Dandelot (1959, 1971) discussed the number of species in this group (which he referred to as the *aethiops* superspecies of the genus *Cercopithecus*). In his 1959 paper he recognized *C. sabaeus, C. aethiops* (with subspecies-groups *aethiops* and *tantalus*), and *C. pygerythrus* (with subspecies-groups *pygerythrus* and *cynosurus*). In 1971 he raised *tantalus* to specific rank and intimated that probably *cynosurus* should be also. The whole genus is in urgent need of revision.

Chlorocebus sabaeus (Linnaeus, 1766)

Green Monkey

1766 *Simia sabaea* Linnaeus. "Cape Verde Islands" (probably Senegal).
1845 *Cercopithecus chrysurus* Blyth. No locality. (Usually placed in synonymy of *tantalus,* but regarded as a synonym of *sabaeus* by Napier [1981]).
1850 *Cercopithecus werneri* I. Geoffroy. Africa.
1851 *Cercopithecus callitrichus* I. Geoffroy. West Africa.

DIAGNOSIS. Long-legged; dorsum grizzled golden green, extending down limbs and to tail base; hands and feet pale; underparts off white; tail tip golden yellow; backs of thighs yellow; cheek whiskers yellow, directed upward in front of ears and over temples from a whorl just in front of ears, sharply distinct from speckled greenish crown. Light brow band poorly expressed or absent. Scrotum very pale blue.

DISTRIBUTION. Senegal east to about the Volta River.

NOTES. In Ghana it occurs (BM[NH] collection) at Damongo, Mandara, Bamboi, and Bugu, around 8°09' to 9°05'N, 2°35' to 1°49'W. Napier (1981) noted that another Ghanaian specimen, from Bole (9°03'N, 2°23'W, so within the range of *C. sabaeus*), has a white brow band, suggesting limited interbreeding with *C. tantalus*.

Chlorocebus aethiops (Linnaeus, 1758)
Grivet

1758 *Simia aethiops* Linnaeus. Sudan: Sennar (fixed by Schwarz [1928a]).
1804 *Simia engytithia* Hermann. No locality.
1819 *Cercopithecus griseus* F. Cuvier. Africa.
1820 *Cercopithecus griseo-viridis* Desmarest. Africa.
1821 *Simia subviridis* F. Cuvier. Africa.
1843 *C[ercopithecus] cano-viridis* Gray. Ascribed to Rüppell: hence, probably Ethiopia.
?1843 *C[ercopithecus] cinereo-viridis* Gray. Ascribed to Temminck.
1902 *Cercopithecus matschiei* Neumann. Ethiopia: Omo River, Malo.
1916 *Cercopithecus (Chlorocebus) toldti* Wettstein. Sudan: Kordofan, Jebel Riha near Kadugli, Nuba Mountains.
1918 *Cercopithecus (Chlorocebus) cailliaudi* Wettstein. Sudan: Blue Nile.
1922 *Lasiopyga (Cercopithecus) weidholzi* Lorenz. Egypt.
1943 *Cercopithecus aethiops zavattarii* de Beaux. Ethiopia: Murle, River Omo, 5°09'N, 36°13'E.

DIAGNOSIS (partly after Kingdon [1997]). Distinguished by having a warm olive grizzled back and yellower crown, gray limbs with light brown digits, and white underside; a narrow white brow band, continuous with prominent, laterally elongated ("falciform" of Dandelot [1971]) white cheek whiskers; face black, with a fine white moustache; hands and feet pale; white tuft at base of tail; scrotum sky blue.

DISTRIBUTION. East of the White Nile, in Sudan from Khartoum in the north (and probably into Egypt) to Mongalla in the south, in Eritrea, and in Ethiopia south to the Omo, where it hybridizes with *C. pygerythrus,* and east as far as the Ethiopian Rift Valley.

NOTES. Dandelot and Prévost (1972) and Kingdon (1997) recognized a subspecies *C. a. matschiei* with a more fawn to russet back, internal aspect of limbs more gray, shorter tail with reduced terminal tuft, and reduced tufts at tail base; this came from the western highlands of Ethiopia. The (slight) color differences, according to Kingdon (1997), might suggest that it may have hybridized extensively with *C. djamdjamensis.* Napier (1981) placed *zavattarii* in the "southern" *(pygerythrus)* group, but as a probable intermediate with the *aethiops* group.

Chlorocebus djamdjamensis (Neumann, 1902)
Bale Mountains Vervet

1902 *Cercopithecus djamdjamensis* Neumann. Ethiopia: bamboo forest near Abera, east of Lake Abaya, 3,300 m.

DIAGNOSIS. A short tail with tuft very reduced or virtually absent; long, thick, deep grizzled brown fur; gray limbs and tail; hands and feet dark gray; underparts creamy white, mixed with gray on the chest; no white on face except for a short but very bushy white beard and cheek ruff and a fine white moustache; white brow band barely indicated, separated from white cheek ruff by a broad black band from eye to ear; inconspicuous red-brown tufts at base of tail. Scrotum blue.

DISTRIBUTION. Highlands of Ethiopia, east of the Rift Valley, in bamboo and forest-edge habitats. Kingdon (1997:62) mapped it in the Bale Mountains and in the mountains east of Lakes Abiata, Shalla, and Zway.

NOTES. Dandelot and Prévost (1972) resurrected the name *djamdjamensis,* previously regarded as a synonym of *matschiei* (itself here regarded as a synonym of *C. aethiops*), and described some specimens from the region of the headwaters of the Webi Shebeyli River. Kingdon (1997) was the first to recognize this as a full species. It has characters of both "northern" and "southern" taxa (here, groups of species), as noted by Napier (1981), but is distinctively different from all others.

Chlorocebus tantalus (Ogilby, 1841)
Tantalus Monkey

DIAGNOSIS.   Like *C. aethiops,* has a grizzled, gold to greenish back and crown, gray limbs, white underside, tail with creamy white tip and a white basal tuft, sky blue scrotum, and black face. It differs in having a black line from corner of eye running back along temples, separating cheek whiskers, which are long, stiff, and yellowish with short black tips, from the sinuous, tapered white brow band; and scrotum is surrounded by a zone of long, orange hairs.

DISTRIBUTION.   From about the Volta River east to Sudan, Uganda, and the Lake Turkana district of Kenya. In Ghana, BM(NH) specimens come from the Accra Plain and Salaga, 8°33′N, 0°31′W. In Sudan, it is found from the Imatong Mountains in the south to Jebel Marra in the north, and intermediates with *C. aethiops,* which is found east of the Nile, are not recorded. In Uganda, it interbreeds with *C. pygerythrus,* as recounted in detail below under *C. t. budgetti.*

Chlorocebus tantalus tantalus (Ogilby, 1841)

1841 *Cercopithecus tantalus* Ogilby. No locality.
1897 *Cercopithecus passargei* Matschie. Nigeria: Yola.
1905 *Cercopithecus pousarguei* Mitchell. Lapsus for *passargei.*
1909 *Cercopithecus tantalus alexandri* Pocock. Lake Chad.
1910 *Cercopithecus viridis* Schultze. Nigeria: Bornu. Nomen nudum.
1914 *Lasiopyga tantalus graueri* Lorenz. Congo-Zaire: Baraka, northwestern shore of Lake Tanganyika.

DIAGNOSIS.   Tends to be olive green in dorsal color, hairs multibanded gray/yellow.

Chlorocebus tantalus budgetti (Pocock, 1907)

1907 *Cercopithecus tantalus budgetti* Pocock. Uganda: Butiaba, Lake Albert.
1909 *Cercopithecus tantalus griseistictus* Elliot. Congo-Zaire: Uele River, Bambara.
1912 *Cercopithecus (Chlorocebus) cynosurus itimbiriensis* Matschie and Dubois. Congo-Zaire: Itimbiri River.
1914 *Lasiopyga tantalus beniana* Lorenz. Congo-Zaire: Beni.

DIAGNOSIS.   Color is more olive brown; cheek whiskers more yellowish with more extensive speckling; hands, feet, and tail tip darker; digits blackish.

DISTRIBUTION.   This subspecies forms a hybrid zone with *C. pygerythrus* along the northern and western shores of Lake Victoria. Dandelot (1959:366, fig. 7) mapped the hybrid zone, as follows:

1. From northeastern Congo-Zaire, as far southeast as Mawambi (on the Ituri River), comes *C. t. budgetti* only, and there are also specimens from Mabenga on the Rutshuru plains (south of Lake Edward); Masindi and Butiaba, in Uganda to the east of Lake Albert; and, curiously, from Entebbe and from Kerio River, in Kenya west of Lake Turkana. The type of *budgetti,* as Napier (1981) pointed out, in fact has some hybrid features.

2. From the latitude of Lake Kivu southward comes *C. p. centralis* only, including Kabare and other localities southwest of Lake Kivu in Congo-Zaire; Rwanda and Burundi; and Bukoba and Kakindu, in Tanzania on the western shore of Lake Victoria. Farther northeast, around Entebbe, come more specimens of this form.

3. In between, there are hybrids. From Abok, Beni, Kaniki, and Kasindi, all just north and northwest of Lake Edward (all in Congo-Zaire) come "hybrids of tantalus predominance," along with "pure" *C. t. budgetti.* From Kazinga, Katebe, and Burumba (all in southwestern Uganda) come "hybrids of vervet predominance," together with "pure" *C. p. centralis.* It is interesting that another "hybrid of vervet predominance" comes from South Kavirondo, on the eastern side of Lake Victoria in Kenya.

Tantalus therefore seem to extend through southern Sudan, northern Congo-Zaire, and northern Uganda east nearly to Lake Turkana in Kenya, whereas vervets extend north to southern Uganda to the west of Lake Victoria and to northern Kenya to the east of it. The two apparently manage to coexist in the Entebbe region (where the vervet in question is *C. p. centralis*) and west of Lake Turkana (where the vervet is *C. p. arenarius*). Hybridization is in fact restricted to a zone east, southeast, and northwest of Lake Albert, in the southern Ituri district of Zaire (where there are backcrosses with tantalus) and in southwestern Uganda (where there are backcrosses with vervets).

Chlorocebus tantalus marrensis (Thomas and Hinton, 1923)

1923 *Cercopithecus tantalus marrensis* Thomas and Hinton. Sudan: foothills south of Jebel Marra.

DIAGNOSIS.   Like *C. t. tantalus* but color tending to be lighter olive-fawn. Distribution is isolated by desert from the rest of the species.

Chlorocebus pygerythrus (F. Cuvier, 1821)
  Vervet Monkey

DIAGNOSIS.   Like previous species, grizzled gray or olive, but this color extends to outer surface of limbs, which are not gray; hands, feet, and tail tip are darker, not lighter; hair at base of tail and in anal region bright red and does not form tufts; white brow band and (very short) cheek whiskers broadly fuse to form a complete face-ring, which grades into greenish speckled crown and neck; and scrotum is turquoise blue.

Chlorocebus pygerythrus hilgerti (Neumann, 1902)

1902 *Cercopithecus hilgerti* Neumann. Ethiopia: Webi Shebeyli, Gobele River.
1902 *Cercopithecus ellenbecki* Neumann. Ethiopia: Lake Zwai, Suksuki and Maki Rivers.
1907 *[Cercopithecus pygerythrus] johnstoni* Pocock. Tanzania: Old Moshi, Mount Kilimanjaro District, 1,500 m.
1909 *Cercopithecus rubellus* Elliot. Kenya: Fort Hall.
1910 *Cercopithecus centralis luteus* Elliot. Kenya: southwest of Mount Kenya, Wambugu.
1912 *Lasiopyga pygerythra callida* Heller. Kenya: south side of Lake Naivasha.
1913 *Lasiopyga pygerythra arenaria* Heller. Kenya: Merille water holes, Marsabit Road.
1913 *Lasiopyga pygerythra tumbili* Heller. Kenya: Taita Hills, Ndi.
1920 *Lasiopyga pygerythra contigua* Hollister. Kenya: inland of Mombasa, Changamwe.

DIAGNOSIS.   Color generally paler brownish yellow than other forms.

DISTRIBUTION.   From the Ethiopian Rift Valley and into parts of the highlands east of the Rift, Somalia, through Kenya and eastern lowlands of Ethiopia to northern Tanzania, west to eastern Uganda.

NOTES.   Napier (1981) noted that the type specimen was described as having long, white cheek whiskers and a white tail tip like *C. aethiops*, indicating interbreeding between the two in this region as in Uganda; she retained *arenarius, callidus, rubellus,* and *johnstoni* as valid subspecies, albeit weakly marked.

For Dandelot and Prévost (1972), the names *hilgerti* and *ellenbecki* actually referred to representatives of *C. aethiops*. Certainly the specimens they described, from the area between Lakes Awusa and Shala, in the Rift Valley, were clearly *C. aethiops* and, in some cases, exactly similar to some from Kaffa Province in the western highlands, but they had not examined the type specimens and relied purely on the type descriptions. Between Lake Shala and Lake Zwai, then, appears to be a boundary between *C. aethiops* and *C. pygerythrus*.

Ethiopian specimens in the BM(NH) are from Shoa, including the region around Addis Ababa, the middle Awash at 1,800 m, and the Bale Mountains up to 1,300 m. *Chlorocebus djamdjamensis* is known from higher altitudes than this, and the separation between them seems to be altitudinal.

Chlorocebus pygerythrus excubitor (Schwarz, 1926)

1923 *Chlorocebus voeltzkowi* Matschie. Nomen nudum.
1926 *Cercopithecus aethiops excubitor* Schwarz. Kenya: Manda Island.

DIAGNOSIS.   Resembling *C. p. hilgerti* but smaller in size.

DISTRIBUTION.   From the Witu Islands, northern Kenya.

Chlorocebus pygerythrus nesiotes (Schwarz, 1926)

1923 *Chlorocebus pembae* Matschie. Pemba Island. Nomen nudum.
1926 *Cercopithecus aethiops nesiotes* Schwarz. Tanzania: Pemba, Chake Chake Island.

DIAGNOSIS. Resembling *C. p. rufoviridis* but smaller in size.

DISTRIBUTION. From Pemba and Mafia Islands, Tanzania.

Chlorocebus pygerythrus rufoviridis (I. Geoffroy, 1843)

1843 *Cercopithecus rufo-viridis* I. Geoffroy. Africa.
1852 *Cercopithecus flavidus* Peters. Mozambique, Quitangonha, 15°S.
?1862 *Cercopithecus circumcinctus* Reichenbach. "W. Afrika."
1870 *Chlorocebus rufoniger* Gray. Error for *rufoviridis.*
1900 *Cercopithecus centralis* Neumann. Tanzania: Bukoba.
1907 *Cercopithecus pygerythrus whytei* Pocock. Malawi: Mount Chiradzulu.
1909 *Cercopithecus silaceus* Elliot. Zambia: south bank of Luangwa River.

DIAGNOSIS. Color more fawn or orange yellow with long, speckled whiskers; underside often reddish infused.

DISTRIBUTION. From Zambezi, east of the Luangwa Valley, north to Lake Victoria, where it hybridizes with *C. tantalus budgetti* as described under that heading.

Chlorocebus pygerythrus pygerythrus (F. Cuvier, 1821)

1811 *Cercopithecus glaucus* Lichtenstein. South Africa. Nomen nudum.
1821 [*Simia*] *pygerythra* F. Cuvier. "Africa": if *pusillus* may be regarded as a substitute name, then type locality may be regarded as Keiskama (Schwarz 1928a).
1825 *Cercopithecus pusillus* Desmoulins. South Africa: Kwazulu-Natal, Keiskama, near Great Fish River.
1829 *S[imia] erythropyga* G. Cuvier. Substitute for *pygerythra.*
1841 *Cercopithecus lalandei* I. Geoffroy. Substitute for *pusillus.*
1931 *Cercopithecus aethiops cloeteii* Roberts. South Africa: Transvaal, Pilgrim's Rest, Mariepskop.
1932 *Cercopithecus aethiops ngamiensis* Roberts. Botswana: Ngamiland, Toten-Maun Road.
1936 *Cerpithecus [sic] aethiops marjoriae* Bradfield. South Africa: Transvaal, Kuruman, Zoetvlei.

DIAGNOSIS. Color ashy gray to olive-green, tending to be grayer in west, more greenish in east.

DISTRIBUTION. South of Zambezi. Napier (1981) noted the strong differentiation between this form, from Beira, and *C. p. rufoviridis* from Gorongoza to

the north; the color becomes more dark gray and less yellow-gray to the south.

Chlorocebus cynosuros (Scopoli, 1786)
Malbrouck

1786 *Simia cynosuros* Scopoli. Congo-Zaire: Banana, Lower Congo (fixed by Schwarz [1928a]).
1833 *Cercopithecus tephrops* Bennett. No locality.
1912 *Cercopithecus (Chlorocebus) cynosurus weynsi* Dubois and Matschie. Congo-Zaire: Banana.
1912 *Cercopithecus (Chlorocebus) aethiops lukonzolwae* Matschie. Congo-Zaire: Lake Mweru, Lukonzolwa.
1912 *Cercopithecus (Chlorocebus) cynosurus tholloni* Matschie. Congo-Zaire: Stanley Pool.
1919 *Cercopithecus pygerythrus katangensis* Lönnberg. Congo-Zaire: Funda Biabo, Shaba.
1926 *Cercopithecus pygerythrus helvescens* Thomas. Namibia: Cunene Falls.

DIAGNOSIS. Similar in most respects to *C. pygerythrus,* but paler, olive-gray, with a pale, blotched face instead of the black face of all other species, and bare skin of palms and soles also pale; ischial callosities "rose pur" (Dandelot 1959); cheek whiskers longer, directed upward and backward; scrotum lapis blue.

DISTRIBUTION. Northern Namibia, Angola, southern Congo-Zaire, and Zambia east to the Luangwa Valley, south to the Zambezi.

Genus *Cercopithecus* Linnaeus, 1758
Guenons

There is no molar flare nor sexual swellings. The ischial callosities are not fused across the midline in either sex, although Napier (1981) mentioned a case of a male *C. neglectus* in which they are contiguous. According to Soligo and Müller (1999), the nails on the toes in *C. cephus* have a two-layered structure, which differs from that described for *Chlorocebus* cf. *aethiops.*

According to Ruvolo's (1988) protein electrophoresis study, *C. hamlyni* branched off first, followed by *C. lhoesti* (note that this was the only species of its group to be studied); there was then a split between a *mitis/cephus* clade and a *neglectus/diana/mona* clade.

Quite contrary to this, Dutrillaux et al. (1988) in their chromosome study placed the *C. lhoesti* group (all three species were studied) in a clade with *Miopithecus, Erythrocebus,* and *Chlorocebus,* outside the main *Cercopithecus* branch altogether. In the main *Cercopithecus* clade, *C. diana* branched off first; the rest formed a dubious cluster, with much parallelism, but the arrangement that makes most sense had a *hamlyni* clade, a *neglectus/mona* clade, and a *mitis/cephus* clade. The sudy of vocalizations by Gautier (1988) reached strikingly similar conclusions, finding exactly the same clades except that the *mitis* group was not associated with the *cephus* group.

Martin and MacLarnon (1988) similarly separated *C. lhoesti* from the rest, but their discriminant analysis of cranial and dental measurements gave quite a different picture. For them, *C. diana* was most distinct, then successively *C. nictitans, neglectus, cephus,* and a *mitis/ascanius* group (*C. hamlyni* was not studied).

The facial patterns and associated gestures of the species and species-groups of *Cercopithecus* were described and illustrated by Kingdon (1980, 1988), who used them to argue taxonomic and phylogenetic questions; for example, the tiny range of *C. sclateri* was seen as correlated with its plesiomorphic facial pattern.

I take the opportunity to report here results of my survey of the banding patterns, conducted many years ago in the BM(NH) collection.

Cercopithecus diana group:

1758 *Cercopitheci* Linnaeus. Stiles and Orleman (1927) argued that this should be accepted as an available name in its singular form, with type *Simia diana* Linnaeus, 1758, and this was accepted by the International Commission on Zoological Nomenclature (1954a), Opinion 238.
1878 *Diana* Trouessart. *Simia diana* Linnaeus, 1758. Not of Risso, 1826 (Osteichthyes).
1904 *Pogonocebus* Trouessart. *Simia diana* Linnaeus, 1758 (fixed by Pocock [1907]).

Cercopithecus mitis group:

1811 *Lasiopyga* Illiger. *Simia nictitans* Linnaeus, 1766 (fixed by G. M. Allen [1939a]).
1862 *Diademia* Reichenbach. *Simia leucampyx* J. B. Fischer, 1829.
1913 *Insignicebus* Elliot. *Semnopithecus albogularis* Sykes, 1831.
1913 *Melanocebus* Elliot. To replace *Diademia* Reichenbach, 1862.

Cercopithecus mona group:

1816 *Monichus* Oken. *Simia mona* Schreber, 1775 (fixed by J. A. Allen [1925]). Unavailable.
1862 *Mona* Reichenbach. *Simia mona* Schreber, 1775.
1897 *Otopithecus* Trouessart. *Cercopithecus pogonias* Bennett, 1833.

Cercopithecus cephus group:

1853 *Cercocephalus* Temminck. *Simia cephus* Linnaeus, 1758.
1862 *Petaurista* Reichenbach. *Simia petaurista* Schreber, 1774. Not of Link, 1795 (Rodentia).
1897 *Rhinostictus* Trouessart. *Simia petaurista* Schreber, 1774.
1913 *Neocebus* Elliot. *Simia cephus* Linnaeus, 1758.

Cercopithecus lhoesti group:

1913 *Allochrocebus* Elliot. *Cercopithecus l'hoesti* Sclater, 1899.

Cercopithecus hamlyni group:

1913 *Rhinostigma* Elliot. *Cercopithecus hamlyni* Pocock, 1907 (but using skull characters derived from type skull of *Cercocebus hamlyni* Pocock, 1906 (= *Lophocebus aterrimus*).

Cercopithecus dryas group

Cercopithecus dryas Schwarz, 1932

**Dryas Monkey or
Salongo Monkey**

1932 *Cercopithecus dryas* Schwarz. Congo-Zaire: Sankuru District, Lomela.
1977 *Cercopithecus salongo* Thys van den Audenaerde. Congo-Zaire: near Wamba, Djolu region.

DIAGNOSIS. Size very small; chin white, continued as a white facial seam, separated from sinuous white brow band on either side by a posterio-superiorly directed bare black line coming from facial skin; elongated tufts between crown and temples. Adults have jet black limbs, inside and out, and a white patch on buttocks and posterior thighs, extending along tail and restricting dark color on tail to a thin line along upper surface; in juveniles these contrasts are absent, with no black on the limbs and no white on the tail and very little on the buttocks.

DISTRIBUTION. Known only from Lomela (2°20′S, 23°12′E), the type locality of *dryas,* and Wamba (0°01′N, 22°32′E), the locality of the type of *salongo* and of specimens obtained (purchased from hunters) by the Kyoto University–led field team working on *Pan paniscus.* It is unknown to people north of the upper Lopori (Colyn 1988).

NOTES. The juvenile and adult coats are so different that they were described as different species. Study of the Kyoto material showed that, exactly as in *C. neglectus,* this species undergoes a dramatic color change as the limbs darken and the white spreads on the buttocks, thighs, and tail (Colyn et al. 1991).

Cercopithecus diana group

Dark gray with white banding and a deep maroon lumbar saddle; underside black, except for anterior chest, throat, and inner side of arms, which are sharply white, but inner side of legs either red or yellow; a white stripe across haunch; face jet black; a pale brow band; prominent, narrow chin beard. Chromosomes: $2n = 58$.

Cercopithecus diana (Linnaeus, 1758)
   Diana Monkey

1758 *Simia diana* Linnaeus. "Guinea" (probably Liberia).
1766 *Simia faunus* Linnaeus. No locality.
1870 *Cercopithecus diana* var. *ignita* Gray. "West Africa."

DIAGNOSIS. Beard short, square, mostly white but black in front; forehead band a small, white crescent with some reddish hairs; buttocks and inner surfaces of thighs deep orange-red. Hairs are gray at base with four pairs of alternating light and dark bands, the first light band being partly merged with the gray base.

DISTRIBUTION. From Sierra Leone to the River Sassandra in Ivory Coast.

Cercopithecus roloway (Schreber, 1774)
   Roloway Monkey

1774 *Simia roloway* Schreber. "Guinea" (i.e., West Africa).
1855 *C[ercopithecus] palatinus* Wagner. "Guinea."

DIAGNOSIS. Beard long, pointed, entirely white; frontal band broader, entirely white; buttocks and inner surfaces of thighs creamy white or with orange tinge. Usually more than four pairs of bands on hairs.

DISTRIBUTION. From the River Sassandra in Ivory Coast east to the River Pra in Ghana.

NOTES. The orange tone of the buttocks and inner thighs becomes deeper at the River Sassandra boundary with *C. diana,* as noted by Oates (1988), but there is no intermediacy in beard or brow band (Booth 1956). The difference between *C. diana* and *C. roloway,* found respectively west and east of the Sassandra River in Ivory Coast, is absolutely clear-cut, and they should stand as distinct species.

Cercopithecus mitis group

Chromosomes: $2n = 70$ or 72.

Cercopithecus nictitans (Linnaeus, 1766)
   Greater Spot-nosed Monkey

DIAGNOSIS. Body deep gray, becoming more olive on back and black on limbs and distal half of tail; arms black, also hands and feet; terminal half of tail black; an oval white spot on nose. The hairs on the body are gray at base, with three or four pairs of alternating white and black bands; on the neck and crown, just two or three band pairs, separated by a zone on the occiput with only one pair; the black tip is very long on the head and neck.

DISTRIBUTION. According to Oates (1988), occurs sporadically in Liberia and western Ivory Coast, along the Ouémé River in Benin, and then quasi-continuously from the Osse River in western Nigeria south and east to the Congo River, across the Oubangui as far east as Buta and Budjala between the Congo and Oubangui in Congo-Zaire (where it is separated from *C. mitis* by the River Itimbiri, according to Colyn [1988]).

Cercopithecus nictitans nictitans (Linnaeus, 1766)

1766 *Simia nictitans* Linnaeus. "Guinea."
1907 *[Cercopithecus nictitans] laglaizei* Pocock. Gabon.
1909 *Cercopithecus sticticeps* Elliot. Congo-Zaire: upper Oubangui River, Ndongoleti.

DIAGNOSIS. Color tends to be grayish olive on back and head; underside as dark as upper, with a black band on chest between arms; interramal region buff; cheek whiskers more bushy.

DISTRIBUTION.   Mainly south of the Sanaga River in Cameroon (though there is a population in the Bamenda highlands [Oates 1988]), through Gabon and Congo-Brazzaville to the mouth of the River Congo, across the Sangha and then the Oubangui into Central African Republic and Congo-Zaire north of the Congo River.

Cercopithecus nictitans martini Waterhouse, 1838

1838 *Cercopithecus martini* Waterhouse. Equatorial Guinea: Bioko Island, but more probably Nigeria according to Pocock (1907).
1849 *Cercopithecus ludio* Gray. West Africa.
1888 *Cercopithecus stampflii* Jentink. Liberia: Pessi County.
1909 *Cercopithecus insolitus* Elliot. Northern Nigeria.

DIAGNOSIS.   Less gray-toned, more greenish; chest, interramal region, throat, and inner surfaces of upper arms white, with only a faint black line across upper chest; rest of underparts lighter than upper; cheek whiskers shorter, crown hair longer; reddish patch around callosities (Oates [1988] stated that this is often absent in western specimens).

DISTRIBUTION.   Cameroon northwest of the lower Sanaga River, into Nigeria; sporadically through southern Nigeria into Benin; Ivory Coast and Liberia. The Bioko Island specimens may represent a separate subspecies (see Oates [1988]). It is extraordinary that, as far as can be detected, there is no even approximately consistent difference between these two widely separated populations; rather, the characters by which they are supposed to differ are distributed sporadically among the various population isolates (Oates 1988). Quite in contrast to the common and continuously distributed *C. n. nictitans,* the range of this subspecies is patchy, with a number of isolates mainly along rivers.

Cercopithecus mitis Wolf, 1822
   Blue Monkey

DIAGNOSIS.   Body speckled blue-gray with blackish limbs (especially arms) and a broad band joining them across scapular region; nape and crown dark gray or black, with pale diadem in front; chin hair pale; long but tidy (rounded), pale cheek whiskers.

DISTRIBUTION.   From the Congo-Oubangui to the East African Rift Valley, south to northern Angola and northwestern Zambia. The boundary between this species and *C. nictitans* is actually the River Itimbiri (Colyn and Verheyen 1987b).

NOTES.   Descriptions in part are after Colyn (1991). Peter Grubb (personal communication) is in the process of producing a detailed description of this species and the next.

Cercopithecus mitis mitis Wolf, 1822

1822 *Cercopithecus mitis* Wolf. "Guinea": probably Angola (Napier 1981).
1829 *Simia leucampyx* J. B. Fischer. "Guinea": Angola (see Schwarz 1928a).
1834 *Cercopithecus diadematus* I. Geoffroy. West coast of Africa.
1838 *Cercopithecus dilophos* Ogilby. No locality.
1848 *Cercopithecus pluto* Gray. Angola.
1907 *[Cercopithecus stuhlmanni] nigrigenis* Pocock. West Africa.

DIAGNOSIS.   Dark gray above, with whitish or gray speckling; black below; crown and nape dark, agouti-banded, but not black, often continued as a broad, black interscapular band and contrasting sharply with white frontal band, which is broad centrally and narrows laterally; cheek whiskers dark gray (with alternating black and white or buff bands); limbs and proximal half of tail black. Hair bases gray with four pairs of alternating white and black bands.

DISTRIBUTION.   Angola, known from Dundo region; not reported in northern Angola or in Congo-Zaire.

Cercopithecus mitis heymansi Colyn and Verheyen, 1987

1987 *Cercopithecus mitis heymansi* Colyn and Verheyen. Congo-Zaire: Yaenero, 0°12′N, 24°47′E.

DIAGNOSIS.   Frontal diadem white, narrow, contrasting; crown fully annulated, continuous with similarly colored mantle; arms black; underside lighter than upper side.

DISTRIBUTION.   According to Colyn and Verheyen (1987b) it is this, not the misallocated *maesi* (which is an intergrade between *stuhlmanni* and *schoutedeni*), that is the taxon from between the Rivers Lualaba and Lomami, where it reaches only to about 2°S, though it extends, or may extend (Colyn [1988]) a little west of the Lomami. A specimen from Lusambo (5°S, 23°27'E) is intermediate between *C. m. heymansi* and *opisthostictus* according to Colyn and Verheyen (1987b).

Cercopithecus mitis schoutedeni Schwarz, 1928

1928 *Cercopithecus leucampyx schoutedeni* Schwarz. Congo-Zaire: Idjwi Island, Lake Kivu.

DIAGNOSIS. Distinguished from *C. m. stuhlmanni* by its gray-olive color; has a black crown and nape and white-speckled frontal band; terminal half of tail black.

DISTRIBUTION. Typically occurs on Idjwi and Shushu Islands, in Lake Kivu; on the mainland, specimens best assigned to this subspecies (but strongly tending toward *C. m. stuhlmanni*) are from the western Virunga Volcanoes and southwest as far as Bobandana, 1°41'S, 29°E.

Cercopithecus mitis stuhlmanni Matschie, 1893

1893 *Cercopithecus stuhlmanni* Matschie. Congo-Zaire: north of Kinyawanga, northwest of Lake Albert.
1902 *Cercopithecus otoleucus* Sclater. Sudan: Mount Lotuke, Didinga Hills.
1906 *Cercopithecus neumanni* Matschie. Tanzania: Kwa Kitoto, Kavirondo.
1907 *[Cercopithecus stuhlmanni] carruthersi* Pocock. Uganda: Rwenzori East, 3,000 m.
1909 *Cercopithecus princeps* Elliot. Uganda: Fort Portal, Mpanga Forest, Toro.
1913 *Cercopithecus (Mona) leucampyx schubotzi* Matschie. Congo-Zaire: Mawambi, Ituri River.
1913 *Lasiopyga leucampix [sic] mauae* Heller. Kenya: Mau Escarpment, between Londiani and Sirgoit.
1919 *Cercopithecus leucampyx maesi* Lönnberg. Congo-Zaire: supposedly from Lake Mai-Ndombe (Leopold II), Kutu; but see Notes.

DIAGNOSIS. Crown black, strongly contrasting with diadem, which is gray, flecked with black, and of even width across forehead; arms black, legs darker than mantle and sometimes partially black; underside lighter than upper side. There are three to five white/black band pairs; the white bands may be yellowish white.

DISTRIBUTION. From region between Uele and Congo Rivers, east of the Itimbiri River, to the Ituri and Semliki Forests and south to about 6°S (eastern Congo-Zaire) east of the Lualaba, into southern Sudan (Didinga Hills and Imatong Mountains), northern Uganda (Bunyoro, Toro), and at least parts of Kenya west of the Rift Valley. The intergradation zone with *C. m. opisthostictus* is north of Lake Lukuga; with *C. m. elgonis* is on the eastern flank of the Rwenzoris and east to the Kakamega Forest.

NOTES. Colyn and Verheyen (1987b) and Colyn (1991) described the variations that occur in this subspecies. Even among members of a single troop, the diadem varies greatly in color and thickness. There are also differences according to altitude: in the mountains west of Lake Edward, the interorbital region is blackish, but this is rare in those from lower altitudes; in high-altitude specimens the cheek whiskers are more developed, and the zone from the chin to the nose is generally white, whereas in the lowlands this tone is much less marked. In the mountains and in the Semliki Valley, the ear tufts are usually white, whereas they are always ochre in more westerly populations.

Cercopithecus l. maesi was said to be in color like *C. m. mitis* but with the frontal band speckled gray, tail black only at tip, and dark reddish brown along its underside. Colyn and Verheyen (1987b) argued that the three specimens on which it was based were actually *stuhlmanni/schoutedeni* hybrids and so must have been from the Sake region, northwest of Lake Kivu.

Cercopithecus mitis elgonis Lönnberg, 1919

1919 *Cercopithecus leucampyx elgonis* Lönnberg. Kenya: Mount Elgon.

DIAGNOSIS. Colyn (1991) distinguished this from *C. m. stuhlmanni*, with which is has usually been combined: the ventral color is black, there are no ear tufts, the legs are very dark, and there is a black collar on the neck.

DISTRIBUTION. The subspecies is restricted to Mount Elgon, but intermediates with *stuhlmanni* are found in Kakamega Forest (closer to *elgonis*) and on the eastern flanks of the Rwenzoris (closer to *stuhlmanni*).

Cercopithecus mitis boutourlinii Giglioli, 1887

1887 *Cercopithecus boutourlinii* Giglioli. Ethiopia: Kaffa.
1901 *Cercopithecus omensis* Thomas. Ethiopia: Omo River, Mursi.

DIAGNOSIS.   Differs from *C. m. stuhlmanni* in that the frontal band blends with crown; arms and legs black; underside black; terminal three-fourths of tail black. The light bands on the hairs are yellow, not white, and there may be either four or five band pairs.

DISTRIBUTION.   Southern Ethiopia: Lake Turkana north to Lake Tana, along the western side of the Ethiopian Rift.

NOTES.   In his discriminant analysis, Colyn (1991) found this subspecies to be quite well separated from the others. Together with its isolated distribution, this might suggest that it should be recognized as a full species; the pelage is, in the opinion of Dandelot and Prévost (1972), if anything intermediate between that of *C. m. stuhlmanni* and *C. m. opisthostictus,* but Peter Grubb (personal communication) finds it strongly distinct.

Cercopithecus mitis opisthostictus Sclater, 1894

1894 *Cercopithecus opisthostictus* Sclater. Zambia: Lake Mweru.

DIAGNOSIS.   Light gray, becoming dark gray on crown and nape, with only slightly lighter gray frontal band; arms black, legs colored as body or slightly darker; underside black; terminal half to three-fourths of tail black, sometimes reddened below especially in juveniles. There are only three white/black band pairs. Dandelot and Prévost (1972) remarked on the resemblance of this subspecies to *C. m. boutourlinii,* from which it differs in the lighter olive-gray tone of the body, so contrasting more with the black areas.

DISTRIBUTION.   From Shaba, north to nearly 6°N on the left bank of the River Lualaba, and northwestern Zambia, west of the Luangwa River. In the Itombwe Mountains and on Mount Kabobo, according to Colyn (1991), it approaches *C. m. stuhlmanni,* especially in having a more contrasted crown.

Cercopithecus doggetti Pocock, 1907
Silver Monkey

1907 *[Cercopithecus stuhlmanni] doggetti* Pocock. Uganda: southwestern Ankole.
1913 *Lasiopyga leucampyx sibatoi* Lorenz. Congo-Zaire: mountain forest, 2,000 m, northwest of Lake Tanganyika.

DIAGNOSIS.   Light silvery brown, becoming black on crown and nape, with sharply contrasting buff-speckled frontal band; only terminal one-third of tail black; tail often reddish underneath. Hairs of body are gray at base and there are then from five to nine pairs of alternating yellow and black bands; on throat there are four band pairs.

DISTRIBUTION.   Parts of the highlands of eastern Congo-Zaire: the mountains west of Lake Albert and those west of Lake Tanganyika (southernmost locality given by Colyn [1991] is Sibato's, 3°52′S, 28°55′E) and extends well west of the main highlands at Nyakanyendje, 3°24′S, 28°18′E. Extends east between Lakes Kivu and Tanganyika into southern Burundi and the Bukoba District of northwestern Tanzania, through Rwanda and north into parts of southern Uganda (Ankole, Busenya, Kaiso [Colyn 1991]).

NOTES.   The wide distribution, mapped by Colyn (1991), appears to adjoin or even interdigitate with that of *C. m. stuhlmanni* in many areas along the mountains of the western Rift, but hybrids are reported from the Bwindi Forest. It appears to extend to the eastern Virunga Volcanoes: it has been recorded from Mount Sabinio and Mount Mgahinga where *C. kandti* has also been recorded (Colyn 1991), but their respective distributions are unclear. Field observations may help to elucidate the distributions and genetic relationships of this species, *C. m. stuhlmanni,* and *C. kandti.*

Cercopithecus kandti Matschie, 1905

Golden Monkey

1905 *Cercopithecus kandti* Matschie. Congo-Zaire: Virunga Volcanoes.
1909 *Cercopithecus insignis* Elliot. "Congo forest."

DIAGNOSIS.   Bright golden body, cheeks, and frontal diadem, contrasting with black limbs, crown, and terminal one-third of tail; perineum and underside of base of tail red. Hairs are yellow at the base, followed by four to seven pairs of alternating red and black bands; the yellow bands are very broad on the flanks, the black broader on the back.

DISTRIBUTION.   Known for certain only from the Virunga Volcanoes and the Nyungwe Forest (southern Rwanda).

NOTES.   The distribution of this highly distinctive species, relative to *C. mitis* and *C. doggetti,* is not clear. In the eastern Virunga Volcanoes both it and *C. doggetti* have been recorded on Mount Mgahinga and Mount Sabinio, and in the western Virungas it and *C. mitis schoutedeni* have been supposed to occur at Burunga and Rubengera. A locality listed by Colyn (1991), Kabare, is so far from other localities that it is surely in error or perhaps merely a place of purchase. Field observations are crucial to resolving these points.

Cercopithecus albogularis (Sykes, 1831)

Sykes's Monkey

DIAGNOSIS.   Body and crown relatively light, speckled yellow-olive or ochery toned, contrasting with dark limbs (especially arms); no scapular band; white chin, cheeks, and throat; long, pale or reddish tufts on ears; a long brow fringe projects forward above the eyes.

DISTRIBUTION.   From Somalia southward, east of the Rift Valley and the Luangwa Valley, to the Eastern Cape.

NOTES.   In Kenya, this species and *C. mitis* are separated by unsuitable habitat in the Rift Valley. Farther south, there are apparent hybrids in Ngorongoro and at Lake Manyara; for this reason C. P. Booth (1968) united them into a single species, and in this she was followed by Napier (1981) and others. This is still the only region where hybrids are definitely known; Ansell (1960) stated that hybrids are not known in the Luangwa Valley of Zambia, where their ranges approach each other.

Cercopithecus albogularis albogularis (Sykes, 1831)

1831 *Semno[opithecus] ?albogularis* Sykes. Tanzania: Zanzibar (Schwarz 1928a).

DIAGNOSIS.   Brow fringe, cheek whiskers, and shoulders speckled yellowish gray, becoming reddish brown on dorsum; crown slightly darker than body; 60% of terminal part of tail black; some red hairs under tail; throat patch extended around neck to form a nearly complete collar. Small in size. Hairs yellowish gray at base followed by three pairs of yellow and black bands.

DISTRIBUTION.   Zanzibar.

Cercopithecus albogularis albotorquatus Pousargues, 1896

1896 *Cercopithecus albotorquatus* Pousargues. No locality.
1907 *Cercopithecus rufotinctus* Pocock. Kenya: ?Mombasa; later (Pocock 1927) changed to Somalia: Juba River, but there seems no reason for the change. Type was a zoo animal.

DIAGNOSIS.   Darker than nominotypical *albogularis* and rump with orange suffusion; collar more nearly complete; red under tail, and sometimes red tones on inner thighs. Hairs are yellow-gray at the base, followed by three pairs of alternating reddish yellow and black bands.

DISTRIBUTION.   From the Tana River forests.

Cercopithecus albogularis zammaranoi de Beaux, 1924

1924 *Cercopithecus (Insignicebus) albogularis zammaranoi* de Beaux. Somalia: Bidi Scionde, lower Juba.

DIAGNOSIS.   Overall dark olive, becoming yellower posteriorly; nape darker than body or crown; no collar, and no red under tail.

DISTRIBUTION. From southern Somalia.

Cercopithecus albogularis kolbi Neumann, 1902

1902 *Cercopithecus kolbi* Neumann. Kenya: Kedong Escarpment; probably Rift Escarpment, southwest of Mount Kenya.
1907 *Cercopithecus kolbi hindei* Pocock. Kenya: Tutha (= River Tusu), Mount Kenya, 2,400 m.
1910 *Cercopithecus kolbi nubilus* Dollman. Kenya: Nairobi forest.

DIAGNOSIS. Like nominotypical *albogularis* but slightly darker overall and dorsum redder; white collar nearly complete. Hairs are yellow-gray at base, with four pairs of alternating reddish yellow and black bands.

DISTRIBUTION. From the highlands of Kenya east of the Rift Valley.

Cercopithecus albogularis kibonotensis Lönnberg, 1908

1908 *Cercopithecus albogularis kibonotensis* Lönnberg. Tanzania: Kibonoto (= Kibongoto), near Mount Kilimanjaro.
1913 *Lasiopyga albogularis maritima* Heller. Kenya: Mazeras.
1913 *Lasiopyga albogularis kima* Heller. Kenya: Taita, Mount Mbololo.

DIAGNOSIS. Hardly different from nominotypical *albogularis;* ends of collar farther apart on nape, and tail black for only its terminal half. Hairs gray at base, with five pairs of alternating reddish yellow and black bands.

DISTRIBUTION. From the mainland of southern Kenya, north to Kilifi Creek, and northern Tanzania inland to Mount Kilimanjaro and Mount Meru.

Cercopithecus albogularis phylax Schwarz, 1927

1927 *Cercopithecus leucampyx phylax* Schwarz. Kenya: Patta Island.

DIAGNOSIS. More reddish overall and darker on crown; collar extends farther around neck. Small in size.

DISTRIBUTION. Patta and Witu Islands.

Cercopithecus albogularis monoides I. Geoffroy, 1841

1841 *C[ercopithecus] monoides* I. Geoffroy. Tanzania: Rufiji River.
1907 *[Cercopithecus albogularis] rufilatus* Pocock. Tanzania: Rufiji River.

DIAGNOSIS. Crown, brow fringe, cheek whiskers, and shoulders speckled yellowish gray, becoming reddish brown on dorsum; only 40% of terminal part of tail black; some red hairs under tail; throat patch little or not extended into a collar.

DISTRIBUTION. Eastern seaboard of Tanzania, from Morogoro (6°49′S, 37°40′E) to the Newala District near the Mozambique border.

Cercopithecus albogularis moloneyi Sclater, 1893

1893 *Cercopithecus moloneyi* Sclater. Mozambique: Karonga, Lake Malawi.

DIAGNOSIS. Speckled gray with dark reddish saddle; distal two-thirds of tail black; tail often reddish beneath; crown and nape dark gray; whiskers and brow fringe light speckled gray; throat patch creamy, extending around neck as a collar. Hairs gray at base with four pairs of red and black bands.

DISTRIBUTION. From Zambia east of the Luangwa River, northernmost Malawi (to 9°30′S), and the Poroto Mountains (9°S, 33°45′E), Southern Highlands of Tanzania.

Cercopithecus albogularis francescae Thomas, 1902

1902 *Cercopithecus francescae* Thomas. Malawi: Mount Waller.

DIAGNOSIS. Darker gray with short gray collar and red ear tufts; hairs under tail reddish laterally, white along midline.

DISTRIBUTION. Considered a synonym of *C. a. moloneyi* by Schwarz (1928a), but regarded as a distinctive subspecies of restricted distribution (Mount Waller and Vipya Plateau, 10°40′–11°50′S) by Ansell (1960).

Cercopithecus albogularis erythrarchus Peters, 1852

1852 *Cercopithecus erythrarchus* Peters. Mozambique: Inhambane.
1892 *Cercopithecus stairsi* Sclater. Mozambique: Chinde, Zambesi delta.
1907 *Cercopithecus albogularis beirensis* Pocock. Mozambique: Beira.
1907 *[Cercopithecus stairsi] mossambicus* Pocock. Mozambique.

1928 *Cercopithecus leucampyx nyasae* Schwarz. Malawi: Mlanje, Fort Lister, 1,065 m.
1948 *Cercopithecus mitis stevensi* Roberts. Zimbabwe: Mount Selinda, Melsetter.

DIAGNOSIS.   Somewhat more yellowish than *C. a. moloneyi,* and posterior dorsum yellow-brown rather than reddish; throat patch whitish but not extended to form a collar; caudal patch, which sometimes extends all around tail base, reddish laterally, white in midline. Hairs gray at base, followed by three or four yellow/black band pairs.

DISTRIBUTION.   Mlanje Plateau, southern Malawi, to Zimbabwe, and most of Mozambique.

Cercopithecus albogularis schwarzi Roberts, 1931

1931 *Cercopithecus leucampyx schwarzi* Roberts. South Africa: Transvaal, Pilgrim's Rest District, Mariepskop.

DIAGNOSIS.   Less dark and has a less extensive white stripe, and some red, under tail base compared with *C. a. labiatus.*

DISTRIBUTION.   Eastern Transvaal.

Cercopithecus albogularis labiatus I. Geoffroy, 1842

1842 *Cercopithecus labiatus* I. Geoffroy. South Africa.
1845 *Cercopithecus samango* Wahlberg. South Africa: inland of Port Natal (= Durban).
1853 *Cercopithecus chimango* Temminck. Emendation of *samango.*

DIAGNOSIS.   A dark gray form, slightly yellower on posterior dorsum, with no red but a creamy white stripe under tail base; frontal fringe dark; a partial white collar. Sometimes a white mustache. Hairs gray at base, with only three pairs of white and black bands.

DISTRIBUTION.   Kwazulu-Natal and Eastern Cape.

Cercopithecus mona group

Chromosomes: $2n = 66$ or 68. The chromosome studies of Dutrillaux et al. (1988) separated *C. campbelli* strongly from *C. mona, pogonias,* and *wolfi;* the latter two are closer together than to *C. mona.*

Cercopithecus mona (Schreber, 1775)
    Mona Monkey

1775 *Simia mona* Schreber. "Barbary."
1804 *Simia monacha* Schreber. Emendation?

DIAGNOSIS.   Dark brown, with gray or black limbs and yellowish head; underside and inner side of limbs white; oval white spot on haunch on each side. Hair bases gray, with four pairs of light/dark bands.

DISTRIBUTION.   Ghana to Cameroon, overlapping with *C. lowei* in the River Afram region, west of the River Volta, and in Cameroon extending somewhat south of the River Sanaga.

Cercopithecus campbelli Waterhouse, 1838
    Campbell's Mona

1838 *Cercopithecus campbelli* Waterhouse. Sierra Leone.
?1838 *Cercopithecus temminckii* Ogilby. Coast of Guinea.
1842 *Cercopithecus burnetti* Gray. Supposedly from Fernando Póo (= Bioko).
1870 *Cercopithecus mona* var. *monella* Gray. Senegal.

DIAGNOSIS.   Dark yellowish gray with somewhat darker gray rump and gray limbs, with yellow tips on thighs; underside and inner side of limbs whitish; hairs of brow band white with black tips; black line from eye to ear; crown speckled with yellow and black bands; cheek whiskers speckled yellow, with whitish bases. Hair bases gray, then four pairs of light/dark bands.

DISTRIBUTION.   From Senegal, just north of the River Casamance, to the River Cavally, on the Liberia–Ivory Coast border.

Cercopithecus lowei Thomas, 1923
    Lowe's Mona

1923 *Cercopithecus campbelli lowei* Thomas. Ivory Coast: Bandama.

DIAGNOSIS.   Brighter colored than *C. campbelli,* more orange, and rump contrastingly dark gray to blackish; thighs black; hairs of brow band yellow with black tips; crown more yellow; cheek whiskers speckled yellow, with gray bases. Hair bases gray, then three pairs of light/dark bands.

DISTRIBUTION. From the River Cavally (Ivory Coast) to the River Volta (Ghana).

NOTES. Kingdon (1997) separated this as a full species from *C. campbelli,* and I agree.

Cercopithecus pogonias Bennett, 1833
Crested Mona

DIAGNOSIS. Agouti gray, generally black on back; arms black, legs gray; underside and inner side of limbs orange; tail black on terminal one-third; crown speckled yellow with three longitudinal black stripes; orange ear tufts.

DISTRIBUTION. From the Cross River in Nigeria to the Congo, and somewhat east of the Oubangui; Bioko.

Cercopithecus pogonias pogonias Bennett, 1833

1833 *Cercopithecus pogonias* Bennett. Equatorial Guinea: Bioko Island.

DIAGNOSIS. Dark gray, with yellowish speckling on crown and nape and white speckling on flanks; a sharply defined black saddle; lower limb segments yellowish, only toes are black; underside orange or yellow; yellow or orange ear tufts; on crown, median black stripe extends forward to brows. Hairs gray at base with five alternating light/dark bands.

DISTRIBUTION. From the Cross River across the River Sanaga to Río Muni; Bioko.

Cercopithecus pogonias nigripes Du Chaillu, 1860

1860 *Cercopithecus nigripes* Du Chaillu. Gabon: Ofoubou River.

DIAGNOSIS. Dark gray, with yellowish speckling on crown and nape and white speckling on flanks; a sharply defined black saddle; lower limb segments buffy gray, hands and feet entirely black; underside orange or yellow; yellow or orange ear tufts; on crown, median black stripe ends well above brows, so yellow lateral areas meet in midline, forming a

U shape. Hairs with only three or four light/dark bands.

DISTRIBUTION. From Río Muni, Gabon, and Congo, south to Cabinda.

Cercopithecus pogonias grayi Fraser, 1850

1850 *Cercopithecus grayi* Fraser. West Africa.
1856 *Cercopithecus erxlebenii* Dallbet and Pucheran. West Africa? Napier (1981) suggested that, according to the type description, this could be a synonym of *C. p. nigripes.*
1909 *Cercopithecus pogonias pallidus* Elliot. Gabon.
1911 *Cercopithecus petronellae* Büttikofer. Upper Congo.

DIAGNOSIS. Dark chestnut-red, speckled with yellow on crown and nape, red on back, and orange on flanks and thighs; saddle dark red, blending with orange of flanks; lower limb segments buffy gray, only toes are black; underside and ear tufts yellow or whitish; on crown, median black stripe extends forward to brows.

DISTRIBUTION. From the middle River Sanaga east to the Central African Republic, the inland region of Congo, and, in Congo-Zaire, east of the River Oubangui to about the Itimbiri River, north of the River Congo.

Cercopithecus pogonias schwarzianus Schouteden, 1946

1944 *Cercopithecus mona schwarzi* Schouteden. Zaire: Mayumbe, Maduda. Not of Roberts, 1931.
1946 *Cercopithecus mona schwarzianus* Schouteden. Replacement for *schwarzi.*

DIAGNOSIS. Described as like *C. p. grayi* but darker, more yellowish red above, and thighs gray instead of orange.

DISTRIBUTION. Known only from the type locality at about 5°S, 13°E.

Cercopithecus wolfi Meyer, 1891
Wolf's Mona

DIAGNOSIS. Dark gray, generally redder on back; arms black, legs reddish; underside sharply demarcated white, yellow, or red; terminal third to half of tail black; pale brow band; only the two lateral black crown stripes; red or white ear tufts. Hairs gray at base, with three or four alternating light/dark bands. Scrotum blue.

DISTRIBUTION. South of the arc of the River Congo, to about 7°30′S.

Cercopithecus wolfi wolfi Meyer, 1891

1891 *Cercopithecus wolfi* Meyer. "Central West Africa."

DIAGNOSIS. A median dorsal chestnut patch; arms black; legs red; underside yellow, sometimes with an orange stripe along flanks; cheek whiskers yellow, speckled with black; ear tufts red.

DISTRIBUTION. From between the Congo and Sankuru Rivers. Colyn (1988) briefly described a few average differences between populations west and east of the River Lomami and noted that east of it the subspecies reaches only to 2°S and is then replaced by *C. w. elegans*.

Cercopithecus wolfi pyrogaster Lönnberg, 1919

1919 *Cercopithecus pyrogaster* Lönnberg. Congo-Zaire: Kwango River, Atene.

DIAGNOSIS. Color as in *C. w. wolfi*, but arms speckled with yellowish white; underside red; cheek whiskers buffy, speckled with black.

DISTRIBUTION. From between the Rivers Kwango and Kasai-Lulua.

Cercopithecus wolfi elegans Dubois and Matschie, 1912

1912 *Cercopithecus (Otopithecus) elegans* Dubois and Matschie. Congo-Zaire: probably Lomami River.

DIAGNOSIS. Dorsal color becoming browner toward rump; forearms black, upper arms with pale speckling; legs light gray; underside white; cheek whiskers white, with dark speckling increasing posteriorly; ear tufts white.

DISTRIBUTION. From between the Rivers Lomami and Lualaba, but only south of 2°S; it is apparently separated from *C. w. wolfi*, which occurs farther north between these two rivers, by a large area of swamp forest (Colyn 1988).

Cercopithecus denti Thomas, 1907
Dent's Mona

1907 *Cercopithecus denti* Thomas. Congo-Zaire: between Mawambi and Avakubi, Ituri Forest.
1912 *Cercopithecus (Otopithecus) denti liebrechtsi* Dubois and Matschie. Congo-Zaire: Stanley Falls.

DIAGNOSIS. Dark brown with yellowish head and dark to black limbs; underside and inner side of limbs sharply white; a pale brow band; black lateral crown stripes. Hairs gray at base, with three or four alternating light/dark bands.

DISTRIBUTION. Eastern Congo-Zaire, east and north of the Congo-Lualaba, in Congo-Zaire not west of the River Itimbiri, but extending into the Central African Republic to M'Brés at 6°49′N, 19°46′E, and east into Rwanda.

NOTES. Napier (1981) pointed out the anomaly that this species is nearly identical to *C. mona*, except that it has no haunch marks and has sharper contrast between upper parts and white underparts.

Cercopithecus cephus group

Chromosomes: $2n = 66$. Kingdon (1980) illustrated and described this species-group in special detail.

Cercopithecus petaurista (Schreber, 1774)
Lesser Spot-nosed Monkey

DIAGNOSIS. Agouti brown; underside white; limbs pale on inner sides; tail whitish below; brow band and upper cheeks black, lower cheeks with bushy white whiskers; a white heart-shaped nose spot.

DISTRIBUTION. From Guinea-Bissau to Togo.

Cercopithecus petaurista petaurista (Schreber, 1774)

1774 *Simia petaurista* Schreber. "Guinea."
1863 *Simia albinasus* Reichenbach. No locality.
1893 *Cercopithecus fantiensis* Matschie. Ghana: Fanti, Rio Boutry.
1923 *Cercopithecus büttikeri pygrius* Thomas. Ivory Coast: Bandama.

DIAGNOSIS. Black brow band continues around face and behind ears to form a complete border to crown, which is much brighter, brindled yellow and black. Lower segment of limbs is speckled like body. Hairs with gray bases and three pairs of alternating yellow and black bands; a line around the occiput has only one yellow band, and the crown has two.

DISTRIBUTION. East of the Sassandra in Ivory Coast.

Cercopithecus petaurista buettikoferi Jentink, 1886

1886 *Cercopithecus büttikoferi* Jentink. Liberia: Soforé Place, St. Paul's River.

DIAGNOSIS. Black brow band does not continue around face or behind ears; crown dully yellow/black agouti; a black/yellow agouti region below eye. Lower segment of limbs gray. Hairs gray at base, with only two alternating yellow/black bands; the head and neck have only two band pairs.

DISTRIBUTION. West of the Nzo-Sassandra system; according to A. H. Booth (1956), both this and *petaurista* are found, along with hybrids, on the right bank of the River Nzo. A skin from Tissalé, Ivory Coast, is intermediate, with only two bands on the body like *buettikoferi* but the distinct crown-cap, black around face, and speckled legs of *petaurista*.

NOTES. There are a few individual variants in both subspecies that bridge the gap and make it impossible to separate them sharply as distinct species.

Cercopithecus erythrogaster Gray, 1866
White-throated Guenon

DIAGNOSIS. Dark agouti red-brown; crown with a triangular black patch, with a single wide gold band anteriorly, the gold contrasting with solid black frontal and parietal bands; arms dark to black; legs paler; underside gray or red; tail as back above with dark tip, whitish below; white cheek and throat whiskers, with zone of yellow-black banded hairs on upper cheeks; heart-shaped nose patch may be brown, black, or with white hairs. Hairs of body gray at base with three pairs of alternating yellow and black bands; on crown with only one, very broad, yellow band and a short black tip; round face is a zone with two band pairs.

DISTRIBUTION. The confirmed distribution was given by Oates (1985) as from 6°02' to 6°56'N and from 4°27' to 6°11'E, in Nigeria, but he mentioned and mapped reports from hunters (later confirmed [Oates et al. 1992]) farther southeast, including slightly to the east of the Niger in the delta region at about 5°N; and (personal communication) Benin.

NOTES. Napier (1981) described the variability of the nose spot in the specimens in the BM(NH) collection. In two the spot is white; in four, black; in two, dark brown; in the type, black with white bases.

Cercopithecus erythrogaster erythrogaster Gray, 1866

1866 *Cercopithecus erythrogaster* Gray. "West Africa." Here restricted, on the evidence of Grubb et al. (1999) to Benin: Lama Forest.

DIAGNOSIS. Underside maroon-red, except for groin, underside of tail base, and inner side of thighs, which are white. Black fronto-parietal band reaches to just behind ears.

DISTRIBUTION. Known only from the Forêt Classée de Lama, in Benin, and presumptively from "a forested area on the Nigerian border north of Porto Novo" in Benin (Grubb et al. 1999:389).

NOTES. Red-bellied guenons have been exported from Lagos, Nigeria, but this seems unlikely to be an actual locality of origin (Grubb et al. 1999). My description is based solely on the type and one other specimen, in the BM(NH).

Cercopithecus erythrogaster pococki Grubb, Lernould, and
Oates, 1999

1999 *Cercopithecus erythrogaster pococki* Grubb, Lernould, and
Oates. Nigeria: "Lagos" (probably meaning inland from
Lagos).

DIAGNOSIS.   Underside pale brownish gray where
nominotypical subspecies is red; black fronto-
parietal band reaches farther around behind crown,
and may join its opposite number at the back.

DISTRIBUTION.   Nigerian part of the range.

NOTES.   Grubb et al. (1999) resurrected a manuscript name of Dollman's and made it available for
the Nigerian (gray-bellied) form of the species. At
the same time, they implied that other color differences may exist between the two, and the one con-
cerning the crown band is obvious in the (admittedly
limited) BM(NH) material. If these differences are
consistent, it may be that full specific status is indicated. The two forms are both living in the zoo directed by J.-M. Lernould at Mulhouse, France.

Cercopithecus sclateri Pocock, 1904
    Sclater's Guenon

1904 *Cercopithecus sclateri* Pocock. Nigeria: Benin City.

DIAGNOSIS.   Gray-brown above; front of crown
brindled black and yellow, forming a gold-flecked diadem, outlined by a black band from eye to ear and
around back of crown; throat with whitish hairs, not
forming a long, clearly white ruff; a narrow, pale
brow band; forearms gray; tail like body above, below red for one-third, then whitish, darkening toward tip; white nose spot, sometimes with a reddish
edge; white ear tufts. Facial skin pale bluish or pink
around eyes. Hairs very light gray at base, with only
two pairs of alternating yellow and black bands.

DISTRIBUTION.   Known from a small area in southeastern Nigeria, from west of Oguta at 5°42′N, just
east of the Niger (Oates and Anadu 1989), east to the
Cross River. Along the Niger, it comes to within 10
to 20 km of the known distribution of *C. erythrogaster*
(Oates et al. 1992).

NOTES.   Kingdon (1980) first proposed that this is a
valid species of the *C. cephus* group, and its limits
(both taxonomic and distributional) have since been
confirmed and delimited by Oates and Anadu (1989)
and Oates et al. (1992).

Cercopithecus erythrotis Waterhouse, 1838
    Red-eared Guenon

DIAGNOSIS.   Gray-brown above; crown black/buff
agouti; underside and inner side of limbs dark gray;
tail red above and below, except for proximal one-
fifth; a black brow band, extending to ears; triangular
red nose spot; red ear tufts. Hairs gray at base with
three pairs of alternating red and black bands.

DISTRIBUTION.   From the Cross River in Nigeria to
the River Sanaga in Cameroon, and at one point just
south of the Sanaga; Bioko.

NOTES.   Occurs south of the Sanaga in the Tinaso,
Lake Tisongo, and Ongue Creek areas, 9°53′ to
10°E, where it interbreeds with *C. cephus*.

Cercopithecus erythrotis erythrotis Waterhouse, 1838

1838 *Cercopithecus erythrotis* Waterhouse. Equatorial Guinea:
Bioko Island.

DIAGNOSIS.   Size smaller; tail shorter; pelage
longer; color darker.

DISTRIBUTION.   Bioko.

Cercopithecus erythrotis camerunensis Hayman, 1940

1940 *Cercopithecus erythrotis camerunensis* Hayman. Cameroon:
Badchama, Mamfe.

DIAGNOSIS.   Size larger; tail longer; pelage shorter;
color lighter, yellower.

DISTRIBUTION.   From the mainland part of the
range.

Cercopithecus ?signatus Jentink, 1886

1886 *Cercopithecus signatus* Jentink. "Perhaps from Banana"
(Congo-Zaire).

Oates (1985) described this monkey, which I have not examined, as follows. Reddish brown above, underside creamy white. Forearms brown or black. Cheek whiskers speckled black and yellow; a black brow band, extending somewhat onto temples; throat gray or silvery white, not forming a ruff; an indistinct white nose spot; creamy yellow ear tufts. Tail colored similarly above and below or underside slightly lighter. The type is unusually large for this species-group. Unfortunately, none of the four specimens referable to this species has a precisely known locality.

Although it has been considered synonymous with *C. erythrogaster* in the past, this is not correct; nor did Oates (1985) agree that it is a synonym of *C. nictitans,* which is much darker and grayer (among other differences). It may, he suggested, represent hybrids between *C. nictitans* and some member of the *C. cephus* group.

To draw attention to this taxon, I here listed it with a query as a species.

Cercopithecus cephus (Linnaeus, 1758)
Mustached Guenon

DIAGNOSIS.   Reddish brown; underside and inner side of limbs gray; a blue chevron on nose and black mustache below; cheek whiskers and throat patch white. Hairs, as in *C. erythrotis,* gray at base with three pairs of alternating red and black bands.

DISTRIBUTION.   From the Sanaga River in Cameroon south and east to the Congo River, and just south of the lower Congo in the northwestern corner of Angola (Machado 1969); extends east of the River Sangha, where it is said to be sympatric with *C. ascanius,* but less common (Lernould 1988), but Colyn (1999) did not find any sympatry.

Cercopithecus cephus cephus (Linnaeus, 1758)

1758 *Simia cephus* Linnaeus. "America" (= West Africa!).
1857 *Cercopithecus buccalis* Leconte. Gabon.
1915 *Cercopithecus pulcher* Lorenz. Cameroon.
1927 *Cercopithecus inobservatus* Elliot. West Africa.

DIAGNOSIS.   Distal three-fourths of tail bright red.

DISTRIBUTION.   The species's range except for coastal Gabon and Congo where *C. c. cephodes* is found. According to Colyn (1999), it ranges east to the Kadei-Mambéré system.

Cercopithecus cephus cephodes Pocock, 1907

1907 *[Cercopithecus cephus] cephodes* Pocock. Gabon.
1947 *Cercopithecus cephus gabonensis* Maclatchy and Malbrant. Gabon: Doumé-Ogooué (fixed by Hill [1966]).

DIAGNOSIS.   Distal three-fourths of tail dark brown above, light grayish brown below, darkening toward tip.

DISTRIBUTION.   From coastal Gabon and Congo, bounded by the Ogooué and Kouilou-Niari.

Cercopithecus cephus ngottoensis Colyn, 1999

1999 *Cercopithecus cephus ngottoensis* Colyn. Central African Republic: Sangora, near Boyali (18°13′E, 4°04′N), 66 km from Bangui on the Bangui-Mbaiki road.

DIAGNOSIS.   Red-tailed, with a white labio-nasal region.

DISTRIBUTION.   From the Kadei-Mambéré region east to the Oubangui, where it occurs as far north as Bangui.

Cercopithecus ascanius (Audebert, 1799)
Red-tailed Monkey

DIAGNOSIS.   Agouti brown; underside and inner side of limbs whitish; at least part of tail red; a heart-shaped nose spot, usually white or yellow, but may be red or black; bushy, adpressed white cheek whiskers. The hair bases on the body are gray, and there are two to four alternating red/black bands, the first pair separated from the light base by an indistinct black band; the subterminal black band is longer than the rest. On the head, the hairs have just two band pairs; around the face and on the bridge of the nose the hairs have just a single straw-colored band and long, black tips.

DISTRIBUTION.   Mainly south and east of the Congo-Oubangui system, through Uganda to the

Rift Valley in Kenya, and western Tanzania; northern Angola and northwestern Zambia.

NOTES.   In most of the subspecies, there is a whorl in front of the ears, from which the white cheek hairs radiate in a sort of spiral: nearly vertically implanted in front, but sweeping down and back below. These long cheek whiskers are bordered inferiorly by a black band, which curves up medially alongside the nose spot and becomes lost in the black or blue circumocular skin. The exception to this is *C. a. schmidti*, which has no whorl; it is not clear whether the lack of the whorl is fully consistent, but (as noted below) *schmidti* might be a candidate for species rank.

Cercopithecus ascanius ascanius (Audebert, 1799)

1799 *Simia ascanius* Audebert. No locality.
1845 *Cercopithecus melanogenys* Gray. Africa.
1863 *Cercopithecus histrio* Reichenbach. No locality.
1886 *Cercopithecus picturatus* Santos. Angola: Quimpalala, near Ambriz, northern coast.

DIAGNOSIS.   Nose spot white; ear tufts reddish. Temporal whorl present. Underside gray, extending under base of tail; tail reddish above, its tip darker; black front band and inferior lateral black band wide; cheeks with extensive black area. Periocular skin light blue. Hairs gray at base with four pairs of alternating red and black bands.

DISTRIBUTION.   From Angola and southwestern Congo-Zaire, south of the Congo and Kasai Rivers. Machado (1969) mapped it in Angola south to somewhat beyond 8°S and east to more than 15°E. Farther east there is apparently a hybrid zone with *C. a. atrinasus* (see below).

Cercopithecus ascanius atrinasus Machado, 1965

1965 *Cercopithecus ascanius atrinasus* Machado. Angola: Zovo, 8°07'S, 18°04'E.

DIAGNOSIS.   Nose spot black; otherwise closely resembling nominotypical *ascanius*. The face, with its light blue circumocular rings, black nose and maxilla, and restricted white hairs emerging from the cheek whorl, has a very distinctive appearance. The cheek whiskers flare laterally and are white above, black in the middle (a lateral extension from the black maxillary zone), and grayish below, extending to the throat.

DISTRIBUTION.   Known only from a small area around the type locality. Machado (1969) mapped it in Angola from 8 to 9°S, 18°00' to 30'E. From three localities to the east and south of this area he reported specimens that he considered hybrids with *C. a. ascanius*, which is unexpected because the nominotypical *ascanius* occurs mainly to the north, across the border in Congo-Zaire. From still farther east, at 8°S, 19°40'E, came a specimen said to be a hybrid with *C. a. katangae*.

Cercopithecus ascanius whitesidei Thomas, 1909

1909 *Cercopithecus ascanius whitesidei* Thomas. Congo-Zaire: upper Lulanga River, Ikau, about 1°N, 22°E.
1913 *Cercopithecus (Rhinostictus) ascanius kassaicus* Matschie. Congo-Zaire: Kasai, Pogge Falls.
1913 *Cercopithecus (Rhinostictus) ascanius omissus* Matschie. Congo-Zaire: obtained from a party coming from Maniema, but doubtless originating elsewhere.
1913 *Cercopithecus (Rhinostictus) ascanius cirrhorhinus* Matschie. Congo-Zaire: probably lower Lomami.
1913 *Cercopithecus (Rhinostictus) ascanius pelorhinus* Matschie. Congo-Zaire: bought at Yambuya, but doubtless originating elsewhere.

DIAGNOSIS.   Nose spot creamy or yellow to deep orange; ear tufts reddish. Temporal whorl present. Underside whitish, extending under base of tail; tail dark above, its distal one-third redder; black frontal band narrow, inferior lateral black band reduced; cheeks with reduced black area, the whiskers much less laterally flaring than in *C. a. ascanius* and *atrinasus*, and white superiorly, agouti-brown inferiorly, with a gray throat. Periocular skin nearly black. Hairs gray at base with three pairs of alternating red and black bands.

DISTRIBUTION.   From between the Congo and the Kasai and Sankuru Rivers; Machado (1969) placed the border with *C. a. katangae* a little south of the Sankuru, and Colyn (1988) found intermediates between the two (usually having the nose spot white) between the Rivers Lualaba and Lomami. The variability among populations ascribed to this subspecies was emphasized by Colyn (1988), who noted a cline of increasing erythrism from southeast to west.

NOTES. The type locality of Matschie's *kassaicus* is within the range of *C. a. katangae,* which is evidently incorrect (see Napier 1981); Schwarz (1928a), who examined the type, confirmed it as a synonym of *C. a. whitesidei.* The type of Matschie's *omissus* is, strictly speaking, according to Machado (1969), transitional between this subspecies and *C. a. katangae.*

Cercopithecus ascanius katangae Lönnberg, 1919

1919 *Cercopithecus ascanius katangae* Lönnberg. Congo-Zaire: Katanga, Kinda.

DIAGNOSIS. Nose spot white; ear tufts reddish. Temporal whorl present. Underside white, extending under base of tail; tail very dark above, its distal one-third blackish; black front band narrow, inferior lateral black band reduced; cheeks with reduced black area. Periocular skin violet in color (Machado 1969). Hairs white at base (not gray) with three pairs of alternating red and black bands.

DISTRIBUTION. From southern Congo-Zaire and Angola, between the Kasai and Sankuru Rivers. According to Machado's (1969) map, it occurs well into Angola, to 11°S, 17°E, except for the small enclave where *C. a. atrinasus* occurs.

Cercopithecus ascanius schmidti Matschie, 1892

1892 *Cercopithecus schmidti* Matschie. Supposedly Congo-Zaire: Maniema; but Uganda, on the northern shore of Lake Victoria, according to Machado (1969).
1913 *Lasiopyga ascanius kaimosae* Heller. Kenya: upper Lukosa River, near Kaimosi.
1913 *Cercopithecus (Rhinostictus) schmidti mpangae* Matschie. Uganda: Mpanga Forest.
1913 *Cercopithecus (Rhinostictus) schmidti sassae* Matschie. Congo-Zaire: Sassa, west of Lake Edward (= Ishasha River).
1913 *Cercopithecus (Rhinostictus) schmidti enkamer* Matschie. Congo-Zaire: Chima Kilima, north of Mawambi, upper Ituri.
1914 *Cercopithecus schmidti montanus* Lorenz. Congo-Zaire: Wabembe country, northwest of Lake Tanganyika.

1914 *Cercopithecus schmidti ituriensis* Lorenz. Congo-Zaire: Ituri Forest.
1917 *Cercopithecus schmidti rutschuricus* Lorenz. Congo-Zaire: mountains on east side of Rutshuru Plains.
1919 *Cercopithecus schmidti orientalis* Lorenz. Kenya: obtained at Simba, Zuwani River, but doubtless originating elsewhere.

DIAGNOSIS. Nose spot white; ear tufts white. No temporal whorl. Underside whitish but base of tail gray below; tail reddish above, its tip darker; black front band and inferior lateral black band narrow; cheeks with extensive whitish area, extending back to partially cover the ears, and margined with black below. Periocular skin slaty blue to nearly black (Machado 1969). Hairs gray at base with two, occasionally three, pairs of red/black bands. The appearance of the face (see Kingdon 1980) is quite different from that of other subspecies, and close study of the consistency of the pattern is necessary to show whether it is consistent; if so, the taxon should rank as a full species. According to Colyn (1988), the color of the temporal and cheek hairs varies, but that of the body and limbs does not, except that in the southeast of the Zairean range individuals have dark gray legs, sometimes extending to the back.

DISTRIBUTION. From east of the Lualaba River, into Uganda, Tanzania, and Kenya; north of the River Congo, it extends north into Sudan (Lernould 1988), and, according to Colyn (1999), it extends west of the Oubangui north of the town of Bangui.

Cercopithecus lhoesti group

Dark gray with reddish zone on back; limbs black; underside dark gray or black; tail black toward tip; white on throat; scrotum light blue. Chromosomes: $2n = 60$.

Cercopithecus lhoesti Sclater, 1899

L'Hoest's Monkey

1899 *Cercopithecus l'hoesti* Sclater. "Congo": restricted to Tschopo River near Kisangani by Pocock (1907).
1905 *Cercopithecus thomasi* Matschie. Rwanda: east shore of Lake Kivu.
1915 *Cercopithecus thomasi rutshuricus* Lorenz. Congo-Zaire: mountains on east side of Rutshuru Plains.

DIAGNOSIS. Crown and nape black with some gray ticking; saddle short, oval, orange-speckled; limbs black; flanks and underside black; tail lighter beneath, above gray except for terminal one-fourth, which is black and tufted; face black; bushy white cheek whiskers and throat ruff. Ears visible. Face gray, with muzzle black. Hairs gray at base with four pairs of alternating white and black bands.

DISTRIBUTION. Eastern Congo-Zaire, from near the Lualaba to Ituri Forest and east into Burundi, Rwanda, and western Uganda; south of the lower River Lindi. The range, as remarked by Colyn (1988), is much the same as that of *C. hamlyni,* but it is found in more of the montane regions along the western Rift.

Cercopithecus preussi Matschie, 1898
Preuss's Monkey

DIAGNOSIS. Crown and nape black, ticked with gray; saddle solidly chestnut-red; flanks gray with buff banding; underside dark gray; limbs black; tail like back proximally, then dark gray with terminal one-fourth black; upper chest and throat white, merging upward with gray cheek whiskers. Face uniformly gray. Ears hidden by crown hair. Hairs gray at base with only two pairs of white/black bands.

DISTRIBUTION. Highlands of Cameroon-Nigeria border region (Mount Cameroon, Bamenda highlands, Korup Reserve, Cross River highlands); Bioko Island.

Cercopithecus preussi preussi Matschie, 1898

1898 *Cercopithecus preussi* Matschie. Cameroon: Victoria.
1905 *Cercopithecus crossi* Forbes. Cameroon.

DIAGNOSIS. Larger; tail longer than head plus body; dorsal saddle brighter chestnut and extends more posteriorly; tail lighter gray.

DISTRIBUTION. Mainland part of range.

Cercopithecus preussi insularis Thomas, 1910

1910 *Cercopithecus preussi insularis* Thomas. Equatorial Guinea: Bioko Island, northern Buntabiri.

DIAGNOSIS. Smaller; tail about equal in length to head plus body; dorsal saddle less bright, does not extend so far onto tail; tail darker gray.

DISTRIBUTION. Bioko Island.

Cercopithecus solatus Harrison, 1988
Sun-tailed Monkey

1988 *Cercopithecus solatus* Harrison. Gabon: River Bali, Forêt des Abeilles, southeast of Booué, 0°14′S, 12°15′E.

DIAGNOSIS. Crown and nape dark with chestnut-orange speckling; saddle chestnut- orange, extending to flanks, shoulders, and upper thighs; rest of limbs black; tail whitish below, above speckled gray proximally, becoming orange-yellow at tip; upper chest and throat ruff white, merging into agouti-gray cheek whiskers; ears not hidden. Face dark, with muzzle black.

DISTRIBUTION. The Forêt des Abeilles, south and west of the great bend of the River Ogooué and east of the River Offoué, in Gabon.

Cercopithecus hamlyni group

Dark olive-gray; limbs black; underside black; tail black on terminal one-fourth; scrotum blue. Chromosomes: $2n = 64$.

Cercopithecus hamlyni Pocock, 1907
**Hamlyn's Monkey or
Owl-faced Monkey**

DIAGNOSIS. Dense, very dark pelage, elongated to form a hood over crown and cheeks. Hairs gray at base with four or five light/dark band pairs.

DISTRIBUTION. Eastern Congo-Zaire: from lowlands near the River Lualaba to Ituri Forest and Rwanda, north to the Lindi-Nepoko system, south to the limit of the rain forest. It extends into the mountains west of Lake Kivu, but elsewhere is more of a lowland species than the otherwise similarly distributed *C. lhoesti* (Colyn 1988).

Cercopithecus hamlyni hamlyni Pocock, 1907

1907 *Cercopithecus hamlyni* Pocock. Congo-Zaire: Ituri Forest.
1910 *Cercopithecus leucampyx aurora* Thomas and Wroughton. Congo-Zaire: south end of Lake Kivu.

DIAGNOSIS. A narrow white stripe running down nose from interocular region to border of upper lip, and a white brow band, the two streaks together forming a characteristic T of white; body olive green to yellow-gray, varying in tone, with silvery gray hairs on backs of thighs, perianal region, and base of tail; underside, limbs, and distal half of tail black.

DISTRIBUTION. From the full range of the species except, it is said, the bamboo zone of Mount Kahuzi; specimens from Kisangani, on the western edge of the range, and from Ruhengeri (Rwanda), on its eastern edge, are alike (Colyn and Rahm 1987).

Cercopithecus hamlyni kahuziensis Colyn and Rahm, 1987

1987 *Cercopithecus hamlyni kahuziensis* Colyn and Rahm. Congo-Zaire: Musisi, 2°18′S, 28°42′E, a marshy zone at base of bamboo forest between Mount Kahuzi and Mount Biega.

DIAGNOSIS. Body darker; facial hairs entirely black, except around eyes where lighter, grayish; no white nasal band or white brow band. Tail shorter than head plus body.

DISTRIBUTION. Said to be from the bamboo zone of Mount Kahuzi.

NOTES. Colyn and Rahm (1987) based their new subspecies on three specimens and a photo of a living animal, only one apparently having a precisely known locality, though all are from Mount Kahuzi. They mentioned two intermediate specimens, one from Mulungu, southeast of Mount Biega, and one known to be only from Mount Kahuzi; and also a typical *C. hamlyni* from the transition between montane and bamboo forest zones on Mount Kahuzi. If the extrapolated distributional data given by Colyn and Rahm (1987) are correct, this is a taxon restricted to the bamboo and intermediate marshy zones on Kahuzi-Biega, with a narrow zone where there may be both hybrids and the two parental forms. More definite locality data are needed (especially field observations); it could turn out to be either a morph of very restricted distribution or a sharply bounded full species.

Cercopithecus neglectus group

Gray with white bands; limbs dark, with arms black; a white haunch stripe; underside dark; tail black; a broad orange frontal band margined with black; a white beard on chin and around mouth. Scrotum blue. Chromosomes: $2n = 58$–62.

Cercopithecus neglectus Schlegel, 1876
De Brazza's Monkey

1870 *Cercopithecus leucampyx* Gray. White Nile. Not of J. B. Fischer, 1829 (= *mitis*).
1876 *Cercopithecus neglectus* Schlegel. Sudan: White Nile; according to Schwarz (1928a), more probably Niam Niam or Monbuttu country.
1886 *Cercopithecus brazzae* Milne-Edwards. Congo-Brazzaville.
1907 *C[ercopithecus] neglectus brazziformis* Pocock. Said to be from "French Congo" (= Congo-Brazzaville).
1908 *Cercopithecus ezrae* Pocock. Probably "Upper Congo."
1919 *Cercopithecus brazzae uellensis* Lönnberg. Congo-Zaire: Uele, Poko.

DIAGNOSIS. Robust, short-tailed. Hairs in adult have a white base and six or seven pairs of alternate white and black bands. There are striking age changes in this species. Newborns are yellow-brown, with the limbs and underside yellow. Infants have hairs that are gray at the base, yellow at the tip; the crown and median dorsal line are dark, the brow band pale yellow, the cheeks pale; the tail yellow; a beard begins to appear. Juveniles are agouti grayish brown with dark hair bands, chestnut brow band, rump, and tail base, and an occasional individual retains this pelage into adulthood (*"ezrae"* morph).

DISTRIBUTION. From northeastern Angola, Río Muni, Gabon, and Cameroon east into Uganda, Mount Elgon and elsewhere in western Kenya, and southwestern Ethiopia.

Tribe Papionini Burnett, 1828

Synonyms are Cynocephalina Gray, 1825 (whose genotype, *Cynocephalus* [baboons], is preoccupied by the name for the colugo [order Dermoptera]); Macacidae Owen, 1843; Cynopithecinae O. Hill, 1966; and Cercocebini and Theropithecini Jolly, 1966.

The third lower molar nearly always has a hypoconulid (may be absent in some small individuals in *Macaca*); molar flare is present except in *Theropithecus;* ischial callosities are fused across the midline in adult male, except in some *Macaca;* there are periodic sexual swellings, except in some *Macaca;* the diploid chromosome number is always 42. Strasser and Delson (1987) added that the pyriform aperture is wider compared with Cercopithecini, there is a steep anteorbital drop (it slopes in Cercopithecini), and facial fossae are present. The fossae are variable in development (relatively poorly in *Macaca*) and occur in both maxilla and mandible.

According to Soligo and Müller (1999), *Macaca mulatta* has a double-layered structure to the nails of the toes, whereas *Papio anubis* has a single layer.

Genus *Macaca* Lacépède, 1799

Macaques

I questioned (Groves 1989) the monophyly of the genus altogether, but this question now appears to be resolved by molecular studies (for example, Morales and Melnick 1998). At any rate the genus is very old and separated from other Papionini 7 million years ago (Stewart and Disotell 1998).

Fooden (1976) divided the genus into four species-groups, based mainly on the shape of the male genitalia. These were the *sylvanus-silenus* group (including *M. sylvanus, silenus, nemestrina,* and the Sulawesi species), the *fascicularis* group (including *M. fascicularis, mulatta, cyclopis,* and *fuscata*), the *sinica* group (including *M. sinica, radiata, assamensis,* and *thibetana*), and *M. arctoides* in a group all by itself. Delson (1980) suggested that *M. sylvanus* should be removed from association with the others of that group and be placed in a group by itself; and *M. arctoides* could be seen as a highly derived member of the *sinica* group.

Hair banding was studied by Inagaki (1996). Band number is extremely variable, but the Sulawesi and *M. nemestrina* groups and *M. thibetana* have saturated or single-banded hair, contrasting with the multi-

banded states usual in other taxa. *Macaca arctoides* has up to 10 bands.

Zhang and Shi (1993) analyzed mitochondrial DNA in Chinese and Vietnamese macaque species using 16 restriction endonucleases, and incorporated some Japanese data; they recovered Fooden's four species-groups, among which the *fascicularis* group was sister group to the rest. *Macaca assamensis* was associated with *M. thibetana* (= Fooden's *M. sinica* group); *M. arctoides* and *M. nemestrina* formed clades of their own.

Morales and Melnick (1998) studied the mitochondrial ribosomal genes (12S and 16S), also by restriction site mapping, but in almost all species, with many specimens of most of them. They affirmed the monophyly of *Macaca* with 99% bootstrap support. Their phylogeny was as follows: (1) *M. sylvanus* was sister to the rest of the genus; (2) the next split was between the *M. nemestrina* + Sulawesi groups and the rest; (3) in the first branch the Sulawesi macaques formed a monophyletic clade as did the *nemestrina* group, but none (!) of the Sulawesi species was monophyletic; (4) the residue, united by only 60% bootstrap support, divided into *fascicularis* + *arctoides* vs. the rest, which in turn divided into a *mulatta* + *cyclopis* + *fuscata* clade and a *radiata* + *sinica* + *assamensis* + *thibetana* group. Although only the last of these corresponds precisely to any of Fooden's groups, these latter were explicitly phenotypic, and indications were given of probable polarities in the morphological characters used (primarily the genitalia).

I propose here to divide the genus into six species-groups, as follows:

1. *M. sylvanus* group: monotypic.
2. *M. nemestrina* group: *M. nemestrina, leonina, silenus, pagensis.*
3. Sulawesi group.
4. *M. fascicularis* group: *M. fascicularis, arctoides.*
5. *M. mulatta* group: *M. mulatta, cyclopis, fuscata.*
6. *M. sinica* group: *M. sinica, radiata, assamensis, thibetana.*

Stewart and Disotell (1998) proposed that the genus evolved in Africa, where its sister clade lives, and consequently *M. sylvanus* is surviving in the ancestral range of the genus.

Macaca sylvanus group:

1758 *Simia* Linnaeus. *Simia sylvanus* Linnaeus, 1758. See Stiles and Orleman (1927); suppressed by the International Commission on Zoological Nomenclature (1929), Opinion 114.
1799 *Macaca* Lacépède. *Simia inuus* Linnaeus, 1758.
1812 *Inuus* E. Geoffroy. *Innus ecaudatus* E. Geoffroy, 1812.
1816 *Sylvanus* Oken. Unavailable.
1827 *Magus* Lesson. *Simia sylvanus* Linnaeus, 1758.
1828 *Pithes* Burnett. *Simia sylvanus* Linnaeus, 1758.
1841 *Salmacis* Gloger. Replacement for *Macaca*.

Macaca nemestrina group:

1820 *Silenus* Goldfuss. *Simia silenus* Linnaeus, 1758.
1839 *Maimon* Wagner. *Simia silenus* Linnaeus, 1758.
1862 *Vetulus* Reichenbach. Replacement for *Silenus*.
1862 *Nemestrinus* Reichenbach. *Simia nemestrina* Linnaeus, 1766. Not of Latreille, 1802 (Diptera).

Sulawesi group:

1835 *Cynopithecus* E. Geoffroy. *Cynocephalus niger* Desmarest, 1822.

Macaca mulatta group:

1840 *Rhesus* Lesson. *Cercopithecus mulatta* Zimmermann, 1780.
1848 *Lyssodes* Gistel. *Macaca speciosus* F. Cuvier, 1825.

Macaca fascicularis group:

1862 *Cynamolgus* Reichenbach. *Macaca irus* F. Cuvier, 1818.

Macaca sinica group:

1862 *Zati* Reichenbach. *Macaca radiata* E. Geoffroy, 1812.

Macaca sylvanus group

Apex of glans penis bluntly bilobed; breadth of glans 59–89% of its length. Tail reduced to at most a nubbin, which contains no vertebrae. Sexual swelling at midcycle, circular, blue-gray in color.

Macaca sylvanus (Linnaeus, 1758)

Barbary Macaque

1758 *Simia sylvanus* Linnaeus. "In Africa, Ceylona." Restricted to "Barbary Coast" by G. M. Allen (1939a).
1766 *Simia inuus* Linnaeus. "Africa."
1799 *Simia pithecus* Schreber. Africa.
1812 *Inuus ecaudatus* E. Geoffroy. Mediterranean coast of Africa, and Gibraltar.
1863 *Pithecus pygmaeus* Reichenbach. Gibraltar.

DIAGNOSIS. Pelage coarse, agouti grayish yellow; crown hairs erect, golden brown; face dark pink or brown, freckled, with light eyelids.

DISTRIBUTION. Atlas range of Morocco, Algeria, and presumably Tunisia; Rif Mountains; Gibraltar (where introduced in Roman times and maintained until current day by repeated introductions from Morocco).

Macaca nemestrina group

Apex of glans penis bluntly bilobed; breadth of glans 59–89% of its length. Sexual swelling at midcycle, reddish, extending to tail root. Crown hairs radiate forward and to sides from a central whorl. Tail short.

Fooden (1975) divided this species-group into two species, *M. silenus* and *M. nemestrina*, the latter with subspecies *M. n. nemestrina, leonina,* and *pagensis*. Subsequent authors, such as Wilson and Wilson (1977), preferred to set the third of these apart as a full species, which according to Whitten and Whitten (1982) was divided into subspecies. It is only the pervasiveness of the BSC that has prevented recognition of *leonina* as a further full species.

Macaca silenus (Linnaeus, 1758)

Lion-tailed Macaque

1758 *Simia silenus* Linnaeus. "Ceylon"; corrected to India: Western Ghats, inland from Malabar Coast, by Fooden (1975).
1777 *Cercopithecus vetulus* Erxleben. "Ceylon."
1792 *Simia (Cercopithecus) silenus albibarbatus* Kerr. "Ceylon and the rest of India."
1792 *Simia ferox* Shaw. "East Indies, Ceylon and interior parts of Africa."
1798 *Simia veter* Audebert. Not of Linnaeus, 1766 (unidentifiable).
1822 *Simia silanus* F. Cuvier. Error for *silenus*.

DIAGNOSIS. Color black, except for long, gray cheek and chin ruff; black crown patch narrow anteriorly. Tail carried simply hanging down behind; its length 55–75% of head plus body length. Supraorbital ridges thin (about 5 mm thick [Fooden 1975]), but strongly projecting. There are strongly marked suborbital depressions, whereas in the other species (and all other macaques) the malar region is flat or somewhat convex. Baculum long, 18.4–21.1 mm, with distal process 4.3–5.5 mm (Fooden 1975, table 6).

DISTRIBUTION. Western Ghats of India, from Ka-lakkadu Hills, 8°25′N, north to Anshi Ghat, 14°55′N (Fooden 1975). Now very much reduced, and species is endangered.

Macaca nemestrina (Linnaeus, 1766)
Sunda Pig-tailed Macaque or Beruk

1766 *Simia nemestrina* Linnaeus. Indonesia: Sumatra.
1774 *Simia platypygos* Schreber. No locality.
1795 *Simia longicruris* Link. No locality.
1800 *Simia fusca* Shaw. No locality.
1821 *Simia carpolegus* Raffles. Indonesia: Bencoolen (Bengkulu), Sumatra.
1826 *Macaca libidinosus* I. Geoffroy. No locality.
1839 *Pithecus maimon* de Blainville. No locality.
1842 *Macacus (Maimon) brachyurus* H. Smith. Indonesia: Sumatra.
1906 *Macaca broca* Miller. Malaysia: Sapagaya River, Sabah.
1936 *Macaca nemestrina nucifera* Sody. Indonesia: Bangka.

DIAGNOSIS. Much larger than other three species. Tail well furred, carried in a backward arch with tip pointing down, less than half of head plus body length; median dorsal region blackish, contrasting with agouti brown of flanks; crown blackish, the whorled region broad in front; cheek whiskers short, pale at base with black tips. Supraorbital ridges thin (about 5 mm [Fooden 1975]) and receding and curved down laterally. Baculum short, 14.9–19.1 mm, with distal process 3.0–4.7 mm (Fooden 1975).

DISTRIBUTION. Malay Peninsula from about 7°30′N, Sumatra, Bangka, Borneo.

NOTES. Though overall larger than the other three species of the group, there is some overlap. According to Fooden (1975), greatest skull length is 139.5–174.5 mm in adult males, compared with 124.1–140.1 in the other species, and 115.4–138.9 vs. 102.3–121.1 mm in adult females.

Macaca leonina (Blyth, 1863)
Northern Pig-tailed Macaque

1863 *Inuus leoninus* Blyth. Burma: northern Arakan.
1869 *Macacus andamanensis* Bartlett. India: Port Blair, Andaman Islands (where not indigenous).
1903 *Macacus coininus* Kloss. Lapsus for *leoninus.*
1906 *Macaca adusta* Miller. Burma: Champang, Tenasserim.
1906 *Macaca insulana* Miller. Thailand: Chance Island (= Ko Chan), Mergui Archipelago.

1919 *Macaca nemestrina indochinensis* Kloss. Thailand: Lat Bua Kao.
1931 *Macaca nemestrina blythii* Pocock. Locality unknown, perhaps the Naga Hills according to Pocock.

DIAGNOSIS. Tail less than half of head plus body length, well furred, carried in a forward arch over back, the tip pointing forward and upward; whole body agouti golden brown, with no contrast be-tween back and flanks; crown brown, the whorled area narrow in front; cheek whiskers fairly long, buffy; a red streak extending laterally from outer cor-ner of each eye. Supraorbital ridges thicker (7–8 mm [Fooden 1975]) but receding and downcurved later-ally. Baculum long as in *M. silenus,* 14.9–21.6 mm, but with distal process even longer, 6.0–9.2 mm (Fooden 1975).

DISTRIBUTION. From about 8°N in peninsular Thailand, through Burma and Indochina into Ban-gladesh, India extending north as far as the Brahma-putra (Choudhury 1997), and southernmost Yunnan, China (Zhang et al. 1997). There may be a gap in the distribution in central and northeastern Burma between about 20 and 25°N, whence it has not been recorded except on the coast in Arakan (Fooden 1975).

NOTES. Fooden (1975) found evidence of appar-ently very restricted hybridization between this spe-cies and *M. nemestrina.* One (out of five) *M. leonina* specimens from the islands of Phuket and Yao Yai (both at 8°N, off the west coast) is large like *M. nem-estrina* and has a short baculum but with a typical *leonina* long distal process; of two specimens from the vicinity of Ban Nong Kok on the opposite main-land, one is *M. leonina* and the other is a clear hybrid. Two of five specimens from the region around Chong (7°33′N, 99°47′E) are *M. nemestrina,* though one is small like *M. leonina;* the other three are clear hybrids. The hybrid zone seems thus very restricted in extent, over about 100 km north-south and span-ning the narrowest part of the aptly named Isthmus of Kra, and even within this zone out of 12 speci-mens only 5 (and one other questionably) show hy-brid features.

Macaca pagensis (Miller, 1903)
**Mentawai Macaque or
Bokkoi**

DIAGNOSIS.   Tail thinly furred, less than 35% of head plus body length; nonagouti dark brown or black, becoming pale brown below; whorled hair on crown broad in front; cheek whiskers short, with white tones. Supraorbital ridges as in *M. leonina*. Baculum unknown.

DISTRIBUTION.   Mentawai Islands.

NOTES.   The two subspecies ascribed to this species are strongly and, apparently, consistently different and might perhaps be more properly ranked as full species.

Macaca pagensis pagensis (Miller, 1903)

1903 *Macacus pagensis* Miller. Indonesia: South Pagai Island.
1923 *Rhesus nemestrina mentaveensis* de Beaux. Indonesia: Sioban, Sipora Island.

DIAGNOSIS.   Color dark brown above; sides of neck and front of shoulders contrastingly pale ochery; legs brown, arms reddish brown.

Macaca pagensis siberu Fuentes and Olson, 1995

1995 *M[acaca] p[agensis] siberu* Fuentes and Olson. Siberut.

DIAGNOSIS.   Described by Whitten and Whitten (1982) as being much darker than *M. p. pagensis,* lacking the light-colored sides of the neck and front of shoulders, and with a contrasting white cheek patch. A specimen in the BM(NH) collection, said to have come from Padang, is identified by them with this subspecies. I described the circumstances surrounding the possibly inadvertent introduction of this name (Groves 1996b).

Sulawesi group

Apex of glans penis bluntly bilobed; breadth of glans 59–89% of its length.
Molecular data (Morales and Melnick 1998) suggest that the Sulawesi macaques separated from the

M. nemestrina group some 4.5 million years ago, which is in excellent agreement with the palaeontologically based reconstruction I proposed (Groves 1976). More recently, Evans et al. (1999) even suggested that the southern and the northern/central groups of species may have separate origins from *M. nemestrina,* but according to John Trueman (personal communication), this proposal does not survive t-PTP testing.

Macaca maura (F. Cuvier, 1823)
Moor Macaque

1823 *Macacus maurus* F. Cuvier. "India."
1829 *Simia cuvieri* Fischer. "Indes Orientales."
1844 *Macacus fusco-ater* Schinz. Celebes.
1866 *Macacus (Gymnopyga) inornatus* Gray. "Borneo."

DIAGNOSIS.   Body brown or black, with smaller gray rump patch.

DISTRIBUTION.   Southern peninsula of Sulawesi, reaching north as far as base of Toraja highlands.

Macaca ochreata (Ogilby, 1840)
Booted Macaque

DIAGNOSIS.   Forearms, shanks, and rump whitish or light brown, paler than trunk; hair on crown darker than surrounding hair and stands erect, forming a rectangular cap. Ischial callosities oval, but lower part often bends laterally (Watanabe et al. 1991b).

DISTRIBUTION.   The whole of the southeastern peninsula of Sulawesi (extending to the north of the lakes region) and large islands off the southern coast.

NOTES.   In the east, the border between *M. o. ochreata* and *M. tonkeana* is the La River; in the west, *M. o. ochreata* extends west along the coast, across the Karaena River in its lower course, but does not reach into the uplands farther inland, where *M. tonkeana* is found (Watanabe et al. 1991b). A specimen from somewhere in the Karaena River region was identified by Watanabe et al. (1991b) as a probable hybrid between the two, and borderland populations showed a certain degree of intermediacy in external features (such as callosity shape, crown hair pattern, and color of limbs), suggesting gene flow.

Macaca ochreata ochreata (Ogilby, 1840)

1840 *Papio ochreatus* Ogilby. Locality unknown.

DIAGNOSIS. Darker: body generally blackish; forearms, shanks, and rump whitish.

DISTRIBUTION. Mainland part of range.

Macaca ochreata brunnescens (Matschie, 1901)

1901 *Papio (Inuus) brunnescens* Matschie. Indonesia: Butung Island.

DIAGNOSIS. Lighter, body generally brown; forearms, shanks, and rump light brown.

DISTRIBUTION. Muna, Butung Islands.

Macaca tonkeana (Meyer, 1899)
Tonkean Macaque

1899 *Macacus tonkeanus* Meyer. Indonesia: Tonkean, Sulawesi.
1901 *Papio (Inuus) tonsus* Matschie. Locality unknown.
1901 *Papio (Inuus) hypomelas* Matschie. Locality unknown.
1949 *Cynopithecus togeanus* Sody. Indonesia: Malenge Island, Togean group.

DIAGNOSIS. Body black, except for cheek whiskers and a wide, bushy-haired area on rump and back of thighs, which are contrastingly gray; underside not lighter than upper, or somewhat paler, grayer on chest and axillae; bare or sparsely haired gluteal fields on either side of tail; ischial callosities orange-toned, oval; muzzle elongated, with lateral ridges. Tail short but not as reduced as in *M. hecki, nigrescens,* or *nigra.*

DISTRIBUTION. Central Sulawesi, southwest to base of Toraja highlands (where it interbreeds with *M. maura*), southeast to the lakes region of the southeastern peninsula; northwest to isthmus between Palu and Parigi, where it interbreeds with *M. hecki.* Togean Islands.

NOTES. The (very narrow) hybrid zone with *M. hecki* was deduced by Watanabe et al. (1991a) to be between Kebun Kopi and Toboli, at the isthmus on the base of the northern peninsula of Sulawesi. More recently a detailed study by Bynum et al. (1997) confirmed this location and emphasized how very restricted in extent it is, about 15 by 7.5 km.

Froehlich et al. (1998) reported that there is a further, so far uncharacterized, species in the eastern peninsula of Sulawesi, east of the Sungai Bongka, and that unfortunately the type series of *Macacus tonkeanus* belongs to it. *Macaca tonkeana* may therefore have to be redesignated. Froehlich et al. (1998) considered that the macaques of the Togean Islands may actually be a hybrid population between the undescribed species and *M. tonkeana.*

Macaca hecki (Matschie, 1901)
Heck's Macaque

1901 *Papio (Inuus) hecki* Matschie. Indonesia: Buol, Sulawesi.

DIAGNOSIS. Black, with brownish ventral surface; forearms dark brown; shanks gray to light brown, paler than trunk; crown hair lengthened, but not forming a distinct crest; cheeks lighter than crown, but without elongated whiskers; ischial callosities gray to yellow, kidney-shaped but with only a partial transverse furrow across them (cf. *M. nigra*). No short-haired gluteal fields. Tail very short, stubby, but not rudimentary as in *M. nigrescens* and *M. nigra.*

DISTRIBUTION. From base of northern peninsula to just east of Gorontalo.

NOTES. This species may hybridize with *M. nigrescens* on the eastern edge of its range (Watanabe and Matsumura 1991).

Macaca nigrescens (Temminck, 1849)
Gorontalo Macaque

1849 *Papio nigrescens* Temminck. Celebes.

DIAGNOSIS. Median crown hairs forming a long, backward-directed crest; ischial callosities gray or brown, very prominent, transversely long-oval; body brown with a thin, black dorsal stripe, but crown crest, limbs, and crown black. Short-haired gluteal fields extend diagonally upward from ischial callosities (Watanabe and Matsumura 1991). Tail rudimentary, nubbinlike.

DISTRIBUTION. A section of the northern peninsula of Sulawesi, from Gorontalo in the south and between Baroko and Bolaangitang in the north, east to the Sungai Onggak Dumoga and the Mount Padang region.

NOTES. Watanabe and Matsumura (1991) recorded probable hybrids with *M. nigra* from Tambun, on the western side of Mount Padang, on the eastern border of the range, and with *M. hecki* east of the Sungai Bolango and near Bolaangitang, on the western border; but individual variation within the species made it difficult to be certain of this, at least in the case of the supposed hybrids with *M. hecki*.

Macaca nigra (Desmarest, 1822)

**Celebes Crested Macaque or
Black "Ape"**

1822 *Cynocephalus niger* Desmarest. "One of the islands of the
    Indian archipelago."
1824 *Cynocephalus malayanus* Desmoulins. "Philippines."
1863 *Cynopithecus aethiops* Reichenbach. "Celebes, Philippines,
    Moluccas."
1931 *Cynopithecus lembicus* Miller. Indonesia: Lembeh Island.

DIAGNOSIS. Median crown hairs forming a long, backward- and upward-directed crest; ischial callosities bright pink, prominent, kidney-shaped, divided by a median groove; color blackish all over. No short-haired gluteal fields. Tail rudimentary, nubbinlike.

DISTRIBUTION. East of the Sungai Onggak Dumoga and Mount Padang, to the tip of the northern peninsula of Sulawesi; Pulau Lembeh; and Pulau Bacan, where doubtless introduced.

NOTES. Individuals with intermediate characteristics, presumably hybrids, were seen along the border of this species with *M. nigrescens* (Watanabe and Matsumura 1991).

Macaca fascicularis group

Glans penis narrow, its shape varies. Face pink or brown. No midcycle sexual swelling, but some swelling and reddening of anogenital skin at puberty.

Macaca fascicularis (Raffles, 1821)

**Crab-eating Macaque or
Long-tailed Macaque or
Kera**

DIAGNOSIS. Apex of glans penis bluntly bilobed; breadth of glans 41–55% of length. Tail at least as long as head plus body. Crown hair directed backward and outward, sometimes forming a small central crest. Face with light spot at inner corner of eyelid.

NOTES. The earliest name for the crab-eating macaque is certainly *Simia aygula* Linnaeus, 1758, which had previously been used for a langur (now known as *Presbytis comata*). This Linnaean name has been suppressed (International Commission on Zoological Nomenclature 1986) and the name *fascicularis* placed on the Official List of Specific Names in Zoology.

For geographic variation in this species, see Fooden (1991, 1995). Geographically varying characters are color, whether lighter (brown or grayish) or darker (nearly black); whether crown or legs contrast with general tone; form of cheek whiskers; and relative length of tail. Of interest is that tail length differentiates the restricted peripheral and insular populations from each other, but its variation in the widespread *M. f. fascicularis* (67–150% of head plus body) covers the entire range. Cheek whiskers are transzygomatic (sweeping upward from near jaw angle to crown, between eye and ear) or infrazygomatic (restricted to mandibular region, ending in a whorl low on cheek, the hairs of the temples running smoothly backward).

Macaca fascicularis fascicularis (Raffles, 1821)

1758 *Simia aygula* Linnaeus. "India."
1775 *Simia cynamolgus* Schreber. Not of Linnaeus, 1758 (a
    baboon).

1821 *Simia fascicularis* Raffles. Indonesia: Sumatra.

1825 *Macacus carbonarius* F. Cuvier. Indonesia: Sumatra.

1826 *Macaca irus* I. Geoffroy. Probably Malaysia: Melaka (Fooden 1995).

1830 *Semnopithecus kra* Lesson. Replacement for *fascicularis*.

1838 *Semnopithecus buku* Martin. For the "Kra Buku" of Raffles, from Sumatra.

1862 *Cynamolgus cynocephalus* Reichenbach. Not of Linnaeus, 1766 (a baboon).

1900 *Macacus pumilus* Miller. Indonesia: Pulau Bunoa (= Benua), Tambelan Islands.

1903 *Macacus phaeura* Miller. Indonesia: Teluk Siaba, Pulau Nias.

1905 *Cynomolgus suluensis* Mearns. Philippines: foot of Crater Lake Mountain, Jolo.

1905 *Cynomolgus cagayanus* Mearns. Philippines: Cagayan Sulu.

1909 *Macaca mordax* Thomas and Wroughton. Indonesia: Cilacap, West Java.

1909 *Macac resima* Thomas and Wroughton. Indonesia: Tasikmalaya, West Java.

1909 *Pithecus validus* Elliot. Vietnam: "Cochin-China."

1909 *Pithecus alacer* Elliot. Indonesia: Bliah (= Selat Bliat), Pulau Kundur, Riau Archipelago.

1909 *Pithecus karimoni* Elliot. Indonesia: Monos, Pulau Karimon, Riau Archipelago.

1909 *Pithecus laetus* Elliot. Malaysia: Pulau Tinggi.

1909 *Pithecus dollmani* Elliot. Singapore: Changi.

1909 *Pithecus bintangensis* Elliot. Indonesia: Sungai Biru, Pulau Bintang, Riau Archipelago.

1910 *Pithecus lapsus* Elliot. Indonesia: Tanjung Pamuja, Bangka.

1910 *Pithecus agnatus* Elliot. Indonesia: Pulau Tuangku, Batu Islands.

1910 *Pithecus lingungensis* Elliot. Indonesia. Pulau Lingung, Natuna Islands.

1910 *Pithecus lautensis* Elliot. Indonesia: Pulau Laut, Natuna Islands.

1910 *Pithecus sirhassensis* Elliot. Indonesia: Pulau Sirhassen (= Serasan), Natuna Islands.

1910 *Pithecus carimatae* Elliot. Indonesia: Teluk Pai, Pulau Karimata.

1910 *Pithecus mandibularis* Elliot. Indonesia: Sungai Sama (= Sungai Ambawang), Pontianak, West Kalimantan.

1910 *Pithecus baweanus* Elliot. Indonesia: Pulau Bawean.

1910 *Pithecus cupidus* Elliot. Indonesia: Pulau Matasiri.

1910 *Pithecus lingae* Elliot. Indonesia: Pulau Lingga.

1910 *Pithecus impudens* Elliot. Indonesia: Pulau Sugi, Riau Archipelago.

1910 *Pithecus capitalis* Elliot. Thailand: Trong.

1911 *Macaca irus argentimembris* Kloss. Malaysia: Penang.

1913 *Pithecus fascicularis limitis* Schwarz. Indonesia: Lelogama, Timor.

1916 *Pithecus mansalaris* Lyon. Indonesia: Pulau Mansalar.

1932 *Macaca irus sublimitis* Sody. Indonesia: Pulau Sumba.

1933 *Macaca irus sumbae* Sody. Lapsus for *sublimitis*.

1949 *Macaca irus sublimitus* Sody. Lapsus for *sublimitis*.

1949 *Macaca irus submordax* Sody. Indonesia: Pucang, Gunung Agung, Bali.

1974 *Simia mauritius* O. Hill. Not of Griffith, 1821 (a langur).

DIAGNOSIS. Pale yellowish gray to golden brown agouti, varyingly erythristic; crown as back, or brighter; cheek whiskers bushy, curving forward, transzygomatic, partly covering ears; tail length averaging 118% of head plus body. Head plus body length of adult male averaging 462 mm, which is about the same as most other subspecies.

DISTRIBUTION. Southern Vietnam, Cambodia, southern Thailand and offshore islands, south to Malay Peninsula, Sumatra, Bangka, the Natuna Islands, Java, Bali, Nusatenggara to Timor, Borneo, Bawean, most but not all offshore islands (Fooden 1995), and extends into the Sulu Archipelago and Zamboanga Peninsula of western Mindanao (Fooden 1991). The range is interrupted in peninsular Thailand by a fairly extensive area of intergradation with *M. f. aurea.*

Macaca fascicularis aurea (E. Geoffroy, 1831)

1831 *Macacus aureus* E. Geoffroy. Burma: Pegu.

1910 *Pithecus vitiis* Elliot. Burma: Domel Island (= Letsok-aw Kyun), Mergui Archipelago.

DIAGNOSIS. Buffy to medium brown; crown usually brighter than back, golden brown; lateral facial crest infrazygomatic, partly covering ears; tail short, averaging 95% of head plus body. Size large, males average 560 mm in head plus body length.

DISTRIBUTION. From southernmost Bangladesh, through Burmese part of range and into west-central Thailand; the boundary with *M. f. fascicularis* is the Thai-Burmese border range, but there is a wide area around the Isthmus of Kra where the two intergrade.

Macaca fascicularis umbrosa (Miller, 1902)

1902 *Macacus umbrosus* Miller. India: Little Nicobar Island.

DIAGNOSIS. Blackish color; crown yellowish brown; legs pale brownish gray; usually forward-directed cheek whiskers, not elongated below ears; tail 115–116% of head plus body length. Rather large; adult males average 502 mm in head plus body length.

DISTRIBUTION. From Great and Little Nicobar and Katchall Islands.

Macaca fascicularis atriceps Kloss, 1919

1919 *Macaca irus atriceps* Kloss. Thailand: Koh Kram (= Ko Khram Yai).

DIAGNOSIS. Buffy to medium brown; middle of crown dark brown or blackish from forehead to nape; cheek whiskers anteriorly directed, transzygomatic; tail length averaging 118% of head plus body.

DISTRIBUTION. Confined to Ko Khram Yai, southeastern Thailand.

Macaca fascicularis condorensis Kloss, 1926

1926 *Macaca irus condorensis* Kloss. Vietnam: Pulo Condore (= Con Son).

DIAGNOSIS. Like *M. f. atriceps* but dark crown patch much broader and grading into general color.

DISTRIBUTION. Confined to Con Son.

Macaca fascicularis fusca (Miller, 1903)

1903 *Macacus fuscus* Miller. Indonesia: Teluk Dalam, Simalar (= Pulau Simeulue).

DIAGNOSIS. Blackish color, including legs; crown blackish with conspicuous yellowish bands; cheek whiskers usually swept back, partly covering ears, not elongated below ears; tail averaging only 97% of head plus body length.

DISTRIBUTION. Confined to Pulau Simeulue.

Macaca fascicularis lasiae (Lyon, 1916)

1916 *Pithecus fuscus lasiae* Lyon. Indonesia: Pulau Lasia, southeast of Pulau Simeulue.

DIAGNOSIS. Color blackish; pale bands on crown inconspicuous; otherwise like *M. f. fusca,* but tail longer, its length 118% of head plus body.

DISTRIBUTION. Confined to Pulau Lasia.

Macaca fascicularis tua Kellogg, 1944

1944 *Macaca irus tua* Kellogg. Indonesia: Pulau Maratua, east of Borneo.

DIAGNOSIS. Color blackish; legs brownish gray, becoming paler distally; crown pale yellowish brown; cheek whiskers directed anteriorly, extending well back under ears; tail length 131% of head plus body. The single adult male is only 440 mm in head plus body length.

DISTRIBUTION. Confined to Pulau Maratua.

Macaca fascicularis karimondjawae Sody, 1949

1949 *Macaca irus karimondjawae* Sody. Indonesia: Pulau Karimunjawa.

DIAGNOSIS. Dark grayish brown; crown often blackish; tail averaging 108% of head plus body length.

DISTRIBUTION. From Pulau Karimunjawa and probably nearby Pulau Kemujan.

Macaca fascicularis philippinensis (I. Geoffroy, 1843)

1843 *Macacus philippinensis* I. Geoffroy. Philippines: Manila.
1851 *Macacus palpebrosus* I. Geoffroy. Philippines: Manila.
1867 *Macacus fur* Slack. Philippines: Luzon.
1870 *Macacus cynomolgus* var. *cumingii* Gray. Philippines.
1905 *Cynomolgus mindanensis* Mearns. Philippines: Pantar, Davao, Mindanao.
1905 *Cynomolgus mindanensis apoensis* Mearns. Philippines: Mount Apo, Mindanao.
1913 *Pithecus mindorus* Hollister. Philippines: Alag River, Mindoro.

DIAGNOSIS. Color dark brown agouti, variably erythristic; crown slightly more golden than body; cheek hairs transzygomatic; tail length averaging 114% of head plus body; averaging larger than *M. f. fascicularis* (adult male mean head plus body length 480 mm).

DISTRIBUTION. From Balabac, Palawan, Culion, Mindoro, Luzon, Samar, Leyte, and northeastern Mindanao.

NOTES. Strictly speaking, the names *mindanensis* and *apoensis* are from the contact zone between this subspecies and *M. f. fascicularis;* they are placed here because the saturate color typical of *philippinensis* occurs throughout Mindanao except for the Zamboanga Peninsula, and it is at this point that average size changes (Fooden 1991, fig. 3, tables 2, 5).

Macaca arctoides (I. Geoffroy, 1831)

Stump-tailed Macaque or
Bear Macaque

1831 *Macacus arctoides* I. Geoffroy. Vietnam: Cochin-China.
1839 *Papio melanotus* Ogilby. "Said to be [from] Madras."
1854 *Macacus ursinus* Gervais. Substitute for *arctoides.*
1871 *Macacus brunneus* Anderson. Burma: Kakhyen Hills, east of Bhamo.
1872 *Macacus rufescens* Anderson. Said to be from Singapore.
1873 *Macacus speciosus* Murie. Not of I. Geoffroy, 1826 (= *M. fuscata*).
1897 *Macacus harmandi* Trouessart. Thailand: Chuntabun.
1912 *Macacus (Magus) arctoides melli* Matschie. China: mountains west of Lochangho (= Lianshan), Guangdong.

DIAGNOSIS. Glans penis greatly elongated, more than 40 mm long, and narrowed from corona to tip. Baculum extremely long and tapering. Face bright pink or red, darkening (and becoming browner) with age and exposure to sunlight. Crown hair long at back and sides, radiating from center; forehead becoming bald with maturity. Pelage long and shaggy, dark brown in adult, white in neonate. Tail very short, about 8% of head plus body length. Cranially, rostrum is relatively short compared with similar-sized (or only slightly larger) *M. nemestrina* or *M. thibetana,* and supraorbital torus is rounded laterally; malar broad, laterally projecting, the anterior surface facing anteriorly.

DISTRIBUTION. Northeastern India and southern China south into the northwestern tip of West Malaysia. In China, it occurs south of 25°N (Fooden 1990; Zhang et al. 1997). In Assam, it is found only south of the Brahmaputra (Choudhury 1997). It appears to be absent from most of Burma and Thailand: records are only from the far north of Burma and from the border ranges between the two countries south into the peninsula, with a few dubious records in central and northwestern Thailand (Fooden 1990).

NOTES. The association of this species in a clade with *M. fascicularis* is surprising, but, as discussed by Morales and Melnick (1998), several lines of (molecular) evidence support it and suggest a separation date of about 1.6 million years ago. Delson (1980) and Fooden (1990) associated it in a clade with the *M. sinica* group, citing such morphological characters as genitalia, copulatory behavior, and facial skin color. It is striking, however, that Fooden (1990:681) stressed that the baculum of *M. arctoides* is very different from that of the *M. sinica* group, instead resembling, in its "gentle curvature and absence of a well-defined distal process," that of the *M. fascicularis* group.

Macaca mulatta group

Apex of glans penis bluntly bilobed; breadth of glans 41–55% of length (as in *M. fascicularis*). Crown hair directed simply backward. Face pink, becoming red in female in midcycle; no midcycle sexual swelling, but at puberty sexual skin swells and reddens from tail base to backs of thighs. Tail shorter than head plus body.

Melnick et al. (1993) and Morales and Melnick (1998) found that Indian and Chinese mitochondrial samples of *M. mulatta* are paraphyletic with respect to *M. cyclopis* and *M. fuscata*, but nuclear genes clearly assort the three species correctly. This, as they pointed out, is a case where mitochondrial lineages have not assorted according to overall phyletic relationships and offers an antidote to the assumption that mitochondrial DNA invariably gives the "right answer" to problems of phylogeny.

Macaca mulatta (Zimmermann, 1780)

Rhesus Monkey

DIAGNOSIS. Brown with reddish-toned hindparts and hindlimbs; tail about half head plus body length and not above 60%.

DISTRIBUTION.   Mapped by Fooden (1989) from the Krishna River in central and eastern India and the lower Tapti River in western India, north into Afghanistan, Kashmir, Kumaun, Nepal, Sikkim, Bhutan, and northeast into China, where, according to Jiang et al. (1991) and Zhang et al. (1997), it extends to the Yangtse and north of its middle course to about 33°N, 110°E. It is absent north of the lower Yangtse, but occurs again north of the lower Hwang He, formerly as far north as Beijing.

NOTES.   While awaiting Jack Fooden's anticipated revision of this species, I have no option but to try to synthesize previous, more restricted, attempts: Pocock (1932, 1939); G. M. Allen (1939b); Feng et al. (1986); Jiang et al. (1991); Pan et al. (1992); Peng et al. (1993). Pocock (1932) had little material from outside South Asia, and the other authors were restricted to study of specimens from China. Pocock (1932) recognized three subspecies in South Asia and (provisionally) three in China, citing the shorter tail as the only feature he could find to differentiate Chinese from South Asian forms. Peng et al. (1993), however, drew a broad distinction between Indian and Chinese samples on both craniometric variables and blood proteins; the Indian population had a narrower postorbital constriction and much narrower biorbital width (mean 51.0 mm, vs. 59.3 for China). The greater cranial width of the endemic Chinese forms is noticeable in general. In discriminant analysis, the Indian group was distinct at a high level from all Chinese samples, even with its small sample size. It would be interesting to place M. cyclopis and M. fuscata in such an analysis, to see whether they are different from any M. mulatta or whether they resemble Chinese forms craniometrically as in mitochondrial DNA (see above).

There is much hybridization between this species and M. fascicularis where their ranges meet; the details have been given by Fooden (1997). The boundary between the two species runs from the southern tip of Bangladesh south to the Irrawaddy delta just inland from the Burma coast, then across the Thai border to Umphang (15°28'N, 99°04'E), north to Ban Mae Na Ree (16°25'N, 99°23'E), then east to Phu Phan near the Laos border at 16°42'N, 104°25'E, southeast to the Thateng District (15°26'N, 106°23'E), and southeast again to Dak Sut (14°56'N, 107°44'E) and Krong Pach (12°43'N, 108°19'E). At all the named localities intermediates were recorded, with tail length intermediate (i.e., between 60 and 90% of head plus body length) and the posterior body at most slightly brighter in tone than the anterior. It should be added that unmixed examples of the two species were recorded at most of these localities in addition to the hybrids.

What is especially noteworthy about this hybrid zone is that the two species are not sister taxa and according to the mitochondrial DNA evidence not even part of sister clades (the M. mulatta group forms the sister clade of the M. sinica group, and the M. fascicularis group is sister to the M. mulatta + M. sinica groups). No hybrids occur between M. mulatta and M. sinica, which meet along a quite similar border in India, nor between the externally very similar M. mulatta and M. assamensis, which are even sympatric in Nepal and northeastern India. It may be the similarity of genital armature between M. mulatta and M. fascicularis that makes the difference; whatever the reason, this constitutes an outstanding example of the principle that ability to hybridize is a plesiomorphic state, rather than a sign of conspecificity.

Macaca mulatta mulatta (Zimmermann, 1780)

1780 Cercopithecus mulatta Zimmermann. "India"; fixed by Pocock (1932) as Nepal Terai.
1792 Simia (Cercopithecus) fulvus Kerr. India.
1798 Simia rhesus Audebert. Locality unknown.
1800 Simia erythraea Shaw. Locality unknown.
1840 Macaca (Pithex) oinops Hodgson. Nepal: "the terai" (tarai).
1840 Macaca (Pithex) nipalensis Hodgson. Nepal: "the terai."
1917 Macaca siamica Kloss. Thailand: Mae Ping rapids, below Chiengmai.

DIAGNOSIS.   Size small: head plus body length of male 483–584 mm (n = 10), of female 437–533 mm (11); greatest skull length of male 112–122 mm (8), of female 102–112 mm (10) (both Pocock [1932, 1939]); skull relatively narrow; tail length 36.4–58.3% of head plus body (n = 21) (Pocock 1932).

DISTRIBUTION. From most of the Indian and Nepalese portion of the range: Pocock (1932) cited it north to the Nepal tarai, the Bhutan Duars, and the Mishmi Hills, and east through Burma to Vietnam. Mapped by Jiang et al. (1991) in China south and west of the Pearl River and south of the upper Yangtse. Mainly from lower altitudes, but up to 2,400 m at Nagarcot, Nepal (Pocock 1932). Measurements of specimens from Burma, Thailand, and Vietnam fall within the range of those from India and Nepal, though perhaps averaging slightly smaller.

Macaca mulatta villosa (True, 1894)

1894 *Macacus rhesus villosus* True. India: Lolab, northern end of Lake Wular, 65 km northwest of Srinagar, Kashmir.

DIAGNOSIS. Medium-sized: head plus body length in male 508–635 mm (*n* = 9), greatest skull length in male according to Pocock (1939) is 131–136 mm in Kashmir and northern Panjab specimens, 125–127 mm in those from Kumaun, and Elliot (1913) gave the length of the type of *villosus* as 131.5 mm; but female is no larger than *M. m. mulatta* (Pocock [1932] gave 107–112 mm for three skulls). Skull relatively narrow, tail relatively long, as in *M. m. mulatta*. Stated to be a little brighter and redder, with the tail, hands, and feet darker above, and the pelage longer, 65–90 mm vs. under 65 mm in *M. m. mulatta;* Pocock (1932:534) stated that in an October skin of *M. m. mulatta* from Nagarcot, Nepal, at 2,400 m, the pelage "is shorter than in October skins of *villosa* from Kumaon at much lower elevations," so the differences between them obviously do not depend just on season or altitude.

DISTRIBUTION. Range is from Kumaun through northern Panjab to southern Kashmir.

Macaca mulatta vestita (Milne-Edwards, 1892)

1892 *Macacus vestitus* Milne-Edwards. China: Tibet, Tengri Nor.
1932 *Macaca mulatta mcmahoni* Pocock. India: Kootai, lower Chitral, Kashmir, 1,100 m.

DIAGNOSIS. Size very large: head plus body length of male unknown, of female 510–630 mm (*n* = 2) (Feng et al. 1986); greatest skull length in male over 140 mm (but a skull, age unstated, in Feng et al. [1986] is only 127.2 mm), in female 114.2–137.0 mm (*n* = 4) (Feng et al. 1986); skull narrow as in *M. mulatta* and *villasa*. Color duller, darker, less tawny; pelage long, thick, shaggy, 60–100 mm according to Pocock (1939). Tail 39.5–41% of head plus body length in two Chitral specimens (Pocock 1932). The measurements of the type and referred specimens of Pocock's *mcmahoni* fall within the range given for this subspecies, and direct comparisons of skins are necessary to confirm the proposed synonymy.

DISTRIBUTION. Pocock (1932) cited descriptions of extremely large rhesus from Kafiristan and Chitral; Feng et al. (1986) and Jiang et al. (1991) mapped it from the valleys of the Ngagong, Salween, and Mekong Rivers in Tibet, east to the gorge of the upper Yangtse.

Macaca mulatta lasiota (Gray, 1868)

1868 *Macacus lasiotus* Gray. China: Sichuan.

DIAGNOSIS. Medium-sized: head and body length of male 610 mm, greatest skull length of male 131–136 mm (*n* = 1) (Pocock 1932), of female 119 mm (Pocock 1932); skull relatively broad; color dark, gray-toned except on the hindquarters; pelage not especially long or thick, but well haired on tail. Measurements taken in the flesh, giving a tail length 53% of head plus body, were quoted by Pocock (1932), who implied, however, that there must have been some mistake because in the dried skin the tail was much shorter than this. Using only Chinese material, Pan et al. (1992) distinguished what was effectively this subspecies (which they called *M. m. mulatta*) from the two endemic Chinese subspecies by the significantly greater anteroposterior diameter of the glenoid fossa, the length of the scar of the origin of the masseter, and especially the moment arm of the temporal muscle on the mandible (i.e., it has a more developed masticatory system).

DISTRIBUTION. According to Jiang et al. (1991), China, north and east of the Yangtse, north to the upper Hwang He, and east to about 104°E.

Macaca mulatta sanctijohannis (Swinhoe, 1866)

1866 *Innus sancti-johannis* Swinhoe. China: North Lena Island, Hong Kong.
1872 *Macacus tcheliensis* Milne-Edwards. China: eastern Chili.
1909 *Pithecus littoralis* Elliot. China: Kuatun, Fujian.

DIAGNOSIS. Size medium: head plus body length of male 610 mm; greatest skull length of male 118–130 mm (*n* = 2) (G. M. Allen 1939b); of female 103–121.5 mm (*n* = 6) (Elliot 1913; Pocock 1932, G. M. Allen 1939b); skull relatively broad. Color apparently similar to that of *M. m. mulatta.* Tail length 39.3% of head plus body in the type, a zoo specimen that suffered from rickets and so was unnaturally short in the body according to Pocock (1932); 32.8% in the type of *littoralis* (Elliot 1913), and 29% in a specimen ascribed to *tcheliensis* by Pocock (1932). In Pan et al.'s (1992) discriminant analysis, this subspecies is distinguished as a whole from both *M. m. mulatta* and *brevicaudata,* albeit with considerable overlap.

DISTRIBUTION. Jiang et al. (1991) and Pan et al. (1992) mapped this subspecies, which they called *M. m. littoralis,* from eastern Guangxi north to the River Yangtze and somewhat across it to about 33°N, and again (as *M. m. tcheliensis*) north of the lower Hwang He (Yellow River).

Macaca mulatta brevicauda (Elliot, 1913)

1909 *Pithecus brachyurus* Elliot. China: Hainan. Not of H. Smith, 1842.
1913 *Pithecus brevicaudus* Elliot. Replacement for *brachyurus* (= *M. nemestrina*).

DIAGNOSIS. Slightly smaller than *M. m. sanctijohannis* and *lasiota* (head plus body length of female 510 mm; skull of female 102–116 mm [*n* = 4] [Elliot 1913; G. M. Allen 1939b]), but similarly broad-skulled; tail length in the type 43% of head plus body (Elliot 1913). On this evidence it is difficult to defend its distinction from *M. m. sanctijohannis,* but Pan et al. (1992) separated it, at least as a whole, on discriminant analysis of cranial measurements.

DISTRIBUTION. Hainan Island.

Macaca cyclopis (Swinhoe, 1862)

Formosan Rock Macaque

1862 *Macacus cyclopis* Swinhoe. Formosa (= Taiwan).
1863 *Macacus (radiatus) affinis* Blyth. Formosa.

DIAGNOSIS. Darker brown than *M. mulatta,* with no contrast between fore- and hindparts; bushy cheek whiskers; tail up to two-thirds head plus body length.

DISTRIBUTION. Taiwan.

NOTES. The mitochondrial DNA findings of Morales and Melnick (1998) and the analysis of blood proteins by Fooden and Lanyon (1989) both imply that this species separated only a quarter of a million years ago from Chinese *M. mulatta.*

Macaca fuscata (Blyth, 1875)

Japanese Macaque

DIAGNOSIS. Yellowish brown, with no contrast between fore- and hindparts; tail very short, well furred.

DISTRIBUTION. Main Japanese islands of Honshu, Shikoku, and Kyushu, and many offshore islets; Yakushima.

NOTES. Molecular (Morales and Melnick 1998) and fossil (Delson 1980) evidence alike suggest that this species diverged from *M. mulatta* only half a million years ago.

The traditional two-subspecies arrangement is maintained here, but there are no modern morphological studies to substantiate it. In fact the genetic evidence (Nozawa et al. 1996) suggests that the really distinctive population is the northernmost one from Wakinozawa (the Shimokita Peninsula, in the far north of Honshu).

Macaca fuscata fuscata (Blyth, 1875)

1825 *Macacus speciosus* F. Cuvier. "East Indies."
1875 *Macacus fuscatus* Blyth. Japan.
1909 *Inuus speciosus japanensis* Schweyer. Japan.

Macaca fuscata yakui Kuroda, 1941

1941 *Macaca fuscata yakui* Kuroda. Japan: Yakushima.

Macaca sinica group

Apex of glans penis sagittate, narrowed toward tip; length of glans less than 30 mm. Facial skin pink or red. As far as is known, no sexual swellings, but a reddening of genital area at midcycle in female.

Fooden (1988) compared the four species of this group in detail.

Macaca sinica (Linnaeus, 1771)
Toque Macaque

DIAGNOSIS (after Fooden [1979]). Tail length at least equal to head plus body, up to 144%. Body yellow-brown or golden brown; crown hairs elongated, radiate from a central whorl to form a cap extending forward as far as brows and backward to nape. Size very small: head plus body length 418–528 mm in adult male, 400–447 mm in adult female.

DISTRIBUTION. Sri Lanka.

NOTES. This species was revised by Fooden (1979).

Macaca sinica sinica (Linnaeus, 1771)

1771 *Simia sinica* Linnaeus. Locality unknown.
1838 *Cercopithecus pileatus* Ogilby. Not of Kerr, 1792 (unidentifiable, but not a macaque).
1862 *Cynamolgus (Zati) audeberti* Reichenbach. "Bengal."
1931 *Macaca sinica inaurea* Pocock. Sri Lanka: Cheddikulam, Northern Province.
1942 *Macaca (Zati) sinica opisthomelas* Hill, 1942. Sri Lanka: Horton Plains.
1965 *Macaca sinica longicaudata* Deraniyagala. Sri Lanka: Menik Ganga, Kataragama District, Northern Province.

DIAGNOSIS. Whole cap is golden brown.

DISTRIBUTION. Fooden (1979) gave the range of this subspecies as the northern half of Sri Lanka, approximately from 8°N on the west coast to 7°30′N in the east, with most of the rest of the island being a contact zone with *M. f. aurifrons*. The contact zone has mainly individuals with the *sinica* pattern, with

at most an "indistinctly paler" anterior crown; the "distinct" and "extreme" contrast types are nearly restricted to the wet zone. Given this, the nominotypical subspecies is here deemed a general dry-zone and hill-zone form, and the two names (*opisthomelas* and *longicaudata*) from the "contact zone" are placed in its synonymy.

Macaca sinica aurifrons Pocock, 1931

1931 *Macaca sinica aurifrons* Pocock. Sri Lanka: Roygam Korale, Western Province.

DIAGNOSIS. Anterior part of cap strongly yellowish, contrasting with darker, browner posterior part.

Macaca radiata (E. Geoffroy, 1812)
Bonnet Macaque

DIAGNOSIS. Tail length 88–136% of head plus body. Gray-brown to golden agouti brown above, becoming duller posteriorly, slightly paler on limbs and darker along dorsal surface of tail; ventrally with hair buffy to whitish, but very sparsely haired, the color of the skin showing through; crown hairs radiate from a central whorl, forming a cap, with posterior hairs longer, extending to occiput, than anterior; very short forehead hairs, anterior to cap, parted in midline. Skin of face and ears brown to pinkish or (in some females) bright scarlet. Head plus body length 450–590 mm (male), 375–515 mm (female).

DISTRIBUTION. Southern India. The range was mapped by Fooden (1989), who recorded it north along the Western Ghats to about the Tapti River and in the center and east to the Krishna River.

Macaca radiata radiata (E. Geoffroy, 1812)

1812 *Cercocebus radiatus* E. Geoffroy. India.
1821 *Simia sinica* Griffith. No locality. Not of Linnaeus, 1771.

DIAGNOSIS (after Fooden [1981]). Crown grayish brown, hairs of cap often black-tipped; body dull gray-brown above with yellow tones, the hairs weakly banded, with a dark median streak posteriorly; limbs ochery gray; ventral skin dark bluish gray.

DISTRIBUTION.   From the major portion of the species's range, south to the Palni Hills, southeast as far
as "Tenmali" (= Timbale), inland of Pondicherry.

Macaca radiata diluta Pocock, 1931

1931 *Macaca radiata diluta* Pocock. India: Boothapundy (=
  Bhutapandi), north of Aramboly, Travancore.

DIAGNOSIS   (after Fooden [1981]). Crown pale yellowish brown; body bright yellow or golden brown,
the hairs conspicuously banded, slightly duller posteriorly; limbs pale gray-brown; ventral skin whitish,
largely unpigmented.

DISTRIBUTION.   From the south tip and southeastern coast of India, north to Kambam (9°44′N,
77°18′E) at the southwestern foot of the Palni Hills
in the west and Pondicherry in the east.

Macaca assamensis (McClelland, 1839)
Assam Macaque

DIAGNOSIS   (after Fooden [1982]). Tail at most 56%
of head plus body length. Golden brown to dark
chocolate agouti brown above, including limbs and
upper side of tail; ventral hair pale buffy or whitish,
but underside very sparsely haired, the pale bluish
skin showing through; crown may be smooth (26%),
tufted (37%), or with a median whorl to form a cap
(37%); cheek and chin whiskers prominent, buffy
to whitish, in adults. Facial skin dark brownish to
purplish, except for very pale supraorbital area.
Head plus body length 538–730 mm in male, 437–
587 mm in female, so much larger than *M. sinica* and
M. radiata and more sexually dimorphic; skull very
prognathous.

DISTRIBUTION.   Himalayan foothills to 2,750 m,
from central Nepal east into northern Burma and
southeast through southernmost China (to the upper Mekong in Tibet, and in the east into southern
Guizhou [Zhang et al. 1997]) to Hoi Xuan in
Vietnam and Thateng in Laos; south through the
Burma/Thailand border ranges as far as Chong-
krong, 14°41′N. Also the Sundarbans in Bangladesh.
A hiatus between central Bhutan and the south side

of the Brahmaputra and the east bank of its upper
course, the Dhibang, marks the division between the
two subspecies.

NOTES.   The mitochondrial ribosomal genes of this
species studied by Morales and Melnick (1998) fell
into two groups: three formed their own clade,
closely associated with *M. thibetana,* and the other
two (thought to be possibly *M. a. pelops*) formed a
clade with *M. radiata.*

   Though separated by such a wide distributional
gap, and possibly (see above) so distinct mito-
chondrially, the two subspecies are very poorly
distinguished.

Macaca assamensis assamensis (McClelland, 1839)

1839 *Macacus assamensis* McClelland. India: Assam.
1872 *Macacus rheso-similis* Sclater. "East Indies."
1932 *Macaca assamensis coolidgei* Osgood. Vietnam: Hoi Xuan.

DIAGNOSIS   (after Fooden [1982]). Size of female
small, head plus body length 437–523 mm, skull
length 113.1–126.8 mm; tail short, 29–47% of head
plus body length.

DISTRIBUTION.   From eastern part of range; north
along the River Dihang (= middle Brahmaputra)
and its Tibetan tributary the Ngagong, according to
the map in Feng et al. (1986).

Macaca assamensis pelops (Hodgson, 1840)

1840 *Macacus (Pithex) pelops* Hodgson. Nepal: the Kachar (i.e.,
  above the timberline).
1846 *Macacus macclellandii* Gray. Nomen nudum.
1867 *Macacus sikimensis* Hodgson. Nomen nudum.
1870 *Macacus problematicus* Gray. India: Dalamcote Fort (=
  Daling), northern West Bengal.

DIAGNOSIS   (after Fooden [1982]). Size of female
large, head plus body length 530–587 mm, skull
length 116–131.5 mm, but male similar in size to
M. a. assamensis, hence less sexually dimorphic; tail
longer, 50–63% of head plus body length.

DISTRIBUTION.   From central Nepal through Sikkim and northernmost West Bengal into central
Bhutan, and the Sundarbans. Apparently occurs

north of the Himalayan crest in the Bhong Valley (Feng et al. 1986).

Macaca thibetana (Milne-Edwards, 1870)
**Tibetan Macaque or
Milne-Edwards's Macaque**

1870 *Macacus thibetanus* Milne-Edwards. China: near Moupin (= Baoxing), Sichuan.
1872 *Macacus tibetanus* Milne-Edwards. Unjustified emendation.
1912 *Macacus (Magus) arctoides esau* Matschie. China: Yao-tze Mountains, west of Lochangho, Guangdong.
1928 *Pithecus pullus* Howell. China: near Kuatun, Fujian.

DIAGNOSIS (after Fooden [1983]). Tail only 7–9% (adult male) and 11–13% (female) of head plus body length. Body dark brown above, weakly or not at all banded, pale buffy or gray below, where sparser than upper side but not very sparse; limbs becoming paler distally to pale brown on hands and feet; crown paler brown, with a small whorl on vertex with short hairs radiating from it; general pelage long and dense; prominent, bushy, pale buffy beard and full cheek whiskers; facial skin pale brownish on muzzle, whitish around eyes but becoming pink in adult females. Size very large, head plus body length in adult male 613–710 mm, adult female 507–630 mm; skull length 147.7–164.0 mm (adult male), 120.7–140.0 mm (adult female).

DISTRIBUTION. Fooden (1983) gave the range as east-central China, 25–33°N, 102°30′–119°30′E, at altitudes of 1,000 to 2,500 m. Later Fooden et al. (1994) gave a range extension south into Guangxi at 23°48′N, about 110°E, and Zhang et al. (1997) mapped it as far west as the Yangtze Gorge in western and northwestern Sechuan; recently Choudhury (1998) recorded this species, on the basis of field observations including photographs, in West Kameng, Arunachal Pradesh, at 2,500 m a.s.l., extending the range by about 1,000 km.

Genus *Lophocebus* Palmer, 1903
Crested Mangabeys

1870 *Semnocebus* Gray. *Presbytis albigena* Gray, 1850. Not of Lesson, 1840 (Indridae).
1903 *Lophocebus* Palmer. To replace *Semnocebus* Gray, 1870.
1914 *Cercolophocebus* Matschie. *Cercopithecus aterrimus* Oudemans, 1890.

The species of this genus had been invariably placed in *Cercocebus* until I separated them (Groves 1978), arguing that *Lophocebus* was close to *Papio* and that the only real similarity, the depth of the suborbital fossae, is convergent, a result perhaps of facial shortening. Mangabeys have sometimes been called "white-eyelid monkeys," but *Lophocebus* do not have white or even especially pale eyelids.

Lophocebus albigena (Gray, 1850)
Gray-cheeked Mangabey

DIAGNOSIS. Pelage long, loose, blackish brown on body and tail, and on crown where it forms a short, partially upright tuft or pair of tufts. A longer, paler cape on shoulders, more prominent in male, continuous with the rather thin, pale, grayish or whitish long cheek whiskers.

DISTRIBUTION. From the Cross River, Nigeria, southeast to the Gabon coast at 2°14′S and the Alima River, Congo-Brazzaville (1°36′S, 16°36′E), east across the Oubangui through the whole rain forest zone north of the Congo and east of the Lualaba, into Uganda (east to Busoga) and Burundi.

Lophocebus albigena albigena (Gray, 1850)

1850 *Presbytis albigena* Gray. West Africa.
1910 *Cercocebus albigena zenkeri* Schwarz. Cameroon: Lokundje River, Bipindi.
1913 *Cercocebus (Leptocebus) albigena weynsi* Matschie. Congo-Brazzaville: Sangha River.

DIAGNOSIS. Mantle light gray, sometimes with fawn tones; midline of nape and withers usually noticeably browner, darker; black of crown and body mat grayish; underside brown; cheeks light creamy white. Crown hair is long and scruffy, often forming two little tufts above brows.

DISTRIBUTION. From the coastal regions of Cameroon, inland to the Oubangui, and south into Gabon.

NOTES. The validity of Schwarz's subspecies *zenkeri* is questionable. I noted (Groves 1978) that there is a tendency for Cameroon coastal specimens to be grayer, with less midline darkening, than those from Congo and Gabon, but the variation is clinal and there is great variability.

Lophocebus albigena osmani Groves, 1978

1978 *Lophocebus albigena osmani* Groves. Cameroon: Edea.

DIAGNOSIS. Mantle rusty brown or tobacco brown, with midline not much darker; black of body with brownish tinge, only crown being jet black; underside yellowish gray, with a yellow tinge anteriorly; arms tend to be paler than body, but hands black; cheeks bright gray white or golden white; crown hair less scruffy, more swept back, without hornlike tufts above brows.

DISTRIBUTION. From Cross River, across Sanaga River to Edea and inland to the Batouri District.

Lophocebus albigena johnstoni (Lydekker, 1900)

1900 *Semnocebus albigena johnstoni* Lydekker. Congo-Zaire: "Near Lake Tanganyika," but Semliki or Ituri Forest according to Lorenz (1917).
1906 *Cercocebus jamrachi* Pocock. Congo-Zaire: Lake Mweru, Molinga (= Mlunga or Kilwa).
1913 *Cercocebus (Leptocebus) albigena ituricus* Matschie. Congo-Zaire: Mokoko and other localities, between Beni and Irumu.
1913 *Cercocebus albigena ugandae* Matschie. Uganda: Chagwe, Nile mouth at Lake Albert.
1917 *C[ercocebus] albigena mawambicus* Lorenz. Congo-Zaire: Mawambi.

DIAGNOSIS. Mantle darkish brown, distinct from the jet black of the crown but not always from the body tone; withers hardly or not darkened; arms blackish; underside dark brown; cheeks light gray-brown, passing to white inferiorly, but very thinly haired. Crown hair backswept but with long eyebrow tufts.

DISTRIBUTION. From Congo-Zaire, from the Oubangui (north of the River Congo) southeast to Kabambare (4°13′S, 27°07′E) and the Burundi, and east to Busoga, at the source of the Victoria Nile, Uganda.

NOTES. There is much cranial variation in this subspecies: thus those from Uganda are small in size but broad. Albinism and partial albinism seem relatively frequent.

Lophocebus aterrimus (Oudemans, 1890)

Black Crested Mangabey

1890 *Cercopithecus aterrimus* Oudemans. Congo-Zaire: Stanley Falls.
1900 *Cercocebus congicus* Sclater. Congo-Zaire: Stanley Falls.
1900 *Semnocebus albigena rothschildi* Lydekker. Congo-Zaire: Bena Dibele, Lukenye River.
1906 *Cercocebus hamlyni* Pocock. "Upper Congo."
1914 *Cercolophocebus coelognathus* Matschie. Congo-Zaire: Lualaba River, Kindu.

DIAGNOSIS. Fur coarse, entirely black, with no shoulder cape; no brow fringe or eyebrow tufts; cheek whiskers thick, elongated, swept back with a slight outward curve, and gray in color contrasting with black of body; a tall, thin central tuft on crown. Skull tends to be rounded, more gracile.

DISTRIBUTION. South of the great bend of the Congo, except the southwesternmost region where *L. opdenboschi* is found.

NOTES. I discussed (Groves 1978) whether this species and *L. opdenboschi* should be recognized as distinct species or included within *L. albigena*. Reasons for coming down on the side of the single-species solution included the occasional occurrence of long, backswept, light cheek whiskers or of a vaguely *albigena*-like mantle, with annulations, in *L. aterrimus*. These rarities reinforce the closeness of the relationship between the two, but do not bridge the gap. One of the intermediates, from Kindu on the River Lualaba, may be a hybrid.

Albinism and partial albinism seem to be unusually common (7 of 112 skins studied by Groves [1978], including the types of *congicus* and *hamlyni*).

Lophocebus opdenboschi (Schouteden, 1944)

Opdenbosch's Mangabey

1944 *Cercocebus opdenboschi* Schouteden. Congo-Zaire: Mwilambongo (= Mwiliambongo, 4°53′S, 19°42′E).

DIAGNOSIS. Fur longer, more lax than in *L. aterrimus;* cheek whiskers relatively short, but thick, not curved, not lighter than body; crown crest broad, pyramidal, laid back. Skull very small, narrow compared with *L. aterrimus,* but within the range of *L. albigena johnstoni.*

DISTRIBUTION. Forests along the Kwilu, Wamba, and Kwango Rivers in southwestern Congo-Zaire and across the border into Angola (Machado 1969). The range may interdigitate with that of *L. aterrimus,* which ranges to Kahemba on the River Kwilu farther upstream than the type locality of *L. opdenboschi* on the same river.

NOTES. The skull differences between this species and *L. aterrimus* are paralleled by those between different populations of *L. albigena johnstoni;* this circumstance was an argument to unite them specifically (and both with *L. albigena*) by Groves (1978). If a contrary view on the relationship between *L. albigena* and *aterrimus* is adopted, as here, then *opdenboschi* must also be regarded as a full species; it is as distinct from *L. aterrimus* as is *L. albigena* (Groves 1978).

Genus *Papio* Erxleben, 1777

Baboons

1777 *Papio* Erxleben. *Cynocephalus papio* Desmarest, 1820 (designated by International Commission on Zoological Nomenclature [1982a], Opinion 1199). Antedated by *Papio* P. L. S. Müller, 1776 (= *Mandrillus*), but that usage was suppressed at the same time as the redesignation of the type species.
1795 *Cynocephalus* G. Cuvier and E. Geoffroy. *Simia cynocephalus* Linnaeus, 1766. Not of Boddaert, 1768 (Dermoptera).
1839 *Chaeropithecus* Gervais. *Simia cynocephalus* Linnaeus, 1766.
1839 *Choeropithecus* de Blainville. *Simia cynocephalus* Linnaeus, 1766.
1840 *Hamadryas* Lesson. *Hamadryas chaeropithecus* Lesson. Not of Hübner, 1806 (Insecta).
1862 *Choiropithecus* Reichenbach. Emendation of *Chaeropithecus.*
1925 *Comopithecus* J. A. Allen. Replacement for *Hamadryas.*

The history of the nomenclature of this genus was recounted by Delson and Napier (1976, 1977). Briefly, the long-used name *Papio* Erxleben, 1777, was found to be antedated by *Papio* Müller, 1776, for the mandrill. Of two alternative solutions proposed (to uphold priority, use *Papio* as a senior synonym for *Mandrillus* and bring in the next available name, *Chaeropithecus,* for the baboons; or to maintain existing usage by suppressing Müller's name), the Commission preferred the latter course (International Commission on Zoological Nomenclature 1982a).

The five species of this genus are extremely distinctive physically (mainly externally), but there are hybrid zones between at least some of the adjacent pairs. This led Groves and Thorington (1970) to unite all but *P. hamadryas* into a single species (for which the prior available name is *P. cynocephalus*), and I myself indicated a preference to unite all of them. Such is the strife into which a strict BSC leads the perplexed adherent.

Jolly (1993) recently surveyed this genus and noted that here, perhaps to a greater extent than in any other Primate genus, there is an offset between phenostructure and zygostructure. If we were presented with a series of modern baboon skulls in a fossil context, the only taxon that would be instantly distinguished as a separate species would be *Papio cynocephalus kindae,* one that is most clearly connected to its closest allies by a graded series of geographically and morphologically intermediate populations. Jolly considered the different options for resolving the taxonomy of the genus, including a sort of "phylogenetic subspecies" concept, which he eventually suggested might be the most viable solution.

I know exactly what he meant. In the early stages of Jolly's baboon study, he and I examined skins and measured skulls together, and concluded together that, under the BSC, it is most likely that only a single species would be recognized, although even this is not absolutely clear-cut: hybridization is not known for sure between *P. papio* and *P. anubis,* and is even thought not to occur between *P. cynocephalus* and *P. ursinus* where these two come into close parapatry despite the undeniable fact that the populations of *P. ursinus* that approach *P. cynocephalus* geographically are the ones that also approach it phenotypically. Baboon systematics is a tangle. I am confident, however, that the five species traditionally recognized are genuinely diagnosable entities; what I am not so confident about is that the geographic divisions within *P. ursinus* are not also diagnosable entities. If this is so, then there would be more species of baboons than recognized here. Yet, Jolly's (1993) final cautionary points must be taken on board: we must not make the mistake of thinking that by proposing a taxonomic scheme we have solved all problems relating to the phenostructural diversity and interrelationships of baboons, nor even begun to resolve their zygostructural complexities. Jolly's (1993) paper is essential reading if one wishes to be aware of the complexities involved in the nature of classification at and below the species level, in baboons in particular and in Primates in general.

Papio hamadryas (Linnaeus, 1758)

Hamadryas Baboon or
Sacred Baboon or
Mantled Baboon

1758 *Simia hamadryas* Linnaeus. Egypt.
1758 *Simia cynamolgus* Linnaeus. Upper Egypt.
1828 *Cynocephalus wagleri* Agassiz. No locality.
1840 *Hamadryas chaeropithecus* Lesson. "Abyssinia, Arabia and Egypt."
1863 *Theropithecus nedjo* Reichenbach. Ethiopia. Considered by Napier (1981) to refer to this species, not to *Theropithecus gelada*.
1870 *Hamadryas aegyptiaca* Gray. New name for *Simia hamadryas*.
1900 *Papio arabicus* Thomas. Yemen: Subaihi country, 100 km northwest of Aden.
1909 *Papio brockmani* Elliot. Somalia: Diredawa, 1,060 m.

DIAGNOSIS. Adult male silvery gray with huge mane, reaching back to rump and contrasting with very short body and limb hair; mane bushing out to form white cheek whiskers, offset by the darker, shorter crown hair; tail tuft white; hands and feet dark. Female a plain olive brown, with no mane. Hairs multibanded, black and gray-white. In male, face and cushionlike skin around callosities bright pinkish red; in female, face gray, skin not swollen around callosities. Tail held in a simple backward-pointing curve.

DISTRIBUTION. Arid zone of northern Ethiopian lowlands, extending east into northern Somalia and west into the Red Sea Hills of Sudan; and along the Red Sea coast of the Arabian Peninsula, from Yemen an unknown distance north, but well into Saudi Arabia.

NOTES. I know of no consistent differences between African and Arabian representatives of this species.

Papio papio (Desmarest, 1820)

Guinea Baboon

1777 *Papio sphinx* Erxleben. "Tropical parts of Africa and India." In part (see Delson and Napier [1976]); not *Simia sphinx* Linnaeus, 1758.
1820 *Cynocephalus papio* Desmarest. "Coast of Guinea."
1851 *Cynocephalus olivaceus* I. Geoffroy. Supposedly, Gulf of Benin.
1853 *Papio rubescens* Temminck. West Africa.

DIAGNOSIS. Brownish red with large mane in adult male, less full around head than in *P. hamadryas*, but again contrasting with shorter rump and limb hair. Hairs with multiple alternating bands of black and yellowish red. Face blackish red, skin around callosities lighter, more pinkish. Tail loops up and back evenly. Nostrils protruding beyond end of snout.

DISTRIBUTION. A small area in far western Africa, from Guinea, Guinea-Bissau, and Senegal and unknown distance into Mauretania and Mali.

Papio anubis (Lesson, 1827)

Olive Baboon or
Anubis Baboon

1827 *S[imia] anubis* Lesson. Africa: fixed by Anderson (1902) as Upper Nile.
1843 *C[ynocephalus] choras* Ogilby. West Africa.
1856 *Cynocephalus doguera* Pucheran and Schimper. Ethiopia: "Abyssinia."
1897 *Papio neumanni* Matschie. Tanzania: Ol Doinyo Lengai, Lake Natron.
1898 *Papio heuglini* Matschie. Sudan: Shilluk Islands, White Nile.
1900 *Papio yokoensis* Matschie. Cameroon: Sanaga River, Yoko.
1902 *Papio lydekkeri* Rothschild. Ethiopia or Sudan: upper Blue Nile.
1902 *P[apio] anubis* subsp. *olivaceus* De Winton. Sudan. Not *Cynocephalus olivaceus* I. Geoffroy, 1851 (= *Papio papio*).
1907 *Papio furax* Elliot. Kenya: Baringo, 1,200 m.
1909 *Papio nigeriae* Elliot. Nigeria: Ibi.
1909 *Papio tessellatum* Elliot. Uganda: Ankole, Mulema, 1,500 m.

1913 *Papio anubis lestes* Heller. Kenya: Athi Plains, Ulukenia Hills.

1913 *Papio anubis vigilis* Heller. Kenya: Lakiundu River near its confluence with northern Guaso Nyiro.

1915 *Papio silvestris* Lorenz. Congo-Zaire: Mawambi, Ituri Forest.

1916 *Papio werneri* Wettstein. Sudan: Kordofan, Nuba Mountains, Gebel Talodi.

1917 *Papio graueri* Lorenz. Congo-Zaire: Rutshuru Plains.

1960 *Papio anubis tibestianus* Dekeyser and Derivot. Chad: Enneri Debassar, Tibesti.

1965 *P[apio] anubis niloticus* Roth. Nomen nudum.

DIAGNOSIS. Olive brown; adult male with large mane, restricted to foreparts and grading into shorter body hair, not sharply set off. Hairs with one or two alternating pairs of rings, black and yellowish brown. Face and skin around callosities dark gray to black; on rump, the bare area is much smaller than in *P. hamadryas* or *P. papio*. Nostrils project forward of snout. Tail directed up for proximal one-fourth, then falls away as if broken.

DISTRIBUTION. From Mali, where its range presumably meets that of *P. papio*, east to the Ethiopian highlands, southeast into Kenya and northwestern Tanzania. It forms a narrow hybrid zone with *P. hamadryas* below the Awash Falls and elsewhere in northern Ethiopia. It hybridizes with *P. cynocephalus* in the eastern part of Tsavo National Park in Kenya, in Manyara National Park in Tanzania, and elsewhere sporadically along a northeast/southwest trending line across this region. In northeastern Congo-Zaire, it enters true rain forest in some regions.

NOTES. It is very doubtful whether any subspecies can be recognized in this species. The rain forest–edge baboons described as *tessellatum* Elliot, 1909, have, as their name indicates, a "tessellated" dorsal pelage, with small patches of black due to the long, black hair bands coinciding over contiguous areas (and with pale cheeks); but tessellated specimens are fairly frequent in the Ethiopian highlands as well (described as *heuglini* Matschie, 1898) and occur in lower frequencies elsewhere. The two skulls on which *tibestianus* Dekeyser and Derivot, 1960, was based (these are in Paris [MNP]) are indeed extremely small, but there are almost equally small skulls (in the London

[BM(NH)] collection) from the Shai Hills, on the coast of Ghana, another isolated population.

In external appearance this species in many respects resembles *P. ursinus* as much as the geographically intervening *P. cynocephalus*. Albrecht and Miller (1993), in their interpretation of the findings of Hayes et al. (1990), found that this is the case in skull measurements as well; their *P. anubis* sample overlaps with both *P. ursinus* and *P. cynocephalus* from Malawi (wrongly reported as *cynocephalus/ursinus* hybrids).

Papio cynocephalus (Linnaeus, 1766)
Yellow Baboon

DIAGNOSIS. Dorsally yellow to yellow-brown, contrasting with white underparts, inner surfaces of limbs, cheeks and lateral patches on muzzle, and fringing hands and feet. Hairs mostly uniformly yellow or yellow-brown, with a black tip. Male has no mane, or hardly any, but a median nuchal crest of long hair; flank hairs long, forming an inconspicuous fringe. Face and (small) bare areas around callosities black. Tail varying, from simply curved in *P. c. kindae* to more usually "broken" as in *P. anubis* but held low. Nostrils set back from snout.

DISTRIBUTION. From the Zambezi and central Angola north and northeast to northern Tanzania, coastal Kenya, and an unknown distance north into Somalia.

NOTES. Hayes et al. (1990) and Albrecht and Miller (1993) confirmed that the skulls of the southern populations of this species are extremely diverse. A sample from Malawi (incorrectly reported as *cynocephalus/ursinus* hybrids) overlaps with *P. anubis,* and samples from western Zambia and from Angola are progressively more distinct (their 1 SD circles only just overlap).

There is a restricted hybrid zone with *P. anubis* near Sultan Hamud (2°02′S, 37°23′E) in Kenya (Maples 1972), and the baboons of Lake Manyara National Park, Tanzania, also appear to be hybrids.

Papio cynocephalus cynocephalus (Linnaeus, 1766)

1766 *Simia cynocephalus* Linnaeus. "Africa," restricted to Kenya: inland from Mombasa (J. A. Allen 1925:315).

?1792 *S[imia] (Papio) variegata* Kerr. "Probably Africa." (May refer to *Mandrillus leucophaeus*). Suppressed by International Commission on Zoological Nomenclature (1970), Opinion 935.

1800 *Simia basiliscus* Schreber. No locality.

1800 *Simia sublutea* Shaw. Africa.

1820 *Cynocephalus babouin* Desmarest. "Northern Africa."

1821 *Cynocephalus antiquorum* Schinz. No locality.

1843 *Cynocephalus thoth* Ogilby. No locality.

1852 *Cercopithecus ochraceus* Peters. Mozambique: Querimba, 10–13°S.

1852 *Cercopithecus flavidus* Peters. Mozambique: Quitangonha, 15°S, mainland north of Mozambique Island.

1892 *Cynocephalus langheldi* Matschie. Tanzania: eastern slope of Nguru Mountains.

1897 *Papio pruinosus* Thomas. Malawi: Fort Johnston (Lesumbwe), Monkey Bay, south end of Lake Malawi.

1907 *Papio strepitus* Elliot. Malawi: Fort Johnston, south end of Lake Malawi.

?1918 *Choiropithecus rhodesiae* Hagner. ?Zambia: "Central Rhodesia."

1928 *Papio cynocephalus jubilaeus* Schwarz. Zambia: Missale, 14°S, 33°10′E (on border with Mozambique).

DIAGNOSIS. No mane in male; pelage on body straight. Natal coat black as in most baboons. Tail in most cases held almost horizontally back, then falling downward, "broken."

DISTRIBUTION. From the central and southeastern part of the range: east of the Luangwa in Zambia, Malawi, northern Mozambique, and most of Tanzania. Jolly (1993) dispelled the commonly held misapprehension that baboons from Mozambique north of the Zambezi and neighboring areas such as Malawi are intermediates between *P. cynocephalus* and *P. ursinus*: they are typical *cynocephalus*, except for the aberrant type of *pruinosus*, an animal that lived a long time in captivity, evidently under poor conditions.

Papio cynocephalus ibeanus Thomas, 1893

1893 *Papio thoth ibeanus* Thomas. Kenya: Lamu.

1942 *Papio ruhei* Zukowsky. Somalia: Webbi Shebeli, hinterland of Mogadishu.

DIAGNOSIS. Jolly (1993) distinguished this subspecies as having a trace of mane, and wavy instead of straight body pelage. The differences are actually quite easily visible in the field; I was struck by the difference between baboons seen at lower altitudes in the eastern Usambaras and those on the lower Tana (which appeared more fawn, less yellow, probably by virtue of the different hair texture), but did not see any in intervening regions.

DISTRIBUTION. Southern Somalia and southeastern and coastal Kenya.

Papio cynocephalus kindae Lönnberg, 1919

1919 *Papio kindae* Lönnberg. Congo-Zaire: Lulua District, Kinda.

DIAGNOSIS. Strikingly smaller in size than other subspecies, though those from western Zambia and southern Congo-Zaire are larger than those from Angola. Uniquely among baboons, the neonatal coat appears to be reddish rather than black.

DISTRIBUTION. On the upper Zambesi in southwestern Zambia, the ranges of this taxon and *P. ursinus griseipes* approach each other closely, but there is no record of interbreeding (Ansell 1978). Similarly, in southern Angola *P. c. kindae* and *P. ursinus ruacana* seem to abut (the former inland, the latter on the coastal strip), with no evidence of interbreeding (Machado 1969).

Papio ursinus (Kerr, 1792)

Chacma Baboon

DIAGNOSIS. Black to brown or gray-buff dorsally, with paler underparts and sides of muzzle. Male without mane, but with elongated hair tufts along nape. Hairs with a single band (either black on brown or yellow-brown on black) and, if the hair is light, a black tip. Face and (small) area around ischial callosities black. Nostrils do not protrude forward of snout. Tail held up and then drops down, "broken," as in *P. anubis*. Facial skeleton, unlike other baboons, points downward as well as forward.

DISTRIBUTION. South of the Zambezi; and reaching north into southern Angola and southwestern Zambia.

Papio ursinus ursinus (Kerr, 1792)

1787 *Simia porcaria* Boddaert. Africa. Not of Brünnich, 1782 (= *Macaca nemestrina*).

1792 *S[imia] (Cercopithecus) hamadryas ursinus* Kerr. Cape of Good Hope.

1804 *Simia sphingiola* Hermann. No locality.

1812 *Papio comatus* E. Geoffroy. South Africa: Cape of Good Hope.

1826 *Cynocephalus capensis* A. Smith. South Africa: Table Mountain to Cape Point, the Paardeberg and the Paarl.

1926 *Papio porcarius orientalis* Goldblatt. Eastern South Africa; restricted by Schwarz (1934b) to Queenstown, Eastern Cape, but corrected by Shortridge (1942) to Atherstone, Grahamstown.

1926 *Papio porcarius occidentalis* Goldblatt. South Africa: Western Transvaal and Cape Province; restricted to Rustenberg, Transvaal, by Schwarz (1934b).

1932 *Papio porcarius nigripes* Roberts. South Africa: northern Transvaal, Magalakuin River.

DIAGNOSIS.   In the southern Cape, members of this subspecies are nearly black in color, with black hands, feet, and tail. Color becomes lighter clinally toward the northeast (Northern Province, formerly northern Transvaal), but still with black extremities.

Papio ursinus griseipes Pocock, 1911

1911 *Papio porcarius griseipes* Pocock. South Africa: Transvaal, supposedly Potchefstroom, but more probably northeastern Transvaal, near Messina (Schwarz 1934b).

1927 *Cynocephalus transvaalensis* Zukowsky. South Africa: Northern Transvaal.

1932 *Papio porcarius chobiensis* Roberts. Botswana: Chobe River, Kabulabula.

1932 *Papio porcarius ngamiensis* Roberts. Botswana: Maun.

DIAGNOSIS.   This subspecies continues the color lightening begun in northerly populations of nominotypical *P. u. ursinus,* being fawn in tone, but the hands, feet, and tail are grayish rather than black.

DISTRIBUTION.   Far southwestern Zambia, Zimbabwe, in Mozambique south of the Zambezi, and in parts of northwestern (former) Transvaal; where they intergrade with *P. u. ursinus,* if they do, is unclear.

Papio ursinus ruacana Shortridge, 1942

1942 *Papio comatus ruacana* Shortridge. Namibia: Otjiwau, 16 km north of Kaoto-Otavi (designated by Meester et al. [1986]).

1965 *P[apio] ursinus chacamensis* Roth. Namibia: Namaland, Damaraland. Nomen nudum.

DIAGNOSIS.   These are small baboons, black-footed like *P. u. ursinus,* darker than *P. u. griseipes,* but tending to be especially dark on the back and crown, where the blackened tones may contrast with the lighter flanks and limbs.

DISTRIBUTION.   Northern Namibia and southwestern Angola, north along the coast to about 11°30′S (Machado 1969).

Genus *Theropithecus* I. Geoffroy, 1843
  Gelada

1843 *Theropithecus* I. Geoffroy. *Macacus gelada* Rüppell, 1835.

1843 *Gelada* Gray. *Gelada rüppelli* Gray, 1843.

1916 *Simopithecus* Andrews. *Simopithecus oswaldi* Andrews, 1916 (fossil species).

Differentiated from *Papio* by small incisors and larger, high-crowned cheekteeth with accessory cusps and very reduced molar flare; relatively shorter but deep face; ascending ramus of mandible vertical; very short index finger (Jolly 1972).

According to Goodman et al. (1998), this genus separated from *Papio* only 4 million years ago; they proposed downgrading it to subgeneric status.

Theropithecus gelada (Rüppell, 1835)
  Gelada

DIAGNOSIS.   (Characters observable only in the living species.) Adult male with enormous cape over head and shoulders, and recurved cheek whiskers; tail shorter than head plus body, tufted; pelage yellow-brown; chest patch edged with white. Nose upturned; a bare red hourglass-shaped pectoral patch, with cyclically prominent marginal vesicles in adult female; mammae close to midline; ischial callosities do not fuse across midline even in adult males; skin below callosities bare, edged with vesicles in female, and incorporating a pair of fatty pads below callosities.

DISTRIBUTION.   High grasslands at 2,350 to 4,400 m in Tigre, Begemdir, Wolle, and Shoa Provinces, Ethiopia.

Theropithecus gelada gelada (Rüppell, 1835)

1835 *Macacus gelada* Rüppell. Ethiopia: Haremat, Semyen, and Gojjam (but species does not occur in Gojjam according to Napier [1981]).
1857 *Theropithecus senex* Pucheran. Ethiopia: Noari region, Semyen.
1843 *Gelada rüppelli* Gray. New name for *Macacus gelada*.

DIAGNOSIS. Mane dark brown; circumpectoral fur not extensive, iron gray in color.

DISTRIBUTION. From Begemdir, Tigre, and Wolle Provinces, west of the Rift Valley.

Theropithecus gelada obscurus Heuglin, 1863

1863 *Theropithecus obscurus* Heuglin. Ethiopia: sources of Takazze River.

DIAGNOSIS. Darker mane; circumpectoral fur extensive and pure white.

DISTRIBUTION. From Wolle and Shoa Provinces, east of the Rift Valley.

Genus *Cercocebus* E. Geoffroy, 1812
White-eyelid Mangabeys

1812 *Cercocebus* E. Geoffroy. *Cercocebus fuliginosus* E. Geoffroy, 1812.
1841 *Aethiops* Martin. Based on "White-eyelid monkeys."
1904 *Leptocebus* Trouessart. *Cercocebus galeritus* Peters, 1879 (here designated).

I summarized (Groves 1978) the differences between this genus and *Lophocebus,* concluding that its affinities lie away from the baboons and gelada. Comparison with *Mandrillus* (Groves 2000a) has shown, in agreement with molecular data (Goodman et al. 1998; Disotell 2000), that *Cercocebus* is very close to *Mandrillus.*

Cercocebus atys (Audebert, 1797)
Sooty Mangabey

DIAGNOSIS. Gray or brown-gray, lighter ventrally and on the elongated cheek whiskers, with somewhat darker hands and feet. Face is dark grayish pinkish colored, with blackish muzzle and ears. Judg-

ing by the skull, this is less sexually dimorphic than *C. torquatus.*

DISTRIBUTION. From Senegal around the coast to Ghana.

NOTES. Previously (Groves 1978) I retained this species in *C. torquatus,* but this now seems ill-advised, because the two are sharply and consistently distinct.

Cercocebus atys atys (Audebert, 1797)

1775 *Simia aethiops* Schreber. West Africa. Not of Linnaeus, 1758 (*Chlorocebus*).
1797 *Simia atys* Audebert. "Afrique occidentale" (Schwarz 1928a).
1812 *Cercocebus fuliginosus* E. Geoffroy. No locality.
1821 *[Cercocebus] aethiopicus* F. Cuvier. West Africa, south of Cape Verde.

DIAGNOSIS. Smoky gray, with only occasionally a trace of a dorsal stripe; generally no whorl on crown; crown hairs have a straw-colored band and black tip. The nuchal markings more typical of *C. a. lunulatus* may be weakly expressed.

DISTRIBUTION. The western part of the range, east to the Nzo-Sassandra system in Ivory Coast; I recorded (Groves 1978) a hybrid specimen from between the Nzo and the Sassandra.

Cercocebus atys lunulatus (Temminck, 1853)

1853 *Cercopithecus lunulatus* Temminck. Ghana: River Boutry.

DIAGNOSIS. A whorl on crown; crown hairs lack the straw-colored band. Dorsal stripe is well expressed and there is a white oval mark, bordered with black, on nape. Hands and feet colored more as body; underside white; face darker than in *C. a. atys.*

DISTRIBUTION. The eastern part of the range, from the Nzo-Sassandra system to the Volta River.

NOTES. It is possible that this subspecies should be elevated to species rank, but the museum material of both subspecies is very poor, and it is not possible to say whether their characters overlap. To a certain degree, it is intermediate between *C. a. atys* and *C. torquatus*.

Cercocebus torquatus (Kerr, 1792)

**Collared Mangabey or
Red-capped Mangabey**

1792 *Simia (Cercopithecus) aethiops torquatus* Kerr. "Ethiopia" = West Africa.
1821 *Cercocebus aethiopicus* F. Cuvier. West Africa, south of Cape Verde.
1843 *Cercocebus collaris* Gray. "Africa."
1843 *Cercocebus crossii* Gray. "Africa."

DIAGNOSIS. Dark gray, with sharply demarcated white underside and inner surfaces of limbs, this zone extending forward to chin, sides of neck, and cheeks. Tail with white tuft. Crown dark red, outlined by a white collar and temporal line. Eyelids bright white. A strong size difference between males and females.

DISTRIBUTION. Rain forest regions from western Nigeria through Cameroon (east as far as the East Dja Reserve) and Río Muni to at least Sette Camma (2°31′S) on the Gabon coast.

NOTES. There are marked geographic variations in this species in the skull (larger north of the River Sanaga and east of the Niger, narrow in Gabon) and length of white tail tip (longer south of the Sanaga), but in a mosaic fashion that makes it impossible to recognize any subspecies.

Cercocebus agilis Milne-Edwards, 1886

Agile Mangabey

1886 *Cercocebus agilis* Milne-Edwards. Congo-Brazzaville: Poste des Ouaddas, Oubangui-Congo confluence.
1900 *Cercocebus hagenbecki* Lydekker. Congo-Zaire: Oubangui River, 480 km above its junction with the Congo.
1914 *Cercocebus fumosus* Matschie. Congo-Zaire: Semliki River, northwest of Beni.
1915 *Cercocebus oberlaenderi* Lorenz. Congo-Zaire: Mawambi.

DIAGNOSIS. Brownish olive to gray-olive, clearly speckled, hairs with two light bands on foreparts, often fading out on hindparts; median dorsal zone tending to be darker; hands very dark brown; tail be-

comes lighter distally and is light below; underside, to chin, and inner surfaces of limbs paler, unspeckled. Cheeks white, because of light bases of back-swept cheek hairs. Crown slightly darker, nearly always with a whorl or center parting in front, bordered in front with a short fringe. Face black; eyelids pale, not white. Sexual size difference as in *C. atys*.

DISTRIBUTION. Río Muni east via the Makokou District (northeastern Gabon) across the Oubangui to the Garamba National Park and the Semliki River, northeastern Congo-Zaire.

NOTES. There are fairly well separated dark and light morphs, the former being much more common. Size is larger to the west.

Cercocebus chrysogaster Lydekker, 1900

Golden-bellied Mangabey

1900 *Cercocebus chrysogaster* Lydekker. Congo-Zaire: Upper Congo.

DIAGNOSIS. Robustly built; rich, dark speckled reddish brown, sharply set off from yellowish creamy cheeks, throat, and inner surfaces of limbs, becoming bright red-gold on chest and belly; tail speckled at root only. Usually no whorl or parting on crown; cheek whiskers long, swept back. Eyelids pale but not white. Sexual size difference marked.

DISTRIBUTION. Rain forests south of the great bend of the Congo River.

NOTES. I suggested (Groves 1978) that the occasional occurrence of an anterior crown whorl may be a character of immaturity. I placed this in *C. agilis* as a subspecies, but it clearly has the criteria of species status under the PSC.

Cercocebus galeritus Peters, 1879

Tana River Mangabey

1879 *Cercocebus galeritus* Peters. Kenya: Tana River, Mitole.

DIAGNOSIS. Inconspicuously speckled gray-yellow with long, loose, wavy pelage; limbs unspeckled;

forearms, hands, and feet dark; underside yellowish white, fluffy-haired. A center parting on crown beginning immediately behind forehead, with very long, dark hair diverging from it on either side; this hair becomes very long, >100 mm, back toward middle of crown. Cheeks and temples whitish. Tail with a slight, pale tuft. Hands and feet dark brown. Face black; eyelids bright white. Size small in male, but female not much smaller than male; skull broad, high-crowned, with deeper suborbital fossa and small teeth compared with *C. agilis*.

DISTRIBUTION. Gallery forests along the Tana River and some of its former courses, between Wenje and Garsen.

NOTES. In some respects this species resembles *C. torquatus* more than the two central African species with which it has commonly been united.

Cercocebus sanjei Mittermeier, 1986

Sanje Mangabey

1985 *Cercocebus galeritus sanjei* Wasser. Nomen nudum.
1986 *Cercocebus galeritus sanjei* Mittermeier. Tanzania: Sanje Waterfall, Mwanihana Forest.

DIAGNOSIS. Speckled gray; underside, but not inner surfaces of limbs, pale orange. Tail with pale tuft. Hands and feet darker. Long crown hairs swept back, up, and sideways to give a "bouffant" appearance, set off in front by thin black brow seam. Face pale grayish, becoming pink around eyes and on nose. Eyelids not strikingly white.

DISTRIBUTION. Known from the Mwanihana Forest Reserve, on the eastern slopes of the Uzungwa Mountains, Tanzania.

NOTES. The circumstances surrounding the introduction of the scientific name of this species, and why Mittermeier (1986) is its author and date, are recounted by Groves (1996b).

Genus *Mandrillus* Ritgen, 1824

Mandrills

1773 *Papio* P. L. S. Müller. *Simia sphinx* Linnaeus, 1758. Suppressed (International Commission on Zoological Nomenclature [1982a], Opinion 1199).
1824 *Mandrillus* Ritgen. *Simia sphinx* Linnaeus, 1758 (see G. M. Allen [1939a:157]). Placed on the Official List of Generic Names in Zoology (International Commission on Zoological Nomenclature 1982a).
1831 *Mandril* Voigt. *Simia sphinx* Linnaeus, 1758.
1839 *Mormon* Wagner. *Simia sphinx* Linnaeus, 1758. Not of Illiger, 1811 (Aves).
1862 *Drill* Reichenbach. *Simia leucophaea* F. Cuvier, 1807.
1870 *Chaeropithecus* Gray. *Simia leucophaea* F. Cuvier, 1807. Not of Blainville, 1839 (= *Papio*).
1904 *Maimon* Trouessart. *Simia sphinx* Linnaeus, 1758.

This taxon is distinguished from *Papio*, which it superficially resembles in its long face, among other features by its very short tail, bright perianal coloration, paranasal ridges, combination of large incisors and relatively small cheekteeth, posteriorly convergent toothrows, shelflike superior temporal lines overlying *M. temporalis* origins, and sternal gland. It is actually much closer to *Cercocebus*; indeed Goodman et al. (1998) calculated that it separated from the latter only 4 million years ago and placed it as a subgenus of *Cercocebus*.

Mandrillus sphinx (Linnaeus, 1758)

Mandrill

1758 *Simia sphinx* Linnaeus. "Borneo." Fixed as Cameroon: Bitye, Dja River, by Delson and Napier (1976).
1766 *Simia maimon* Linnaeus. "Ceylon."
1766 *Simia mormon* Alströmer. No locality.
1780 *Simia madarogaster* Zimmermann. No locality.
1792 *Simia suilla* Kerr. No locality.
1799 *Simia latidens* Bechstein. No locality.
1827 *Simia (Papio) pennanti* Griffith. No locality.
1909 *Papio planirostris* Elliot. Cameroon: Fan country.
1917 *Mandrillus schreberi* Matschie. New name for *Simia maimon* "Schreber" (recte Linnaeus).
1917 *Mandrillus tessmanni* Matschie and Zukowsky. Equatorial Guinea: middle Benito River, near Alen, Okak-land.
1917 *Mandrillus escherichi* Matschie and Zukowsky. Equatorial Guinea: Río Muni, Temboni River, Ekododo.
1917 *Mandrillus zenkeri* Matschie and Zukowsky. Cameroon: between Bipindi and Yaounde.
1917 *Mandrillus hagenbecki* Matschie and Zukowsky. ?Nigeria: Lagos (certainly in error: zoo specimen, bought in Bordeaux).
1917 *[Mandrillus zenkeri]* var. *ebolowae* Matschie and Zukowsky. Cameroon: Ebolowa.

1922 *Mandrillus insularis* Zukowsky. Supposedly from Equatorial Guinea: Bioko; but locality is in error.
1922 *Maimon burlacei* Rothschild. Cameroon: Dja River, Bitye.

DIAGNOSIS. Nose and lips red, paranasal ridges blue, longitudinally grooved; cheek whiskers and beard orange; hairs of body multibanded black and reddish yellow. In skull, paranasal ridges of adult male broad, converging gradually anteriorly.

DISTRIBUTION. South of River Sanaga in Cameroon, through Río Muni, Gabon, and Congo to River Kouilou (Napier 1981).

NOTES. Linnaeus's description was based on a monkey illustrated by Gesner in 1554 in his *Historia Animalium de Quadrupedibus Viviparis*. Gesner's figure is clearly a *Mandrillus* but, as Delson and Napier (1976) pointed out, it could as well have been a drill as a mandrill. To fix the name, they nominated as neotype the specimen that is holotype of *Maimon burlacei* Rothschild, which definitely is a mandrill.

Grubb (1973) and Napier (1981) reviewed the evidence for the distribution of this species and concluded that there is no evidence that it occurs north of the River Sanaga or on Bioko. Delson and Napier (1976) concurred and doubted the real existence of any of the taxa erected by Matschie, Zukowsky, or Rothschild.

Mandrillus leucophaeus (F. Cuvier, 1807)
Drill

DIAGNOSIS. Face black, except for lower lip, which is red; paranasal ridges smooth; cheek whiskers and beard white; hairs of body brown with (usually) a single straw-colored or yellow band and black tip; underside white or gray-white In skull, paranasal ridges converge sharply anteriorly.

DISTRIBUTION. Southeastern Nigeria and northwestern Cameroon, between the Cross River and Sanaga River, and slightly south of the lower Sanaga at Ongue Creek (Grubb 1973); Bioko Island.

Mandrillus leucophaeus leucophaeus (F. Cuvier, 1807)

1792 *Simia (Papio) sylvicola* Kerr. Suppressed (International Commission on Zoological Nomenclature [1970], Opinion 935).
?1792 *[Simia] (Papio) variegata* Kerr. "Probably Africa." May refer to *Papio cynocephalus*. Suppressed (International Commission on Zoological Nomenclature [1970], Opinion 935).
1792 *Simia (Papio) cinerea* Kerr. Suppressed (International Commission on Zoological Nomenclature [1970], Opinion 935).
1792 *Simia (Papio) livea* Kerr. Suppressed (International Commission on Zoological Nomenclature [1970], Opinion 935).
1795 *Simia sylvestris* Link. Suppressed (International Commission on Zoological Nomenclature [1970], Opinion 935).
1800 *Simia silvestris* Schreber. Suppressed (International Commission on Zoological Nomenclature [1970], Opinion 935).
1800 *Simia sylvicola* Shaw. "Guinea." Suppressed (International Commission on Zoological Nomenclature [1970], Opinion 935).
1807 *Simia leucophaea* F. Cuvier. "Probably coasts of Africa." Placed on the Official List of Specific Names in Zoology (International Commission on Zoological Nomenclature [1970]).
1838 *Cynocephalus drill* Lesson. New name for *Simia leucophaea*.
1906 *Papio mundamensis* Hilzheimer. Cameroon: Mundame, Mukonje Farm.

DIAGNOSIS. Skull averaging larger than in *M. l. poensis* (greatest length 192–231 mm), narrower across zygomatic arches (120–130 mm), but braincase wider (84–102 mm); toothrow longer (54–56 mm); frontal region rises steeply above nasion. Hairs on sides of crown ringed with yellow and black; legs less buffy than arms.

DISTRIBUTION. Mainland part of species range.

NOTES. There are slight differences between different populations. Male skulls from the Cross River District are 205–235 mm long ($n = 23$); from Edea, just north of the mouth of the River Sanaga, 199–213 ($n = 7$); and from the Yabassi (Korup) region, 192–204 ($n = 2$). Skins from the Cross River have body hairs with a long, light gray-brown base, followed by a straw-colored band and a black tip; those from Yabassi have a lighter gray base and the band is yellower. Yabassi skins are whiter below, rather than gray-white.

Mandrillus leucophaeus poensis Zukowsky, 1922

1922 *Mandrillus poensis* Zukowsky. Equatorial Guinea: Bioko.

DIAGNOSIS. Size slightly smaller than *M. l. leucophaeus* (skull length 186–207 mm, *n* = 6) but skull wider across zygomatic arches (128–135 mm), braincase narrower (86–94 mm at meatus level); toothrow shorter (49–50 mm); frontal strongly slopes back from nasion to glabella. Body more grayish toned. Hairs on sides of crown yellow-brown with only a black tip; legs more buffy than arms.

SUBFAMILY COLOBINAE JERDON, 1867

Strasser and Delson (1987) defined the subfamily as having a complex (at least three-chambered) stomach, a short pollex, reduced molar flare (but this is parallel to most Cercopithecini), shortened trigonid, high molar relief, posteriorly deepened mandibular corpus, and some pedal specializations. Colobine feet are consistently paraxonic (that is, with the second digit as short as the fifth and the third and fourth digits subequal and forming the axis of the foot [Strasser 1994]). The entire foot is modified to increase the span of the grasp and for more efficient supination: elongated, well-curved astragalocalcaneal facet, strong lateral rotation of astragalar head, more vertically oriented navicular, shortened distal tarsus, and associated muscular dispositions (Strasser 1988).

Trachypithecus obscurus has two-layered nails on the feet (Soligo and Müller 1999), but because of the variation within the Cercopithecinae this condition cannot be assumed to characterize all the Colobinae.

Delson (1976) pointed out that two other names, Presbytina Gray, 1825, and Semnopithecidae Owen, 1843, antedate Colobidae and would thus upset common usage were strict priority to be maintained. Delson (1976) ascribed the name Colobidae to Blyth, 1875, but D. Brandon-Jones (1978) pointed out that the correct attribution is to Jerdon, 1867.

I proposed (Groves 1989) that the basic division is between *Nasalis* (including *Simias*) and the rest, but Strasser and Delson (1987) preferred an Asia/Africa split, although admitting that the Asian group would

be hard to diagnose. Genetic studies tend to separate the Colobidae into an African and an Asian group (Wang, Forstner et al. 1997; Zhang and Ryder 1998). In some cases, *Nasalis* is the first branch in the Asian group, but in others (Zhang and Ryder 1998) it assorts with *Rhinopithecus* and *Pygathrix*.

A general overview of the subfamily was given by Oates et al. (1996).

African Group

In the African genera, the pollex is absent (or just a nubbin), and there are certain pedal specializations (Strasser and Delson 1987).

Genus *Colobus* Illiger, 1811
Black-and-white Colobus

1811 *Colobus* Illiger. *Simia polycomos* Schreber, 1800.
1821 *Colobolus* Gray. Emendation.
1870 *Guereza* Gray. *Guereza rüppellii* Gray, 1870.
1887 *Stachycolobus* Rochebrune. *Colobus satanas* Waterhouse, 1838.
1887 *Pterycolobus* Rochebrune. *Semnopithecus vellerosus* I. Geoffroy, 1834.

Ischial callosities of males are confluent across the midline; there is no sexual swelling in female or perineal organ in male; the subhyoid sac is present and the larynx is enlarged; and the stomach is three-chambered. Oates et al. (1996) discussed the polarity of these characters. The intermembral index is 79. Infant is white at birth, except in *C. satanas*.

In characterizing some of the subspecies of *C. angolensis* and *C. guereza,* I have had the benefit of an unpublished study by D. L. Hull, for which I am most grateful.

Colobus satanas Waterhouse, 1838
Black Colobus

1838 *Colobus satanas* Waterhouse. Equatorial Guinea: Bioko Island.
1857 *Semnopithecus anthracinus* Leconte. Gabon.
1917 *Stachycolobus municus* Matschie. Equatorial Guinea: Wurminsog.
1917 *Stachycolobus limbarenicus* Matschie. Gabon: Lambaréné.
1917 *Stachycolobus zenkeri* Matschie. Cameroon: Bipindi.
1943 *Colobus metternichi* Krumbiegel. Equatorial Guinea: Bioko Island, northern slopes of Santa Isabel mountain.

DIAGNOSIS. Completely black, lacking zones of elongated hair on body or tail; the neonate is brown, not white like other species (Oates et al. 1996). The dorsal outline of the braincase in lateral view is curiously saddle-shaped. Skull is less prognathous than in other species, and the external nose is comparatively short; vocalizations are less derived (Oates and Trocco 1983).

DISTRIBUTION. Coastal regions of Cameroon south of the Sanaga River, Equatorial Guinea (including Bioko Island), and Gabon, inland to the Lopé National Park.

NOTES. The Bioko form of this species may be subspecifically distinct (D. Brandon-Jones and T. Butynski, personal communication).

Colobus angolensis Sclater, 1860
Angola Colobus

DIAGNOSIS. Body black; white circumfacial hair, forming laterally elongated tufts on cheeks; long white epaulettes; tail tufted, sometimes entirely white or white restricted to tuft. Nose is longer than in *C. satanas,* and skull is more prognathous; vocalizations (Oates and Trocco 1983) are more derived.

DISTRIBUTION. From Angola north to the Congo-Oubangui system and east through the montane and coastal forests of Tanzania to southeastern Kenya.

NOTES. Descriptions of the subspecies mainly follow Colyn (1991) and an unpublished study by D. L. Hull. They divide into two groups, one with large epaulettes and a white pubic band *(angolensis, ruwenzorii, palliatus),* and one with thin epaulettes and no white in the pubic region *(cottoni, cordieri, prigoginei);* they are linked by "Nkungwe's Angolan Colobus," a subspecies not formally named (Nishida et al. 1981).

Colobus angolensis angolensis Sclater, 1860

1860 *Colobus angolensis* Sclater. Angola: inland from Bembe.
1908 *Colobus angolensis sandbergi* Lönnberg. Angola: Lifizi River near junction with Zambezi River.

1913 *Colobus (Colobus) palliatus weynsi* Matschie. "Lower Congo."
1914 *Colobus maniemae* Matschie. Congo-Zaire: Sankuru River, Bena Makima.
1914 *Colobus benamakimae* Matschie. Congo-Zaire: Sankuru River, Bena Makima.

DIAGNOSIS. Cheek whiskers and epaulettes white, broad, forming a continuous zone on either side, sometimes overlain by long, black hairs; pubic region with a narrow median white stripe; 30 to 70% of distal part of tail white.

DISTRIBUTION. From south of the great bend of the River Congo, south to about 12°S and west to about 16°W, in both Congo-Zaire and Angola.

Colobus angolensis cottoni Lydekker, 1905

1905 *Colobus palliatus cottoni* Lydekker. Congo-Zaire: Zokwa, Ituri River.
1913 *Colobus (Colobus) palliatus mawambicus* Matschie. Congo-Zaire: northern Pemba, Ituri.
1914 *Colobus mawambicus nahani* Matschie. Congo-Zaire: Panga, Aruwimi River.

DIAGNOSIS. White cheek whiskers well developed, more so than epaulettes, with which they form a continuous but narrow band on either side; no white in pubic region; distal half or so of tail grayish.

DISTRIBUTION. From northeast of the River Congo, west to about 20°E along the River Congo, east to Lake Albert, north to the Uele-Oubangui, and in the southeast hybridizing with *C. a. ruwenzorii* on the equator around 28°E.

Colobus angolensis ruwenzorii Thomas, 1901

1901 *Colobus ruwenzorii* Thomas. Congo-Zaire: Bwamba country, northwestern flank of Mount Rwenzori.
1914 *Colobus adolfi-frederici* Matschie. Rwanda: Rugege Forest.

DIAGNOSIS. Cheek whiskers and epaulettes forming a broad, continuous white band, sometimes overlain with long black hairs; a 6- to 10-cm-wide band of white or grayish in pubic region; distal 5–10 cm of tail grayish.

DISTRIBUTION. Occurs along both sides of the western Rift, from the Semliki Valley and the

Rwenzoris through southwestern Uganda, Rwanda, Mount Kahuzi, and Burundi to the eastern shores of Lake Tanganyika in Tanzania at about 4°S.

Colobus angolensis cordieri Rahm, 1959

1959 *Colobus polykomos cordieri* Rahm. Congo-Zaire: Kampunzu, Pangi Territory.

DIAGNOSIS. Cheek whiskers poorly developed, forming a continuous band with epaulettes; no white in pubic region; tail wholly grayish except for proximal 5–8 cm.

DISTRIBUTION. Congo-Zaire between the Rivers Lowa and Luama, from the Lualaba in the west to about 28°E, where it hybridizes with *C. a. ruwenzorii.*

Colobus angolensis prigoginei Verheyen, 1959

1959 *Colobus polykomos prigoginei* Verheyen. Congo-Zaire: Mount Kabobo, 2,400 m.

DIAGNOSIS. Similar to *C. a. cordieri* but tail yellowish white instead of grayish; pelage long and silky.

DISTRIBUTION. Known only from Mount Kabobo.

Colobus angolensis subsp.

This unnamed population (Nishida et al. 1981) has large epaulettes like the *angolensis-palliatus-ruwenzorii* group, but unlike them lacks a white pubic band, in which it resembles the *cottoni-prigoginei-cordieri* group. No white forehead band. Tail grayish only at tip, with a slight brush in males but not in females. It has been found on Mount Nkungwe and the ridges of the northern Mahale Mountains as far as Mount Pasagula to the north and Mount Kahoko and Mount Sibindi to the south. These areas are more than 300 km from the nearest localities of other subspecies.

Colobus angolensis palliatus Peters, 1868

1868 *Colobus palliatus* Peters. Tanzania: Pangani River.
1902 *Colobus sharpei* Thomas. Malawi: Fort Hill.
1914 *Colobus langheldi* Matschie. Tanzania: Ujiji.

DIAGNOSIS. Epaulettes large; a white pubic band, broad and oval in males, narrow in females; tail tip white, bushy, occupying about one-third of tail length; white forehead band fairly broad, broadly connected via full cheek whiskers with epaulettes; occipital hairs lengthened; coat long, thick, and soft; a whorl on withers.

DISTRIBUTION. Discontinuously distributed through the southern highlands and coastal and gallery forests of southern and eastern Tanzania (Rungwe Mountains, Lake Rukwa, Ulugurus, Ngurus, Uzungwas, Usambaras, Pangani River, Umba River) into southeastern Kenya (Diani Beach, Shimba Hills, Mrima Hill).

Colobus polykomos (Zimmermann, 1780)
King Colobus

1780 *Cebus polykomos* Zimmermann. Sierra Leone.
1792 *Simia (Cercopithecus) regalis* Kerr. Sierra Leone.
1795 *Simia tetradactyla* Link.
1800 *Simia polykomos* Schreber. Sierra Leone and Guinea.
1800 *Simia comosa* Shaw. Sierra Leone.

DIAGNOSIS. Body, crown, and limbs black; tail wholly white, without a tuft; long, straggly epaulettes and circumfacial hair (extending back onto anterior half of crown and under throat to sternal region) gray-white; hair on throat sparse. Ischial callosities are margined with white fur ventrally in both sexes; in male, the white extends up between the callosities and down toward the genitalia. Tail is more than 160% of head plus body length. Facial skeleton short, palate small; frontal process of malar curves laterally toward zygoma; supraorbital torus interrupted at glabella. Nose somewhat longer than in *C. angolensis,* and vocalizations are more derived.

DISTRIBUTION. Far western Africa, from Sierra Leone and Guinea east to the Sassandra River and its tributary, the Nzo, in Ivory Coast.

Colobus vellerosus (I. Geoffroy, 1834)
Ursine Colobus

1834 *Semnopithecus vellerosus* I. Geoffroy. "Africa."
1835 *Semnopithecus bicolor* Wesmael. "Coasts of Africa."

1835 *Co[obus] ursinus* Ogilby. "Algoa Bay."
1838 *Colobus leucomeros* Ogilby. Supposedly from Gambia River.
1927 *Colobus polykomos dollmani* Schwarz. Ivory Coast: Bandama. See under Notes.

DIAGNOSIS.   Black, with broad, bushy white cir-cumfacial ruff (not extending back onto crown or far back under throat), white patch on thigh, and wholly white tail with slight tuft; hair on throat thick; no epaulettes or with a few white hairs on shoulder. White fur extends all around the ischial callosities in males and around their ventral and lateral margins in females. Tail is less than 160% of head plus body length. Facial skeleton longer, palate larger than in *C. polykomos;* frontal process of malar curves medi-ally toward junction with zygomatic process; supra-orbital torus straight from side to side. Nose is long and convex, and vocalizations are strongly derived (Oates and Trocco 1983).

DISTRIBUTION.   From the Bandama River and its tributary, the Nzi, east into western Nigeria.

NOTES.   Hybridization between *C. vellerosus* and *C. polykomos* was analyzed by Groves et al. (1993). Hybrids occur between the Sassandra-Nzo system and the Bandama-Nzi, but are in all cases closer to *C. vellerosus,* suggesting that this more apomorphic species crossed the Bandama from the east and ge-netically swamped a preexisting population of *C. po-lykomos.* The type of *dollmani* is such a hybrid; it is listed in the synonymy because it is phenetically so much closer to *C. vellerosus.*

Colobus guereza Rüppell, 1835
Mantled Guereza

DIAGNOSIS.   Black, with elongated white veil or mantle from shoulder to haunch, curving across lumbar region to join veil of opposite side; short but bushy white cheek hairs, forming part of circumfa-cial ruff; white thigh stripe; tail with long, full white tuft, either restricted to terminal zone or occupying nearly all of tail. Facial appearance and vocalizations

are similar to those of *C. vellerosus;* the two are sister species (Oates and Trocco 1983).

DISTRIBUTION.   From the Donga River region of Nigeria and the Yabassi District of Cameroon, east through the Batouri region and across the Oubangui River through northern Zaire into southern Sudan, Uganda, and the Kenyan and Ethiopian highlands, and south to Mount Kilimanjaro and Mount Meru and the Kahé District in Tanzania.

NOTES.   An unpublished survey of the geographic variation in pelage of this species (and *C. angolensis*) by David Hull has kindly been made available to me; he previously studied craniometric variation by dis-criminant analysis (Hull 1979). In pelage characters, the subspecies of *C. guereza* seem not as distinct as those of *C. angolensis.*

Colobus guereza guereza Rüppell, 1835

1816 *Lemur abyssinicus* Oken. "Abyssinia." Unavailable.
1835 *Colobus guereza* Rüppell. Ethiopia: Gojjam Province.
1870 *Guereza rüppellii* Gray. Replacement for *guereza.*
1901 *Colobus abyssinicus poliurus* Thomas. Ethiopia: Omo River.
1913 *Colobus (Guereza) poliurus managaschae* Matschie. Ethiopia: Managasha Forest, upper Awash River.

DIAGNOSIS.   Proximal half of tail generally gray, distal half with a long, white tuft forming on average 52% of tail length (unpublished study by D. L. Hull); white mantle relatively long, extending onto dor-sum, shorter on flanks but long behind, covering one-fifth of the length of the tail; thigh stripe some-what diffuse. Tail longer than head plus body.

DISTRIBUTION.   Forested areas of the Ethiopian highlands west of the Rift Valley and along the Awash River, extending into lowland forests along the Omo River and in the Blue Nile gorge. Hull (in an unpublished study) found quite well-marked, but average, differences between northern and southern (Omo River) samples, and it may be that the name *poliurus* should be revived for the latter. The skull (Hull 1979) resembles that of *C. g. caudatus.*

Colobus guereza gallarum Neumann, 1902

1902 *Colobus gallarum* Neumann. Ethiopia: Garar Mulata and Djaffa Mountains, headwaters of Webi Shebeyli.

DIAGNOSIS. Tail tuft very bushy, white, occupying on average 40% of the tail length (unpublished study by D. L. Hull); proximal part of tail black; mantle relatively short, thin, but covering base of tail, and more developed on shoulders than *C. g. guereza,* but no white sprinkling on shoulders below veil.

DISTRIBUTION. The Ethiopian highlands east of the Rift Valley.

NOTES. Craniometrically, this seems to be the most distinctive member of the species (Hull 1979).

Colobus guereza occidentalis (Rochebrune, 1887)

1887 *Guereza occidentalis* Rochebrune. Congo-Brazzaville: Noki.
1913 *Colobus (Guereza) matschiei uellensis* Matschie. Congo-Zaire: Uele River.
1913 *Colobus (Guereza) matschiei ituricus* Matschie. Congo-Zaire: Ituri.
1913 *Colobus (Guereza) matschiei dianae* Matschie. Congo-Zaire: Kissenge, northeastern shore of Lake Edward.
1913 *Colobus (Guereza) matschiei brachychaites* Matschie. Sudan: Lado.
1913 *Colobus abyssinicus terrestris* Heller. Uganda: Rhino Camp.
1914 *Colobus (Guereza) escherichi* Matschie. Congo-Brazzaville: Gombe, Sangha River.
1914 *Colobus occidentalis ituricus* Lorenz. Congo-Zaire: Mawambi, Ituri Forest.
1914 *Colobus occidentalis rutschuricus* Lorenz. Congo-Zaire: Sassa (= Ishasha) River, southeast of Lake Edward.

DIAGNOSIS. White tail tuft occupying only one-third of tail length, the shortest in the species; tail longer than head plus body; flank veil creamy, not extending onto back, long on shoulders but covering only root of tail; white sprinkling on shoulders. Hair relatively short.

DISTRIBUTION. From Uganda, west of the Nile, and southwestern Sudan, south to about 1°25′S in the Ituri Forest and west (north of the River Ouban-gui) into Cameroon, as far west as Batouri and Lomie, and across the River Sanaga to Yabassi, south to the Makokou/Belinga District in northeastern Gabon.

Colobus guereza dodingae Matschie, 1913

1913 *Colobus (Guereza) matschiei dodingae* Matschie. Sudan: southwestern Didinga Hills, 4°10′N, 33°42′E.

DIAGNOSIS. Tail tip (which is not very bushy) is white for only 40% of tail length; tail much longer than head plus body; flank veil slightly creamy, not extending up onto back, but covering part of tail behind; hair fairly short, coarse; no white sprinkling on shoulders.

DISTRIBUTION. Known only from the Didinga Hills, southeastern Sudan.

NOTES. Craniometrically (Hull 1979) this subspecies is very close to *C. g. occidentalis.*

Colobus guereza percivali Heller, 1913

1913 *Colobus abyssinicus percivali* Heller. Kenya: Mathews Range, Mount Gargues.

DIAGNOSIS. White tail tuft extends over 63% of tail on average; tail about as long as head plus body; flank veil very long, creamy yellow, extending well up onto flanks and loins, but shortened posteriorly so only just reaching tail base; no white sprinkling on thighs or shoulders below veil. Hair very long, more than 400 mm on loins (cf. only 300 mm in *C. g. gallarum*).

DISTRIBUTION. Restricted to Mount Gargues (or Varagess).

NOTES. This subspecies is close to *C. g. matschiei* (Hull 1979, and an unpublished study).

Colobus guereza matschiei Neumann, 1899

1899 *Colobus matschiei* Neumann. Kenya: Kwa Kitoto, Kisumu.
1913 *Colobus abyssinicus roosevelti* Heller. Kenya: Mau Forest, Njoro.
1925 *Colobus abyssinicus elgonis* Granvik. Kenya: east slope of Mount Elgon.

DIAGNOSIS. White tail tuft short, occupying 45% of tail length on average; rest of tail black; tail much longer than head plus body; flank veil much yellowed, not extending up onto back, but covering base of tail; white hairs sprinkled over shoulders, nearly (or just) linking veil to white of throat. Hair short.

DISTRIBUTION. Kenya west of the Rift Valley and some forests within the Rift itself, such as Lake Naivasha; west to Mount Elgon, south to Ngorongoro Crater and the Grumeti River in Tanzania.

NOTES. The southwestern populations are rather distinct from those from Uganda and western Kenya, in both cranial (Hull 1979) and pelage features (an unpublished study by Hull). Cranially it is close to *C. g. kikuyuensis*, but in external appearance these two Kenyan subspecies are strikingly different.

Colobus guereza kikuyuensis Lönnberg, 1912

1912 *Colobus abyssinicus kikuyuensis* Lönnberg. Kenya: Escarpment Station.
1913 *Colobus (Guereza) caudatus thikae* Matschie. Kenya: west slope of Mount Kenya.
1913 *Colobus (Guereza) caudatus laticeps* Matschie. Kenya: west slope of Mount Kenya.

DIAGNOSIS. White tail tuft very bushy, extending over 72% of tail length on average; tail proximally gray; tail short, about equal to head plus body; flank veil very long, extending well up on back and beyond base of tail tuft; hair over 400 mm long on loins; thigh stripe short.

DISTRIBUTION. From Ngong Escarpment to Mount Kenya and the Aberdare Range.

Colobus guereza caudatus Thomas, 1885

1885 *Colobus guereza caudatus* Thomas. Tanzania: Mount Kilimanjaro (northeast slope), Useri, 900 m.
1906 *Colobus albocaudatus* Lydekker. Replacement for *caudatus*.

DIAGNOSIS. Flank veil and tuft even longer than in *C. g. kikuyuensis*, the tuft making up over 80% of tail length (an unpublished study by D. L. Hull); only extreme proximal part of tail black; fur on underside less woolly.

DISTRIBUTION. Mount Kilimanjaro, Mount Meru, Momela Lakes, and Kahé.

Genus *Piliocolobus* Rochebrune, 1887
Red Colobus

1887 *Piliocolobus* Rochebrune. *Simia (Cercopithecus) badius* Kerr, 1792.
1887 *Tropicolobus* Rochebrune. *Colobus rufomitratus* Peters, 1879.

Members of this genus share with *Procolobus* a small larynx, loss of subhyoid sac, ischial callosities that remain separate in both sexes, a four-chambered stomach (possessing a praesaccus in addition to other compartments), periodic sexual swellings in adult females and a perineal organ in subadult males, and almost consistent presence of a sagittal crest in adult males. The intermembral index is 86. Many features of the skull (Verheyen 1962) distinguish the genus from both *Colobus* and *Procolobus*.

The taxonomy of the red colobus has long been a headache. The usual practice has been to "give up," and unite them all into one species, except perhaps for one or two outstandingly different taxa *(kirkii, tholloni)*—a clearly unsatisfactory course of action. My approach here has been to try to break up the group into its more clearly distinct groups; it is to be taken as an attempt to bring some incipient sense into it, certainly not as definitive.

Piliocolobus badius (Kerr, 1792)
Western Red Colobus

DIAGNOSIS. Sharply distinct black or gray dorsally, extending onto upper part of limbs, and bright red ventrally, extending to rest of limbs and to cheeks. Cranially, it differs from other species, except *P. preussi*, in less prognathous face, with the lower orbital margin not standing forward of the upper, and low degree of sexual dimorphism; the temporal lines meet farther forward on the braincase.

DISTRIBUTION. West African region, from Senegal coast (formerly) to Ghana.

Piliocolobus badius badius (Kerr, 1792)

1792 *Simia (Cercopithecus) badius* Kerr. Sierra Leone.
1800 *Simia ferruginea* Shaw. Sierra Leone.
1812 *Colobus ferruginosus* E. Geoffroy. New name for *ferruginea*.
1838 *Colobus rufoniger* Ogilby. Sierra Leone.

DIAGNOSIS. Glossy black above, bright red below and on limbs. Nostrils raised on a swelling of nasal septum.

DISTRIBUTION. Rain forest from Sierra Leone east to the Nzi-Bandama system in Ivory Coast.

Piliocolobus badius temminckii (Kuhl, 1820)

1820 *Colobus temminkii* [sic] Kuhl. No locality.
1835 *Colobus fuliginosus* Ogilby. Gambia.
1838 *Colobus rufo-fuliginus* Ogilby. Replacement for *fuliginosus*.

DIAGNOSIS. Essentially a pale version of *P. b. badius:* back dark gray, underside and limbs light orange-red.

DISTRIBUTION. Open forest and savanna woodland areas of Senegal and Guinea.

Piliocolobus badius waldronae (Hayman, 1936)

1936 *Colobus badius waldroni* Hayman. Ghana: Ashanti, Goaso.

DIAGNOSIS. Differs from *P. b. badius* by the greater extent of the black zone on the limbs, and by the flat nose.

DISTRIBUTION. Formerly, from the Nzi-Bandama system in Ivory Coast to the Volta River in Ghana.

NOTES. This taxon is the only Primate taxon known with fair certainty to have become extinct in modern times (John Oates, personal communication).

Piliocolobus pennantii (Waterhouse, 1838)
   Pennant's Colobus

DIAGNOSIS. Dorsally red, with black tones on shoulders (sometimes infusing entire back), extending onto tail and usually onto upper arms; underside pale red or even whitish; hands and feet black, contrasting with general tone of limbs; long cheek hairs, whitish or buffy in color, this tone continuous with pale tones of underside. Prominent tufts at bases of ears.

DISTRIBUTION. Bioko; the western part of the Niger delta; and a small area in Congo, about 0 to 1°S, on the banks of the Rivers Likouala and lower Sangha.

NOTES. The three subspecies appear very close, but each is known from rather few specimens and the distinguishing characters may prove constant (diagnostic), in which case they will rate as full species. They are clearly distinct from *P. preussi* on the one side and *P. foai oustaleti* on the other. The reasons for the extinction of geographically intervening forms between the three, and the insertion of *P. preussi* between them, are unclear.

Piliocolobus pennantii pennantii (Waterhouse, 1838)

1838 *Colobus pennantii* Waterhouse. Equatorial Guinea: Bioko Island.

DIAGNOSIS. Very slightly darker than *P. p. bouvieri;* hairs of mantle black with orange bands. Whitish color on outer side of arm restricted to a fringe extending from chest, rest of outer side of arm orange-brown. Black extends from hands and feet some way up outer side of limbs. Crown black; no whorls.

DISTRIBUTION. Bioko.

Piliocolobus pennantii epieni (Grubb and Powell, 1999)

1999 *Procolobus badius epieni* Grubb and Powell. Nigeria: Sampou-Apoi, 4°55′N, 6°00′E, Bayelsa State, delta of the Niger.

DIAGNOSIS. Like *P. p. pennantii* with black crown and banded black and orange mantle, but whitish tone of underparts "intruding onto outer side of arm, restricting orange-brown tones to a narrow band on trailing edge of outer arm, or completely excluding these tones" (Grubb and Powell 1999:71). Black of hands and feet not extending up limbs. Crown black; hairs over ears form a whorl.

DISTRIBUTION. The range appears restricted to the Niger delta, between the Forcados-Nikrogha Creek and the Sagbama-Osiama-Agboi Creek, in the marsh forest, inland of the mangrove zone; it may occur sporadically farther east, almost to 7°E (Grubb and Powell 1999).

Piliocolobus pennantii bouvieri Rochebrune, 1887

1887 *Piliocolobus bouvieri* Rochebrune. Supposedly from Gambia.
1914 *Colobus (Piliocolobus) likualae* Matschie. Congo-Brazzaville: Sangha River, opposite mouth of Likouala River.

DIAGNOSIS. Crown red, with no whorls, but a stiff, black superciliary band; glossy red above, with a narrow, blackish, unbanded mantle extending to midback; thighs yellower; tail golden brown. A small tuft behind ear.

DISTRIBUTION. The Sangha-Likouala part of the range.

NOTES. There have been no recent reports of this taxon, and it may be extinct (John Oates, personal communication).

Piliocolobus preussi Matschie, 1900
Preuss's Red Colobus

1900 *Piliocolobus preussi* Matschie. Cameroon: Barombi, Elephant Lake.

DIAGNOSIS. Hairs ticked red and black, the black predominating on dorsum, the red on flanks. Crown blackish. Cheeks orange. Lower flanks, tail, and limbs (including hands and feet) bright saturated red. Underside white as in *P. pennantii,* but not extending onto arms. Cranially it resembles *P. badius.*

DISTRIBUTION. Korup National Park, Cameroon (near the coast just east of the Nigerian border), and neighboring region, extending into Nigeria in the Ikpan block of the Oban Division of Cross River National Park (Grubb and Powell 1999).

Piliocolobus tholloni (Milne-Edwards, 1886)
Thollon's Red Colobus

1886 *Colobus tholloni* Milne-Edwards. Congo-Zaire: lower Congo.
1913 *Colobus (Piliocolobus) lovizettii* Matschie. Congo-Zaire: Lake Leopold II, Kutu.

DIAGNOSIS. Uniform foxy red, darker on shoulders and lighter, yellower on underside. Circumfacial hair short, black; only posterior part of cheeks is light like underparts. Hands, feet, and tail tip dark. Long, black tufts at base of tail. Skull very elongated rostrally.

DISTRIBUTION. South of the great bend of the River Congo; its eastern limit is not the Lualaba (= Upper Congo) itself, but the River Lomami.

NOTES. This species is obviously related to other central African forms (here, *P. foai*), but the abundant material in the MRAC (Tervuren) makes it clear that it is always distinct and recognizable.

Piliocolobus foai (Pousargues, 1899)
Central African Red Colobus

DIAGNOSIS. Variable, but all having dark to black hands and feet, red on crown, black brow band, and light cheeks.

DISTRIBUTION. The central region, from Congo-Brazzaville through Congo-Zaire, north and east of the Congo River, into the southernmost forests of Sudan and the Central African Republic.

NOTES. This species, as recognized here, contains all those forms that are not outstanding and diagnostic in any way, and is very likely a paraphyletic "rump," in the words of Peter Grubb (personal communication).

Piliocolobus foai foai (Pousargues, 1899)

1899 *Colobus foai* Pousargues. Congo-Zaire: Ouroua, between southwest of Lake Tanganyika and upper Congo.
1909 *Colobus graueri* Dollman. Congo-Zaire: Wabembe country, 80 km west of north end of Lake Tanganyika.
1914 *Piliocolobus kabambarei* Matschie. Congo-Zaire: Kabambare, Lualaba River.
1914 *Piliocolobus lulindicus* Matschie. Congo-Zaire: Lulindi River.

DIAGNOSIS. Prominent tufts at bases of ears; forehead or anterior crown crest present; pelage tending to be black on the proximal dorsum, red elsewhere, long in highland populations, short in riverine ones.

DISTRIBUTION. This subspecies is isolated between the River Lowa-Oso and about 6°S; north of its range, between the Rivers Lowa-Oso and Maiko, there are no red colobus.

NOTES. Colyn (1993) recognized the riverine *lulindicus* as separate from montane *foai* because of its shorter pelage, poorly developed crest, and smaller size.

Piliocolobus foai ellioti (Dollman, 1909)

1909 *Colobus ellioti* Dollman. Congo-Zaire: 90 km west of south end of Lake Edward.
1914 *Piliocolobus anzeliusi* Matschie. Congo-Zaire: upper Ituri.
1914 *Piliocolobus ellioti melanochir* Matschie. Congo-Zaire: between Beni and Irumu.
1914 *Colobus variabilis* Lorenz. Congo-Zaire: Ituri Forest.
1914 *Colobus multicolor* Lorenz. Congo-Zaire: Ituri District, Mawambi.
1925 *Colobus langi* J. A. Allen. Congo-Zaire: Risimu, between Kisangani and Bafwaboli.

DIAGNOSIS. It is hardly possible to characterize this very variable subspecies, except that the black shoulders vs. red saddle distribution of *P. f. foai* is generally reversed, but even this is inconsistent.

DISTRIBUTION. West of Lake Edward, south of the Aruwimi-Ituri, as far as the Zaire-Lualaba. The southern border is the River Maiko, south of which there is a gap in the range of red colobus until the River Lowa is reached, beyond which *P. f. foai* is found.

NOTES. At the western end of the range, a pattern with red anterior parts (including arms) and blackish hindparts (including legs and tail) predominates; farther east, there is a great variability of patterns. Colyn (1993) ascribed this to intermixture between a formerly separate riverine form *(langi)* and *P. f. semlikiensis,* and it may indeed reflect an ancient mixture between two separate forms; but the population is now, except perhaps at its extreme western end, inextricably variable, and the name *ellioti* must be used for the entire ensemble. I have found that east of 29°E specimens are predominately blackened on the foreparts, with the underside gray; west of this longitude they are more usually darkened on the hindparts too (but less than on the foreparts), and the underside is yellowish. I agree with Colyn (1993) that the *langi* color predominates in the Kisangani region, and that there body size averages smaller. A chestnut-brown color morph, golden-toned below, occurs rarely throughout the range.

Piliocolobus foai oustaleti (Trouessart, 1906)

1906 *Colobus oustaleti* Trouessart. Congo-Zaire: Oubangui, Youmba.
1906 *Colobus nigrimanus* Trouessart. Congo-Brazzaville: Liranga.
1913 *Colobus (Piliocolobus) powelli* Matschie. Congo-Zaire: Zokwa, upper Ituri.
1914 *Colobus (Tropicolobus) umbrinus* Matschie. Congo-Brazzaville: Sangha River, Bungi, between Ouesso and Ikelemba.
1914 *Colobus (Tropicolobus) schubotzi* Matschie. Congo-Zaire: Koloka, between Likati and Bima Rivers.
1919 *Colobus (Piliocolobus) brunneus* Lönnberg. Congo-Zaire: upper Uele, Sili.

DIAGNOSIS. Coat agouti-banded, dark brown to coppery red, with dark to black hands and feet. The most common color is dark smoky brown, a little paler below, but red morphs occur in the far west of the range *("nigrimanus")* and the far east; to the north in the gallery forests of the Uele tributaries occur lighter, brownish fawn animals with light forearms and shanks *("brunneus");* and a raw sienna type, with pale buffy or golden underparts, and coppery red under the tail root, is restricted to the Oubangui/Lobaye area.

DISTRIBUTION. This subspecies occupies a huge range, from the River Sangha in the west across the Oubangui east to Lake Albert; the southern border is the River Congo-Zaire and, in the east, its tributary, the Aruwimi; to the north, it occurs well out into the savanna woodlands north of the River Uele. The boundary between it and *P. f. ellioti* is vaguely defined, and gene flow from the latter probably accounts for the red specimens in the east of the range; Colyn (1991) discussed the question, concluding that in the east the River Ituri is the approximate boundary, whereas farther west it is the River Aruwimi. It approaches the range of *P. pennantii bouvieri* closely in the west, but there is no sign of hybridization.

Piliocolobus foai semlikiensis (Colyn, 1991)

1991 *Colobus badius semlikiensis* Colyn. Congo-Zaire: Tungula, right bank of River Semliki.

DIAGNOSIS.   Prominent tufts at bases of ears; arms brick red with forearms bordered with dark gray and hands black, legs blackish gray with reddish lights, tail black; body (above and below) anteriorly gray with dark to blackish mantle.

DISTRIBUTION.   The *Cynometra* forests east of the Middle Semliki Valley. Colyn (1991) located it on both banks, but it is clear from his list of localities that west of the river the *semlikiensis*-like specimens occur as extremes of the phenotypic range of *P. f. el-lioti*. Even east of it, *P. f. ellioti* is said to occur at Mut-sora and Talya, about 0°20'N, 29°45'E, so that *P. f. semlikiensis* is confined to a very small region indeed, about 0°20' to 0°50'N, between the River Semliki and the Rwenzori massif (where there are no red col-obus, forming a gap between the ranges of *P. foai* and *P. tephrosceles*).

NOTES.   For Colyn (1991), this subspecies was one of the two parent populations of the extremely vari-able mixture called *ellioti,* and indeed some speci-mens from within the range of *P. f. ellioti* are *semli-kiensis*-like. Whether or not it is true that present-day *P. f. ellioti* is actually a hybrid swarm, *P. f. semlikiensis* seems now to retain some degree of homogeneity by being "protected" from *ellioti* gene flow by the Semliki River.

Piliocolobus foai parmentierorum Colyn and Verheyen, 1987

1987 *Colobus rufomitratus parmentieri* Colyn and Verheyen. Congo-Zaire: Mabobi, left bank of Congo River.

DIAGNOSIS.   Prominent tufts at bases of ears; no crest on crown or forehead; upper side black on mid-back, shoulders, and upper arms, rest brick red; un-derside whitish, including throat (continuous with epaulettes) and inner surfaces of limbs; crown agouti brownish red, sharply contrasting with black of neck and bordered anteriorly by a black frontal band that becomes thicker on temples; hands and feet black, contrasting with limbs; tail becomes progressively

darker along its length, but not quite black at tip. Facial skin black except for depigmented lips, chin, and bases of nose.

DISTRIBUTION.   Between the Lomami and Lualaba Rivers, south to about the Rivers Ruiki and Lutanga.

NOTES.   Despite its distribution west of the Lua-laba, separated from *P. tholloni* by the River Lomami, its affinities, as noted by Colyn and Verheyen (1987a), lie not with *P. tholloni* but very clearly with *P. foai,* especially nominotypical *P. f. foai*.

Piliocolobus tephrosceles (Elliot, 1907)

Ugandan Red Colobus

1907 *Colobus tephrosceles* Elliot. Uganda: Toro, Ruahara River, east side of Mount Rwenzori, 1,200 m.
1914 *Tropicolobus gudoviusi* Matschie. Tanzania: west of Lake Victoria, between Ussuwi and Ihangiro.

DIAGNOSIS.   Prominent tufts at bases of ears; pel-age long, glossy black above including tail, light to white below; crown red; forehead crest marked, dark reddish brown, bordered by a superciliary-temporal black stripe; forearms and legs dull light gray; long, lighter-toned tufts at base of tail. Skull has a distinc-tive transverse groove across nasion from one orbit to the other; this appears to be constant in occur-rence in adults.

DISTRIBUTION.   Eastern border of the western Rift Valley, from about 2°N to 8°S, in Uganda, Rwanda, Burundi, and Tanzania.

NOTES.   There is geographic variation in this spe-cies: (1) Ugandan populations are dull black or brown-black with a red tinge, lighter gray-brown rump, and much lighter, often straw gray, forearms and legs; underparts light gray to whitish; tail brown, darkening toward tip. I have seen 21 skins from this region. (2) Those from the Uganda-Tanzania border (6 skins seen) are black dorsally with a deep red tinge on rump and loins, lighter red-brown forearms, brown legs and tail, and white underparts. (3) Far-ther south in Tanzania, to the Mahali Mountains, I have seen 6 skins, which are jet black, the red-brown restricted to the croup and haunch; arms and legs mixed with brown; the tail becomes black distally; underside white. (4) Finally, 2 skins from Lake Rukwa are lighter, blackish gray, with the tail shorter and colored like the body, the arms and legs dull straw gray, and the underside light gray; as noticed by Rodgers et al. (1984) in the field, the cheek whiskers are longer and the red cap may extend down to form sideburns.

Piliocolobus gordonorum Matschie, 1900

Uzungwa Red Colobus

1900 *Piliocolobus gordonorum* Matschie. Tanzania: Uzungwa Mountains.

DIAGNOSIS.   Dorsum either wholly deep shining black, or black with brick red lumbar region and rump; arms and thighs black, but shanks mixed with silvery; tail short, slightly bushy, black or mixed with ochraceous above, but tending to off white below; underside and inner surfaces of limbs white. Crown shining red, with laterally directed hairs forming a toupee. Cheeks white, this zone separated from red of crown by a thick black line along temples. Face depigmented around nose and mouth. Skull broad, short-faced, small-toothed, lacking a nasion groove. Sexual swelling larger than in *P. tephrosceles*.

DISTRIBUTION.   Uzungwa Mountains and forests between Little Ruaha and Ulanga Rivers, south of Iringa, Tanzania, ranging from 200 to 1,623 m.

NOTES.   The red-backed morph is rare, about 2% of the population (Struhsaker and Leland 1980); the hairs on the loins have unusually long, reddish bases, and the red tone may be caused simply by abrasion of the black terminal part. In some respects the color, with its bicolored tail, white underside, black arms, and silvery white tones on the legs, resembles that of a juvenile *P. kirkii,* in transition between the infant white coat and the adult coat. Struhsaker (1981) found that the vocalizations resemble those of *P. kirkii.*

Piliocolobus kirkii (Gray, 1868)

Zanzibar Red Colobus

1868 *Colobus kirkii* Gray. Tanzania: Zanzibar.

DIAGNOSIS.   Arms, shoulders, and forepart of back black; remainder of back, crown, and base of tail brick red; circumfacial hair, underparts, inner surface of forelimbs, whole of hindlimbs, and terminal part of tail white. Feet dark gray. Face black with pink zone around mouth and lower part of nose. Hair on cheeks and forepart of crown shaggy, framing face in a halo. Infant is white. Cranium resembles those of *P. tephrosceles* and *P. foai.*

DISTRIBUTION.   Zanzibar (Unguja) Island. A specimen in the BM(NH), collected by Kirk, is labeled "Pangani River, opposite Zanzibar"; on my visit to this region in 1971 I found only *Colobus angolensis,* and widespread replacement of native vegetation by exotic flora along the coasts south of the river mouth.

Piliocolobus rufomitratus (Peters, 1879)

Tana River Red Colobus

1879 *Colobus rufomitratus* Peters. Kenya: Tana River, Muniuni.

DIAGNOSIS.   Body gray with a slight buffy tinge; tail darker, becoming nearly black toward tip; limbs lighter; underparts grayish white; crown brick red, surrounded by a dark gray rim, and with a whorl behind each ear leading to a low transverse crown crest. Neonates are completely blackish, like most other species, but with long white bases to hairs. Clitoris is large, prominent; sexual swellings large. Size very small, like *P. kirkii.* The skull is distinctive: very broad across orbits but narrow across canines, with long basicranium and short palate; nasals triangular, prominent. Very concave behind supraorbital torus. Coronal suture runs straight across; in all other species it is somewhat V-shaped backward. Torus is interrupted by a depression over glabella. Lateral orbital margins merge smoothly with zygomata, with no constriction at junction. Suborbital fossae deep. Occiput steep, rounded. Teeth large. The brachial index is higher than in other species.

DISTRIBUTION. Tana River gallery forests, from Garsen north nearly to Wenje.

NOTES. Although externally rather similar to *P. tephrosceles,* cranially it is strongly different from all other red colobus.

Genus *Procolobus* Rochebrune, 1887
Olive Colobus

1887 *Procolobus* Rochebrune. *Colobus verus* van Beneden, 1838.
1895 *Lophocolobus* Pousargues. *Colobus verus* van Beneden, 1838.

This genus shares several probably synapomorphic features with *Piliocolobus,* especially the perineal organ; but its skull and teeth are distinctive, differing strongly from those of both *Piliocolobus* and *Colobus* by the presence of a mediosagittal mental foramen, usually six cusps on M_3, and the presence of a prominent lingual cingulum, with a small cingular cusp, on incisors (Verheyen 1962). Oates et al. (1996) pointed out other unique features: the graded color pattern, the small crest on the crown, the hairy ear pinna, and the minutely papillated glans penis.

Procolobus verus (van Beneden, 1838)
Olive Colobus

1838 *Colobus verus* van Beneden. "Africa."
1840 *Semnopithecus (Colobus) olivaceus* Wagner. Replacement for *verus.*
1866 *Colobus?? chrysurus* Gray. West Africa.
1866 *Colobus cristatus* Gray. West Africa.

DIAGNOSIS. Dorsally a dull olive-brown, becoming white below; an off white circumfacial ruff; crown with a low longitudinal crest, formed by a pair of lateral pale whorls.

DISTRIBUTION. Mapped by Oates (1981) with a more or less continuous range from Sierra Leone to Ghana, extending just east of the River Volta, and again in Nigeria on the south bank of the River Benue just above its confluence with the Niger. He considered that the distributional gap is likely to be real rather than an artifact of collecting or observation.

Langur Group

Until the early years of the twentieth century, all langurs were commonly referred to a single genus, called either *Semnopithecus* or *Presbytis.* The name *Pithecus* came into general use early in that century, until it was banned (International Commission on Zoological Nomenclature 1929). Pocock (1928) first divided the langurs into three "sections," the *Entellus, Pyrrhus,* and *Aygula* groups, distinguished by the color of the neonate: black, golden red, and "cruciger," respectively (we now know that it is not quite as simple as this). Later he recognized (Pocock 1935) these three groups as genera, calling them, respectively, *Semnopithecus, Trachypithecus,* and *Presbytis,* and Hill (1934) in addition recognized a fourth, *Kasi,* for two species that Pocock had placed in *Semnopithecus.*

The problem is that, just as the langur group as a whole is defined mainly (entirely?) by plesiomorphic traits, as the more obviously apomorphic groups, *Presbytis* and *Semnopithecus,* are separated out, still a plesiomorphic rump remains, *Trachypithecus.* A future phylogenetic analysis of the group should not take the genus *Trachypithecus* for granted but should keep the species-groups separate in case one or more of them should turn out to be on a clade with one of the other two genera.

Genus *Semnopithecus* Desmarest, 1822
Sacred Langurs or
Gray Langurs or
Indian Langurs

1822 *Semnopithecus* Desmarest. *Simia entellus* Dufresne, 1797.

The neonate is black; adults are gray or gray-fawn, with contrasting jet black skin of face and ears; there is a whorl behind brow from which hair radiates; and long, stiff, forward-pointing brow fringe. The skull is strongly prognathous, with well-developed supraorbital torus and muscle attachments including temporal and nuchal lines; basicranial flexion is reduced; the pyriform aperture is evenly narrow; there is no marked arch under malar-maxillary suture; and no infralambdoidal swelling; the mandibular corpus deepens markedly posteriorly, with expanded jaw angle. The intermembral index is 79; the brachial index is 100. The clitoris is small and contained within the vulva.

Hill (1939) divided this genus into four species:

1. *S. schistaceus:* including both montane forms and *hector* of the tarai. This species was defined as having long whiskers, concealing the ears, and the color of the crown and whiskers strongly contrasted with that of the body.

2. *S. entellus* (monotypic). This was not defined.

3. *S. priam:* including *anchises* and *thersites.* This species was described as being crested, with little or no contrast between the color of the extremities and the rest of the limbs.

4. *S. hypoleucos:* including *aeneas, iulus, dussumieri, achates,* and *elissa.* This species was described as having an ashy gray body contrasting with black extremities.

Roonwal (1981; Roonwal et al. 1984) brought a new character into the taxonomy of this group: tail carriage. This very neatly cuts the genus into two groups: a northern group, in which the tail is looped forward, and a southern group, in which it is looped back. There is some variation in the northern group: it remains well above the back in *hector* and *schistaceus,* but reaches well below (the tip falling to the left or right of the body in different individuals) in *ajax* and *entellus.* The border between the two styles was found (Roonwal et al. 1984) to be from between Bharuch and Surat, 21°50′–21°10′N, then east-southeast to the Arni-Hewari area, about 20°N, 78°E, then more southeast to 16°43′N in the Godavari delta. In only two troops were there individuals with both styles: a single juvenile male had the southern carriage in an otherwise northern-style troop at Bharuch in Gujarat, and both types were about even in a troop near Arni in Maharashtra.

I examined the skins in the Natural History Museum (BM[NH]), London, in company with Doug Brandon-Jones. I am grateful for the insights gained in discussion with him; the arrangement that follows is based on our discussions and joint examination, and I owe him a deep debt, though he may well disagree with the details of my conclusions.

Body measurements given below are from Roonwal (1981) and Roonwal et al. (1984).

Semnopithecus schistaceus Hodgson, 1840
Nepal Gray Langur

1840 *Semnopithecus schistaceus* Hodgson. Nepal.
1840 *Semnopithecus nipalensis* Hodgson. Nepal.
1909 *Presbytis lania* Elliot. China: Chumbi Valley.
1928 *Pithecus entellus achilles* Pocock. Nepal: Satthar Hill, Gorkha, 3,600 m.

DIAGNOSIS.   Tail carriage of northern type. Coat thick, plush. Very large in size, adult male head plus body length over 760 mm; female rather smaller, head plus body length averaging 84% of male's; tail very short, 106–145% (usually <127) of head plus body length (rarely over 125%). Roonwal (1981) gave weights of 14.5–19.5 kg for three males and 15.4–16.8 kg for three females. Brownish toned, generally dark sepia brown; crown creamy, sharply contrasting with body; underside white, restricted to ventral surface (not reaching flanks), but inner sides of limbs very broadly white; rump white, but not extensively; whiskers very long, creamy. Hands not black or at most with black patches; feet with pale tone on ankle and pale edging to soles. Tail tip is white.

DISTRIBUTION.   From the Himalayan slopes from at least as far west as Gorkha, central Nepal, to Sikkim, at 1,500–3,500 m; occurs in Tibet in Bo Qu Valley and Ji Long Zang Bu Valley, in the Mount Everest region, and at 3,000 m to below 2,800 m in the Chumbi (or Chumvi) Valley, the Tibetan region that wedges between Sikkim and Bhutan (Feng et al. 1986).

NOTES.   Napier (1985) sorted out the correct nomenclature of this taxon. She pointed out that *schistaceus* is the high-altitude langur of Nepal, and that it is Pocock's (1928) *achilles,* not his *hector* as he later (1939) surmised, that is effectively a redescription of Hodgson's taxon. The type of Elliot's *lania* (in the BM[NH]) falls well within the range of this species.

Semnopithecus ajax (Pocock, 1928)

Kashmir Gray Langur

1928 *Pithecus entellus ajax* Pocock. India: Deolah, Chamba, 1,800 m.

DIAGNOSIS. Lighter, yellower in color than *S. schistaceus*, with less sharply defined white on head; underside slightly yellowish-tinged white; loose, shaggy coat, forming a kind of mane over the foreparts; arms dark below elbows, and hands with dark patches, but legs more as body with feet not much darker; tail tip extensively white. Cheek whiskers bush out to the side to a much greater degree than in other highland taxa. Extremely large like *S. schistaceus* and even more sexually dimorphic, head plus body length of female averaging 79% of male's.

DISTRIBUTION. The western representative of the species, living on the Himalayan slopes from Dehra Dun west into Pakistani Kashmir, at about 2,000–3,000 m.

Semnopithecus hector (Pocock, 1928)

Tarai Gray Langur

1928 *Pithecus entellus hector* Pocock. India: Sitabani, Ramnagar, Kumaon, 600 m.

DIAGNOSIS. Tail carriage of northern type. Pelage thick, plush. Light grayish yellow, "mauve-blond," with a very slightly darker line down back; slightly larger than *S. entellus,* head plus body length in male 660–762 mm, with large teeth; head strikingly white, whiskers very long, whitish, forming a bushy halo of even width all around face; limbs browner than body below elbows and knees, with a few black hairs intermixed on hands and feet; underside buffy white, not pure white; tail tip white.

DISTRIBUTION. The Himalayan foothills, from Kumaun east beyond Katmandu, to at least the Hazaria District, at 600–1,800 m. It is not known what happens where this taxon meets the true montane *S. schistaceus,* but they appear to maintain their separate existence along a wide swath from Kumaun to Bhutan. Hill (1939) included it in *S. schistaceus,* but it seems to me that despite the large size it is somewhat

more like *S. entellus,* without the extreme head/body contrasts, extremely large size, and strong sexual dimorphism of *S. schistaceus.*

Semnopithecus entellus (Dufresne, 1797)

Northern Plains Gray Langur

DIAGNOSIS. Tail carriage of northern type. Head plus body length in male up to 740 mm, female only a little smaller (head plus body length averaging at least 88% of male's); tail usually more than 150% of head plus body length, occasionally down to 120%; no tuft on crown; dorsum darker than flanks, often with quite noticeable contrast; tail only somewhat darker, becoming whiter or fully white at tip; hands and feet suddenly black, contrasting with limbs, which are only slightly darker than body; head creamy, lighter than head and body; whiskers whitish.

DISTRIBUTION. From Pakistan through the lowlands of India north of the Godavari and Krishna Rivers, and south of the Ganges.

Semnopithecus entellus entellus (Dufresne, 1797)

1797 *Simia entellus* Dufresne. India: Bengal.

DIAGNOSIS. A somewhat erythristic form, creamy yellow on flanks, browner on dorsum, and red-gold on underside; coat short, wavy; whiskers long. Head plus body length 610–690 mm, not much different in males and females.

DISTRIBUTION. East from Nimar District through Orissa, Bihar, and Bengal, and south into northern Andhra Pradesh.

NOTES. The syntypes of *Simia entellus* are in the Paris (MNP) collection and are still in good enough condition to place them with confidence in this taxon.

Semnopithecus entellus subsp.

D. Brandon-Jones (personal communication) has pointed out that the western representatives of what have customarily been referred to *S. e. entellus* are less reddish, creamy below, longer-coated with shorter whiskers, and slightly smaller than the eastern ones, and may merit subspecific rank. They occur from about the Krishna River northwest into lowland Pakistan, including Rajasthan and Gujarat. Hrdy (1977) gave weights of up to 13.6 kg.

Semnopithecus hypoleucos Blyth, 1841
Black-footed Gray Langur

1841 *Semnopithecus hypoleucos* Blyth. India: Travancore.
1928 *Pithecus entellus aeneas* Pocock. India: Makut, South Coorg, 76 m.

DIAGNOSIS. Tail carriage of southern type. Deep mauve brown, with pale orange underside; yellow-brown head, with short, black-infused whiskers and crown; hair simply slopes up and back in middle of crown, but no upright tuft. Limbs below elbow and knee are black even on their inner surfaces, contrasting strongly with body color, and the contrast is further enhanced by the noticeably longer body hair above the elbows. Tail likewise black, with a few white hairs at tip. In the type the head plus body length is 559 mm, the tail 145% of head plus body length.

DISTRIBUTION. Restricted to the South Coorg region of Kerala; known localities are Markut and Wotekolli, at 600 m (2,000 ft). The dark color reflects the humid coastal habitat (Gloger's Rule).

NOTES. The type of *hypoleucos* is in the Zoological Survey of India collection (ZSI) in Calcutta. It is of the sharply black-legged form like the type of *aeneas* and the other four South Coorg skins in the BM(NH).

Semnopithecus dussumieri I. Geoffroy, 1843
Southern Plains Gray Langur

1843 *Semnopithecus dussumieri* I. Geoffroy. India: Malabar coast, Mahé.
1844 *Presbytis anchises* Blyth. India: Deccan.
1928 *Pithecus entellus priamellus* Pocock. India: Shernelly, Cochin.
1928 *Pithecus entellus iulus* Pocock. India: Jog, Gersoppa Falls, Kanara-Mysore boundary, 400 m.

1928 *Pithecus entellus elissa* Pocock. India: Nagarhole.
1928 *Pithecus entellus achates* Pocock. India: Haunsbhavi, Dharwar, 600 m.

DIAGNOSIS. Tail carriage of southern type. Gray-brown, often somewhat mauve in tone; underside creamy yellow, often darkened on the chest; very small in size (head plus body length 580–640 mm); tail very long, about 160% of head plus body length; head creamy, often not strongly distinct from body; whiskers short, whitish with a few black hairs; crown tuft very small if present at all, very little darker than rest of head; rump whitish; limbs below elbow and knee grayer, but digits, and sometimes entire hands and feet, to wrists and ankles, black, this tone occasionally going up forearms (but not legs); tail dark with tip paler brown or white. The contrast between the black on the extremities and the grayer limbs may become more marked with age (Agrawal 1974).

DISTRIBUTION. This is the widespread species of southwestern and west-central India, probably from the Godavari (northernmost locality is Helwak, Satara District) east to Diguvamatta and south through the foothills of the Nilgiri and other hill systems to Shernelly, but not including the enclave where *S. hypoleucos* is found.

NOTES. Hill (1939) was almost surely correct to synonymize *iulus* with *dussumieri;* the type skin of the latter, he suggested, was faded, making it appear paler than the type of *iulus;* but the two adult cotypes (there is also a juvenile) are actually quite well preserved and vary in tone of body and even of hands and feet. The type of *priamellus,* in the BM(NH), is a very poor specimen; it is not even possible to be sure if it had any crown tuft, but it has black extremities, as does the type of *elissa.* The type of *achates* is also black-footed and indeed beyond its lesser degree of black extension can hardly be distinguished from that of *elissa.*

The exact meaning of "Deccan" in the type locality of *anchises* is obscure, and different authors have had different concepts of the "real" *anchises,* but Roonwal et al. (1984) were almost surely incorrect when they used the name to indicate the langurs of Orissa, northeastern Maharashtra, southern Madhya Pradesh, and northern Andhra Pradesh (east-central India); these probably are representatives of *S. e. entellus.*

Over this wide area, there is much variation, but much of it is purely individual; the main clinal change is in the color of the tail tip, which is generally pale brown in southern specimens, becoming briefly white farther north and extensively white in the northernmost extremes of the range. In the far north, the light color of the underside advances well up the flanks, and there is only a difference in degree from *S. entellus.* Finally, although *S. hypoleucos* is strikingly different, the black of the hands is most extensive at localities such as Nagarhole (type locality of *elissa*), close to the restricted region where *S. hypoleucos* occurs, suggesting a limited amount of outward gene flow.

Semnopithecus priam Blyth, 1844

Tufted Gray Langur

1844 *Semnopithecus priam* Blyth. India: Coromandel coast.
1844 *Semnopithecus pallipes* Blyth. "Southern India."
1847 *Semnopithecus priamus* Blyth. Emendation.
1847 *Presbytis thersites* Blyth. Sri Lanka: Trincomalee.

DIAGNOSIS. Tail carriage of southern type. Brownish gray, variable in tone but usually grayer than *S. dussumieri;* small in size like *S. hypoleucos,* tail even longer (up to 174% of head plus body length); head, including cheek whiskers, creamy white, but this is not all that distinctively set off from body color; crown with a coconutlike crest, which (together with surrounding region of central crown) is darker, browner than rest of head; underside and inner sides of limbs paler than upper side, creamy yellow; rump whitish; tail darker, with white tip; hands paler than body, and feet often nearly white.

DISTRIBUTION. Sri Lanka, except the rain forest and hill zones, and southeastern India. In India, from Aramboli, on the southern tip of India, through the Dharmapuri Range to the Palkonda Hills (i.e., north to nearly 15°N and west to about 78°W).

NOTES. Phillips (1935) gave a mean weight of 12.5 kg for six males and 7 kg for seven females; this is therefore the only southern species in which there is notable sexual dimorphism. Hill (1939) thought that *thersites* was "doubtfully separable" from *priam,* and Brandon-Jones and I could not distinguish southeast Indian from Sri Lankan specimens.

The types of *priam* and *thersites* are in the Zoological Survey of India collection (ZSI), Calcutta. Though rather poorly preserved (there has never been any possibility, because of obvious financial constraints, of keeping types in climate-controlled conditions), both are clearly pale-footed and have small crown tufts, and indeed they seem rather alike.

Genus *Trachypithecus* Reichenbach, 1862

Lutungs

1862 *Trachypithecus* Reichenbach. *Semnopithecus pyrrhus* Horsfield, 1823.
1862 *Kasi* Reichenbach. *Cercopithecus johnii* Fischer, 1829.

The neonate is usually golden colored; the frontal whorl is poorly or not developed; brow fringe is developed, but not as much as in *Semnopithecus.* The canine/sectorial complex is strongly sexually dimorphic; the cranial vault is low, but gracile; the supraorbital torus is not strong, is rounded above each orbit and depressed at glabella; the muscular markings are less developed than in *Semnopithecus.* The ascending ramus of mandible is high. The intermembral index is 82 in most species, but 76 in the *T. vetulus* group; the brachial index is 94, but 101 in *T. vetulus.*

As noted above, the monophyly of this genus is not unquestionable. Zhang and Ryder (1998) found that of the four species whose mitochondrial tRNA and cytochrome *b* sequences they studied, two (*T. vetulus* and *T. johnii*) shared a clade with *Semnopithecus,* and the other two (*T. francoisi* and *T. phayrei*) together formed a clade that was sister to all other Asian colobines. Christian Roos (personal communication) found a similar set of relationships. Brandon-Jones (1984) combined the two genera (the prior name is *Semnopithecus*), and this move may prove ultimately correct when molecular clock dates become clearer.

Trachypithecus vetulus group

Neonate not invariably golden; sacral hair much shorter than hair of mantle, with a gray or white triangular patch at tail root; female has white pubic patch; cheeks and crown lighter than body. Intermembral index lower, brachial index higher than in other species-groups. As already noted, this group may actually be closer to the other South Asian langurs (*Semnopithecus*) than to Southeast Asian species.

Trachypithecus vetulus (Erxleben, 1777)
Purple-faced Langur

DIAGNOSIS. Neonate black or gray, with white cheek hairs. In adult, long, backswept white (sometimes brown-tipped) cheek whiskers, contrasting with crown, which is brownish; color brown or blackish brown, with whitish throat and interramal region; tail white, continued back from sacral patch.

DISTRIBUTION. Sri Lanka.

NOTES. The four subspecies seem very clearly separated on the evidence of the specimens in the BM(NH), but further material, in museums worldwide, needs to be examined; finally, field surveys should enable their distinctiveness to be assessed and at what level.

Trachypithecus vetulus vetulus (Erxleben, 1777)

1777 *Cercopithecus vetulus* Erxleben. "Ceylon" (Sri Lanka).
1780 *Cercopithecus kephalopterus* Zimmermann. Sri Lanka.
1785 *Cercopithecus cephalopterus* Boddaert.
1800 *Simia veter* Shaw.
1812 *Cercopithecus latibarbatus* E. Geoffroy.
1825 *Cercopithecus leucoprymnus* Otto.
1825 *Semnopithecus fulvogriseus* Desmoulins.
1876 *Semnopithecus kelaarti* Schlegel.

DIAGNOSIS. Gray-black, with light-tipped hairs; hairs of lumbar region, haunches, and distal part of tail light brown with long creamy tips, making these areas overall white; head, including cheek whiskers, brown or gray-brown, contrasting with blackish body; cheek whiskers conceal lower half of ears only; small in size and strongly sexually dimorphic (head plus body length 540–648 mm in males, 492–525 mm

in females; weight 6.6–7.7 kg in males, 5.0–5.2 kg in females, according to Phillips [1935]).

DISTRIBUTION. This is the southerly wet-zone form, in rain forest from south of the Kalu Ganga to about Rama, ascending to nearly 1,000 m inland.

Trachypithecus vetulus nestor (Bennett, 1833)

1833 *Semnopithecus nestor* Bennett. Sri Lanka: "probably Rayigam" (Phillips 1935).
1923 *Pithecus vetulus phillipsi* Hinton. Sri Lanka: Gonapola, Panadura.

DIAGNOSIS. Lighter, more gray-brown than *T. v. vetulus,* with less contrasting (silvery gray) rump patch; forearms and shanks darker, nearly black; crown and nape paler brown; tail slightly tufted at tip; smaller, with little or no sexual dimorphism (head plus body length 504–560 mm in both sexes, weight 3.4–3.8 kg in males, 3.1 kg in females).

DISTRIBUTION. Another wet-zone subspecies, from north of the Kalu Ganga as far as the rain forest extends.

NOTES. There is some variability in the gray tone of the body; in general, the consistency of the differences from *T. v. vetulus* needs to be explored further.

Trachypithecus vetulus philbricki (Phillips, 1927)

1927 *Pithecus philbricki* Phillips. Sri Lanka: Kantalai, Eastern Province.
1955 *Presbytis senex harti* Deraniyagala. Thunakai, Northern Province.

DIAGNOSIS. Overall brown, like *T. v. nestor,* but the rump patch is much less conspicuous; lower part of limbs black; tail tip is pale, tawny to white, not tufted; larger, head plus body length 552–640 mm in both sexes, weight of males 6.0–7.7 kg, of females 4.76–7.25 kg according to Hill (1936), but "about 11.4 kg" according to Deraniyagala (1955); hair short, silky; whiskers very long.

DISTRIBUTION. Found throughout the dry zone of the north and east of Sri Lanka, and reaching as high as 1,500 m in the East Matale and Madukelle Hills (Deraniyagala 1955).

NOTES. The northern form described as *harti* was said to be darker, more reddish brown, with golden tips to the hairs, more prominent light rump patch, and averaging smaller in size (male head plus body length 465–605 mm, weight 8.8 kg). It is at most the northern end of a cline in size and color.

This is the subspecies formerly called *"Presbytis senex senex"* until Napier (1985) revised and corrected the nomenclature.

Trachypithecus vetulus monticola (Kelaart, 1850)

1850 *Presbytis cephalopterus* var. *monticola* Kelaart. Sri Lanka: Nuwara Eliya.
1851 *Presbytis ursinus* Blyth. Sri Lanka: Nuwara Eliya.

DIAGNOSIS. Brown, with only slightly contrasting crown and rump patch, but long, white cheek whiskers that hide ears; very thick pelage; short tail. Linear measurements as in *T. v. philbricki* (head plus body length 552–590 mm), but build more robust and weighing 9.0–9.3 kg in both sexes. Hill (1936) stated that the hairs, which are thick, are sparsely pigmented in the cortex but almost continuously pigmented in the medulla, whereas in *T. v. nestor* the pigment is in scattered granules in the cortex and either continuous or confined to localized patches in the medulla.

DISTRIBUTION. A large high-mountain subspecies, living at altitudes from 1,200 to 2,000 m.

Trachypithecus johnii (Fischer, 1829)
Nilgiri Langur

?1800 *Simia leonina* Shaw. On this name, see D. Brandon-Jones (1995b).
1829 *Cercopithecus johnii* Fischer. India: Tellicherry (but see D. Brandon-Jones 1995b).
1834 *Semnopithecus cucullatus* I. Geoffroy. India: Western Ghats, probably Nilgiri Hills according to D. Brandon-Jones (1995b).
1839 *Semnopithecus jubatus* Wagner. Southern India, probably Nilgiri Hills according to D. Brandon-Jones (1995b).

DIAGNOSIS (partly after D. Brandon-Jones [1995b]). Black, hairs with brown bases; whole of head and neck pale brown, becoming yellowish on nape and sometimes on brow. Crown hairs posteriorly directed; nuchal hairs elongated. Tail black, or only partly gray.

DISTRIBUTION. Western Ghats of India, from Aramboli Pass (at 8°16′N near the southern tip of India) north to Srimangala (12°01′N, 75°58′E); see map in D. Brandon-Jones (1995b:26).

NOTES. D. Brandon-Jones (1995b) noted apparent geographical variation in the presence and extent of silvering on the rump, which is conspicuous, even extending forward onto the back or backward along the tail, in the north of the range, such as the Nilgiri Hills, whereas to the south of about 10°35′N this is generally very poorly expressed, and farther south pale hairs are even confined to the pubic region (of both sexes). This is obviously a case for further research.

Hohmann and Herzog (1985) observed some odd-colored langurs living in troops of this species in the Nilgiri Hills. These were brown, with black tail, distal limbs, hands and feet, and exposed skin, and the head creamy, this tone extending back to the shoulders and around under the jaw; one individual, presumed to be an adult male, had a small crest on the crown, and an adult female had a white patch on the inner side of the thighs. These individuals were explained as probable hybrids between *T. johnii* and *Semnopithecus entellus*. The latter species was rare in the Nilgiris, but one was observed in a *T. johnii* troop in which three "brown" langurs were seen. Later Hohmann (1988) gave evidence that the loud calls of the "brown" langurs were likewise intermediate between the two, reinforcing the hypothesis that they are hybrids.

Trachypithecus cristatus group

Face entirely slaty black; female with white pubic patch; underside as dark as upper.

Trachypithecus auratus (E. Geoffroy, 1812)
Javan Lutung

DIAGNOSIS. Larger than *T. cristatus*, lacking extensive gray tipping; teeth and jaws exceptionally large compared with size of skull.

DISTRIBUTION. Java, Bali, and Lombok.

NOTES. D. Brandon-Jones (1995b) discussed the nomenclature of this species and gave a detailed description, characterizing it as having erect, forward-curled circumfacial hair; hair on ear pinna whitish with tinges of yellowish to blackish; female with a pale (usually yellowish white) pubic patch. The descriptions and distributions of the two subspecies are based on his paper.

The following body measurements of adults or presumed adults, in mm, with number of specimens in parentheses, are taken from D. Brandon-Jones's (1995b:35–36) compilation: head plus body length 440–650 (58), tail 610–865 (58), hindfoot 143–180 (55), ear 30–45 (51).

Study of mtDNA restriction site polymorphisms in this species from West, Central, and East Java showed no geographical structure; West Javan haplotypes were more diverse than those from elsewhere, and those of *T. cristatus* were nested within those of *T. auratus* (Rosenblum et al. 1997).

Trachypithecus auratus auratus (E. Geoffroy, 1812)

1812 *Cercopithecus auratus* E. Geoffroy. Indonesia: Semarang according to Müller (1839), but this is a place of acquisition, and D. Brandon-Jones (1995b) restricted it to Batu District, 7°25′S, 112°31′E.

1823 *Semnopithecus pyrrhus* Horsfield. Indonesia: Province of Pasuruan, East Java (see Chasen [1940]), further restricted by D. Brandon-Jones (1995b) to Batu District.

1823 *Semnopithecus maurus* Horsfield. Indonesia: West Java.

1919 *Pithecus pyrrhus sondaicus* Robinson and Kloss. Indonesia: Cibodas, West Java.

1931 *Pithecus pyrrhus kohlbruggei* Sody. Indonesia: Sendang, Bali.

1934 *[Trachypithecus pyrrhus] stresemanni* Pocock. Indonesia: Danau Bratan, 760 m, or Celukanbawang, Bali.

DIAGNOSIS. The common morph is glossy black with a slight brownish tinge especially on venter, sideburns, and legs; but hairs are light-tipped on arms, legs, head, flanks, and sometimes back. The rarer erythristic morph is deep orange, becoming more yellowish on limbs, hairs around ears, and on venter, often with dorsal black tinge. Skin of face, palms, and soles black in common morph, depigmented or freckled in red morph. The type series of *sondaicus* are somewhat light-tipped and are at most intergrades between the two subspecies.

DISTRIBUTION. According to D. Brandon-Jones (1995b), the boundary between this subspecies and the next runs from the south coast of Java at 109°E approximately northwestward to the vicinity of Jakarta, though approaching the west coast at Gunung Ujungtebu, 6°15′S, 106°00′E. The red morph has a very restricted distribution, between Blitar, Ijen, and Pugeran.

Trachypithecus auratus mauritius (Griffith, 1821)

1812 *Cercopithecus maurus* E. Geoffroy. Not *Simia maura* Schreber, 1774, which was regarded by Weitzel and Groves (1984) and by D. Brandon-Jones (1995b) as probably indeterminable.

1821 *Simia mauritius* Griffith, 1821. "Mauritius": probably West Java, restricted by D. Brandon-Jones (1995b) to Jasinga.

DIAGNOSIS. Pelage is glossy black with a very slight brownish tinge especially on venter, sideburns, and legs; hairs lack light tips.

DISTRIBUTION. Very restricted in West Java to the north coast from Jakarta and inland to Bogor, Cisalak, and Jasinga, southwest to Ujung Kulon, then along the south coast as far as Cikaso and possibly Ciwangi.

Trachypithecus cristatus (Raffles, 1821)
  Silvery Leaf Monkey or
  Silvery Lutung

DIAGNOSIS. Teeth and jaws relatively small. Lighter, grayer than *T. auratus,* with long gray-white hair tips. Circumfacial hair as in *T. auratus.*

DISTRIBUTION. Borneo, Sumatra, Natuna Islands, Bangka, Belitung, many islands of Riau Archipelago, and a strip along west coast of peninsular Malaysia.

NOTES. Restriction site polymorphisms of this species were nested within those of *T. auratus,* indicating that the separation of the two species is rather recent (Rosenblum et al. 1997).

Trachypithecus cristatus cristatus (Raffles, 1821)

1821 *Simia cristata* Raffles. Indonesia: Bencoolen (Bintuhan), Sumatra.

1822 *Semnopithecus pruinosus* Desmarest. Sumatra.
1878 *Semnopithecus rutledgii* Anderson.
1909 *Presbytis cristata pullata* Thomas and Wroughton.
    Indonesia: Batam Island, Riau Archipelago.
1910 *Pygathrix ultima* Elliot. Malaysia: Mount Dulit, Sarawak.

DIAGNOSIS. Some tone of gray, with long lightened hair tips, evenly scattered over body and crown. A rare red morph occurs in Borneo.

DISTRIBUTION. Found on Sumatra, some islands in the Riau Archipelago (Lingga, Bintang, Sugi, Jombol, Bakang), Bangka, Belitung, Borneo, and the western coastal strip of peninsular Malaysia.

NOTES. This subspecies has often been divided into two: *cristatus* and *ultimus*. In the former, the hair tips are shorter, especially on the back, giving an overall almost gray-green color; in the latter, the tips are very long and very blond, making the general tone silvery gray. The two morphs are certainly very distinct, but are not geographically segregated in Borneo though they are somewhat more so in Sumatra. The *ultimus* morph predominates in Aceh and the Malay Peninsula, the *cristatus* morph farther south (Palembang, Telukbetung, Bengkulu, Padang, Lotabumi, Teluk Panji, Kerinci, Tapanuli Bay) and on Bangka, Belitung, and the Riau Islands; they occur at almost equal frequencies throughout Borneo, with the *ultimus* form being perhaps somewhat more common.

The mtDNA haplotype of a single Sumatran specimen studied by Rosenblum et al. (1997) was not identical to any from West Malaysia, but not more divergent than were some of the Malaysian haplotypes from each other.

Trachypithecus cristatus vigilans (Miller, 1913)

1913 *Presbytis vigilans* Miller. Indonesia: Sirhassen Island, South Natuna Islands.

DIAGNOSIS. Somewhat darker than *T. c. cristatus,* with less lightened hair tips; crown dark, surrounded by a well-defined zone of long hair tips. Skull has a more sloping lower frontal region.

DISTRIBUTION. From the Natuna Islands.

Trachypithecus germaini (Milne-Edwards, 1876)
Indochinese Lutung

DIAGNOSIS. Distinguished from *T. cristatus* by its larger size and paler color; by the light hair tips being very short or absent, making the body color evenly pale gray; and by the very long silvery or yellowish circumfacial hair, forming a distinctive "halo" all around. Tail and forearms are darker gray. Tail is very long, 720–838 mm ($n = 6$), compared with 560–751 mm ($n = 13$), whereas the two species have much the same head plus body length.

DISTRIBUTION. Central Thailand and Burma, from the northern end of the peninsula, into northeastern Thailand and at least southern Vietnam (north to 15°N [Fooden 1996]). As shown by Fooden (1971), its range runs west through Kanchanaburi Province into Burma and to the Bay of Bengal, inserting between the distributions of the two white-eyed species, *T. obscurus* and *T. phayrei*. Its range is, in turn, separated (but much more widely) from that of the related *T. cristatus* by *T. obscurus,* which occurs all up the peninsula to the latitude of Bangkok.

Choudhury (1997) mentioned "many reports" from rain forest districts in Assam and Mizoram.

Trachypithecus germaini germaini (Milne-Edwards, 1876)

1876 *Semnopithecus germani [sic]* Milne-Edwards. Cochin China and Cambodia.
1909 *Presbytis margarita* Elliot. Vietnam: Langbian.
1916 *Presbytis germaini mandibularis* Kloss. Thailand: Koh Chang.
1919 *Presbytis cristatus koratensis* Kloss. Thailand: Lat Bua Kao, 50 km west of Korat.

DIAGNOSIS. Color medium gray, caused by short creamy tips on dark gray or brown-black hairs; underside, throat, and shanks creamy, grading into upper side; hands and feet black; hairs form a creamy yellow halo around face; tail nearly black above, lighter below.

DISTRIBUTION. From Kanchanaburi west to the sea in Burma in a narrow strip; southeastern Thailand, including Koh (island) Chang, as far north as Loei; Cambodia; southern Vietnam (known from Bien Hoa). To the north of this range, *T. phayrei* is found; to the south, after a narrow strip of *T. barbei*, is found *T. obscurus*.

Trachypithecus germaini caudalis (Dao, 1977)

1977 *Presbytis cristata caudalis* Dao Van Tien. Probably North Vietnam.

DIAGNOSIS. Body dark, with a black tone to dorsum, becoming brown-gray with silvery tips on flanks; the long cheek hairs pale yellow; tail not distally black; legs pale yellowish outside; tail at base bicolored, with a sharply marked yellowish stripe underneath, contrasting with the black upper side.

DISTRIBUTION. The syntypes of this subspecies had been living in the Botanic Gardens of Hanoi and were of unknown origin. I have seen similar specimens in the USNM from Nongkhor, Sriracha, and also a skin labeled "Hainan?."

Trachypithecus barbei (Blyth, 1847)

Tenasserim Lutung

1847 *Presbytis barbei* Blyth. Burma: Ye, south of Moulmein.
1928 *Pithecus pyrrhus atrior* Pocock. Burma: Ye.

DIAGNOSIS. General color dark, nearly black, with no silvering; tail dark gray, slightly paler than body, with no black on distal part. To judge by preserved skins, the upper eyelids are probably pale, and the mouth has white hairs around it.

DISTRIBUTION. A very small region in the northern peninsular areas of Burma and Thailand, 14°20′–15°10′N, 98°30′–98°55′E (Fooden 1976). To the south, the range borders on that of *T. obscurus*.

NOTES. Pocock (1939) was the first to draw attention to the confusion surrounding the name *barbei*. Blyth (1847) first said that the two specimens on which the name is based came from Ye, in Tenasserim, Burma, and that they had pale face markings like *T. obscurus;* later (1863) he changed the locality, on Mr. M. Barbe's information, to Tipperah (now Tripura), northeastern India, and also changed the

description, stating then that the face was black without white markings.

Khajuria and Agrawal (1979), after earlier disagreement as to the correct allocation of the name, made an examination of two specimens, male and female, labeled *T. barbei* in the collection of the Zoological Survey of India (ZSI) (which had inherited, via the Indian Museum, the old collection of the Asiatic Society of Bengal, of which Blyth was curator). One of these skins, the female, has an odd asymmetrical white patch on the inside of the base of the thigh, exactly as mentioned by Blyth (1847), and Khajuria and Agrawal (1979) logically concluded that the two skins were indeed Blyth's (1847) syntypes of *barbei*. The specimens do not now have the white color of the lips mentioned by Blyth, but according to Khajuria and Agrawal (1979) this could have been obliterated by preservative chemicals, and this might explain why later (1863) Blyth said that the face was entirely black.

As for Mr. M. Barbe, on whose information Blyth (1863) changed the type locality to Tipperah, this was not the Rev. J. Barbe who collected the original specimens (Khajuria and Agrawal 1979).

My examination of the two skins in the early 1980s led me to agree with Khajuria and Agrawal (1979) that they represent the same taxon as that later described as *atrior* by Pocock. Napier (1985) likewise accepted that the two seem similar but, without comparing them directly, was unwilling to synonymize them.

The types of *barbei,* the type of *atrior,* and other skins from the same area are all so characteristically different from *T. germaini* that I am sure they represent a distinct species, which (if I have correctly interpreted the color of the facial skin) may be as closely related to the *T. obscurus* group as to the *T. cristatus* group.

Trachypithecus obscurus group

No white pubic patch in female; face dark gray, with white eye-rings (not always fully encircling eyes) and depigmented area on mouth and nasal septum.

Trachypithecus obscurus (Reid, 1837)

**Dusky Leaf Monkey or
Spectacled Leaf Monkey**

DIAGNOSIS. Legs and tail whitish or gray, generally contrasting with body tone; a light or whitish cap on crown.

DISTRIBUTION. Malay Peninsula, north to 15°10′N in Thailand and Burma.

Trachypithecus obscurus obscurus (Reid, 1837)

1837 *Semnopithecus obscurus* Reid. Malaysia: restricted to Malacca (Melaka) by Chasen (1940).
1841 *Semnopithecus leucomystax* Müller and Schlegel. "Malacca."

DIAGNOSIS. Body bronze to medium drab brown; limbs and tail contrastingly pale; posterior crown patch creamy or yellow-brown; dorsal stripe distinct. There is some variability in the darkness of tone of the body.

DISTRIBUTION. Peninsular Malaysia, north about to Perlis.

Trachypithecus obscurus flavicauda (Elliot, 1910)

1910 *Pygathrix flavicauda* Elliot. Thailand: Trang.
1916 *Presbytis obscura smithii* Kloss. Thailand: Klong Bang Lai, Patiyu.
1933 *Presbytis ruhei* Knottnerus-Meyer. Thailand: Songkhla.
1934 *Trachypithecus obscurus corax* Pocock. Burma: Tenasserim town.

DIAGNOSIS. Generally much darker than *T. o. obscurus,* more blackened on body, so that the creamy yellow or brownish gray legs and tail are more contrasting. Fooden (1971) described the very considerable range of variation in this subspecies: some specimens are much less contrastingly patterned than others; females are paler, browner than males; and subadult males are more blackish.

DISTRIBUTION. Found from Perlis north to the northern border of the species's range in peninsular Thailand and Burma; also Koh Lak Island (Thailand) and James, Kisseraing, and King Islands, Mergui Archipelago, Burma.

Trachypithecus obscurus halonifer (Cantor, 1845)

1845 *Semnopithecus halonifer* Cantor. Malaysia: Penang Island.

DIAGNOSIS. The general tone is brownish, with a distinct dorsal stripe; limbs and tail do not greatly contrast with body; crown is drab creamy buff.

DISTRIBUTION. Restricted to Penang Island.

Trachypithecus obscurus carbo (Thomas and Wroughton, 1909)

1909 *Presbytis obscura carbo* Thomas and Wroughton. Thailand: Terutau Island (not Langkawi in Malaysia, according to Chasen [1940]).
1913 *Presbytis corvus* Miller. Thailand: Terutau Island.

DIAGNOSIS. This taxon closely resembles *T. o. halonifer,* but is slightly darker, with the dorsal stripe less clearly marked. Actually, it barely differs from *T. o. halonifer;* moreover those on Terutau are slightly darker than those on Langkawi and Dayang Bunting, so there is in effect a stepped south-north cline of color darkening up these west-coast islands, from Penang north to Terutau.

DISTRIBUTION. Found on Terutau, Langkawi, and Dayang Bunting Islands, off the west coast of the peninsula on either side of the Thai-Malaysian border.

Trachypithecus obscurus styx (Kloss, 1911)

1911 *Presbytis obscura styx* Kloss. Malaysia: East Perhentian Island, Trengganu.

DIAGNOSIS. Color is deep brown, with no gray or black tones, and almost no contrasts of color between body and limbs or tail, which are only slightly lighter and grayer; crown is buffy; no trace of a dorsal stripe.

DISTRIBUTION. Restricted to East Perhentian Island.

NOTES. The resemblance of this east-coast island form to those of the west coast *(T. o. halonifer* and *carbo)* is striking. A possible explanation might be that these are the remnants of a dark, uncontrasty taxon that formerly inhabited the southern peninsula, and that after the terminal Pleistocene sea level rise the northerly, more contrasting *T. o. flavicauda* spread south, swamping the mainland populations but unable to spread to what by that time were offshore islands.

Trachypithecus obscurus seimundi (Chasen, 1940)

1940 *Pithecus obscurus seimundi* Chasen. Thailand: Pennan Island.

DIAGNOSIS. This is another dark insular form, resembling *T. o. carbo* in its dark overall tone, but frontal band is browner, less black; limbs are slightly paler.

DISTRIBUTION. Restricted to Pennan Island.

Trachypithecus obscurus sanctorum (Elliot, 1910)

1910 *Pygathrix sanctorum* Elliot. Burma: St. Matthew Island, Mergui Archipelago.

DIAGNOSIS. Black body and arms, with yellow-gray tail, dark smoky gray legs; feet black; chest creamy colored posteriorly.

DISTRIBUTION. Apparently this is restricted to St. Matthew Island, with the nearby islands having *T. o. flavicauda*.

NOTES. This is simply a more blackened version of *T. o. flavicauda*; it bears no relation to the more southerly island forms.

Trachypithecus phayrei (Blyth, 1847)
Phayre's Leaf Monkey

DIAGNOSIS. Very similar to *T. obscurus*, but legs, tail, and crown-cap are not contrastingly pale colored.

DISTRIBUTION. From Burma and Thailand, north of the peninsular zone, to southern Assam, northern Burma, southern Yunnan, and northern Vietnam, where it goes east to about 106°E and thence south to about 17°N (Fooden 1996).

Trachypithecus phayrei phayrei (Blyth, 1847)

1847 *Presbytis phayrei* Blyth. Burma: Arakan.
1863 *Presbytis barbei* Blyth. India: Tipperah (Tripura). Not of Blyth, 1847 (see under Notes and under *T. barbei*).
1878 *Semnopithecus holotephreus* Anderson. Locality unknown.
1909 *Presbytis melamera* Elliot. Burma: Cadu Ciaung, near Bhamo.

DIAGNOSIS. Dark brown, buffy, or gray-brown above, with sharply contrasting gray or whitish underparts; white ring encircles eye; no whorl behind brow.

DISTRIBUTION. From Pegu north through Arakan to Tripura, southern Assam (Choudhury 1997) and eastern Bangladesh.

NOTES. The name *barbei* Blyth, 1847, has often been used for this species in the past, because of confusion engendered by Blyth himself (compare Blyth [1863] with Blyth [1847]). The subject is discussed under *T. barbei*.

Trachypithecus phayrei crepusculus (Elliot, 1909)

1909 *Presbytis crepuscula* Elliot. Burma: Mount Muleiyit, 1,500 m.
1909 *Presbytis crepuscula wroughtoni* Elliot. Thailand, Pachebon.
1919 *Presbytis argenteus* Kloss. Thailand: Lat Bua Kao, west of Korat.

DIAGNOSIS. Body light gray to gray-fawn or silvery brown to nearly yellow, paler than in *T. p. phayrei*; forehead brown, gradually becoming lighter farther back over crown, sometimes creamy on nape; hands and feet black; depigmented area around mouth is very small, but white eye-rings are complete.

DISTRIBUTION. A rain forest form, found from Raheng (central Thailand) and Mae Ping rapids (northwest Thailand) north to Xishuanbanna (Yunnan), east to southwestern Laos and northern Vietnam, and west to the coast of the Bay of Bengal south of the range of *T. p. phayrei*. Northeasterly specimens tend to be darker, with darker limbs and tail.

Trachypithecus phayrei shanicus (Wroughton, 1917)

1917 *Pithecus shanicus* Wroughton. Burma: Se'en, Hsipaw, Shan State.

DIAGNOSIS. General color fawn, with no contrastingly light pelage areas except a trace of lightening on nape; hands and feet slightly darker; depigmented area around mouth is very wide, with white hairs, but white around eyes is restricted to inner side. A whorl or parting usually present behind brows, but no crest on crown. Rather small in size.

DISTRIBUTION. Found in the northern Shan States and neighboring dry zone of northern Burma, into Yunnan in the Yingjiang-Namting River and Tunchong-Homushu Pass Districts (Li and Lin 1983).

Trachypithecus pileatus group

Large sized, with long, backward-swept cheek whiskers contrasting with short, upright crown hair; ventral hair diverges bilaterally, whereas in all other *Trachypithecus* it streams backward or even slightly converges toward the midline.

Trachypithecus pileatus (Blyth, 1843)
Capped Langur

DIAGNOSIS. Contrastingly colored, with gray to brown dorsum, black crown-cap, white to red underparts (belly often redder than chest and throat) and buffy white to red cheek whiskers. Cheek whiskers curl forward behind cheeks.

NOTES. Choudhury (1997) was skeptical of the real existence of most of the described subspecies, stating that there is seasonal variation in pelage, and that he has observed wild individuals with characteristics of more than one race. They are retained here because museum material appears to allocate them to separate geographic regions and altitudes, but fieldwork is necessary to test their validity.

Trachypithecus pileatus pileatus (Blyth, 1843)

1843 *Semnopithecus pileatus* Blyth. "Received from Barrackpore [menagerie], stated to be Malayan."
1851 *Semnopithecus argentatus* Horsfield. Bangladesh: Sylhet.

DIAGNOSIS. Belly yellowish gray-white; chest and throat whitish, tinged with buff; whiskers buffy whit-

ish, sharply contrasting with dark tone of the dorsum; crown-cap black; tail tip dark. Dorsal color varies from iron gray to pale brown; upper arm and whole of legs lighter; hands and feet darker like dorsum.

DISTRIBUTION. This subspecies is widely distributed in the highlands south and east of the Brahmaputra and west of the Chindwin, in Burma, India, and Bangladesh. In Burma, the Chin Hills, south to Mount Victoria; in India, the Khasi, Garo, Naga, and Jaintia Hills. Choudhury (1997) gave also the Karbi Plateau and Barail Range, in Assam. Altitudes range from about 600 to 3,000 m.

Trachypithecus pileatus durga (Wroughton, 1916)

1916 *Presbytis durga* Wroughton. India: Cachar, Assam.
1923 *Pithecus pileatus saturatus* Hinton. India: Bara Hapjan, Lakhimpur, Assam.

DIAGNOSIS. Upper parts tend to be paler, slaty gray; belly, whiskers, throat, and underparts (including inner surfaces of limbs) orange to gingery, this tone extending farther up flanks than in *T. p. pileatus*. Head not as dark. Hair shorter.

DISTRIBUTION. Found adjoining the range of *T. p. pileatus* to the north, but at lower altitudes, from nearly sea level up to 600 m. Known from Naga Hills, at lower altitudes than *T. p. pileatus*, Lakhimpur, Golaghat, Cachar Hills, Samaguting, and Sibsagar.

NOTES. The real existence of this subspecies, and its relationship to *T. p. pileatus*, needs to be tested carefully: both the individual variation within a troop and even whether the darkness of the upper side and redness of the underside are not somehow a purely phenotypic result of temperature and humidity.

Trachypithecus pileatus brahma (Wroughton, 1916)

1916 *Presbytis brahma* Wroughton. India: Seajulia, Dafla Hills, North Lakhimpur, Assam.

DIAGNOSIS. Less contrastingly colored, to some extent, than other subspecies: body dark gray above becoming fawn-gray on lower flanks, rump, and legs; tail black; crown not dark; hands and feet black; inside of limbs pale gray; belly white, tinged with buff; chest and throat white.

DISTRIBUTION.   Known only from the Dafla Hills, north of Brahmaputra.

Trachypithecus pileatus tenebricus (Hinton, 1923)

1923 *Pithecus pileatus tenebricus* Hinton. India: Matunga River, North Kamrup, Assam.

DIAGNOSIS.   Darker than other subspecies, more or less ashy black above, with black crown-cap defined behind by a gray collar; legs grayer; belly whitish in male, rather pale red in the female, and rest of underparts yellowish orange as are bases of whiskers; most of whiskers gray. Rather small in size.

DISTRIBUTION.   The most northwesterly subspecies, from the Manas region (from 100 m up to nearly 1,000 m) into Bhutan, well to the west of *T. p. brahma*.

Trachypithecus shortridgei (Wroughton, 1915)
   Shortridge's Langur

1915 *Presbytis shortridgei* Wroughton. Burma: Homalin, upper Chindwin.
1915 *Presbytis shortridgei belliger* Wroughton. Burma: Hkamti, upper Chindwin.

DIAGNOSIS.   Strikingly uniform in color: silvery gray on head and body (including underside), with legs lighter; tail darkening increasingly toward tip; crown and whiskers colored as body; hands and feet black; whiskers shorter than in *T. pileatus* and *T. geei*; tail lacking long hairs.

DISTRIBUTION.   Mainly from lower altitudes east of the Chindwin from Kachin north to the Myitkyina District; also on the Dulong River in Gongshan District, Yunnan.

NOTES.   This species is at least as different from *T. pileatus* as is *T. geei*, and, like the latter, entirely lacks the parti-colored pelage so characteristic of *T. pileatus*.

Trachypithecus geei Khajuria, 1955
   Gee's Golden Langur

1955 *Trachypithecus geei* Khajuria. India: Jamduar, Sankosh River, Assam.

DIAGNOSIS.   A nearly uniform golden color (tending to whitish in winter), with very long, lax pelage and long cheek whiskers.

DISTRIBUTION.   From the *duars* of northwestern Assam and the area immediately to the south nearly to the Brahmaputra, between the Manas and Sankosh Rivers in India, and across the border into Bhutan, where it extends as far north as Black Mountain (about 27°30′N) and across onto the west bank of the River Sankosh (= Mo) for a short distance. In Bhutan it also extends east of the Manas along either side of the main stream and of its left-bank tributaries the Kur and Dangme as far northeast as Tashigang, only 20 km from the border with Sikkim. It is reported to be marginally sympatric with *T. pileatus* in a small area of Bongaigaon District, Assam (Choudhury 1997).

Trachypithecus francoisi group

Color mainly black, with jet black face; an elongated ridge of hair from lateral corners of upper lip curling upward and ending in front of ears, like a handlebar mustache; a high crest on crown, formed by long hairs directed backward from a frontal whorl, meeting forward-directed hairs from a whorl on nape. There is always some white zone on the pelage, generally involving at least the "mustache."

This distinctive group of "pied leaf monkeys," as D. Brandon-Jones (1995b) called them, is found in a small area through northern Vietnam, northern Laos, and southernmost China, and within this region is restricted to limestone formations, where they are largely terrestrial and seek refuge, at night or when alarmed, in caves and rock shelters. They all have a small erect crest on the crown, which except in *T. ebenus* is demarcated behind by hairs from a pair of whorls on the nape, and all have somewhat elongated hairs in line from mouth corners up cheeks to level of ears.

Table 18
External Measurements for the *T. francoisi* Group

Species (*n*)	Head + body length (mm)	Tail length (mm)	Hindfoot length (mm)	Ear length (mm)
T. francoisi (10)	470–635	740–960	152–175	30–42
T. hatinhensis (2)	500–665	810–870	140–155	28–35
T. laotum (3)	460–535	810–895	155–163	45–45
T. delacouri (5)	570–735	725–970	150–184	40–45
T. p. poliocephalus (7)	492–590	820–887	150–171	33–40
T. p. leucocephalus (6)	470–615	765–890	163–183	23–34

Note: From Dao (1985) and D. Brandon-Jones (1995b:35–36), except for those for *T. p. leucocephalus*, which are from Li and Ma (1980). No data are available for *T. ebenus*.

Inclusion of *T. johnii* and *T. auratus* in this species-group by D. Brandon-Jones (1995b), who even awarded each of these two species an Indochinese subspecies, seems at first sight bizarre, but in fact follows logically from that author's view of pelage evolution in langurs (in my view [Groves 1989] perfectly plausible) combined with his determinedly non-cladistic philosophy of taxonomy. From a PSC point of view all (nearly all?) the taxa are sharply distinct, and are here given species rank, except that *poliocephalus* and *leucocephalus* are combined, with some misgivings.

Table 18 gives external measurements for taxa of this group. Dao (1985) gave weights for two *T. francoisi* as 6.5 and 7.2 kg, for one *T. hatinhensis* as 8 kg, and for one *T. p. poliocephalus* as 6.7 kg. Li and Ma (1980) gave weights of four *T. p. leucocephalus* as 7.7 to 9.45 kg.

It appears from this, if the measuring techniques are comparable, that *T. delacouri* is rather larger than other species, and *T. laotum* averages smaller. *Trachypithecus hatinhensis* has a relatively short hindfoot and small ears; the ears of *T. laotum* and *T. delacouri* are unusually large.

Trachypithecus francoisi (Pousargues, 1898)
François's Langur

1898 *Semnopithecus francoisi* Pousargues. China: Luong-Tchéou (= Longzhou, 22°24′N, 106°50′E), Guangxi.

DIAGNOSIS. A narrow tract of slightly elongated white hair runs from corner of mouth along side of face to upper edge of ear pinna, in a sort of handlebar

mustache. Crown with a tall, pointed crest, set off behind by lateral ridges of hair coming from a pair of whorls on either side of midline of nape. Female with depigmented pubic skin, clothed with white to yellowish hairs.

DISTRIBUTION. From the Red River in Vietnam across the Chinese border as far as the Daming Hills (about 23°30′N, 108°20′E) in Guangxi and Xingyi (about 25°N, 105°E) in Guizhou. In Guangxi the approximate border with *T. p. leucocephalus* is the River Zuo, but some groups seem to have established themselves south and east of this river, and there is interbreeding with *leucocephalus;* the former occurrence of *leucocephalus* to the west of the river is suggested by Li and Ma's (1980) report of *leucocephalus*-like individuals in Daxian County. *Trachypithecus francoisi* is the one member of this species-group that is noticeably widespread geographically, and it may be that it has spread its range at the expense of other taxa and is even today in the process of "swallowing up" *T. p. leucocephalus* by genetic introgression.

Trachypithecus hatinhensis (Dao, 1970)
Hatinh Langur

1970 *Presbytis francoisi hatinhensis* Dao Van Tien. Vietnam: Xom-cuc, Hà-tinh.

DIAGNOSIS. Differs from *T. francoisi* by the white tract extending behind the ear to the lateral part of the nape, sometimes almost reaching the midline; young have a broad, white band across the forehead until the middle juvenile period. Female has a small triangular pubic patch. Tail is held above the back in an S curve.

DISTRIBUTION.   From localities centered on Quang Binh Province, Vietnam, from 19°39′N, 105°29′E south to 16°10′N, 107°40′E (Nadler 1996); in addition, Nadler (1998) mapped a locality about 15°00′N, 108°30′E, and Nhat et al. (1998) mentioned uncertain reports from Gialai and Kontoum Provinces. Although the information transmitted by me, and reported by D. Brandon-Jones (1995b), that it occurs in Cuc Phuong National Park is now known to be inaccurate, this species and *T. delacouri* overlap widely in distribution, though they occur on different limestone blocks. The total population is not more than 200–250 individuals.

NOTES.   The differences between this species and *T. francoisi* are very small, but consistent; especially, the white frontal band restricted to the juvenile is unique (it survives in the adult in *T. laotum* and is not present at any age in *T. francoisi*).

Trachypithecus poliocephalus (Trouessart, 1911)
White-headed Langur

DIAGNOSIS.   Head and neck white to yellowish, this color extending back at least to the nuchal whorls, but on crest black-tipped; white or light-tipped hairs on feet; gray- or white-tipped hairs on rump.

DISTRIBUTION.   Cat Ba Island, Vietnam; and a small area in Guangxi. This presumably represents the remnant of a wider range, over most of which it has been replaced by *T. francoisi*. The Guangxi population, *T. p. leucocephalus*, is itself strongly affected by gene flow from contiguous *T. francoisi*.

NOTES.   The similarity in color pattern between *T. p. poliocephalus* and South Indian *T. johnii*, noted by D. Brandon-Jones (1995b) and leading him to propose that they are conspecific, is striking, but most likely plesiomorphic, as noted by that author himself on pp. 33–34.

Trachypithecus poliocephalus poliocephalus (Trouessart, 1911)

1911 *Semnopithecus (Lophopithecus) poliocephalus* Trouessart. Vietnam: supposed to be Caï-Khin (identified as Cai Kinh limestone massif at about 21°45′N, 106°30′E by D. Brandon-Jones [1995b]), but because this is outside the known range of the taxon Brandon-Jones (1995b:24) proposed to restrict it instead to Cat Ba Island.

DIAGNOSIS.   Head and neck more yellowish to gray or creamy rather than white, this color extending for up to about 40 mm behind nuchal whorls and bordered posteriorly by brownish hairs with yellowish bases; similar brownish, yellow-based hairs also found on the feet. Hairs on crest, and sometimes back to nape, black-tipped. Hairs on sacral region, generally extending to back of thighs, gray-tipped. Pale (orange to yellow) hairs in pubic region in both sexes. Posterior dorsal pelage longer than in *T. francoisi*.

DISTRIBUTION.   Known only from Cat Ba Island, in Ha Long Bay.

Trachypithecus poliocephalus leucocephalus Tan, 1955

1955 *Trachypithecus (Presbytis) leucocephalus* T'an Pang-chi'eh (= Tan Bangjie). China: Sangeng Mountain, Luobai Commune, Zuo County, Guangxi.

DIAGNOSIS.   Head completely white, this color extending down to shoulders and upper chest; crest may be black-tipped; croup and thighs mixed with white and white-tipped hairs, not wholly gray or white; some white hairs on hands and feet; in most specimens, tail is white to brownish for two-thirds to three-fourths of its length. According to Li and Ma (1980) and Tan (1985), the occipital bone is higher than in *T. francoisi*, the nasals are slightly longer, and biorbital width is greater.

DISTRIBUTION.   The known range (Li and Ma 1980) is restricted to Fusui, Chongzuo, Linming, and Longzhou Counties. The range is bordered on the north and west by the River Zuo, on the east by the River Ming; to the southeast it does not extend as far as Mount Xiwan. A skin of this taxon, labeled *"poliocephalus,"* in the USNM, collected by F. R. Wulsin, is ticketed "Kwantung or Hainan." Neither origin is impossible in theory, although it evidently does not occur there today.

NOTES. Li and Ma (1980) noted considerable variability in color: there may be white patches, or patches with white hairs mixed with the black, on backs of limbs, especially arms; various amounts of white on feet and tail. They also reported a group of four, seen on the east side of the River Zuo, that lacked white on the tail; on the west side of the river, in the Liupang region of Daxian County, in the range of *T. francoisi,* they saw some individuals more resembling *T. p. leucocephalus.*

The occurrence, on both sides of the River Zuo, of animals apparently intermediate between *leucocephalus* and *francoisi* led Li and Ma (1980) to propose that the two are conspecific. D. Brandon-Jones (1995b), taking the variability in the known specimens into account, together with reports of apparent sympatry with *T. francoisi* and a report of fully and partially albinistic animals in Daxian, opined that *leucocephalus* actually represents a concentration of semialbinistic *francoisi.* For Tan (1985), however, the evidence was that they are sympatric, but remain in separate groups, despite the occurrence of some interbreeding. To me the available information suggests a northern population of *T. poliocephalus,* which it resembles very closely; it is sympatric over much of its (very small) range with *T. francoisi* and certainly the two do interbreed to some extent, such that some at least of the remaining population of *leucocephalus* is infused with gene flow from *francoisi.*

Genetically, this taxon does separate from *T. francoisi:* at least, it can be said that the mitochondrial ND3 and ND4 sequences of three *leucocephalus* and four *francoisi* individuals do sort cleanly into two separate clades in both neighbor-joining and parsimony analyses, the latter with 98 and 99% bootstrap values (Zhang and Ryder 1998).

Trachypithecus laotum (Thomas, 1911)
Laotian Langur

1911 *Pithecus laotum* Thomas. Laos: Ban Na Sao, on the Mekong at 17°30'N.

DIAGNOSIS (after D. Brandon-Jones [1995b]). Resembles *T. hatinhensis* but retains the broad, white forehead band throughout life, so that the head is mainly white except for black crest, joined by a narrow strip to black of shoulders and back. There are pale hairs around mouth and on chin and throat. Female has a small orange-yellow pubic patch over spottily depigmented skin. May be yellow hairs at tail tip.

DISTRIBUTION. Known from a small region of Laos, about 17°50'–18°30'E, 104°00'–105°00'E, in the Nam Kading and Khammouane National Biodiversity Conservation Areas.

Trachypithecus delacouri (Osgood, 1932)
Delacour's Langur

1932 *Pithecus delacouri* Osgood. Vietnam: Hoi Xuan.

DIAGNOSIS. Overall black, but with white hindparts and legs above the knees and elongated gray cheek whiskers; female with white pubic patch, overlying depigmented skin. Tail hair black, elongated especially in its distal half, contrasting with short body hair. The crown crest is relatively thin and forward-pointing; ears are rather outstanding from the head; tail is held in a backward curve.

DISTRIBUTION. Vietnam south of the Red River, from 21°36'N, 104°31'E south to 18°N, 105°26'E in Ha Tinh Province; range includes the Cuc Phuong National Park. This range is considerably larger than was previously realized, but the remaining populations are isolated and total only 200–250 individuals (Nadler 1996).

Trachypithecus ebenus (Brandon-Jones, 1996)
Indochinese Black Langur

1996 *Semnopithecus auratus ebenus* Brandon-Jones. Vietnam: either Lai Chau or, more probably, the Fan Si Pan mountain chain, 22°30'N, 103°50'E.

DIAGNOSIS (partly after D. Brandon-Jones [1995b]). Brow hairs more erect, less backwardly directed, than in other species of the group; crest and its posterior demarcating ridge less clearly defined, because of erectness of hairs between crest and pinna; no nuchal whorls. Pelage entirely glossy black, becoming browner toward hair bases especially on venter. Faint yellowish tips to cheek hairs; whitish, yellowish, and yellow-tipped hairs sparsely intermingled with black on chin, upper lip, nasal septum, and fringing palms and soles; whitish hairs on anterior part of ear pinna. Pubic patch (female) orange medially to whitish laterally, with (unlike others of the genus) black hairs intermingled.

DISTRIBUTION. Brandon-Jones (1995b) speculated on field reports in northwesternmost Vietnam that could possibly be of this taxon, but Nadler (1998) cited apparently reliable sightings in the Hin Namno National Biodiversity Conservation Area in Laos, on the Vietnam border at about 17°30'N, a limestone region not too far from areas where *T. hatinhensis* and *T. laotum* occur.

NOTES. Brandon-Jones described this as a subspecies of the Javan species, which is here allocated to a different species-group. As he in fact opined, the almost entirely glossy black pelage, the only real resemblance to the Javan form, is likely to be a retained plesiomorphic condition. Nadler (1998) recently illustrated a living captive specimen.

Genus *Presbytis* Eschscholtz, 1821
Surilis

1821 *Presbytis* Eschscholtz. *Presbytis mitratus* Eschscholtz, 1821.
1879 *Corypithecus* Trouessart. *Semnopithecus frontatus* Müller, 1838.
1879 *Lophopithecus* Trouessart. *Semnopithecus rubicundus* Müller, 1838.
1879 *Presbypithecus* Trouessart. Replacement for *Presbytis*.

This genus is distinguished by the predominance of an underbite in incisor occlusion; incisors homodont; the canine/sectorial complex not greatly sexually dimorphic; the skull vault rounded; the cranial base strongly flexed; brow ridges relatively weak, and the upper margin of the orbit horizontal; the nasal profile convex; the pyriform aperture suddenly contracting to a point inferiorly; a well-marked arch under the malar-maxillary suture; a swelling behind

the lambdoid suture; the mandible with a shallow corpus, and the angle not enlarged; the relatively short tail; the longer hindlegs (intermembral index 76) and longer forearm (brachial index 109); the neonate usually white with a thick dorsal stripe and sometimes a cross-stripe over the shoulders, but in *P. comata* it is gray or black. A true pendulous scrotum is present; the clitoris is large and external to labia. The hypoconulid on M_3 is usually absent; there is no functional occlusion to distal surface of distal loph on M^3.

The number of species in this genus is questionable. The definition of the Sumatran species, where the question is most relevant, follows Aimi and Bakar (1996), except for the additional recognition of *P. siamensis*.

Presbytis melalophos (Raffles, 1821)
Sumatran Surili

DIAGNOSIS. Crown has a distinct black crest; tail bicolored, dark above and white below. No pale eyerings. Interorbital region narrow, nasals long. Infant has a thick, dark dorsal stripe, but no shoulder cross-stripe. Vocalizations are distinctive.

DISTRIBUTION. Sumatra, south of the Wampu and Simpang Kiri Rivers, except for the eastern coastal forests; separated from *P. femoralis* by the Rokan River and from *P. siamensis* by the Kampar River and its upper tributaries.

NOTES. This species has proved much more difficult to divide into subspecies than has *P. femoralis*, because of widespread geographic gradients in color and individual variation (often in the form of quite sharply marked morphs). There appears to be sympatry between *P. melalophos* and *P. femoralis* in the Tapanuli Highlands (Wilson and Wilson 1977; J. Vermeer, personal communication).

Presbytis melalophos melalophos (Raffles, 1821)

1821 *Simia melalophos* Raffles. Indonesia: Bencoolen (Bengkulu), Sumatra.
1831 *Semnopithecus flavimanus* I. Geoffroy. Indonesia: Sumatra (restricted to Bengkulu by Chasen [1940]).

1842 *Semnopithecus nobilis* Gray. "India": restricted to Inderapura, Sumatra, by Chasen (1940), and further to the highlands inland of Inderapura by Wilson and Wilson (1977).

1861 *Semnopithecus femoralis* var. *aurata* Müller and Schlegel. Mount Ophir region (= Gunung Talamau).

1876 *Semnopithecus ferrugineus* Schlegel. Indonesia: Padang, Sumatra.

DIAGNOSIS. Usually a single, indistinct whorl on forehead; a U-shaped parting on nape. General color of body and tail pale reddish foxy red, red-orange, or red-brown, generally overlain with black; outer sides of limbs tend to be paler, more orange, than body; tail whitish underneath in proximal half; crown crest black; underside white, yellow, or pale orange, not extensively, and chin and extremities are pale but inner sides of limbs are white; cheeks straw-colored, separated from whitish forehead by a reddish or blackish band; hands and feet may be black but are usually colored like limbs; facial skin is black.

DISTRIBUTION. Western Sumatra, from the upper Sungai Rokan south to the upper Sungai Hari and beyond along the Barisan Range into Lampung (Aimi and Bakar 1996).

NOTES. There is a gradation in this subspecies from very red, with one or occasionally two whorls, in the Bengkulu District, to much paler with no whorls, farther east.

The two forms described as *nobilis* and *ferruginea* have commonly been retained as separate subspecies, but the work, including extensive fieldwork, of Kawamura (1984), Aimi et al. (1986), and Aimi and Bakar (1992) showed that they are in fact color morphs that recur in different localities. The foxy red *nobilis* morph was described as an inland subspecies from south of the Batang Hari River by Wilson and Wilson (1977), but Aimi and Bakar (1992) noted that this region is not at all close to the Lake Singkarak locality from which a specimen was described by Pocock (1935). I have seen a skin of this morph, in the MCZ, from Kayutanan, another rather disparate locality. The less-reddish *ferruginea* morph was thought by Hooijer (1962) and Wilson and Wilson (1977) to be a northerly subspecies, from the Padang High-

lands, but Kawamura (1984) found this morph together with *nobilis* and typical *melalophos* morphs in the same troop.

It may be true that the three morphs predominate in different regions, but their evident widespread coexistence over most of the subspecies's range requires that the names be synonymized.

The *aurata* morph, golden buff in color with a black spot on the chest, black hands and feet, and white tail tip, has a more restricted distribution. It was observed by Kawamura (1984) in the western Barisan Range just north and south of the equator, and by Aimi et al. (1986) from the western coast around Panti; Aimi and Bakar (1992) found it again, on the border of *P. m. melalophos* with *P. m. bicolor*.

Presbytis melalophos bicolor Aimi and Bakar, 1992

1992 *Presbytis melalophos bicolor* Aimi and Bakar. Indonesia: Batang Kering, 0°50′S, 101°23′E, 14 km north of Kiliranjao, West Sumatra.

DIAGNOSIS. Dark gray-black or chocolate dorsum contrasts with white underside and cheek ruff; limbs are dark on outer surfaces (black on the hindlimbs), white on inner; hands and feet black; tail black above, white below, with white tip; crown crest with black median stripe and black tip; a black fringe along forehead; face dark gray with no eye-ring, but muzzle black and chin gray or pinkish colored.

DISTRIBUTION. A montane form, from the middle and lower Sungai Hari, which separates it from *P. m. melalophos* (with which it intergrades in the foothills to the west), to the middle Sungai Inderagiri, which separates it from *P. siamensis catemana*. Three skins in the Leiden collection (RML) (Hooijer's [1962:9–10] nos. 8, 9, and 12), from north of Kerinci, are variants of this subspecies grading toward *P. m. melalophos*.

NOTES. Wilson and Wilson (1977:219) referred to this as "black and white subspecies, not previously described," and it was mentioned in addition by Kawamura (1984:38). It was formally named by Aimi and Bakar (1992).

Presbytis melalophos sumatrana (Müller and Schlegel, 1841)

1841 *Semnopithecus sumatranus* Müller and Schlegel. Indonesia: Mount Ophir (= Gunung Talamau), north of Padang.
1903 *Presbytis batuanus* Miller. Indonesia: Pinie (Pini) Island, Batu Archipelago.
1948 *Presbytis aygula margae* Hooijer. Indonesia: Serdang, Deli, Sumatra.

DIAGNOSIS. Usually no whorls on forehead; crest is gray to dark brown, sometimes indistinctly formed; dorsally very dark gray-brown to nearly black, with outer sides of limbs black, including hands and feet; white of underside extends to chin, cheeks, wrists and ankles, and under tail, but chest is black or dark brown. Face is light, bluish gray to pinkish colored, with mouth black but lower jaw white.

DISTRIBUTION. Found on Pulau Pinie, in the Batu Archipelago; and, on the Sumatran mainland, in the northern highlands and west coast south of the Sungai Simpang Kiri north of Gunung Talamau, the former Mount Ophir, south of which *P. m. melalophos* occurs; southeast to the Rokan River, which separates it from *P. femoralis percura,* and on the east coast from Sungai Wampu to the Sungai Barumun, which separates it from *P. siamensis paenulata* (Aimi and Bakar 1996). There is a clinal darkening from south (*sumatrana*-like) to northeast (*margae*-like). Hooijer's *margae* was described as a subspecies of *P. aygula* (i.e., *thomasi* + *comata* + *hosei*), but it vocalizes like true *P. melalophos* (Wilson and Wilson 1977) and it is clearly distinct from *P. thomasi* by lacking white on the head and by the lack of a posterior component to the crown crest; Aimi and Bakar (1992) showed that it is simply a clinal variant of *P. m. sumatrana.*

Wilson and Wilson (1977:220) stated that in the Tapanuli highlands "local inhabitants reported a grey race that vocalized like *melalophos,* and a dark race that vocalized like *femoralis.*" Jan Vermeer (personal communication) recorded *P. femoralis paenulata* 150 km north of the Barimun River, deep within the range of *P. melalophos sumatrana.*

The insular population described as *batuanus* differs from the more southerly (west-coast) mainland populations in that there is usually a single fron-tal whorl, and the white does not usually extend to the wrists and ankles. Pulau Pini is a shelf island, which was connected to the mainland during periods of low sea level.

Presbytis melalophos mitrata Eschscholtz, 1821

1821 *Presbytis mitrata* Eschscholtz, in Kotzebue. Indonesia: Sumatran mainland opposite Zutphen Islands.
1906 *Presbytis fusco-murina* Elliot. Indonesia: Teluk Betung, Lampung.
1940 *Pithecus femoralis fluviatilis* Chasen. Indonesia: Muaraduwa, Palembang, Sumatra.

DIAGNOSIS. Generally no whorls. Overall from dark mouse brown or gray to very pale red-yellow or yellow-gray, with some black overlay; limbs with very pale-rooted hairs; forehead and crown similar, the bases here being white, this color showing on the crest, but tuft itself is generally black on top and sides; limbs mixed, whitish and gray or red-brown; hands and feet gray. Underside creamy yellow or white, this tone extending well up onto flanks. Tail above redder than body; below pale buffy, tip white. Face gray, muzzle often pink, set off by white cheek ruff; a white or pink crescent around outer corner of each eye.

DISTRIBUTION. The southeasternmost subspecies, from Lampung north to the upper Musi River drainage, west of Palembang, and (according to Aimi and Bakar [1996]) as far north as the Batang Hari River; not extending west to the Barisan Range, where *P. m. melalophos* occurs.

NOTES. The status of Eschscholtz's name *mitrata* as senior synonym of Elliot's *fusco-murina* was affirmed by Aimi and Bakar (1992). It was named from a specimen obtained by Kotzebue (Eschscholtz 1821:353), "brought to us, for sale, by the inhabitants of Sumatra" while his ship was anchored off the Zutphen Islands. The description given by Eschscholtz exactly fits this subspecies.

Chasen (1940) noted considerable color variation and separated it into two subspecies (southerly *fusco-murina* and northeasterly *fluviatilis*); Wilson and Wilson (1977) divided it slightly differently, into a western mountain form with gray face and white cheek ruff, and an eastern lowland one with depigmented muzzle and less-pronounced ruff. Finally Aimi et al. (1986) further described the depigmentation, noting that the skins of the palms and soles is black in northern populations but pinkish colored in the southern.

A further variant is the largely white "mutant" described by Pocock (1935); both Wilson and Wilson (1977) and Aimi et al. (1986) saw it south of Palembang, and there is a skin of it from Kotabumi in the Singapore collection (ZRC).

Presbytis femoralis (Martin, 1838)
Banded Surili

DIAGNOSIS. Interorbital region wider than in *P. melalophos;* nasals short. Crest on crown shaggy, ill defined. Dorsally dark brown or blackish, ventrally white continuous with a prominent white zone on outside of thighs. White eye-rings present. Infant has "cruciger" pattern. Vocalizations distinct from those of *P. melalophos.*

DISTRIBUTION. In the Malay Peninsula, restricted to the far south and to the northwest, extending north throughout peninsular Thailand and Burma; the two parts of the range are separated by *P. siamensis.* In Sumatra, found between the Rokan and Siak Rivers; north of the Rokan, *P. melalophos* is found, and south of the Siak, *P. siamensis.* Also occurs on Singapore.

Presbytis femoralis femoralis (Martin, 1838)

1838 *Semnopithecus femoralis* Martin. Singapore.
1876 *Semnopithecus neglectus* Schlegel. Singapore.
1913 *Presbytis australis* Miller. Malaysia: Jambu Luang, Johor.

DIAGNOSIS. A pair of whorls on crown, with a long crest between, or one of the pair may be suppressed. Body blackish above, the head tending to red-brown; underside sharply gray, with a median white line. There is no white under the tail or on the limbs. Hairs are directed backward on chest (except, sometimes, those on the lateral edges), forming a short crest where they meet the forward-directed hairs of the belly.

DISTRIBUTION. This taxon is found on Singapore and in Johore. Singaporean examples average darker ventrally, and slightly smaller, than those from the mainland.

Presbytis femoralis robinsoni Thomas, 1910

1910 *Presbytis robinsoni* Thomas. Thailand: Ko Khau, Trang.
1911 *Presbytis neglecta keatii* Robinson and Kloss. Thailand: Ko Khau, Trang.

DIAGNOSIS. One or a pair of whorls on crown; gray-brown or blackish brown, with white underside; inner surfaces of limbs white nearly to heel, but chest black; hands and feet black. Fooden (1976) stated that the eyelids are black, with only a patch around the mouth being white. Hairs on sides of chest are directed more outward than backward, but in midline directed back. Whitish morphs are known and were described by Pocock (1935).

DISTRIBUTION. Found from Larut Hills, Perak (4°47′N, 100°45′E), north as far as 13°N in Phet Buri Province, Thailand (Fooden 1976).

Presbytis femoralis percura Lyon, 1908

1908 *Presbytis percura* Lyon. Indonesia: Kompei, East Sumatra.

DIAGNOSIS. One or (occasionally) a pair of whorls, well back on crown; blackish above, except for head, which is gray; white zone of underside is narrow but extends to cheeks, chin, wrists, and ankles, but not onto outer surfaces of limbs and is indistinct under tail, and chest is black. Face is gray, with narrow pale eye-rings; muzzle gray. Hairs on chest are all directed backward.

DISTRIBUTION. A small range in the east of Sumatra, south of the Rokan River, only as far as the Siak River, according to Aimi and Bakar (1996).

NOTES. Chasen (1940) noted the close similarity of this subspecies to *P. f. femoralis,* just across the strait; it differs mainly in the more extensive white of the underside. The caudal shift of the whorls results in the crest being "shaggy" in living animals (Aimi and Bakar 1996).

Presbytis natunae (Thomas and Hartert, 1894)

Natuna Islands Surili

1894 *Semnopithecus natunae* Thomas and Hartert. Indonesia: Bunguran Island, North Natuna Islands.

DIAGNOSIS. Forehead whorl or whorls absent or very indistinct. Body dark grayish brown above, with the head and lower parts of limbs blacker; white of underside extremely widespread, extending to backs of thighs, chest, chin, and often to wrists and ankles, but not under tail. Buccal depigmentation extends to nose; there is a slight depigmented area bordering the frontal hair on and lateral to glabella; and the white eye-rings are very large, especially below the eye, but the upper and lower halves do not connect around the eyes. Hairs are directed outward on chest, except on midline. Cheek whiskers are exceptionally bushy and prominent, white except for their upper margins.

DISTRIBUTION. Bunguran Island.

NOTES. D. Brandon-Jones (1984) considered this a subspecies of *P. siamensis*, to which it is obviously related, but its outstanding facial patterns make it quite distinct.

Presbytis chrysomelas (Müller, 1838)

Sarawak Surili

DIAGNOSIS. Whorls, of which there are nearly always a pair, very small and nearer to brow than in other species; crest high, upstanding, and narrow; no defined fringe over the eyes. A U-shaped parting on nape, rather than a whorl as in other species. Chest hairs uniformly directed backward, meeting forward-directed hairs of belly in a ridge. Skull distinguished by a characteristic swelling on nasion.

DISTRIBUTION. North and northeast of the Kapuas River, in northwestern Kalimantan; throughout Sarawak; as far as Melalap, Sabah.

NOTES. This has generally been included in *P. femoralis* (for example, D. Brandon-Jones [1984]), but it differs consistently.

Presbytis chrysomelas chrysomelas (Müller, 1838)

1838 *Semnopithecus chrysomelas* Müller. Indonesia: Pontianak, West Borneo.

DIAGNOSIS. Color is generally jet black, occasionally with some brown hairs on dorsum. Cheeks brown-gray to white; pale zone of underside extending to cheeks, chin, wrists, halfway down thighs, and underside of tail.

DISTRIBUTION. From the Kapuas River as far as the IV Division of Sarawak.

NOTES. There is a great deal of variability in this subspecies. The underside is generally fawn, darker on chest and upper abdomen, only lower abdomen may be white, though whitish tones may be restricted to patches on chest, inner limb surfaces, and underside of tail; inner sides of limbs may be creamy yellow rather than white, and this tone may go all the way to the wrists (but often very poorly expressed), and variably down inner side of thigh toward knee. The white on the underside of the tail may be restricted to patches on the proximal part or may reach the tip. Skins from Sarawak nearly always have two whorls; those from the Landak region, Kalimantan, are more variable.

Presbytis chrysomelas cruciger (Thomas, 1892)

1892 *Semnopithecus cruciger* Thomas. Malaysia: Miri, Sarawak.
1934 *Presbytis arwasca* Miller. Malaysia: Miri, Sarawak.

DIAGNOSIS. Browner in general than *P. c. chrysomelas*, the hair bases being redder. There tends to be a greater amount of white or creamy fawn on the underside: there is often a whitish streak running forward from the abdomen along the chest midline, and the white always reaches the wrists and ankles and nearly always the tail tip. The thigh stripe is wider: according to measurements by Pocock (1935), it is 15–21 mm wide, cf. 7–13 mm in *P. c. chrysomelas*. The forehead whorls are more elongated or like a parting, and the crest is not quite so high. The nasion swelling is less prominent.

DISTRIBUTION. From the Baram District, northeastern Sarawak, to Sabah.

NOTES. There is a strikingly different black-and-red morph, called the "cruciger" type, which is found in the same troops as the normal brown morph and appears to be quite common: indeed the type of *cruciger* was of this form (the type of *arwasca* was of the "normal" morph). It has crown, flanks, and legs bright red, and a thick, black dorsal band (extending onto tail), crossed by another band down the arms. This pattern represents a persistence, even intensification, of the cross-pattern of the infant coat.

Presbytis chrysomelas subsp.

There is an isolated population in a small area of southeastern Sarawak (Betong, Saribas; Nimong; Batang Lupar) that is entirely "cruciger"-like superficially, but it differs from the "cruciger" morph of *P. c. cruciger* in characteristic ways: the dividing line between the black of the dorsum and red of the flanks is less sharp, and there are some black hairs on the flanks; the underside is dull creamy rather than white, and there is more red on the chest; the black patch on the knee is more conspicuous, and there is always black patching on the shank; there are two small whorls on the forehead, with a tall crest between; the nasion swelling is conspicuous.

A "cruciger" morph is not known for *P. c. chrysomelas,* and this would fill that role (it differs from the cruciger morph of *P. c. cruciger* in the same ways in which their "normal" morphs differ) except that it is geographically separate. It represents a very restricted, but quite distinct, undescribed subspecies.

Presbytis siamensis (Müller and Schlegel, 1841)
White-thighed Surili

DIAGNOSIS. Much paler (grayish brown) than *P. femoralis,* with hands and feet, as well as brow, blackish; underside whitish, continuous with a grayish-whitish zone on outside of thigh.

DISTRIBUTION. Most of peninsular Malaysia and two small sections of the eastern forests of Sumatra, including some islands of the Riau Archipelago.

NOTES. Chasen (1940) noted that what he called *P. femoralis paenulatus* and *P. f. cana* are sharply dis-

tinct in their reddish brown color from their blackish southerly and westerly neighbors, analogous to the pale *P. f. siamensis* replacing a more southerly dark form in the Malay Peninsula. D. Brandon-Jones (1984) proposed that the three pale taxa are rather weakly separated from each other and could together represent a separate species, and this insight is followed here.

Presbytis siamensis siamensis (Müller and Schlegel, 1841)

1841 *Semnopithecus siamensis* Müller and Schlegel. "Malacca."
1843 *Semnopithecus nigrimanus* I. Geoffroy. "Java"; actually, Malay Peninsula (according to Chasen [1940]).
1909 *Presbytis nubigena* Elliot. Malaysia: Malacca (Melaka).
1909 *Presbytis dilecta* Elliot. Malaysia: Dusun Tua, Selangor.

DIAGNOSIS. Body rather light, grayish, varying from medium brown-gray to buff; underside and inner side of legs white, reaching down to wrists and ankles and forward to throat and cheeks; no white under tail except at base; tail, hands, and feet nearly black.

DISTRIBUTION. Most of peninsular Malaysia, except the south and northwest: Perak north to the Piah Valley (5°14′N, 101°07′E); Selangor; Negri Sembilan; Pahang; Melaka. There is also a BM(NH) skin from Mabek, Thailand, 6°22′N, 101°08′E, which is well into the area occupied by *P. femoralis robinsoni.* In Sumatra, too, the species appears to have a population isolate within the range of a different, related species (Vermeer's [1998] record of *P. siamensis paenulata* within the range of *P. melalophos sumatrana* in Sumatra: see under *P. s. paenulata*); it may be that it is a primitive species, formerly more widely distributed but gradually replaced by more derived relatives.

NOTES. Chasen (1940) separated this taxon into two subspecies, a darker one with the thighs and buttocks grayish, from southern localities *(nubigena),* and a lighter one with smaller dark thigh stripe and more whitish thighs and buttocks, from northern localities. Studies by D. Brandon-Jones (personal communication) indicate that the differences are at best clinal.

Presbytis siamensis rhionis Miller, 1903

1903 *Presbytis rhionis* Miller. Indonesia: Bintang Island, Riau Archipelago.

DIAGNOSIS.   Usually a single whorl on forehead; yellow-brown with a gray tinge, but the head and lower limb segments blacker; white extensive on body, reaching ankles but not under tail. Buccal depigmentation extends up to nose. Hairs directed mainly sideways on chest.

DISTRIBUTION.   Known only from Pulau Bintang; Chasen (1940) mentioned that there are langurs on Batam and Galang Islands, but their taxonomic affiliation is unknown.

NOTES.   This subspecies curiously resembles *P. s. siamensis* in color, but has more contrast between the light body and the dark crown and tail, and greater extension of white down the limbs; also the whorl is farther back on the crown.

Presbytis siamensis cana Miller, 1906

1906 *Presbytis cana* Miller. Indonesia: Kundur Island, Riau Archipelago.
1908 *Presbytis catemana* Lyon. Indonesia: Kateman River, East Sumatra.
1984 *Presbytis femoralis amsiri* Kawamura. Nomen nudum. This putative subspecies, never formalized, proved to be a variant of *catemana* (M. Aimi, personal communication).

DIAGNOSIS.   Two, one, or no whorls on forehead; yellow-gray-brown, with the head darker; the shanks colored like the body; the white of the underside extending onto the backs of thighs only as a wide gray area, but extending to chin, cheeks, wrists, and ankles. Tail lighter below, but not white. Hairs on chest directed mainly outward. Face is gray with bluish eye-rings and pinkish-colored muzzle. The ascending ramus of the mandible has a curious abrupt eversion on either side, just above the angle.

DISTRIBUTION.   Pulau Kundur and the opposite mainland of Sumatra, between the Siak and Inderagiri Rivers.

NOTES.   This taxon differs very little from *P. s. rhionis* except for its rather darker limbs and tail, as

noted as long ago as 1940 by Chasen. The form described as *catemana* was supposed to be darker, with less extensive white on the backs of the thighs, but the differences are very slight and they overlap extensively.

Vermeer (1998) reported seeing a white langur, with a little gray on the back and a black crest, near Jambi. Its vocalizations were not those of *P. melalophos*; an extremely pale form of *P. siamensis*, related to *P. s. cana*, which is found not much farther to the north, may be the most plausible explanation.

Presbytis siamensis paenulata (Chasen, 1940)

1940 *Pithecus femoralis paenulatus* Chasen. Indonesia: Teluk Panji, Barumun River, south of Labuhanbilik, northeastern Sumatra.

DIAGNOSIS.   Dorsum and tail darker reddish brown; underside, including whole of inner side of limbs, whitish to pale gray-brown; outer sides of limbs brown, except for gray-white backs of thighs and outer sides of arms; hands and feet black; tail paler below than above, but not sharply bicolored; crest small, black, with pale patch in front; face black with complete white eye-rings and pinkish-colored lips.

DISTRIBUTION.   Found in a small wedge of coastal forest in eastern Sumatra. Stated by Wilson and Wilson (1977) and Aimi and Bakar (1996) to be separated from *P. femoralis percura* by the Rokan River; it was considered by Aimi and Bakar (1996) to be separated from *P. melalophos sumatrana* by the Barimun River, but Vermeer (1998) recorded it near Lake Toba, 150 km north of the Barimun River, deep within the range of the latter.

Presbytis frontata (Müller, 1838)
White-fronted Langur

1838 *Semnopithecus frontatus* Müller. Indonesia: southeastern Borneo.
1909 *Presbytis nudifrons* Elliot. Malaysia: Bajalong, Sarawak.

DIAGNOSIS.   Body pale grayish brown above; tail yellowish gray; hands and feet, brow, crown crest, and cheeks blackish; lower segments of limbs grading toward black of hands and feet; underside yellowish brown. Crown crest tall, partly directed forward; one or two forehead whorls at base of crest, enlarged, making a conspicuous light frontal patch.

DISTRIBUTION. Central and eastern Borneo, from central Sarawak to the southern coast.

Presbytis comata (Desmarest, 1822)
Javan Surili

DIAGNOSIS. Dark gray or black above, with at least some white below; crown crest black, without stripes.

DISTRIBUTION. Western Java, as far east as the Gunung Slamet region.

NOTES. This species was generally known as *Presbytis aygula* (Linnaeus, 1758) until it was shown (Napier 1985) that Linnaeus's *Simia aygula* actually referred to the crab-eating macaque, *Macaca fascicularis*. The Commission suppressed the name *aygula* (International Commission on Zoological Nomenclature [1986], Opinion 1400), so that the next available name, *comata,* must be used for this species.

The proposal by D. Brandon-Jones (1995a) that the eastern subspecies *P. c. fredericae* is a distinct species was shown by Nijman (1997), on the basis of both museum and field observations, to be less clear-cut than originally thought, because of the existence of a chain of intermediate populations. This is yet one further example of the crucial value of good field observations in straightening out taxonomic affinities.

Presbytis comata comata (Desmarest, 1822)
1822 *Semnopithecus comatus* Desmarest. Java.

DIAGNOSIS. Upper parts dark gray; pelage on body longer than on limbs or tail; limbs and tail very dark gray, generally darker than back; underside, including inner aspects of arms and legs, whitish. Crown crest black.

DISTRIBUTION. According to Nijman (1997), the typical coloration as described above occurs in individuals from Mounts Salak, Pancar, and Gede-Pangrango (106°45′–107°00′E, 6°35′–6°45′S). Far-

ther east, on Mount Tilu, south of Bandung (107°30′E, 7°09′S), the underside seems to be more intermingled with gray. Farther east still, on Mount Sawal and in the Ceringin/Cisaga District (108°16′–108°30′E, 7°12′–7°27′S), the whitish or light gray throat is bordered behind (on the upper breast) by a broad gray band, which has a thin, whitish median stripe joining the white areas of throat and belly; and the arms are almost black in tone. This coloration thus approaches that of *P. c. fredericae*. It is possible that this population should be ranked as a third subspecies, intermediate between the otherwise strikingly different *P. c. comata* and *fredericae*.

In addition to these variations in adult coloration, the neonate is medium gray on Gede-Pangrango, but dark gray on Mount Sawal (Nijman 1997).

Presbytis comata fredericae (Sody, 1930)
1930 *Pithecus aygula fredericae* Sody. Indonesia: Mount Slamet, central Java, 1,000 m.

DIAGNOSIS. Upper parts are black, not gray; underside is black except for lower abdomen and inner aspect of legs and tail, which are white, and throat and upper part of chest, which are light gray or whitish, with these two light zones being joined by a thin, whitish streak.

Nijman (1997) described some variation in this general color pattern. The description above characterizes animals of Mount Slamet and Mount Prahu (109°13′–109°55′E, 7°19′–7°20′S). At lower altitudes, near Linggo, general color is less dark, like that of specimens from Mount Sawal farther west (see under *P. c. comata*), and the black zone on the breast is dark gray. In addition, the neonate is black on Mount Slamet but dark gray on the Dieng Range.

Presbytis thomasi (Collett, 1892)
Thomas's Langur
1892 *Semnopithecus thomasi* Collett. Indonesia: Langkat, Sumatra (Aceh).
1942 *Presbytis thomasi nubilus* Miller. Indonesia: Blanganga, 1,100 m, Aceh.

DIAGNOSIS. Iron gray, with creamy white underside, this tone extending to inner sides of limbs and underside of tail. A white patch above each eye, with a black stripe in the midline and on either side, extending up to crest.

DISTRIBUTION. Northern Sumatra, north of the Wampu and Alas (= Simpang Kiri) Rivers; Aimi and Bakar (1996) also found one population south of the latter river.

NOTES. This species is kept separate from *P. comata*, contrary to D. Brandon-Jones (1984), who combined them (also *P. hosei*) because of their gray color.

Presbytis hosei (Thomas, 1889)
Hose's Langur

DIAGNOSIS. Iron gray with whitish tips; hands and feet black, this tone sometimes extending to lower part of legs; underside grayish to buffy white, becoming white on inner sides of limbs. Crown crest black, and sometimes rest of crown blackish. Facial skin pinkish on lower jaw and cheeks, darker (reddish or black) elsewhere.

DISTRIBUTION. Northeastern Borneo: eastern Sarawak, Brunei, and Sabah south as far as the Sungai Karangan in Kalimantan.

NOTES. D. Brandon-Jones (1984, 1996) combined this species with *P. thomasi* and *P. comata* because of their gray color. I keep them separate because not only are they diagnosably different, but the gray color is likely, as explained by Brandon-Jones (1996), to be a metachromatic stage in the evolution of langurs; hence such a composite species would be paraphyletic. Indeed, I have my doubts whether this species ought not really to be divided into three (*canicrus* and *sabana* raised to specific rank).

Measurements for this species, following Brandon-Jones (1996), are given in Table 19.

Presbytis hosei hosei (Thomas, 1889)

1889 *Semnopithecus hosei* Thomas. Malaysia: Niah, Sarawak.

DIAGNOSIS. D. Brandon-Jones (1996) argued that this is a subspecies of very restricted distribution (the type locality, where it is now extinct, and probably the lower Baram River, where it may also be extinct), characterized by lacking sexual dichromatism, both sexes having the black color restricted to the crown, with the rest of the head being white. Legs darkening distally, becoming black on feet; underside buffy white; tail paler below than above. The black on the face is restricted to the upper part, around the eyes, the rest being pink, but the lips bluish.

Presbytis hosei everetti (Thomas, 1892)

1892 *Semnopithecus everetti* Thomas. Malaysia: Mount Kinabalu, Sabah.

DIAGNOSIS. This subspecies has an odd sexual dichromatism, unparalleled among langurs: in the male the black color is, as usual, restricted to the crown, whereas in the adult female it spreads to cover the forehead and nape, extending laterally to the ear, except for a whitish spot above the brows. Color of limbs, underside, tail, and facial skin as in nominotypical *hosei*.

DISTRIBUTION. Extends from about latitude 2°40′N to Mount Kinabalu and along the northwestern coast of Sabah into Brunei; in Sarawak it is confined to hilly country, being replaced on the coast itself by *P. h. hosei* (D. Brandon-Jones 1996). The localities closest to those of *P. h. hosei* given by Bran-

Table 19
Measurements (mm) of Subspecies of *P. hosei*, after D. Brandon-Jones (1996)

Character	P. h. everetti (19)	P. h. sabana (10)	P. h. canicrus (13)
Head + body length	448–648	480–557	430–510
Tail length	600–755	646–840	690–827

don-Jones are Labi Ridge (4°12′N, 114°35′E), Rumah Penghulu Bayak (4°00′N, 114°29′E), and Bukit Selikan (3°31′N, 114°04′E).

Presbytis hosei sabana (Thomas, 1893)

1893 *Semnopithecus sabanus* Thomas. Malaysia: Paitan, Sabah.

DIAGNOSIS. Crown gray, with only crest itself black; black color restricted to hands and feet, contrasting; underside grayish anteriorly, but white on belly and inner side of limbs. Facial skin reddish, except for bluish lips, with a sharply marked black patch on each side, between eye, nose, and mouth. Measurements in Table 19 show this to be about the same size as *P. h. everetti,* but longer-tailed.

DISTRIBUTION. According to D. Brandon-Jones (1996), this subspecies is confined to central and eastern Sabah, as far southwest as Kalabakan at 4°26′N, 117°29′E, although it may occur in far northeastern Kalimantan where its border with *P. h. everetti* is somewhere between 3 and 4°N.

Presbytis hosei canicrus Miller, 1934

1934 *Presbytis canicrus* Miller. "Dutch North-East Borneo."

DIAGNOSIS. Resembles *P. h. sabana* in general color, but crown wholly brownish; underside buffy white; face skin wholly blackish down to a line between mouth and ear, and below this whitish buff. As seen in Table 19, this seems to be a small but long-tailed form.

DISTRIBUTION. According to D. Brandon-Jones (1996), this is a subspecies of very restricted distribution, along the eastern coast from Kutai (0°24′N, 117°16′E) to Gunung Talisayan (1°40′N, 118°10′E), although there is a gap of about 1° between its known northern boundary and the known southern boundary of *P. h. everetti.*

Presbytis rubicunda (Müller, 1838)
Maroon Leaf Monkey

DIAGNOSIS. Strikingly maroon-red to brick red in color, the underside being only a lighter, more yel-

lowish or creamy tone and not sharply white. Facial skin blue-gray, except for upper lip and chin, which are pinkish. A frontal whorl, 200 mm behind brow, from which crown crest sweeps back and frontal fringe sweeps forward.

DISTRIBUTION. The whole of Borneo, and Karimata Islands.

NOTES. My study of the USNM, BM(NH), Bogor (MZB), Singapore (ZRC), and Sarawak Museum (SM) skins shows that the described subspecies are probably worth recognizing, mainly on general body color, and the east/west division on the basis of color of hands and feet, and there is an additional south/north division between "narrow-crested" and "broad-crested" races. The former have a U-shaped parting on the nape, from which the hairs supporting the crest from behind run forward in a narrow stream; in the latter there is only an indistinct whorl on the nape, from which the hairs of the back of the crest radiate out anteriorly.

Presbytis rubicunda rubicunda (Müller, 1838)

1838 *Semnopithecus rubicundus* Müller. Indonesia: Mount Sekumbang, southeast of Banjermasin, South Kalimantan.

DIAGNOSIS. Maroon to mahogany red, slightly paler on underparts and inner surfaces of limbs; hands and feet heavily washed with black, especially on digits; tail with scattered black hairs, especially toward tip. "Narrow-crested."

DISTRIBUTION. Southeastern Kalimantan, approximately east of the Sungai Barito and south of the Sungai Mahakam. Specimens from Samarinda and Longbleh have crests of intermediate type.

Presbytis rubicunda, undescribed subsp.

Like nominotypical *rubicunda,* but "broad-crested." It is found in northeastern Borneo, from about the Sungai Mahakam north; throughout Sabah, except where *P. r. chrysea* is found; west into Sarawak to the Kelabit uplands, Mount Mulud, and the IV Division.

Presbytis rubicunda rubida (Lyon, 1911)

1911 *Pygathrix rubicunda rubida* Lyon. Indonesia: mouth of Kendawangan River, southwestern Kalimantan.

DIAGNOSIS. More yellowed than *P. r. rubicunda*, with hands and feet not or only slightly blackened; "narrow-crested."

DISTRIBUTION. Southwestern Borneo, more or less south of the Sungai Kapuas and west of the Sungai Barito.

Presbytis rubicunda ignita Dollman, 1909

1909 *Presbytis ignita* Dollman. Malaysia: Mount Mulud, Sarawak.

DIAGNOSIS. "Broad-crested" like the undescribed subspecies (above), but more foxy red, with hands and feet light, not washed with black.

DISTRIBUTION. From north of the Kapuas, into Sarawak, as far as the Baram River and the borders of Brunei. Strictly speaking, the type locality is in the (fairly broad) intergradation zone; 14 skins from the Singapore and Kuching collections vary from foxy red to deep maroon, but their hands and feet have at most a slight trace of darkening.

Presbytis rubicunda chrysea Davis, 1962

1962 *Presbytis rubicunda chryseus* Davis. Malaysia: east of Kretam Kecil River, southeast end of Dewhurst Bay, Kinabatangan, Sabah.

DIAGNOSIS. Rich reddish gold, paler on abdomen and sides of thighs; digits buff, washed with brown; tail washed with brown at tip.

DISTRIBUTION. Apparently confined to a very small area of eastern Sabah; the undescribed subspecies (above) occurs at Sandakan, only 80 km away from the type locality.

Presbytis rubicunda carimatae Miller, 1906

1906 *Presbytis carimatae* Miller. Indonesia: Karimata Island.

DIAGNOSIS. Dark brick red, the hands and feet somewhat darkened; crown with a golden tinge; "narrow-crested."

Presbytis potenziani (Bonaparte, 1856)

Mentawai Langur or Joja

DIAGNOSIS. A small, erect, slightly forward-directed crest on crown; a ridge around chin and up to ear, where forward-directed interramal hair stream meets backward-directed cheek hair stream. Blackish above; tail entirely black; underside and inner side of thigh reddish orange, may be overlaid with black; inner side of upper arm whitish or reddish; throat, cheeks, and chin yellowish white, grading backward through gray to the reddish of the underside; brow band whitish; pubic region yellowish white in both sexes. Skin of face black, becoming somewhat depigmented around mouth. Skull more prognathous than other species, narrower interorbitally, and with a deeper postorbital constriction.

DISTRIBUTION. Mentawai Islands (predominantly Siberut, Sipura, Pagai Utara, and Pagai Selatan).

NOTES. This species has been monographed in detail by D. Brandon-Jones (1993). The more robust skull, as described above, is the reason why it has sometimes been thought to be intermediate between *Presbytis* and *Trachypithecus*, together with the misallocation of a golden *Trachypithecus* infant skin to the species (see Brandon-Jones [1993]): the infant is in fact typically *Presbytis*, white with a thick, dark dorsal stripe.

Presbytis potenziani potenziani (Bonaparte, 1856)

1856 *Semnopithecus potenziani* Bonaparte. "Tenasserim"; corrected by Thomas (1895) to the Mentawai Islands, and further restricted by Chasen and Kloss (1927) to Indonesia: Sipora Island, and amended to Pulau Pagai Utara (North Pagai Island) by D. Brandon-Jones (1993).
1867 *Semnopithecus chrysogaster* Peters.

DIAGNOSIS. Underparts bright reddish orange, with black-tipped hairs toward midline of chest.

DISTRIBUTION. Pagai Utara (North Pagai), Pagai Selatan (South Pagai), and Sipura (Sipora); as discussed by Brandon-Jones (1993), there may be slight differences between Sipura and Pagai Islands specimens.

Presbytis potenziani siberu (Chasen and Kloss, 1927)

1927 *Pithecus potenziani siberu* Chasen and Kloss. Indonesia: Siberut Island.

DIAGNOSIS. Underparts overlain with blackish, with long, black hair tips; white pubic patch more sharply marked.

Odd-Nosed Group

Whether this is more than an informal group is unclear; the only unifying features are the high intermembral index (more than 90), and that all have modifications to the nose, but in different ways in the *Pygathrix/Rhinopithecus* group (which has flaps of skin on the upper margins of the nostrils) and the *Nasalis/Simias* group (in which the nostrils are entire). In the first group, the face is short and wide and the nasal bones are reduced; in the second, the face is long and narrow, the interorbital pillar being unusually narrow for the subfamily, and the nasals are lengthened. An extensive analysis by Jablonski and Peng (1993) affirmed the monophyly of the first of these two groups, but Judith Caton (personal communication) doubts their close affinity, because of differences in the histology of the stomach.

Genus *Pygathrix* E. Geoffroy, 1812
  Doucs

1812 *Pygathrix* E. Geoffroy. *Simia nemaeus* Linnaeus, 1771.
1821 *Daunus* Gray. *Simia nemaeus* Linnaeus, 1771.

The facial skeleton is short and wide, with strong supraorbital ridges and a broad interorbital pillar. The body has black and white multibanded hairs, giving a light, almost bluish gray appearance. The tail, which extends to a small triangular patch on the croup, is white, and the male has a white spot at each anterior angle of the croup patch. In the male, a white stripe extends up between the ischial callosities; the neck and cheek whiskers are white.

Pygathrix nemaeus (Linnaeus, 1771)
  Red-shanked Douc

1771 *Simia nemaeus* Linnaeus. Vietnam: supposedly Cochin-China.
1926 *Presbytis nemaeus moi* Kloss. Vietnam: Langbian Peak, 1,650–1,975 m.

DIAGNOSIS. Facial skin yellow-brown, except around mouth and chin, where it is white. Eyes slanting, forming an angle of over 20° from the horizontal. White cheek whiskers long, copious, flaring to side and curving downward. Black frontal band broad. Underside dark with much agouti banding. Tail with copious tassel. Forearms and wrists white; hind shanks maroon red; fingers, toes, and thighs black.

DISTRIBUTION. From 20°N south to at least 14°N (Fooden 1996; Nadler 1997), but Nadler also recorded a hybrid between this and *P. nigripes* at about 13°N.

Pygathrix nigripes (Milne-Edwards, 1871)
  Black-shanked Douc

1871 *Semnopithecus nigripes* Milne-Edwards. Vietnam: Saigon.

DIAGNOSIS. Facial skin blue-gray, only circumocular patches being yellow-brown. Axis of eye at about 11.2° from the horizontal. Black frontal band broad, extending down in front of ears to connect with black of shoulders and forming a wedge from ear toward, but not reaching, mouth corners. White cheek whiskers short, thin. Body and crown dark gray-agouti. Belly dark brown-agouti. Tail white, as usual, with thin tassel. Hind shanks black.

DISTRIBUTION. Fooden (1996) and Nadler (1997) mapped the range between about 10°30'N and 14°30'N.

NOTES. Hybrids with *P. nemaeus* were recorded by Nadler (1997) in the area where their ranges overlap, between about 14°30'N and 13°N.

Pygathrix cinerea Nadler, 1997
  Gray-shanked Douc

1997 *Pygathrix nemaeus cinereus* Nadler. Vietnam: Play Ku, Gia Lai Province, 13°59'N, 108°00'E.

DIAGNOSIS. Facial skin yellow-brown, except around mouth and chin, where it is white. Axis of eye at 14.6° from the horizontal, intermediate between other two species. White cheek whiskers long, copious, flaring to side and curving downward. Black frontal band small, not connected to black of shoulders. Throat white, with a broad orange collar, bordered below by a black line that joins black patches on shoulders and inner surfaces of upper arms. Body, nape, and crown light gray-agouti, this color extending to arms (except inside forearms); legs dark gray-agouti, except for inner surface of thigh, which is black. Rest of underside very pale whitish gray with little banding. Hands and feet black, except that some agouti hairs occur on backs of hands. Tail white, as usual, with thin tassel.

DISTRIBUTION.   Recorded by Nadler (1997) from a few localities in central Vietnam, between 13°59′ and 14°46′N.

NOTES.   Nadler (1997) carefully distinguished this new taxon from the hybrids recorded between *P. nemaeus* and *P. nigripes*. Hybrids, as described by Fooden (1996) and Nadler (1997), are much more variable, and have something of the characters of the two parent species on the limbs and underside (i.e., they do not have gray shanks or whitish underparts), but they do occur in much the same general area as *P. cinerea*. DNA analysis (C. Roos, personal communication) confirmed that Nadler's new taxon is distinct and not merely a stable form of hybrid; in fact *P. cinerea* forms a sister clade to *P. nemaeus* + *P. nigripes*.

Genus *Rhinopithecus* Milne-Edwards, 1872
Snub-nosed Monkeys

1872 *Rhinopithecus* Milne-Edwards. *Semnopithecus roxellana* Milne-Edwards, 1870.
1924 *Presbytiscus* Pocock. *Rhinopithecus avunculus* Dollman, 1912.

The facial skeleton is more prognathous and wider, and the supraorbital torus stronger, than in *Pygathrix,* and the palate is longer. The underparts and inner side of limbs are lighter than the upper parts, as is the circumfacial zone. The ears are tufted. The nose is not merely upturned, but pressed flat against the interorbital pillar, so that the nostrils point directly forward, even somewhat upward.

I placed (Groves 1989 and elsewhere) *Rhinopith-* ecus in *Pygathrix* as a subgenus, but Jablonski and Peng (1993) argued strongly for its recognition as a full genus. This course is followed here, although the sister-group status of the two genera (at least as against *Trachypithecus,* the other odd-nosed monkeys not being tested) was affirmed by Jablonski and Peng (1993) on morphological grounds and by W. Wang et al. (1997) on genetic grounds.

Jablonski and Peng (1993) found that *R. roxellana* was sister species to *R. bieti* + *R. brelichi,* and that *R. avunculus* was sister to them all. Zhang and Ryder (1998), however, concluded that relationships between *R. bieti, R. roxellana,* and *R. avunculus* are more or less three-way. Under these circumstances, it seems inappropriate to divide the genus into subgenera or species-groups.

Rhinopithecus roxellana (Milne-Edwards, 1870)
Golden Snub-nosed Monkey

DIAGNOSIS.   Very large size; body yellowish red, varying from brown-toned to bright orange-red, overlain with black on back; limbs as body, with thick black stripe running down outer side (not to hands and feet), and whitish patch on back of thighs; crown dark in adult; underside yellowish white. Facial skin is white on muzzle (which is sparsely haired), pale blue around eyes; these two zones are separated, trefoil-like, by thickly furred wedges from cheeks across bridge of nose. Males are much more brightly colored, with contrasting black back, than females or young; adult males develop a swollen reddish wart-like excrescence on upper lip near corner of mouth.

DISTRIBUTION.   Mountains of central and western China: Sichuan, Ganssu, Hubei, and Shaanxi.

Rhinopithecus roxellana roxellana (Milne-Edwards, 1870)

1870 *Semnopithecus roxellana* Milne-Edwards. China: near Moupin, Sichuan.
1872 *Semnopithecus roxellanae* Milne-Edwards. Replacement.

DIAGNOSIS   (after Y. Wang et al. [1997]). Tail relatively short, about equal to head plus body length; dorsally dusky brown, especially on shoulders and limbs; braincase long and broad; dental arch wide.

DISTRIBUTION.   Western Sichuan and Ganssu.

Rhinopithecus roxellana qinlingensis Y. Wang, Jiang, and Li,
1998

1998   *Rhinopithecus roxellana qinlingensis* Wang, Jiang, and Li.
China: South of Qinling Mountains, Shaanxi.

DIAGNOSIS   (after Y. Wang et al. [1997]). Tail
longer, about 115% of head plus body length; pelage
golden red; braincase shorter, narrower; dental arch
narrower.

DISTRIBUTION.   The Qinling Range.

Rhinopithecus roxellana hubeiensis Y. Wang, Jiang, and Li,
1998

1998   *Rhinopithecus roxellana hubeiensis* Wang, Jiang, and Li.
China: Shennongjia, Hubei.

DIAGNOSIS   (after Y. Wang et al. [1997]). Tail longest, nearly 130% of head plus body length; dorsal
pelage more gray-brown; braincase long and wide as
in *R. r. roxellana* but dental arch narrow as in *qinlingensis*. No nasal bone: premaxillae meet and fuse
above nasal aperture.

DISTRIBUTION.   Shennongjia, in Hubei, on the borders with Sichuan and Shaanxi.

Rhinopithecus bieti Milne-Edwards, 1897
   Black Snub-nosed Monkey

1897   *Rhinopithecus bieti* Milne-Edwards. China: Kiape, a day's
journey from Atuntze, Yunnan.

DIAGNOSIS.   Upper parts black; underside striking
white, extending high on flanks and sides of neck,
and to backs of thighs and forward to form a circumfacial ring. Hair on thigh backs very long, wavy, especially in adult males; on proximal part of tail, hairs
are wavy and parted in dorsal midline. A thin, high,
forward-curved crest on crown; skin of face pale yellowish or greenish around eyes, bright pink or red on
muzzle, the fur wedges separating these zones not as
striking as in *R. roxellana*. As in *R. roxellana*, pelage is
much more contrastingly colored in adult males than
in females and young, with white thigh backs standing out especially strikingly in the field.

DISTRIBUTION.   Ridge of the Mekong-Salween divide, in Yunnan.

Rhinopithecus brelichi Thomas, 1903
   Gray Snub-nosed Monkey

1903   *Rhinopithecus brelichi* Thomas. China: Van Gin Shan
(Fanjinshan), Guizhou.

DIAGNOSIS.   Dark brown above, with white patch
on withers; underside and inner side of limbs bright
gingery red in adult male, becoming overlaid with
black on chest and belly. Hands and feet blackish.
Hairs of vertex reddish in front, and lie flat. Tail
nearly 150% of head plus body length, black, with
curly hairs at base; tip white. Cheeks and interramal
region black; face white, except for dark bluish zone
around eyes, lacking fur wedges.

DISTRIBUTION.   Restricted to Fanjinshan, south of
the Yangtze in Guizhou.

Rhinopithecus avunculus Dollman, 1912
   Tonkin Snub-nosed Langur

1912   *Rhinopithecus avunculus* Dollman. Vietnam: Yen Bay,
Songkoi River.

DIAGNOSIS.   More slenderly built than other species, with elongated digits. Upper parts black; underside white, this zone advancing high up on flanks,
around face, and down limbs, leaving only a thin,
black stripe down outer sides to hands and feet. An
orange patch on throat. Hairs of vertex black, and lie
flat. Face bluish white; cheeks and muzzle blue-
black; enlarged lips pink. Tail very long, nearly 150%
of head plus body length, with curly, intermixed
black and straw-colored hairs along upper surface,
and a white tip.

DISTRIBUTION.   Restricted to a few areas of far
northwestern Vietnam; now critically endangered.

Genus *Nasalis* E. Geoffroy, 1812
   Proboscis Monkey

1812   *Nasalis* E. Geoffroy. *Cercopithecus larvatus* van Wurmb,
1781.
1821   *Hanno* Gray. *Simia nasica* Schreber (recte Lacépède, 1799).

The skull is long and narrow, with elongated nasals and strong supraorbital ridges. The thumb is much less reduced than in other colobines. The nose turns forward from an early age and becomes prominent in the adult female and enormous and downturned in the adult male. Ischial callosities are united across the midline in the adult male.

Nasalis larvatus (van Wurmb, 1781)

Proboscis Monkey

1781 *Cercopithecus [sic] larvatus* van Wurmb. Indonesia: Pontianak, West Borneo.
1792 *Simia (Cercopithecus) capistratus* Kerr. Borneo.
1799 *Cercopithecus nasica* Lacépède. Borneo.
1828 *Nasalis recurvus* Vigors and Horsfield. Indonesia: Pontianak (see Chasen [1940]).
1940 *Nasalis larvatus orientalis* Chasen. Indonesia: Salim Batu, Bulungan, northeastern Kalimantan.

DIAGNOSIS.   Color brick red, more banded anteriorly; tail and triangular croup patch white, as in *Pygathrix* and *Trachypithecus vetulus*. Cheek whiskers lighter, yellower, adpressed and backswept.

DISTRIBUTION.   Whole of Borneo, in suitable areas (mainly mangrove and some riverine forests).

Genus *Simias* Miller, 1903

Pig-tailed Langur

1903 *Simias* Miller. *Simias concolor* Miller, 1903.

The nose is a smaller version of that of *Nasalis*. The tail is very short, upturned, curly, and nearly hairless.

As in *Nasalis*, the thumb is long, and the ischial callosities of the adult male meet in the midline.

Simias concolor Miller, 1903

Pig-tailed Langur or Simakobu

DIAGNOSIS.   There are two color morphs: black (with some agouti banding on anterior part of body) and creamy; white cheek patches.

DISTRIBUTION.   Mentawai Islands: Siberut, Sipura, Pagai Utara, Pagai Selatan, and some offshore islets.

Simias concolor concolor Miller, 1903

1903 *Simias concolor* Miller. Indonesia: South Pagai Island.

DIAGNOSIS.   Less dark, more brownish toned.

DISTRIBUTION.   Pagai group (including offshore islets) and Sipura.

Simias concolor siberu Chasen and Kloss, 1927

1927 *Simias concolor siberu* Chasen and Kloss. Indonesia: Siberut Island.

DIAGNOSIS.   Black color (in dark morph), with more banding on anterior parts; white cheeks more conspicuous.

DISTRIBUTION.   Restricted to Pulau Siberut.

Hominoids—Superfamily Hominoidea Gray, 1825

Goodman et al. (1998) found that the separation date of Hominoidea and Cercopithecoidea is within the range of many families, rather than superfamilies. If they are combined, the name Cercopithecoidea Gray, 1821, has priority.

Family Hylobatidae Gray, 1870

As recounted earlier (Chapter 2, in the section on Ages of the Different Ranks), the separation date of the family Hylobatidae from the Hominidae is rather recent (about 18 million years ago), but is still within the range of universally recognized families in the other orders of mammals.

Genus *Hylobates* Illiger, 1811
Gibbons

The basic four-way split within this genus is hard to resolve. Haimoff et al. (1982), using morphological and behavioral data, favored an initial separation of *Nomascus*, followed by *Symphalangus*, then *Bunopithecus*. A study on chromosome morphology linked *Symphalangus* to *Nomascus* to the exclusion of *Hylobates* (Tuinen and Ledbetter 1983). Garza and Woodruff (1992), using the mitochondrial cytochrome *b* gene, found a three-way split between subgenera *Hylobates, Symphalangus,* and *Nomascus* (they could not obtain samples of *Bunopithecus*). More recently Hayashi et al. (1995), using mtDNA fragments, and Zhang (1997), using the mitochondrial cytochrome *b* gene, found that *Nomascus* is sister group to the other three subgenera, although in the latter case the non-*Nomascus* clade has a bootstrap value of only 55%. Using molecular clock assumptions, Hayashi et al. (1995) calculated the divergence of subgenera *Hylobates, Symphalangus,* and *Nomascus* as about 6 million years ago (that is, earlier than the *Homo-Pan* split); if this finding is confirmed by further study, then full generic separation for the three subgenera must be considered.

One reason why generic separation is not favored at present is the position (and correct nomenclature!) of what is here still called *Bunopithecus*. Only Hall et al. (1998) have so far examined hoolocks; they found them to be about as different from other subgenera as they were from each other. Hall et al. (1998) confirmed the findings of Hayashi et al. (1995) that the four hylobatid "subgenera" are somewhat more different from each other than are *Homo* and *Pan*.

Marshall and Sugardjito (1986) gave detailed descriptions of gibbon pelage characters and, especially, vocalizations (illustrated with sonograms), and color photos of almost all taxa of gibbons, with descriptive notes, were given by Geissmann (1994).

SUBGENUS *HYLOBATES* ILLIGER, 1811

Lar-group Gibbons

1777 *Gibbon* Zimmermann. *Simia longimana* Schreber, 1775. Rejected by the International Commission on Zoological Nomenclature (1954b), Opinion 257.
1811 *Hylobates* Illiger. *Homo lar* Linnaeus, 1771.
1821 *Laratus* Gray. *Homo lar* Linnaeus, 1771.
1829 *Cheiron* Burnett. *Homo lar* Linnaeus, 1771.
1932 *Brachitanytes* Schultz. *Symphalangus klossii* Miller, 1903.

(After Weitzel et al. [1988].) The cranial length is small, generally 75–82 mm; the vault is low; the orbital rim is thickened; the facial profile is strongly sinuous; the crown hair is directed fanwise from the front of the scalp; there is no laryngeal sac; the intermembral index is below 136; the thoracic vertebrae usually number 13; and there is no color change at sexual maturity in the female (except for loss of pale cheek whiskers in *H. agilis*). The baculum is very short, maximum length 6 mm; uniquely in the genus there is no baubellum. Chromosomes: $2n = 44$.

Interrelationships among the various species of the subgenus seemed clear enough to me (Groves 1972) and to Haimoff et al. (1982): *H. klossii* was clearly, on morphological grounds, the sister species of the others. This is not the finding of molecular studies. For Garza and Woodruff (1992), studying the cytochrome *b* gene, the split is between *H. lar* and *H. agilis*, on the one hand, and *H. muelleri*, *H. pileatus*, and *H. klossii* on the other (*H. moloch* was not studied); for Hayashi et al. (1995), on long mt DNA fragments, *H. pileatus* was sister species to the rest, within which *H. lar* + *klossii*, *H. agilis*, and *H. moloch* form more or less a three-way split (in this case, *H. muelleri* was not studied). Hayashi et al. (1995) dated the splitting up of the subgenus *Hylobates*, on a molecular clock basis, as 3.5 million years ago.

There are hybrid zones between all taxa of subgenus *Hylobates* that have adjoining ranges. I had some thoughts (Groves 1993) on these hybrid zones and what they suggest about mate recognition signals. There is positive assortative mating in areas where *H. lar* has sharply distinct light and dark color

morphs, and this appears (on very slender, but suggestive, evidence at present) to hold in its hybrid zone with *H. pileatus* in Khao Yai National Park. In central Borneo, where there are two taxa that are similar in color but as different in vocalizations as are *H. lar* and *H. pileatus,* the hybrid zone is much wider. Polygynous pairing, extremely rare in purebred gibbons, seems more frequent in hybrid zones (S. Bricknell affirmed that this is the case in the Borneo hybrid zone as it is for Khao Yai).

Hylobates (Hylobates) lar (Linnaeus, 1771)

Lar Gibbon or
White-handed Gibbon

DIAGNOSIS. Hands and feet white, sharply contrasting with body color; a white ring around face; color either dark (brown or black) or pale (buff or creamy). Hair not elongated or laterally projecting over ears.

DISTRIBUTION. Northern Sumatra; Malaysia, except for a strip between the Perak and Mudah Rivers; north through Burma and Thailand, east of the Salween and west of the Mekong, into China. Formerly extended into southeastern Thailand, where it was parapatric with *H. pileatus* (and still forms a narrow hybrid zone with it in Khao Yai National Park).

Hylobates (Hylobates) lar lar (Linnaeus, 1771)

1771 *Homo lar* Linnaeus. Restricted to Malacca by Kloss (1929).
1774 *Simia longimana* Schreber. Malaysia: Malacca (Melaka).
1812 *Pithecus variegatus* E. Geoffroy. Malaysia: Malacca (Melaka).
1828 *Simia albimana* Vigors and Horsfield. Supposedly from Sumatra.

DIAGNOSIS. Pale phase creamy, dark morph medium brown, often with paler lumbar region and darker, browner legs. Light (gray) hair bases form at least half of hair length.

DISTRIBUTION. Malay Peninsula, south of the Perak River, and north of the Mudah River to 9°N.

NOTES. A neotype, BM(NH) 55.1488, has been designated for this taxon (International Commission on Zoological Nomenclature [1982b], Opinion 1219).

Hylobates (Hylobates) lar vestitus Miller, 1942

1929 *Hylobates lar albimanus* Kloss, 1929. Not of Vigors and Horsfield, 1828.
1942 *Hylobates albimanus vestitus* Miller. Indonesia: Blangnanga, 1,100 m, Aceh, Sumatra.

DIAGNOSIS. Always light brown, from golden to fawn to gray-brown; no dark morph. Somewhat darker ventrally and on crown; lumbar region lighter than trunk, limbs darker.

DISTRIBUTION. Sumatra, north of the latitude of Lake Toba.

Hylobates (Hylobates) lar entelloides I. Geoffroy, 1842

1842 *Hylobates entelloides* I. Geoffroy. Malay Peninsula about 12°N. Name placed on the Official List of Specific Names in Zoology (International Commission on Zoological Nomenclature [1982b], Opinion 1219).

DIAGNOSIS. Pale morph honey-colored, usually darker on legs; dark morph black. White of hands and feet not extending onto wrists and ankles. Hair bases only one-third the length of the hair.

DISTRIBUTION. From 10°N to around 15°N, in Thailand and Burma; it is the subspecies found in Khao Yai National Park, separated from the peninsular populations by the Chao Phraya plain (Brockelman 1985).

Hylobates (Hylobates) lar carpenteri Groves, 1968

1968 *Hylobates lar carpenteri* Groves. Thailand: Doi Angka, 1,040 m.

DIAGNOSIS. Pale morph uniform creamy white; dark morph nearly black. White face-ring exceptionally broad, with superolateral "angles" where the cheek and brow components meet. Often a white pubic tuft in dark morph. Hair bases, up to half of hair length, lighter gray. Hair very long, 79–103 mm compared with a maximum of 56 mm in more southerly subspecies.

DISTRIBUTION. From about 16°N (south of which it fuses with *H. l. entelloides*) to Chieng Dao, at 19°22'N, in Thailand.

Hylobates (Hylobates) lar yunnanensis Ma and Wang, 1986

1986 *Hylobates lar yunnanensis* Ma and Wang. China: Menglian, southwestern Yunnan, 2,000 m.

DIAGNOSIS. Hair very long, 120–150 mm; light hair bases short, only one-tenth to one-fifth the length of the hair. Hairs of pubic region dark brown or red brown.

DISTRIBUTION. Range described as from Menglian to Cangyuan (Ma and Wang 1986); the map shows this as in far southwestern Yunnan, on the Burmese and Thai borders from about 22 to 23°30'N.

Hylobates (Hylobates) agilis F. Cuvier, 1821
Agile Gibbon

1821 *Hylobates agilis* F. Cuvier. Indonesia: West Sumatra.
1828 *Hylobates rafflei* E. Geoffroy. Indonesia: West Sumatra.
1829 *Hylobates unko* Lesson. Indonesia: Sumatra. Restricted by Marshall and Sugardjito (1986:142) to "southwest shore of Selat Rupat, 1°40'N, 101°20'E, Riau."
1862 *Hylobates albo nigrescens* Ludeking. Indonesia: Agam, West Sumatra.
1862 *Hylobates albo griseus* Ludeking. Indonesia: Agam, West Sumatra.

DIAGNOSIS. Hands and feet colored as body or somewhat darker; either black (or maroon-brown) or pale (buff, gray, or creamy, with darker ventral surface) in color; a white brow band; males and juveniles have light (white or reddish white) cheek whiskers and partial beard in addition. Hairs somewhat elongated over ears.

DISTRIBUTION. Sumatra, south of the latitude of Lake Toba; Malay Peninsula, north of the Perak River, south of the Mudah River.

NOTES. In Barisan Range, the pale morph is more common; in eastern lowlands of Sumatra, and in the Malay Peninsula, black predominates.

Hylobates (Hylobates) albibarbis Lyon, 1911
Bornean White-bearded Gibbon

1911 *Hylobates mülleri albibarbis* Lyon. Indonesia: near Sukadana, Southwest Borneo.

DIAGNOSIS. Similar to *H. agilis* but with no black morph; pelage with light gray-brown tones, becoming golden toned on rump, with blackened ventral zone, hands, and feet; males often have a light genital tuft; dark crown-cap margined with buff.

DISTRIBUTION. Southwestern Borneo, south of the Sungai Kapuas and west of the Sungai Barito.

NOTES. For me (Groves 1972) the southwestern Bornean gibbon was simply a geographic variant of *H. muelleri;* it certainly differs in its pelage from the other variants of Bornean gibbons no more than they differ from each other. The finding (Marshall and Marshall 1976) that it has the *H. agilis* vocalization raised a dilemma: shall pelage or vocalization take precedence when allocating that population *(albibarbis)* to a species? Marshall and Sugardjito (1986) showed that *albibarbis* does have features like the light (nonblack) morph of *H. agilis,* although there are still differences (black hands and feet, black rather than brown ventral surface; it is "the most colorful and contrastingly patterned of gibbons" [Marshall and Sugardjito 1986:141]). Thus, the southwestern Bornean gibbon differs diagnostically from both *H. muelleri* and *H. agilis.*

According to Zhang (1997), this species is sister group to *H. lar,* but with only 76% bootstrap support; on the other hand, *H. agilis* was not studied.

Hylobates (Hylobates) muelleri Martin, 1841
Müller's Bornean Gibbon

DIAGNOSIS. Brown to gray, with dark cap and ventral surface, varying both individually and with age (becoming more extensive with age); light brow band often not well marked. Laterally directed hair markedly elongated over ears.

DISTRIBUTION. Borneo, except for southwestern portion.

NOTES. The information supplied by J. T. Marshall (personal communication) and reported in Weitzel et al. (1988), especially that some of the USNM specimens are mislabeled, helps to reinstate the described subspecies of *H. muelleri,* which now appear to have more homogeneity than I had hitherto supposed.

Hylobates (Hylobates) muelleri muelleri Martin, 1841
1841 *Hylobates mülleri* Martin. Indonesia: southeastern Borneo.

DIAGNOSIS. Dark color, with darkened ventral surface and crown-cap; hands and feet darker than body or limbs.

DISTRIBUTION. Southeastern Borneo: approximately, south of the Sungai Mahakam and west of the Sungai Barito.

Hylobates (Hylobates) muelleri funereus I. Geoffroy, 1850
1850 *Hylobates funereus* I. Geoffroy. Philippines: Sulu Islands.

DIAGNOSIS. Color very dark, dark brown or gray; hands and feet not darker than limbs, and sometimes lighter.

DISTRIBUTION. Northeastern Borneo: Sabah, south to the Sungai Mahakam and perhaps west to the Baram District and the IV Division of Sarawak; the reputed type locality is obviously in error, which may have arisen because Sabah was formerly part of the Sultanate of Sulu.

Hylobates (Hylobates) muelleri abbotti Kloss, 1929
1929 *Hylobates cinereus abbotti* Kloss. Indonesia: Pontianak, southwestern Borneo.

DIAGNOSIS. Color medium gray, with little ventral darkening or cap on crown; hands and feet not darkened.

DISTRIBUTION. Western Borneo, north of the Sungai Kapuas, and east as far as the Saribas District of Sarawak.

Hylobates (Hylobates) moloch (Audebert, 1797)
Silvery Gibbon

1797 *Simia moloch* Audebert. Indonesia: Java.
1799 *Simia leucisca* Schreber. Indonesia: Java.
1804 *Pithecus cinereus* Latreille. Indonesia: Java.
1949 *Hylobates lar pongoalsoni* Sody. Indonesia: Kali Kidang, Mount Slamet.

DIAGNOSIS. Silvery gray with at least a trace of black cap and venter; poorly expressed light face-ring, but a short, white beard on chin. Hair somewhat elongated over ears. Teeth large, with large cingula, unreduced third molars, unmodified Y-5 pattern, and commonly supernumerary cusps.

DISTRIBUTION. Java.

Hylobates (Hylobates) pileatus Gray, 1861
Pileated Gibbon

1861 *Hylobates pileatus* Gray. Cambodia.

DIAGNOSIS. Hands and feet white as in *H. lar;* female and juvenile buffy, with black on ventral surface, which spreads with age (ultimately down inner surfaces of limbs), and black crown-cap; adult male black; female and juvenile with white face-ring, adult male with white face-ring or just brow band, and white pubic tuft; long light to white hairs on sides of crown. This is a sister group of *H. muelleri* according to Zhang (1997).

DISTRIBUTION. Cambodia and southwestern Laos, west of the Mekong, and southeastern Thailand. The Mun River, a tributary of the Mekong, may have formed the northern barrier of this species, and the Bang Pakong River the eastern (Geissmann 1991).

NOTES. According to Geissmann (1991), in 1925 both this species as well as a single *H. lar* were reported at Sriracha, on the coast 80 km southeast of Bangkok. He suggested that the two existed along a belt of parapatry between this region and the Takhong River, an upper tributary of the Mun River; the Takhong overlap zone, where there is some hybridization, is probably the only part of this belt that still exists.

Hylobates (Hylobates) klossii (Miller, 1903)
Kloss Gibbon or
Bilou

1903 *Symphalangus klossii* Miller. Indonesia: South Pagai Island.

DIAGNOSIS. Color entirely black, with sparser pelage than others of the subgenus; bare area on throat; long pollex and hallux; mandibular symphysis sub-

vertical; nasals flat; foramen magnum facing somewhat backward; zygomata flattened.

DISTRIBUTION. Mentawai Islands: Siberut, Sipura, Pagai Utara, Pagai Selatan.

NOTES. Despite the striking morphological differences between this species and others of the subgenus, molecular data resolutely refuse to put it outside the rest of the group. In Garza and Woodruff's (1992) study and, most lately, in Zhang's (1997) analysis, this species is sister to the *pileatus/muelleri* clade, although with only 77% bootstrap support.

There may be geographic variation in the species, perhaps in size, and in the direction of the hair stream on the outer side of the forearm (Groves 1972).

?SUBGENUS *BUNOPITHECUS* MATTHEW AND GRANGER, 1923
Hoolock Gibbon

1923 *Bunopithecus* Matthew and Granger. *Bunopithecus sericus* Matthew and Granger, 1923, a fossil (middle [?] Pleistocene) form.

Hair is sparse compared with that of most members of subgenus *Hylobates*. Sexual dichromatism is present, as in *Nomascus:* juveniles are black with white brow streaks; this color is retained in the adult male but changes with maturity in the female to buffy brown with a white face-ring, continued around under the eyes as suborbital streaks. The nasals are convex, long, and pointed, giving a hooked external nose. The ears are small. A laryngeal sac is present, small, in both sexes. The baculum is long, 6.4–8.0 mm. Chromosomes: $2n = 38$.

For Zhang (1997), the three-way split between this subgenus, *Symphalangus,* and *Hylobates* could not be resolved.

My examination of the type of *Bunopithecus sericus* leads me to doubt whether it belongs to *Hylobates* at all; it probably represents an extinct genus of the Hylobatidae. The subgenus of the hoolock gibbon may therefore have to be renamed, and this is a major stumbling block to raising the subgenera of *Hylobates* to full generic status.

Hylobates (?Bunopithecus) hoolock (Harlan, 1834)
Hoolock Gibbon

DIAGNOSIS.   As for subgenus.

DISTRIBUTION.   East and south of the Brahmaputra in India, across northern Burma (south, in
Arakan, to Mount Victoria) into China, as far east as
the Salween.

Hylobates (?Bunopithecus) hoolock hoolock (Harlan, 1834)

1795 *Simia golock* Bechstein. Placed on the Official Index of
      Rejected and Invalid Specific Names in Zoology
      (International Commission on Zoological Nomenclature
      [1982b], Opinion 1219).
1834 *Simia hoolock* Harlan. India: Garo Hills, Assam.
1834 *Hylobates fuscus* Winslow Lewis. "Vicinity of Himalaya
      Mountains."
1837 *Hylobates choromandus* Ogilby. Locality unknown.
1840 *Hylobates scyritus* Ogilby. India: Assam.

DIAGNOSIS.   In black specimens, color jet black;
preputial tuft black or only faintly grizzled; white
brow streaks close together, connected by white
hairs; little white on chin or under eyes.

DISTRIBUTION.   West of the Chindwin River.

Hylobates (?Bunopithecus) hoolock leuconedys Groves, 1967

1967 *Hylobates hoolock leuconedys* Groves. Burma: Sumprabum,
      1,200 m.

DIAGNOSIS.   In black morph, back with brown
overlay (Marshall and Sugardjito 1986), preputial tuft
white; brow streaks well separated with no white
hairs between; chin and suborbital zone often with
white hairs. Adult female has somewhat lighter
hands and feet.

DISTRIBUTION.   East of the Chindwin; in China,
Ma and Wang (1986) mapped this taxon as far east as
the Salween River, north to nearly 26°N.

SUBGENUS *SYMPHALANGUS* GLOGER, 1841
Siamang

1841 *Symphalangus* Gloger. *Simia syndactylus* Raffles, 1821.
1843 *Siamanga* Gray. *Simia syndactylus* Raffles, 1821.

Size is very large, with cranial length over 90 mm;
the vault is low; orbital rims are somewhat raised;
the facial profile is sinuous; the crown hair is directed
fanwise from a whorl behind the brows; there is a
large laryngeal sac in both sexes; the intermembral
index is above 136; there are usually 13 thoracic vertebrae; all ages and sexes are black. The baculum in
a single adult male measured 14.5 mm long. Chromosomes: $2n = 50$.

Hylobates (Symphalangus) syndactylus (Raffles, 1821)
Siamang

1779 *Simia gibbon* C. Miller. Indonesia: Fort Marlborough,
      Bencoolen (= Bintuhan), Sumatra.
1821 *Simia syndactyla* Raffles. Indonesia: Bencoolen (Bintuhan),
      Sumatra. Given precedence over *Simia gibbon* (International
      Commission on Zoological Nomenclature [1982c],
      Opinion 1224).
1908 *Symphalangus syndactylus continentis* Thomas. Malaysia:
      Semangko Pass, 900 m, Selangor-Pahang border.
1911 [*Siamanga syndactylus*] *volzi* Pohl. Indonesia: Sago, Padang
      highlands, Sumatra.

DIAGNOSIS.   As for subgenus.

DISTRIBUTION.   Sumatra, Malay Peninsula at least
as far north as Perak.

SUBGENUS *NOMASCUS* MILLER, 1933
**Concolor-group Gibbons or
Crested Indochinese Gibbons**

1933 *Nomascus* Miller. *Hylobates leucogenys* Ogilby, 1840.

Members of this subgenus are slightly larger than
members of the subgenus *Hylobates;* the vault is high
and rounded; orbital rims are flattened; the facial
profile is straight; the crown hair is erect, elongated
in middle (forming a short tuft) in the male, at sides
in the female; there is a small laryngeal sac in the
male; the intermembral index is above 136; there are
usually 14 thoracic vertebrae; the adult male and
juvenile are black, the adult female is buffy with a
blacklongitudinal streak on the crown and black hair
around the face. The baculum is very long, 8.2–12.1
mm. Chromosomes: $2n = 52$.

I gave (Groves 1972) brief descriptions of the taxa
of *Nomascus,* at that time regarded as subspecies of a

single species, *H. concolor*, although its possible future splitting into full species was anticipated. Some of the differences among the taxa are not at all obvious in preserved flat skins, and Schilling (1984) and Geissmann (1994) added further distinguishing features.

Notes on the complexity and partially mosaic nature of character variation in this genus, including chromosome rearrangements, were given by Groves (1993).

Hylobates (Nomascus) concolor (Harlan, 1826)
   Concolor Gibbon or
   Black Crested Gibbon

DIAGNOSIS (partly after Ma and Wang [1986]). Hair fine, almost silky. Male and juvenile entirely black. Adult female with extensive black ventral zone, extending with age along inner aspect of limbs; crown streak black. Crown tuft present in both sexes. Frontals higher; nasals higher, narrower; M^3 subequal to M^1. Upper canine bulky, with long and deep mesial groove. Baculum short, 9.0–9.2 mm long (Groves and Wang 1990), wide at base, without anterior projection. Intermembral index 126–149. Compared with *H. leucogenys*, chromosome 7 apparently has a pericentric inversion (Couturier and Lernould 1991).

DISTRIBUTION. From about 20°N in Vietnam, north into Laos; and a small region in Laos on the east bank of the Mekong at 20°17′–20°25′N.

NOTES. It seems very doubtful that the long-lived male gibbon called "Zombie," of Twycross Zoo, is *H. hainanus* as has commonly been reported. His chromosomes, vocalizations, and so on have been extensively reported in the literature; in fact, our detailed knowledge of all-black crested gibbons is almost entirely based on him. It is therefore very important to deduce exactly what taxon he is; *H. c. concolor* seems, for many reasons, most probable.

Notes on subspeciation in *H. concolor* were given by Ma et al. (1988) and Geissmann (1989).

Hylobates (Nomascus) concolor concolor (Harlan, 1826)

1826 *Simia concolor* Harlan. Locality unknown; fixed as Vietnam: Chapa by Groves (1972).
1827 *Hylobates harlani* Lesson. Replacement for *concolor*.
1840 *Hylobates niger* Ogilby. Error for *concolor*.
1897 *Hylobates henrici* Pousargues. Vietnam: Lai Chau.

DIAGNOSIS. Hair 32–52 mm long on dorsum. Crown hair markedly elongated. Crown patch of adult female relatively short (80–120 mm [Groves 1972]), diamond-shaped with elongated posterior angle, very occasionally narrowly connected to black face-ring; adult female pale brown- or gray-yellow, with no light face-ring, and black ventral hair short and not extending much onto limbs. According to Ma and Wang (1986), ear is small, 28 mm long, and skull is relatively broad.

DISTRIBUTION. Between the Black and Red Rivers in Vietnam and China; Ma and Wang (1986) mapped this taxon north to 23°45′N, and in a couple of places (about 22°55′N and 23°20′N) as going over onto the left bank of the Red River.

Hylobates (Nomascus) concolor nasutus Kunkel d'Herculais, 1884

1884 *Hylobates nasutus* Kunkel d'Herculais. Vietnam: Along (= Ha Long) Bay.

DIAGNOSIS. The only modern evidence for this taxon is a female, "Patzi," formerly in Tierpark Berlin, captured in the hinterland of Hon Gai, east of the Red River in Vietnam at 20°57′N, 107°05′E (Geissmann 1989). She was buff brown, with thinner, much darker brown pelage on chest, separated from general body color on flanks by very pale gray hair, reaching up to form a circumfacial ring. The black face-ring was extended superolaterally into small tufts; crown patch was very broad, 67 mm on crown, joined to the black face-ring, and 300 mm long, extending down to cover shoulders, where it was 98 mm broad. Hair on withers was 68 mm long.

Geissmann (1989) proposed that "Patzi" might represent a separate taxon, from Ha Long Bay north of the Red River delta.

Hylobates (Nomascus) concolor lu Delacour, 1951

1951 *Hylobates concolor lu* Delacour. Laos: Ban Nam Khueng.

DIAGNOSIS. Hair 40–47 mm long on dorsum. Very difficult to separate from nominotypical *concolor,* although the available material of this small, isolated population suggests that some, at least, of the males have a silvery eye-ear streak (though I cannot detect it in all specimens, and Geissmann [1989] argued that such a stripe can occur in any *Nomascus* gibbons); females are grayer-toned, with extensive ventral blackening, if not quite to the extent of *H. c. furvogaster.* Crown patch in an adult female 130 mm long. May be slightly smaller in size than *H. c. concolor.*

DISTRIBUTION. A tiny area in a bend of the Mekong in Laos at about 20°N, completely separated from other subspecies.

Hylobates (Nomascus) concolor jingdongensis Ma and Wang, 1986

1986 *Hylobates concolor jingdongensis* Ma and Wang. China: Wenbu, Jingdong, central Yunnan (Wuliang Mountain), 1,800 m.

DIAGNOSIS. Crown hair less elongated than in nominotypical *concolor.* Crown patch of adult female tadpole-shaped, with a short posterior tail; adult female light golden yellow, with gray-white face-ring; black on ventral region as in *H. c. concolor.* Ear longer, 30–37 mm, and skull slightly narrower.

DISTRIBUTION. According to Ma and Wang (1986), this is found in a small region around Wuliang Mountain, between the Mekong and the Red River, from about 24 to 25°N.

Hylobates (Nomascus) concolor furvogaster Ma and Wang, 1986

1986 *Hylobates concolor furvogaster* Ma and Wang. China: Menglai, Cangyan, Yunnan (2,000 m).

DIAGNOSIS. Small in size. Crown hair, as in *H. c. jingdongensis,* not greatly elongated, but hair very long over ear. Crown patch of adult female figured as tadpole-shaped, 180 mm long by 95 mm wide, much larger than in any other subspecies; adult female

described as gray-yellow-brown (Ma and Wang 1986:400) or dark gray (Ma and Wang 1986:410), with a narrow gray-white face-ring and long, thick, black ventral hair that extends strongly along limbs and under armpits. Ear and skull not described.

DISTRIBUTION. This is the only *Nomascus* taxon occurring west of the Mekong; it occurs in a small region of Yunnan near the Thai border at 23°15′–23°40′N, 99°05′–99°29′E, only just to the north of the northernmost extension of *H. lar.*

NOTES. Thomas Geissmann (personal communication) thinks that the female specimens of this subspecies are subadult, and that this accounts for the unusually extensive black ventral zone.

Hylobates (Nomascus) hainanus Thomas, 1892
Hainan Gibbon

1892 *Hylobates hainanus* Thomas. China: Hainan.

DIAGNOSIS. Hair 35–75 mm long. Crown hair very little elongated. Crown patch of the adult female indistinct, 108–140 mm long; Ma and Wang (1986) illustrated it as short and very wide, oval with a somewhat elongated, pointed posterior end; adult female light brownish gray to brownish yellow, without complete face-ring either white or black, but some white hair on sides of face and chin, and no black on ventral surface at all. Size very small, skull length 102–105 mm, according to Ma and Wang (1986), compared with at least 111 mm in other taxa; juvenile skulls I have seen (no adults available) are, however, rather large.

DISTRIBUTION. Hainan Island; Dao (1983, 1985) reported its existence in Vietnam, northeast of the Red River, in Hoa Binh and Cao Bang Provinces.

NOTES. There seems no doubt about the correct identification of the mainland specimens ascribed to this subspecies. Dao (1985) reported a male and a female from Trung Kanh, Cao Bang Province; the male was entirely black, the female brown gold with no black on ventral surface, a small crown patch, and pale area on the cheek. A second female, from Chi Ne, Hoa Binh Province, was referred to this taxon by Dao (1983). These localities are well inland of the Ha Long Bay presumed distribution of the very different *H. concolor nasutus* (note that, by a failure of communication between the coauthors, Groves and Wang [1990], the name *nasutus* was mistakenly applied to mainland *hainanus*).

Hylobates (Nomascus) leucogenys Ogilby, 1840
  Northern White-cheeked Gibbon

1840 *Hylobates leucogenys* Ogilby. Thailand: "Siam." Restricted by Kloss (1929) to Pak Lay, Laos; amended by Fooden (1997) to Muang Khi, 18°27'N, 101°46'E.

DIAGNOSIS (partly after Ma and Wang [1986]). Hair coarser than in *H. concolor,* its length 40–70 mm. Crown crest very high in male, absent in female. Male and juvenile black with silvery hairs intermixed, and with white cheek whiskers, forming a narrow streak reaching level of tops of ears and not approaching the corners of the mouth. Female is creamy orange, occasionally off white, with few or no black hairs on ventral surface, dark streak on crown is dark brown (68–145 mm), and circumfacial hair is white. Frontals lower; nasals wide, flat; M³ the smallest upper molar; upper canine more slender, pointed, mesial groove at most short and shallow; baculum 9–12 mm long, with projection anteriorly. Some sexual dimorphism in size. Intermembral index 121–140.

DISTRIBUTION. Between the Mekong and the Black River, from 19°N in Laos and Vietnam, into China, just across the border. It overlaps with *H. concolor* in Luchun, Yunnan (Ma and Wang 1986), and, presumptively, the Ma River region, Vietnam (Dao 1983), but Fooden (1996) examined the evidence for sympatry in Vietnam and found it wanting.

Hylobates (Nomascus) siki Delacour, 1951
  Southern White-cheeked Gibbon

1951 *Hylobates concolor siki* Delacour. Vietnam: Thua Luu.

DIAGNOSIS. Hair short, 33–42 mm long. Color as in *H. leucogenys.* Male and juvenile have white cheek whiskers as in *H. leucogenys,* but they reach only to level of halfway up ears, with a pointed upper end, and extend to jaw angles and along the margin of the upper lips, and onto the sides of the chin. Female's crown patch and hair pattern resemble those of *H. gabriellae.* Circumfacial hair in adult female is white. Baculum only 7.5 mm long (but the full maturity of that specimen is not certain). Intermembral index 141 ($n = 1$). There is a reciprocal translocation between chromosomes 1 and 22, not seen in *H. leucogenys* nor apparently in *H. concolor* (Couturier and Lernould 1991).

DISTRIBUTION. Central Vietnam and Laos, between 15°45'N and about 20°N, according to Fooden (1996), who recorded an apparent overlap or interdigitation between the ranges of *H. siki* and *H. leucogenys* between about 19 and 20°N.

NOTES. There is some difference of opinion as to whether this is a subspecies of *H. leucogenys* or of *H. gabriellae.* The chromosomes of this taxon have been reported as different from those of both, and the facial pattern of the male is not precisely like either of them. Garza and Woodruff's (1992) and Zhang's (1997) mtDNA analyses both placed it on a clade with *H. leucogenys,* with a bootstrap value of 97% in the first case but only 86% in the second; Zhang (1997) recommended giving it species rank. To rank it as a full species, despite the probable interbreeding with *H. gabriellae* at Savannakhet (Groves 1974), seems the best way of cutting the Gordian knot.

Hylobates (Nomascus) gabriellae Thomas, 1909
  Red-cheeked Gibbon

1909 *Hylobates gabriellae* Thomas. Vietnam: Langbian, 460 m.

DIAGNOSIS. Hair short but not quite as short as in *H. siki,* 37–45 mm long. Male and juvenile black, with silvery hairs intermixed as in *H. leucogenys,* but have a brownish chest contrasting with the black body; pale reddish or reddish yellow cheek whiskers, reaching less than halfway up ears, with a rounded upper margin; they reach corners of mouth, but do not extend along lips or sides of chin. Adult female more red-gold and rarely has any white around face; crown cap in adult female triangular, pointed behind, only 65–89 mm long; no crown tuft in female. Cheek whiskers are noticeably brushed sideways in both sexes. Baculum 8.2 mm long, in shape like that of *H. concolor.* Very little sexual size difference. Canines slender. Intermembral index 137–150. There is the same reciprocal translocation between chromosomes 1 and 22 as in *H. siki,* and the same pericentric inversion on chromosome 7 as in *H. concolor* (Couturier and Lernould 1991).

DISTRIBUTION.   Eastern Cambodia, southernmost Laos, and southern Vietnam, from 15°30′N to the borders of the Mekong delta.

Family Hominidae Gray, 1825
Great Apes

Synonym is Pithecidae Gray, 1821, based on *Pithecus,* a name now suppressed, so fortunately the family-group name likewise cannot be used; also Simiadae and Simiina Gray, 1872 (based on *Simia,* likewise suppressed).

The sinking of the "family Pongidae" into the Hominidae is now widely adopted, and is obligatory on cladistic grounds unless the Pongidae are restricted to the orangutans (here, subfamily Ponginae).

SUBFAMILY PONGINAE ELLIOT, 1912

Genus *Pongo* Lacépède, 1799
Orangutans (strictly, orang hutan)

1799 *Pongo* Lacépède. *Pongo borneo* Lacépède, 1799.
1813 *Lophotus* Fischer. No type species; name given to flanged males, under the impression that they are a different animal from the (then) more familiar small juveniles.
1816 *Faunus* Oken. Unavailable (International Commission on Zoological Nomenclature 1956).
1828 *Macrobates* Bilberg. No type species.

The external phenotypic differences between Bornean and Sumatran orangutans were given by Weitzel et al. (1988), and cranial and other differences were discussed by Groves (1986); further cranial differences were elucidated by Röhrer-Ertl (1984). The two are clearly diagnosably different.

Uchida (1998b) found significant differences between the teeth of Sumatran and Bornean orangutans, and between those from north and south of the Kapuas River in western Borneo. Sumatran orangutans, compared with Bornean, have a relatively small paracone on both P^3 and M^1, M^1 that is larger than M^2 (instead of equal in size), and broader M_3. In Borneo, the upper molars are longer in the southwestern population compared with the western, M^1 metacone is smaller, P_4 has a larger talonid, and M_1 is longer and narrower, and in these features the southwestern orangs resemble those from Sumatra; the western population resembles the Sumatran in only one feature, the large entoconid on M_1, and is intermediate in a few others. In her canonical analysis (fig. 3), the two Bornean samples overlap, whereas the Sumatran does not overlap with either of the two Borneans.

The differences between Bornean and Sumatran orangutans have been brought to a head by the molecular data, which indicate a very long separation time between them. Protein differences are well above the level usually associated with subspecies difference (Janczewski et al. 1990). There is a fixed difference in chromosome 2, apparently in several amino acids, and in mtDNA cleavage sites (Ryder and Chemnick 1993). The mitochondrial control region and NADH1, COII, ATPase 6, ATPase 8, and cytochrome *b* genes are more different than those of *Pan troglodytes* from *P. paniscus,* for example (Xu and Arnason 1996). Ryder and Chemnick (1993) estimated the time since separation as about 1.5 million years.

The eighteenth-century history of knowledge of the orangutan, and its correct nomenclature, was elucidated by Groves and Holthuis (1985).

Pongo pygmaeus (Linnaeus, 1760)
**Bornean Orangutan or
Maias**

DIAGNOSIS.   Build stouter and more stocky than in the Sumatran species; pelage maroon; face more prognathous, figure-8 shaped; cheek pads of adult male large, forward-curving, sparsely haired; laryngeal sac large, extensive. Skull with a suborbital fossa; compared with *P. abelii*, interorbital pillar more prominent; foramen magnum shorter; brachial index above 100.

DISTRIBUTION. Borneo, but apparently absent from the southeast.

NOTES. The name *Simia satyrus* Linnaeus, 1758, was used as the prime reference for this species by Röhrer-Ertl, who argued that the correct name is therefore *Pongo satyrus*. This name, however, can never be used, because it was suppressed early in the twentieth century (International Commission on Zoological Nomenclature 1929).

I first raised the question (Groves 1986) of whether there is geographic variation within Borneo; I noted that skulls of males from south of the Sungai Kapuas are larger than those from north of it, and that limited evidence indicated that those from Sabah have very narrow biorbital breadth. Groves et al. (1992) found some multivariate separation between samples within Borneo, as much as between any of them and Sumatra specimens. Uchida (1998b) confirmed dental differences between samples from north and south of the Sungai Kapuas. I recently used multivariate analysis on eight craniometric variables to differentiate male samples from northwest Kalimantan (north of the Kapuas), southwest Kalimantan (south of the Kapuas), and Sabah. Anne Russon kindly sent me measurements of skulls confiscated in Samarinda and thought to be from the Kutai National Park; multivariate analysis located them within the Sabah group, which is geographically understandable. I here (Figure 6) reproduce a diagram of the first two discriminant functions for the adult male crania, with the two male Samarinda skulls included with the Sabah group. The three geographic groups are clearly separated. It is interesting that the Northwest Kalimantan group overlaps with the Sumatrans.

There is considerable time depth, as measured by DNA studies, within Bornean orangs (Xu and Arna-

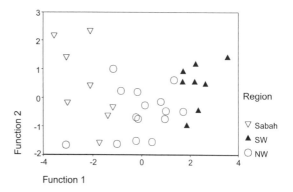

FIGURE 6. Diagram of the first two discriminant functions for adult male crania of *Pongo pygmaeus*. The first function accounts for 70%, the second for 23%, of the total variance.

son 1996), but the relationship of this to geography is unclear.

Pongo pygmaeus pygmaeus (Linnaeus, 1760)

1760 *Simia pygmaeus* Linnaeus. "Habitat in Africa." Fixed by Rothschild as Landak River (Groves et al. 1992).

1896 *Pithecus satyrus landakkensis* Selenka. Indonesia: Landak (Northwest Kalimantan).

1896 *P[ithecus] satyrus batangtuensis* Selenka. Indonesia: From right bank of the Ketungau River, upstream to right bank of Kapuas River, near Batangtu (Northwest Kalimantan).

1896 *P[ithecus] satyrus dadappensis* Selenka. Indonesia: Left bank of Katungau River, in region of Dadap Mountains, north of Genepai (Northwest Kalimantan).

1896 *P[ithecus] satyrus genepaiensis* Selenka. Indonesia: left bank of Katangau River, near Genepai Mountains (Northwest Kalimantan).

1896 *P[ithecus] satyrus skalauensis* Selenka. Indonesia: left bank of the Katangau, west of Dadap region (Northwest Kalimantan).

1896 *P[ithecus] satyrus rantaiensis* Selenka. Indonesia: Rantai, northwest of Skalau region (Northwest Kalimantan).

1896 *P[ithecus] tuakensis* Selenka. Indonesia: Upper Mrekai River (Northwest Kalimantan).

DIAGNOSIS. In general, medium in size, with relatively broad palate: in adult males, mean greatest skull length 236.0 mm (SD 7.5), palate breadth averaging 74.0 mm, bicanine breadth 72.1 mm ($n = 15$ in all cases).

DISTRIBUTION. The Northwest Kalimantan population, north of the Sungai Kapuas and (formerly) going northeast an unknown distance into Sarawak.

Pongo pygmaeus wurmbii (Tiedemann, 1808)

1808 *Simia wurmbii* Tiedemann. Indonesia: Sukadana, Southwest
    Borneo.

DIAGNOSIS. Very large, but with narrow palate: in
adult males, mean skull length 248.6 mm (SD 10.6);
palate breadth 74.4 mm, bicanine breadth 73.6 mm
($n = 7$ in all cases), so about the same as *P. p. pygmaeus* in a much larger skull.

DISTRIBUTION. The Southwest Kalimantan population, south of the Sungai Kapuas, west of the Sungai Barito.

Pongo pygmaeus morio (Owen, 1837)

1837 *Simia morio* Owen. Borneo.
1853 *Pithecus brookei* Blyth. Sarawak.
1853 *Pithecus owenii* Blyth. Sarawak.
1855 *Pithecus curtus* Blyth. Sarawak.

DIAGNOSIS. Very small, proportionally like *P. p.
pygmaeus*: adult males's mean skull length 232.7 mm
(SD 10.3), palate breadth 69.3 mm, bicanine breadth
70.4 mm ($n = 7$ in all cases).

DISTRIBUTION. From Sabah, south to the Sungai
Mahakam.

Pongo abelii Lesson, 1827
    **Sumatran Orangutan or
    Mawas**

1827 *Pongo abelii* Lesson. Sumatra.
1841 *Simia gigantica* Pearson. Sumatra.
1896 *P[ithecus] satyrus deliensis* Selenka. Indonesia: left streams of
    Langkat River and Deli (Sumatra, Aceh).
1896 *P[ithecus] satyrus langkatensis* Selenka. Alternative name for
    deliensis.
1896 *P[ithecus] satyrus abongensis* Selenka. Indonesia: north of
    Langkat River, near Abong-abong Mountains (Sumatra,
    Aceh).

DIAGNOSIS. More linear in build than Bornean
species; pelage more cinnamon-colored; adult male
with prominent beard and moustache, cheek flanges
flat, covered with downy hair; females with beard;
face more orthognathous, oval; skull lacking suborbital fossa; interorbital pillar flatter; foramen magnum longer; brachial index less than 100.

DISTRIBUTION. Sumatra; generally held to be restricted to the north, north of the latitude of Danau
Toba, but it has recently been argued that there is
convincing evidence of its former, and perhaps continuing, occurrence well down the west coast (Rijksen and Meijaard 1999).

SUBFAMILY HOMININAE GRAY, 1825
    African Great Apes

Although the evidence now seems fairly convincing
that *Gorilla* is sister to a *Pan-Homo* clade, the separation times seem to have been close, and I recognize
no tribes here.

Genus *Gorilla* I. Geoffroy, 1853
    Gorillas

1853 *Gorilla* I. Geoffroy. *Troglodytes gorilla* Savage and Wyman,
    1847.
1913 *Pseudogorilla* Elliot. *Gorilla mayema* Alix and Bouvier, 1877.

I recognized (Groves 1986 and elsewhere) three taxa
of gorillas (*gorilla, graueri,* and *beringei*) and classified
all three in the one species, because phenetic relationships among them in craniometrics and other
characters were rather linear: *beringei* and *gorilla*
seemed perfectly distinct, but *graueri* was in most respects intermediate. Uchida (1998a) found that the
relationship between the three taxa is rather more
three-way; in fact in a canonical analysis based on
lower molars (her fig. 7) *beringei* overlaps with both
of the other two, which do not overlap with each
other. The talonid of P_4 is relatively much smaller in
graueri than in the other two. Both *graueri* and *beringei,* however, have much larger cheekteeth, rather
smaller incisors, and more sexual dimorphism in
molar size than *gorilla,* with *beringei* being more extreme than *graueri;* and *beringei* has much more sexual dimorphism in the upper canines. Externally,
eastern and western gorillas are instantly recognizable. Differences between eastern and western gorillas in the mitochondrial D-loop are greater than between common and pygmy chimpanzees (Garner
and Ryder 1996). There seems no reason to retain the
two geographic isolates in the same species.

Gorilla gorilla (Savage, 1847)

Western Gorilla

DIAGNOSIS. Teeth relatively small; incisors relatively large; less sexual dimorphism in molar size; palate short, in adult males maximum length 121.5 mm; hair short, sparse on brows; color brown-gray, often red on crown; adult male becomes light gray on back and thighs; nostrils flared; a prominent "lip" above nasal septum. Mental foramen is usually single.

DISTRIBUTION. Western parts of the range of the genus: southeastern Nigeria, southern Cameroon, Río Muni, Gabon, Congo, Cabinda, Mayombe, southernmost Central African Republic, and an isolate in northern Congo-Zaire.

Gorilla gorilla gorilla (Savage, 1847)

1847 *Troglodytes gorilla* Savage. Gabon: Mpongwe. (Savage and Wyman are joint authors of the paper in which *T. gorilla* was described, but careful reading of the paper shows that Savage alone was responsible for the name.)

1848 *Troglodytes savagei* Owen. Gabon: Mpongwe.

1855 *Gorilla gina* I. Geoffroy. West coast of Africa.

1856 *Satyrus adrotes* Mayer. Replacement for *Troglodytes gorilla*.

1856 *Sat[yrus] africanus* Mayer. Replacement for *Troglodytes gorilla*.

1862 *Gorilla castaneiceps* Slack. Gabon?

1877 *Gorilla mayêma* Alix and Bouvier. Congo-Brazzaville: village of King Mayêma, 4°35′S.

1903 *Gorilla gigas* Haeckel. Cameroon: Yaounde.

1905 *Gorilla gorilla matschiei* Rothschild. Cameroon.

1905 *Gorilla jacobi* Matschie. Cameroon: Lobo-mouth station, Dja-Nyong junction.

1912 *Gorilla gorilla schwarzi* Fritze. Cameroon: Sogema Farm (actually Sogemafam, Dja River).

1914 *Gorilla hansmeyeri* Matschie. Cameroon: Assobam Road, west of Mokbe, south of Dume River.

1914 *Gorilla zenkeri* Matschie. Cameroon: Mbiawe, Lokundje River, six hours upstream from Bipindi.

1927 *G[orilla] uellensis* Schouteden. Congo-Zaire: Bondo region, Djabbir.

1927 *Gorilla gorilla halli* Rothschild. Equatorial Guinea: Río Muni, Punta Mbonda.

1943 *Gorilla (Pseudogorilla) ellioti* Frechkop. Gabon: Fernan Vaz.

DIAGNOSIS. Size larger than *G. g. diehli:* in adult males of coastal sample (smallest-sized), mean greatest skull length is 296.1 mm (SD 16.6), cranial length 196.2 mm (SD 13.7), face height 146.2 mm (SD 10.1); but relatively narrow, mean biorbital breadth 136.4 mm, and mean bizygomatic breadth 173.6 mm ($n = 71$ in all cases). Cameroon plateau gorillas average larger, mean greatest skull length 304.1 mm (SD 16.6) ($n = 112$), and Sangha valley gorillas ($n = 37$) are more like coastal.

DISTRIBUTION. These are the gorillas of Cameroon (south of the Sanaga River), south to the Congo River mouth and east to perhaps the Oubangui River (well across the Sangha River, at any rate); there is, or was, an isolated population at Djabbir, near Bondo, on the north bank of the Uele River.

NOTES. Time depths between mtDNA lineages within this subspecies are very great, indicating a long-standing demic population structure (Garner and Ryder 1996).

Gorilla gorilla diehli Matschie, 1904

1904 *Gorilla diehli* Matschie. Nigeria-Cameroon border: Mun-Aya, Cross River, 6°6′N, 9°20′E.

DIAGNOSIS. Small in size: in adult males mean greatest skull length is 283.2 mm (SD 13.6), cranial length 183.4 mm (SD 13.7), face height 139.5 mm (SD 7.4); but relatively broader, mean biorbital breadth 136.0 mm, bizygomatic breadth 176.4 mm (i.e., about the same as in the coastal gorillas, in the preceding subspecies, in a smaller-skulled form) ($n = 25$ in all cases).

DISTRIBUTION. A small area on the Nigeria-Cameroon border, extending a short distance on either side of the border in the upper Cross River forest highlands.

NOTES. Discriminant analysis of male crania, using 18 variables, separates Cross River gorillas on average, but not absolutely, from other *G. gorilla* (Figure 7). The first discriminant function accounts for 59% of the total variance, the second for 33%. Overall, 84% of Cross River specimens are correctly identified, higher than for any of the other three samples.

The recent rediscovery of the Cross River gorilla population has sparked a lot of interest and conserva-

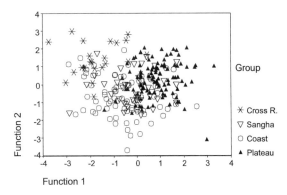

FIGURE 7. Diagram of the first two discriminant functions for adult male crania of *Gorilla gorilla,* using 18 variables. The first function accounts for 59% of the total variance, the second for 33%.

tion activity, and deservedly so; I here recognize the subspecies *diehli,* based on it, more as a recognition of these concerns and as a stimulus to further research than anything else.

Matschie (1904) described this subspecies on much the same cranial differences as I have cited above, but emphasized the low, broad planum nuchale. He had nine skulls from the Cross River, but referred one, which lacked this distinguishing feature, to *gorilla.*

Gorilla beringei Matschie, 1903

Eastern Gorilla

DIAGNOSIS. Cheekteeth large; incisors smaller than in *G. gorilla;* more sexual dimorphism in molar size; palate long, in adult males minimum length 126 mm; hair longer noticeably on brows; color jet black; adult male becomes gray-white on back only, spreading somewhat to thighs only later in life; nostrils narrow; no "lip" above nasal septum. Mental foramen is commonly double or multiple.

DISTRIBUTION. Eastern region of the range of the genus, in eastern Congo-Zaire extending into northern Rwanda and southwestern Uganda.

Gorilla beringei beringei Matschie, 1903

Mountain Gorilla

1903 *Gorilla beringeri* Matschie. Rwanda: Mount Sabinyo. Misspelling.

1905 *Gorilla beringei* Matschie. Justified emendation.
1917 *Gorilla beringei mikenensis* Lönnberg. Congo-Zaire: Mount Mikeno.

DIAGNOSIS. Size is very large; facial skeleton wider than in *G. b. graueri;* ascending ramus high, especially in females; jaw angles strongly flared in adults; mental foramen nearly always multiple; humerus relatively short; clavicle long; vertebral border of scapula sinuous; hallux long, more adducted to sole; pelage on scalp long, shaggy; nostrils angular, clearly outlined above; less padding above mouth. The talonid of P_4 is larger; strong sexual dimorphism in upper canines.

DISTRIBUTION. Virunga Volcanoes, straddling the border between Uganda, Rwanda, and Congo-Zaire; supposedly also the Bwindi-Impenetrable Forest in southwestern Uganda.

NOTES. The Bwindi Forest gorillas are not different from those of the Virunga Volcanoes in their mtDNA (Garner and Ryder 1996), but according to Sarmiento et al. (1996), they differ strongly in morphology: especially, they are smaller and more narrow-bodied with relatively longer limbs, hands, and feet; shorter pollex and hallux; narrower biorbital widths; and smaller teeth. They projected that further study would enable the Bwindi gorillas to be set aside as a distinct subspecies, or possibly they will have to be transferred to *G. b. graueri.*

Gorilla beringei graueri Matschie, 1914

Eastern Lowland Gorilla

1908 *Gorilla manyema* Rothschild. "South Congo" region. Lapsus for *mayêma* Alix and Bouvier, 1877.
1914 *Gorilla graueri* Matschie. Congo-Zaire: 80 km northwest of Boko, Wabembeland.
1927 *Gorilla gorilla rex-pygmaeorum* Schwarz. Congo-Zaire: Luofu, west of Lake Edward.

DIAGNOSIS. Compared with *G. b. beringei,* cranial size not quite as large; facial skeleton narrower; ascending ramus of female lower; jaw angles not strongly flared; mental foramen less frequently multiple; humerus longer; clavicle shorter; vertebral border of scapula straight; hallux shorter, more divergent; pelage on scalp shorter, less shaggy; nostrils rounded, not clearly outlined above; upper lip strongly padded. The talonid of P₄ is smaller; less sexual dimorphism in upper canines. Most of these features resemble those of *G. gorilla* (exceptions: shape of nostrils, shape of face, and premolar talonid size), and this taxon is indeed phenotypically rather intermediate between *G. b. beringei* and *G. gorilla;* but in these characters it is plesiomorphic. Few of the differences from *G. b. beringei* are absolute; indeed one population, that from the mountains west of Lake Albert (today apparently restricted to Mount Tshiaberimu), approaches *G. b. beringei* in several respects, such as flaring jaw angles and more adducted hallux.

DISTRIBUTION. Eastern Congo-Zaire, including both the lowlands east of the Lualaba and the Mitumba Range from Mount Tshiaberimu south to the Itombwe Mountains near Fizi.

NOTES. Mitochondrial D-loop sequences (from wild Kahuzi-Biega and captive samples, the latter apparently from Masisi) differ at 96% bootstrap value from *G. b. beringei* (Garner and Ryder 1996).

Genus *Pan* Oken, 1816

Chimpanzees

1812 *Troglodytes* E. Geoffroy. *Troglodytes niger* E. Geoffroy, 1812. Not of Vieillot, 1806 (Aves).
1816 *Pan* Oken. *Pan africanus* Oken, 1816.
1828 *Theranthropus* Brookes. *Troglodytes niger* E. Geoffroy.
1838 *Anthropopithecus* de Blainville. *Simia troglodytes* Blumenbach, 1799.
1841 *Hylanthropus* Gloger. *Simia troglodytes* Blumenbach, 1799.
1860 *Pseudanthropus* Reichenbach. Replacement for *Troglodytes.*
1866 *Pongo* Haeckel. Replacement for *Troglodytes.* Not of Lacépède, 1799 (Ponginae).
1866 *Engeco* Haeckel. *Simia troglodytes* Blumenbach, 1799.
1895 *Anthropithecus* Haeckel. Emendation of *Anthropopithecus.*
1905 *Fsihego* de Pauw. *Fsihego ituriensis* de Pauw, 1905.
1954 *Bonobo* Tratz and Heck. *Pan satyrus paniscus* Schwarz, 1929.

Should *Pan* be placed in *Homo* as a subgenus? Goodman et al. (1998) said yes: the chimpanzee and human lines separated only 6 million years ago, less than the time they set for the separation of two genera. Takahata and Satta (1997) made this separation even more recent, at 4.5 million years ago, which is well below the usual generic separation level (and Easteal and Herbert [1997] more recent still, at 4 million or less), although I argued earlier (in Chapter 2 in the section on Ages of the Different Ranks) that a mere 4 million would still be acceptable (just!) for generic separation. Nonetheless, the idea that chimpanzees belong to *Homo* has to be taken very seriously indeed. I have argued elsewhere against the lazy acceptance of the pre-Darwinian assumption that chimpanzees are "just animals"; if they were to be called *Homo* then we would be forced to concede that their treatment as anything but quasi-human would be simply immoral. A rose by any other name would smell as sweet? Not necessarily: the names we give to things, and to organisms, really does alter the way we think about them.

The work by Oken in which the name *Pan* first appeared has been suppressed (International Commission on Zoological Nomenclature 1956). This caused some dismay because of the almost universal use of *Pan* with that author and date, so the name was more recently made officially available (International Commission on Zoological Nomenclature 1985b).

Pan troglodytes (Blumenbach, 1799)
**Chimpanzee or
Common Chimpanzee**

DIAGNOSIS. Relatively heavily built, long-armed, short-legged; face pink at birth, turning black by maturity, circumbuccal region not contrastingly pink; a noticeable white chin beard in adults; skull large, vault flattened, with long prognathous face; brow ridges well developed, generally somewhat raised; white anal tuft shed after juvenile phase. Vulva backwardly placed; sexual cycle about 35 days, with sexual swelling occupying only 8–10 days midcycle.

DISTRIBUTION. From Senegal in the west to the Congo River, south to its mouth, and east across the Oubangui (always north of the Congo) to southeastern Congo-Zaire, southern Sudan, western Uganda, Rwanda, Burundi, and Tanzania, extending some distance east of the northern half of Lake Tanganyika.

NOTES. The characterizations of the three traditionally recognized subspecies follow Groves (1986), who was well aware of the limitations of the data. I am not at all confident that these distinctions would hold in large samples, let alone in every individual. A large craniometric data set I collected awaits full analysis and has been shared with a number of colleagues. In preliminary studies using most of this material, Groves et al. (1992) and Shea et al. (1993) found that there is extensive overlap in a multivariate craniometric study of this species, although on size-corrected data (female skulls) *P. t. schweinfurthii* stands out somewhat more from the other two subspecies recognized at that time than they do from each other, and overlaps slightly more with *P. paniscus;* more than half the sample of *P. t. verus* and *P. t. troglodytes* was correctly classified, but in some analyses less than half of the sample of *P. t. troglodytes.* Some other samples, especially within *P. t. schweinfurthii,* were rather distinctive craniometrically (Groves et al. 1992).

And yet Morin et al. (1994), sequencing the mitochondrial control and cytochrome *b* regions, found that the three "traditional" subspecies have their own unique lineages. Those of *P. t. verus* are so different from those of the other two (if the substitution rate is 20.4% per million years, then the *verus* lineage became separate 1.58 ma) that the authors proposed to raise this taxon to specific rank, as *Pan verus;* note, however, that all three are diagnosably different in this region of DNA, so objectively there would be three species, not just two. Gonder et al. (1997) sequenced chimpanzees from western Nigeria and the Nigeria-Cameroon border, finding that they form a fourth clade, closer to *verus.*

Should one contemplate raising to species rank three taxa between which one is hard-pressed to find any phenotypic differences at all (let alone fixed ones!)? I also have in mind the strictures on the subject of mitochondrial DNA that I raised earlier in the book, derived from my reading of Avise (1994). We may discover fixed differences in nuclear DNA, which, of course, if in one or more translated sequences would be expected to show up in some phenotypic character. Meanwhile one acknowledges the intense biogeographic and conservation implications of Morin et al.'s (1994) findings, but it is premature to permit them to affect chimpanzee taxonomy.

Stimulated by the findings of Gonder et al. (1997), I examined the distribution of some cranial nonmetrical features in my database, separating their Nigerian taxon *(vellerosus)* from the others (and, indeed, skulls from both sides of the Niger are alike, with due acknowledgment of small sample sizes). Some are traditionally examined variants (pterion patterns, exsutural bones, and so on); Vandebroek (1969) claimed that there are differences in the pattern of bones on the medial wall of the orbit, so this was examined too. The data for an ethmo-lacrimal suture in the orbit, presence of an extra bone at lambda, and tendency for the infraorbital foramen to be multiple are given in Table 20. Vandebroek's (1969) claim is completely vindicated. Bones along the lambdoid suture vary much as the lambda bones, but less markedly.

One notes in these data that, in agreement with Morin et al. (1994), *P. t. verus* is more different from the other subspecies than they are from each other and, in agreement with Gonder at al. (1997), the Nigerian subspecies is different, but its nonmetrical frequencies are like those of *P. t. troglodytes* and *schweinfurthii.*

According to Braga (1998), the timing of the closure of the incisive (premaxillary-maxillary) suture differs in the three (conventional) subspecies, especially its anterior portion, which closes much later, often after the eruption of the third molars, in *P. t. verus.*

Pan troglodytes troglodytes (Blumenbach, 1799)

1758 *Simia satyrus* Linnaeus (in part). Name suppressed by the International Commission on Zoological Nomenclature (1929), Opinion 114.
?1792 *Simia satyrus pongo* Kerr. "Java and interior parts of Guinea."
?1792 *Simia satyrus jocko* Kerr. No locality.
1799 *Simia troglodytes* Blumenbach. "Angola."
1812 *Troglodytes niger* E. Geoffroy. Supposedly, coast of Angola.
1816 *Pan africanus* Oken. "Congo, Angola, Guinea, Sierra Leone."
1831 *Troglodytes leucoprymnus* Lesson. "Coast of Guinea."
1840 *Anthropopithecus pan* Lesson. No locality.
1855 *Troglodytes tschego* Duvernoy. West coast of Africa.

Table 20

Distribution of Some Cranial Nonmetrical Features in Chimpanzees

Character	P. t. verus	P. t. vellerosus	P. t. troglodytes	P. t. schweinfurthii
Ethmo-lacrimal suture				
Percent	20.5	85.7	82.9	85.9
n (sides)	44	21	211	92
Bone at lambda				
Percent	56.3	21.1	23.7	28.1
n	32	19	93	64
No. of infraorbital foramina				
One	26.4	10.3	18.5	16.7
Two	45.9	41.4	47.6	63.5
Three or more	27.7	48.3	33.9	19.8
n (sides)	148	87	439	263

1856 *Satyrus lagaros* Mayer. No locality.
1856 *Satyrus chimpanse* Mayer. No locality.
1860 *Troglodytes calvus* du Chaillu. Gabon: south of Cape Lopez.
1860 *Troglodytes kooloo-kamba* du Chaillu. Gabon: upper Ovenga River, Ashankolo Mountains.
1866 *Troglodytes aubryi* Gratiolet and Alix. Gabon?
1870 *Pseudanthropus fuliginosus* Schaufuss. Congo-Brazzaville.
1876 *Anthropopithecus angustimanus* Brehm. Nomen nudum.
1895 *Anthropopithecus fuscus* Meyer. No locality.
1899 *Troglodytes livingstonii* Selenka. Nomen nudum.
1903 *Anthropithecus mafuca* Haeckel. Gabon: Loango Coast. Nomen nudum.
1905 *Simia pygmaeus raripilosus* Rothschild. Congo-Brazzaville.
1914 *Anthropopithecus reuteri* Matschie. Cameroon: Kadei, Dume River.
1914 *Anthropopithecus ochroleucus* Matschie. Cameroon: upper Lobo River, north of Sangmelima.
1919 *Anthropopithecus schneideri* Matschie. Gabon: Fernan Vaz.
1919 *Anthropopithecus pusillus* Matschie. Gabon: lower Ogooue River.
1932 *Anthropopithecus heckii* Koch. Cameroon (hinterland).

DIAGNOSIS. Facial color in juvenile pink, then develops tan spots and turns deep black with maturity. Scalp parting poorly marked; midline hair runs caudad; scalp thinly haired from an early age, becoming very bald especially in females, in which the bald patch extends onto temples. White chin beard long, sparse; sideburns long, directed downward. Head very broad; occiput less steeply descending than in other subspecies; brow ridges run straight across, thickened especially at glabella; muzzle less broad; facial profile very concave. Size large; very sexually

dimorphic; limbs long compared with skull size (Groves et al. 1992).

Weights of three adult males from the Bipindi District, Cameroon, according to Powell-Cotton Museum (PCM) records, are 50–70 kg (mean 60); weights of eighteen adult males in the Franceville Medical Research Centre, Gabon, kindly communicated by Caroline Tutin, vary from 41.5 to 61.5 kg (mean 51.79). Both these means are well above most of those recorded for other subspecies, though the ranges overlap. Two females from Bipindi are both 50 kg and one from Batouri is 42.25 kg; nineteen females from the Franceville colony are 34.8–58.8 kg (mean 43.8) (courtesy of Caroline Tutin).

Head plus body length, from Powell-Cotton Museum records, is 800–845 mm (mean 819.2) for five males from the Batouri District, and 879–959 mm (mean 914.3) for three from the Bipindi District. Thirty-one females from Batouri are 702–892 mm (mean 795.6), and five from Bipindi are 838–905 mm (mean 871.2).

This subspecies has a high frequency of an ethmo-lacrimal suture, and low frequencies of lambda and lambdoid bones.

DISTRIBUTION. Central part of range: from Congo River mouth north into Cameroon, conventionally into Nigeria east of the River Niger, and into the

Central African Republic east to the River Oubangui; but boundaries with other subspecies are unclear. It is probable that all the names listed in the synonymy do refer to Central African chimpanzees, but the proposition should be tested if possible.

Pan troglodytes verus Schwarz, 1934

1904 *Simia chimpanse* Matschie. No locality. Not *Satyrus chimpanse* Mayer, 1856.
1934 *Pan satyrus verus* Schwarz. Sierra Leone: Karima District, Sanda Magbolonto chiefdom.

DIAGNOSIS.   Facial color dark around eyes, more pinkish elsewhere in juvenile, retaining some contrast in adult. Scalp parting in juvenile, hair even in adult diverging to either side. Bald spot triangular in male, beginning as widening of parting; in female, very wide, extending down onto temples. White chin beard thick, full, rounded; sideburns long, directed downward. Head broad, flat-topped; occiput steeply descending; brow ridges raised, arched over each eye, and still prominent lateral to eye; muzzle broad; facial profile somewhat concave. Body proportions and size apparently like those of *P. t. schweinfurthii* (Groves et al. 1992).

Two males, according to museum records, weighed 46.3 and 48.5 kg; a single female was only 21.2 kg (Morbeck and Zihlman 1989).

This subspecies has a very low frequency of an ethmo-lacrimal suture and very high frequencies of extra bones at lambda and on the lambdoid suture. The frequency of a single infraorbital foramen bilaterally is higher than in other subspecies.

DISTRIBUTION.   Senegal to the Volta or the Dahomey Gap. The eastern boundary is traditionally given as the Niger River, but Gonder et al. (1997) found that mtDNA sequences from specimens from western Nigeria (west of the river) are different from those of *P. t. verus* (see next subspecies).

Pan troglodytes vellerosus (Gray, 1862)

1862 *Troglodytes vellerosus* Gray. Cameroon: Mount Cameroon.
1914 *Anthropopithecus ellioti* Matschie. Cameroon: Basho.
1914 *Anthropopithecus oertzeni* Matschie. Cameroon: Basho.

1919 *Anthropopithecus papio* Matschie. Cameroon: Barombi, Elephant Lake.

This subspecies is recognized on the evidence of Gonder et al. (1997), who found that sequences from the control region of mtDNA obtained from specimens from Ondo State forests (western Nigeria) and from three localities on the Nigeria-Cameroon border were alike and formed a sister clade to corresponding sequences of *P. t. verus* (from Senegal, Liberia, and Ivory Coast). The authors proposed, on general zoogeographic grounds, that the Sanaga River is the most likely boundary with *P. t. troglodytes*.

From my data, skulls of this subspecies are unlike those of *P. t. verus,* but resemble those of *P. t. troglodytes* and *P. t. schweinfurthii* in their high frequency of an ethmo-lacrimal suture and low frequency of occurrence of bones at lambda and on the lambdoid suture. They have a higher average number of infraorbital foramina than any other subspecies. In these features, skulls from western Nigeria, though few in number, seem to be like those from the Cross River District, Mount Cameroon, and the Edea/Ongue Creek region of the Cameroon coast north of the Sanaga River. I have not seen any of these chimpanzees alive, nor any photographs of them.

Assuming that chimpanzees from Mount Cameroon are of the same subspecies as those from Nigeria, as argued by Gonder et al. (1997), then the prior available name for this subspecies is *vellerosus* Gray, 1862. Basho, type locality of two of Matschie's names, is on the Cross River, so these two names certainly apply to this subspecies. Elephant Lake is in the highlands north of Mount Cameroon, so the name *papio* likewise almost certainly applies to this subspecies.

Pan troglodytes schweinfurthii (Giglioli, 1872)

1872 *Troglodytes schweinfurthii* Giglioli. Congo-Zaire: upper Uele, Niam-niam country.
1887 *Troglodytes niger* var. *marungensis* Noack. Congo-Zaire: Marungu, west of Lake Tanganyika.
1899 *Troglodytes livingstonii* Selenka. Nomen nudum.
1905 *Fsihego ituriensis* de Pauw. Congo-Zaire: Ituri.
1912 *Simia (Anthropopithecus) nahani* Matschie. Congo-Zaire: Aruwimi River, Banalia.

1912 *Simia (Anthropopithecus) ituricus* Matschie. Congo-Zaire: Ituri River, Makala-Avakubi road.

1912 *Simia (Anthropopithecus) cottoni* Matschie. Uganda: Ishasha River.

1912 *Simia (Anthropopithecus) adolfi-friederici* Matschie. Rwanda: Bugoye Forest.

1912 *Simia (Anthropopithecus) kooloo-kamba yambuyae* Matschie. Congo-Zaire: lower Aruwimi River, Yambuya.

1914 *Anthropopithecus purschei* Matschie. Tanzania: between Lake Kivu and Lake Luhondo, Tshingogo Forest.

1914 *Anthropopithecus pfeifferi* Matschie. Tanzania-Burundi border: east of Ruzizi, near source of Akanyaru River.

1914 *Anthropopithecus graueri* Matschie. Congo-Zaire: 80 km northwest of Boko, Wabembe country.

1914 *Anthropopithecus calvescens* Matschie. Congo-Zaire: Luama River, between Niembo and Kabambare.

1914 *Anthropopithecus castanomale* Matschie. Burundi: northeast shore of Lake Tanganyika.

1914 *Anthropopithecus schubotzi* Matschie. Congo-Zaire: upper Ituri, between Kilo and Irumu.

1914 *Anthropopithecus steindachneri* Lorenz. Congo-Zaire: Moera, north of Beni.

DIAGNOSIS. Face in juvenile pink, with no freckling; dark, but not usually deep black, in adult. Scalp parting poor in juvenile; hair in midline runs caudad, in male gradually becoming thinner all over, in female, never a very wide bald spot, never extending down onto temples. White chin beard full but straggly; sideburns full, backswept. Head more rounded than in other subspecies; occiput elongated, not steep; brow ridges thinner, run straight across, with marked glabellar prominence, and not prominent lateral to eye; muzzle narrow; facial profile fairly straight. Less sexually dimorphic than *P. t. troglodytes,* smaller, and shorter limbed (Groves et al. 1992).

According to Morbeck and Zihlman (1989), six males from Mahale weighed 30.3–52.0 kg (mean 42.0), and nine from Gombe 31.8–49.5 kg (39.5). Records attached to a male from Tshibati (north of Lake Kivu) in the Zurich collection (UZ) give the weight as 36.5 kg, well within the Tanzanian ranges, whereas three males from Kivu Province (Tervuren [MRAC] and Harvard [MCZ] collections) are much larger, 52.5 and 61.0 kg. Eight females from Mahale were 30.0–45.5 kg (35.2), and six from Gombe 22.7–35.5 kg (29.8); one from Tshibati is 29.3 kg; but one from Kivu in the Tervuren collection is 45 kg. These figures suggest that there may be noticeable varia-

tions in body size across the range of the subspecies.

Head plus body lengths for seven males from Faradje and Avakubi (on the Uele River) were 770–925 mm (mean 834) (J. A. Allen 1925); three males from Kivu Province (Tervuren and Harvard) were 772–800 mm, and one from Tshibati (Zurich collection) 900 mm. Four females from the Uele District were 700–850 mm (783) (J. A. Allen 1925), and one from Kivu was 785 mm. These figures confirm that this subspecies is somewhat smaller than *P. t. troglodytes.*

In its cranial nonmetrical features, this subspecies tends to resemble *P. t. troglodytes.*

DISTRIBUTION. Congo-Zaire, north of the River Congo and east of the River Lualaba, into Sudan, Uganda, Rwanda, Burundi, and Tanzania. Groves et al. (1992) noted considerable geographic variation in canonical analysis of skull measurements, and it may be that more than one subspecies should really be recognized.

Pan paniscus Schwarz, 1929
Bonobo or
Pygmy Chimpanzee or
Gracile Chimpanzee

1929 *Pan satyrus paniscus* Schwarz. Congo-Zaire: 30 km south of Befale, south of upper Maringa River.

DIAGNOSIS. Relatively slenderly built, short-armed, long-legged, with especially small head; face black from birth, with pink mouth; hardly any chin beard; skull short, very rounded, short-faced; brow ridges slight, not raised; white juvenile tuft retained into adulthood. Vulva more forwardly placed; sexual cycle long, over 40 days, with sexual swelling occupying most of the cycle. Cranial nonmetrical features tend to resemble those of *P. troglodytes* subspecies other than *verus.* The pterion ($n = 35$) is more usually (57.1%) of sphenoparietal pattern, whereas in *P. troglodytes* ($n = 288$) it is overwhelmingly (99.0%) of frontotemporal type. The anterior portion of the incisive suture closes much earlier than in *P. troglodytes* (Braga 1998).

DISTRIBUTION. Congo-Zaire, south of the great bend of the Congo River.

Genus *Homo* Linnaeus, 1758

Humans

1758 *Homo* Linnaeus. *Homo sapiens* Linnaeus, 1758.

1894 *Pithecanthropus* Dubois. *Anthropopithecus erectus* Dubois, 1892. Trinil, Java: early or middle Pleistocene.

1900 *Proanthropus* Wilser. *Anthropopithecus erectus* Dubois, 1892. Trinil, Java: early or middle Pleistocene.

1907 *Palaeanthropus* Bonarelli. *Homo heidelbergensis* Schoetensack, 1908. Mauer, Germany: middle Pleistocene.

1909 *Pseudhomo* Ameghino. *Homo heidelbergensis* Schoetensack, 1908. Mauer, Germany: middle Pleistocene.

1911 *Notanthropus* Sergi. *Notanthropus eurafricanus* Sergi, 1911. Combe Capelle, Eguisheim, Grimaldi, Laugerie-Basse, and Chancelade, France (Combe Capelle selected by Campbell [1965]): late Pleistocene.

1915 *Archanthropus* Arldt. *Homo neanderthalensis* King, 1864. Neandertal, Germany: late Pleistocene.

1926 *Anthropus* Boyd Dawkins. *Homo neanderthalensis* King, 1864. Neandertal, Germany: late Pleistocene.

1927 *Sinanthropus* Black and Zdansky. *Sinanthropus pekinensis* Black and Zdansky, 1927. Zhoukoudian, China: middle Pleistocene.

1928 *Cyphanthropus* Pycraft. *Homo rhodensiensis* Woodward, 1921. Kabwe, Zambia: middle Pleistocene.

1932 *Praehomo* von Eickstedt. *Anthropopithecus erectus* Dubois, 1892. Trinil, Java: early or middle Pleistocene.

1932 *Javanthropus* Oppenoorth. *Homo (Javanthropus) soloensis* Oppenoorth, 1932. Ngandong, Java: middle or late Pleistocene.

1938 *Africanthropus* Weinert. *Palaeoanthropus njarasensis* Reck and Kohl-Larsen, 1936. Eyasi, Tanzania: middle Pleistocene.

1943 *Maueranthropus* Montandon. *Homo heidelbergensis* Schoetensack, 1908. Mauer, Germany: middle Pleistocene.

1944 *Meganthropus* Weidenreich. *Meganthropus palaeojavanicus* Weidenreich, 1944. Sangiran, Java: early Pleistocene.

1948 *Nipponanthropus* Hasebe. *Nipponanthropus akasiensis* Hasebe, 1948. Nishiyagi, Japan: probably middle Pleistocene.

1949 *Telanthropus* Broom and Robinson. *Telanthropus capensis* Broom and Robinson, 1949. Swartkrans, South Africa: early Pleistocene.

1950 *Europanthropus* Wüst. *Homo heidelbergensis* Schoetensack, 1908. Mauer, Germany: middle Pleistocene.

1954 *Atlanthropus* Arambourg. *Atlanthropus mauritanicus* Arambourg, 1954. Tighenif, Algeria: early or middle Pleistocene.

1965 *Tchadanthropus* Coppens. *Tchadanthropus uxoris* Coppens, 1965. Yayo, Chad: middle Pleistocene.

For convenience, all generic synonyms known to me (mostly after Campbell [1965]) are listed here. *Homo* needs no special definition, because I am not dealing here with the related fossil genera *Ardipithecus*, *Australopithecus*, and *Paranthropus*. It may be, how-ever, that ultimately all these genera, and probably *Pan* as well, will be combined into one, in which case the name *Homo* takes priority, and the composite genus will need to be diagnosed with respect to *Gorilla*.

Homo sapiens Linnaeus, 1758

Modern Humans

1758 *Homo sapiens* Linnaeus. Based on humans from Europe, Africa, Asia, and America. No type specimen exists, and no lectotype has ever been proposed.

1758 *Homo troglodytes* Linnaeus. Unavailable: based on a hypothetical concept.

1758 *Homo monstrosus* Linnaeus. Unavailable: based on a hypothetical concept.

1825 *Homo japeticus* Bory de St. Vincent. Europe: Caspian Sea to Cap Finistère.

1825 *Homo arabicus* Bory de St. Vincent. Arabia and North Africa.

1825 *Homo indicus* Bory de St. Vincent. Indian subcontinent.

1825 *Homo scythicus* Bory de St. Vincent. From the Caspian Sea to the Sea of Japan, in Siberia.

1825 *Homo sinicus* Bory de St. Vincent. Korea and Japan to Burma.

1825 *Homo hyperboreus* Bory de St. Vincent. Lappland to Kamchatka.

1825 *Homo neptunianus* Bory de St. Vincent. Madagascar east through the Pacific to western Americas, from California to Chile.

1825 *Homo australasicus* Bory de St. Vincent. Australia.

1825 *Homo colombicus* Bory de St. Vincent. Eastern North America, south to the Caribbean.

1825 *Homo americanus* Bory de St. Vincent. South America, except the west and the south.

1825 *Homo patagonus* Bory de St. Vincent. Southernmost South America.

1825 *Homo aethiopicus* Bory de St. Vincent. Most of subsaharan Africa.

1825 *Homo cafer* Bory de St. Vincent. Southeastern Africa.

1825 *Homo melaninus* Bory de St. Vincent. Melanesia east to Fiji.

1825 *Homo hottentotus* Bory de St. Vincent. Southernmost Africa.

1899 *Homo spelaeus* Lapouge. France: Cro-Magnon (selected by Campbell [1965]).

1899 *Homo priscus* Lapouge. France: Chancelade (selected by Campbell [1965]).

1906 *Homo grimaldii* Lapouge. France: Grimaldi. Late Pleistocene.

1910 *Homo aurignacensis hauseri* Klaatsch and Hauser. France: Combe Capelle (selected by Campbell [1965]). Late Pleistocene.

1911 *Notanthropus eurafricanus recens* Sergi. France: Combe Capelle (selected by Campbell [1965]). Late Pleistocene.

1911 *Notanthropus eurafricanus archaius* Sergi. Czech Republic: Predmost. Late Pleistocene.

1915 *Homo fossilis proto-aethiopicus* Giuffrida-Ruggeri. France: Combe Capelle. Late Pleistocene.

1917 *Homo capensis* Broom. South Africa: Boskop. Late Pleistocene or Holocene.

1921 *Homo sapiens cro-magnonensis* Gregory. France: Cro-Magnon (selected by Campbell [1965]).

1921 *Homo sapiens grimaldiensis* Gregory. France: Grimaldi. Late Pleistocene.

1921 *Homo wadjakensis* Dubois. Indonesia: Wajak. Holocene.

1931 *Homo drennani* Kleinschmidt. South Africa: Cape Flats. Holocene.

1932 *Palaeanthropus palestinus* McCown and Keith. Israel: Skhul (selected by Campbell [1965]). Middle-late Pleistocene boundary.

1940 *Homo leakeyi* Paterson. Kenya: Kanjera. Late Pleistocene or Holocene. Nomen nudum.

I have listed here the names that have been applied to *Homo sapiens* both living and fossil, as far as are known to me. The curious plethora of species described by Bory de St. Vincent (1825) is of historic interest and must be taken into account for nomenclatural purposes. Whether any subspecies do exist in modern *Homo sapiens* is probably not resolvable.

Appendix: A Word about Fossil Primates

Having little experience with fossil "prosimians," and none at all with fossil platyrrhines, in this short final section I will treat only a few of the catarrhines. *Eosimias* would rank as plesion under Simiiformes, and *Amphipithecus*, *Pondaungia*, and *Siamopithecus* under Catarrhini, before the separation between Eocatarrhini and Eucatarrhini.

FOSSIL CERCOPITHECOIDS

The fossil history of the Cercopithecoidea now appears exceedingly simple. Although the earliest known representative, *Prohylobates*, recognizable by its (incomplete) bilophodonty, is known from 18 to 20 ma, a primitive genus, *Victoriapithecus*, is well known from much later deposits (14.7 ma at Maboko). It still lacks full bilophodonty, retaining a crista obliqua and lacking a distal loph on upper molars and deciduous premolars, and still possessing a hypoconulid on M_{12}, and other features; the skull is strikingly like those of the catarrhine plesion *Aegyptopithecus*, the early hominoid *Afropithecus*, and to some extent the middle Miocene hominoids *Sivapithecus* and *Dryopithecus* (Benefit and McCrossin 1997). There is absolutely no trace of colobine/cercopithecine divergence at that time.

FOSSIL HOMINOIDS

The growing number of Miocene hominoid genera was reviewed by Andrews (1992), who, after separating out *Proconsul* and its allies, allocated the rest to the Hominidae. The early and middle Miocene representatives of the latter (sivapithecines aside) he placed in a subfamily Dryopithecinae, with three tribes: Afropithecini, Kenyapithecini, and

Dryopithecini, which he agreed might well be a heterogeneous group. Afropithecini had increased enamel thickness, enlarged premolars, and primitive postcrania; included genera are *Afropithecus* (including the Moroto palate), *Heliopithecus*, *Otavipithecus*, and the Maboko material, all early Miocene. Kenyapithecini had even thicker enamel and included *Kenyapithecus* and *Griphopithecus*, both middle Miocene. Dryopithecini contained only *Dryopithecus* (including *Hispanopithecus* and *Rudapithecus*).

Further discussion of these groupings by Cameron (1998) concluded that: (1) *Afropithecus*, *Heliopithecus*, and the Moroto and Maboko hominoids do indeed form a clade, distinct from the Proconsulidae and Hominidae. (2) *Afropithecus* makes a poor putative ancestor for the Moroto hominoid; such a relationship would require several state reversals. (3) *Heliopithecus* is related to *Afropithecus*, but remains distinct. (4) The Maboko hominoid, usually allocated to *Kenyapithecus*, is generically distinct. (5) *Kenyapithecus* is likely to be a sister taxon to the Proconsulidae and Hominidae, but is not related to the Afropithecini.

In view of the subsequent description, based on much richer material, of the Moroto hominoid as *Morotopithecus*, and its redating at 20 ma, Cameron's (1998) point no. 2, at least, must be regarded as insightful and rather prophetic.

Andrews and Pilbeam (1996) discussed the emerging apparent contradiction between cranial and postcranial evidence for the affinities of Miocene hominoid fossils. *Sivapithecus* has a very oranglike skull, but its postcranial skeleton lacks many synapomorphic states of living Hominidae. The reverse situation seems to be the case with *Dryopithecus laietanus*, from a 9.5-million-year-old Spanish deposit, which has a primitive skull but shortened lumbar vertebrae, apparently a broad thorax, long clavicle, flat deltoid plane of humerus, curved radial shaft, reduced contact of ulnar styloid with triquetrum, long and curved proximal phalanges, and apparently elongated arm and shortened leg; all of these are hominid apomorphies and some specifically oranglike, mainly the long clavicle, long arms, and long, curved proximal phalanges (Moyà-Solà and Köhler 1996).

But help is at hand. The striking facial resemblance of *Sivapithecus* to *Pongo* could actually be plesiomorphic; if so, apparent similarities could be superficial (Benefit and McCrossin 1997). *Dryopithecus* does retain primitive features, such as short metacarpals with small articular surfaces, and lumbars that are longer than those of living hominids (Moyà-Solà and Köhler 1996), and the specifically oranglike features are shared with gibbons too, suggesting that they might be primitive for the Hominoidea as a whole and reversed in the Homininae. We may then interpret *Sivapithecus* and *Dryopithecus* as successive branches off the hominid clade before the separation of Ponginae and Homininae.

The most striking recent advance in understanding of the early history of the Hominoidea concerns the discovery of new material from Moroto and the redating of the site at somewhat over 20 million years old (Gebo et al. 1997). The new material has permitted the description of a new genus and species, *Morotopithecus bishopi*. The genus had a primitively small, asymmetrical femoral head, with a large trochanteric crest and greater trochanteric fossa; robust femoral shaft; broad knee joint; smoothly curved, superiorly widened glenoid articular surface of scapula; shortened lumbars; apparently broad thorax; and primitive craniodental features, such as large molar cingula and small size of maxillary sinus. The animal was large in size, an estimated 36–54 kg. Gebo et al. (1997) said that the genus thus had clear hominoid synapomorphies, with only *Dryopithecus* and *Oreopithecus* being more derived postcranially, but retained primitive features; they considered it likely to be the sister taxon of the crown Hominoidea.

My task is to approach the taxonomy of these fossils, only touching on their evolution in so far as it affects taxonomy. What we have among those Miocene taxa for which we have good evidence is (1) Definite neocatarrhines, but without many derived conditions one way or the other (*Afropithecus, Heliopithecus*) or with a few derived conditions of the Hominoidea (*Griphopithecus, Otavipithecus,*

Kenyapithecus). (2) Hominoids, but neither hylobatids nor hominids, retaining apparently primitive conditions (postcranial, dental) from group 1, but possessing a clear suite of derived character states of their own (beaded cingulum, mandibular symphysis with unusual buttressing): *Proconsul, Rangwapithecus, Kamoyapithecus*. (3) Hominoids, but neither hylobatids nor hominids, without their own apomorphies: *Morotopithecus, Dryopithecus*.

There remains *Sivapithecus*: a probably derived hominoid, but less distinctly so than group 3. This can be ranked currently only as incertae sedis.

Members of group 1 are, for the time being, ranked as plesia under Neocatarrhini; all we really know about them is that they are more derived, and share more crown catarrhine features, than the Oligocene genera. Group 2, with its own internal diversity, ranks as a separate family, Proconsulidae. Members of group 3 rank as plesia under Hominidae. So a classification of the Catarrhini looks like this:

Catarrhini
  Plesion *Amphipithecus*
  Plesion *Pondaungia*
  Plesion *Siamopithecus*
    Eocatarrhini
      Family Parapithecidae
    Eucatarrhini
      Plesion Oligopithecidae
      Plesion *Propliopithecus*
      Plesion *Aegyptopithecus*
      Neocatarrhini
        Plesion *Afropithecus*
        Plesion *Heliopithecus*
        Superfamily Pliopithecoidea
          Family Pliopithecidae
        Superfamily Cercopithecoidea
          Plesion *Prohylobates*
          Plesion *Victoriapithecus*
          Family Cercopithecidae
        Superfamily Hominoidea
        Plesion *Griphopithecus*
        Plesion *Otavipithecus*
        Plesion *Kenyapithecus*
          Family Proconsulidae
          Plesion *Morotopithecus*
          Plesion *Dryopithecus*
          Family Hylobatidae
          Family Oreopithecidae
          Family Hominidae

Glossary

acrocentric Of a chromosome, having a long arm on one side of the centromere only.

allopatric Of two or more populations or taxa, inhabiting different geographic areas.

alpha taxonomy A mode of taxonomy concerned with basic description at the species and genus level.

apomorphic, -ous Derived (q.v.).

ascertainment bias The skewing of a study toward the characters and the sample in which variation had initially been detected.

autapomorphic, -ous Uniquely derived (q.v.).

available Of a scientific name, published in a way that fulfills the requirements of the *Code* and so is potentially usable if the taxon to which it refers is deemed valid.

basilar suture Strictly, the spheno-occipital synchondrosis; the synchondrosis between the basioccipital and basisphenoid, whose final fusion marks the end of the pre-mature growth phase.

beta (and gamma) taxonomy A mode of taxonomy concerned with the detailed elucidation of relationships.

binomial A two-word scientific name for a species, in which the first word is the name of the genus, the second that of the species.

biological species concept (BSC) The concept that species are "groups of actually or potentially interbreeding natural populations which are reproductively isolated from other such groups" (Mayr 1963:19). The key concept is the existence of reproductive isolating mechanisms (q.v.).

bizygomatic width (breadth) Width of cranium across the zygomatic arches.

BSC Biological species concept (q.v.).

canonical variates Uncorrelated variables resulting from transformation of the original variables, in a form of multivariate analysis (q.v.).

centromere The part of a chromosome that joins the two chromatids together.

character Any attribute of an organism.

character states Variants of a character.

choanae Internal nares.

cladistics The principle of deducing the interrelationships of organisms by discovering the polarities (q.v.) of their character states.

cladistic taxonomy (phylogenetic systematics) Taxonomy based on cladistic principles (i.e., strictly on phylogenetic affinities).

cladogram A diagram representing sister-group relationships.

Code *International Code of Zoological Nomenclature* (q.v.).

cohesion species concept The concept that a species is "the most inclusive population of individuals having the potential for phenotypic cohesion" (Templeton 1989:13). The key concept is the intrinsic cohesion mechanism (q.v.).

Commission The International Commission on Zoological Nomenclature (q.v.).

composite species concept (CSC) The concept that a species is "the set of all organisms belonging to an originator internodon, and all organisms belonging to any of its descendant interndons, excluding further originator internodons and their descendant internodons" (Kornet and McAllister 1993:78). The key concept is the originator internodon (q.v.).

condylobasal length Length of cranial base, from prosthion to back of occipital condyles.

convergence The development of apparently similar structures from quite different ancestral conditions.

coordination, principle of The principle that a name given to a species can be applied to a subspecies and vice versa, one given to a genus can be applied to a subgenus and vice versa, and one given to a family can be applied to a superfamily, subfamily, tribe, or subtribe and vice versa, with the same types, priority, and so on.

craniometry Measurements of skulls.

crown-group The lineages terminating in modern taxa, as far back as their last common ancestor.

CSC Composite species concept (q.v.).

dental formula I, incisor; C, canine; P, premolar; M, molar. D in front of I, C, or P indicates a deciduous tooth. A superscript number after I, C, P, or M indicates a maxillary (upper) tooth; a subscript indicates a mandibular (lower) tooth; hence, I^1 = first (central) upper incisor; M_2 = second lower molar.

derived (apomorphic) Of a character state, changed from an ancestral state.

diagnosable Of a species, identifiable 100% of the time; having fixed genetic differences from all others.

discriminant analysis (or canonical analysis) A form of multivariate analysis (q.v.) in which groups are the OTUs, the aim being to maximize between-group difference relative to within-group difference.

eumelanin A form of melanin producing black tones when concentrated and brown when dilute.

evolutionary species concept The concept that a species is "a lineage . . . evolving separately from others and with its own unitary evolutionary role and tendencies" (Simpson 1961:153).

evolutionary taxonomy A taxonomic school purporting to rely on the total evolutionary relationships of organisms.

FAD Of fossils, first appearance datum.

glabella A forward protrusion in the midline between the two supraorbital bars.

Gloger's Rule The observation that, very frequently, taxa in warm, humid environments are more heavily pigmented than their relatives in cooler and/or drier environments.

Gregg's paradox The representation of a monotypic group at several taxonomic levels (e.g., the aardvark is not merely a genus [*Orycteropus*] but also a family [Orycteropidae] and an order [Tubulidentata]).

hamulus A hooklike process, especially that prolonging the posterio-inferior corner of one or both pterygoid plate pairs (q.v.).

heterochrony Evolutionary dissociation between aspects of development and growth. Includes paedomorphosis (retention of juvenile features of the ancestor in adult descendant, by either slowing development with respect to growth [neoteny] or accelerating ontogeny in its entirety [progenesis]) and hypermorphosis (delay in cessation of growth well into maturity).

holophyletic Synonym of monophyletic (q.v.).

holotype The specimen designated as the type of a species (or subspecies) at the time of its original description.

homology Equivalent states of a character in different taxa.

homonyms In nomenclature, two or more names that are the same in spelling but refer to different taxa. Specific names may be primary or secondary homonyms (q.v.).

homoplasy Character incongruence on a cladogram (due to parallelism, convergence, or reversal [q.v.]).

ingroup A group of taxa being analyzed.

inion The point on the midline of the occiput where the superior nuchal lines meet.

International Code of Zoological Nomenclature The book (now in its fourth edition) that lists the rules to be applied when determining the correct names of animals.

International Commission on Zoological Nomenclature A body that lays down the rules for zoological nomenclature and decides difficult cases.

internodon An evolutionary lineage between two nodes or between a node and a termination.

intrinsic cohesion mechanism The (genetic or demographic) means whereby a species maintains its phenotypic cohesion.

junior synonym A name applied to a taxon that is not the earliest available one, and so is not in current use.

lectotype A specimen designated as the type of a species (or subspecies) subsequent to its original description.

lineage An evolutionary line.

mesopterygoid fossa The fossa bounded by the pterygoid plates, continuous with the nares anteriorly.

metacentric Of a chromosome, with a long arm on either side of the centromere.

metaspecies A species that is not known to be monophyletic.

MN stages Neogene biostratigraphic divisions (first proposed by P. Mein and C. Guérin), based on associations and evolutionary stages of fossil mammal species and paleofloras in western Europe and applicable, with due caution, in neighboring regions.

molecular clock The principle that DNA (especially noncoding DNA) evolves stochastically, so that in the long term evolutionary changes are regular, and genetic distance is equivalent to phylogenetic distance.

monophyletic Of a taxonomic group, descended from a common ancestor that was itself a member of that taxonomic group, and including *all* the descendants of that ancestor.

monotypic Of a taxonomic group, with only one representative at the next lowest level; so a monotypic family has only one genus, a monotypic genus has only one species. Of a species, having no subspecies.

morphometrics Study of phenetic relationships, based on measurements.

multivariate analysis (multivariate morphometrics) Morphometrics based on the simultaneous analysis of multiple measurements. *See* discriminant analysis, principal components analysis, and canonical variates.

neogene The second half of the Cenozoic Era: Miocene, Pliocene, and (in some schemes) Pleistocene and Holocene.

neotype A specimen designated as the type of a species (or subspecies) if the holotype or lectotype has been lost or never existed.

new combination A species name consisting of a previously published specific name plus the name of the genus to which it has been newly transferred.

NF (nombre fondamental, or fundamental number) In a karyotype, the number of major chromosome arms.

node A branching point on a cladogram or phylogenetic tree.

nomenclature The system of the naming of organisms.

nomen nudum (plural, nomina nuda) A name published for a given taxon, but without a description or a bibliographic reference to one. A nomen nudum is not available in nomenclature.

nominotypical Bearing the same name: the subspecies that bears the same name as the species, the subgenus that bears the same name as the genus, or the subfamily that bears the same name as the family.

numerical taxonomy *See* phenetics.

objective synonyms Two or more specific names of the same type specimen; or two or more generic names of the same type species.

obligatory ranks The taxonomic levels of the modified Linnaean system: kingdom, phylum, class, order, family, genus, and species.

ontogeny Individual development, from conception to maturity.

operational taxonomic units (OTUs) The units of comparison in a taxonomic study. They may be of any taxonomic rank, or may even be individuals.

ophryonic groove A transverse groove separating the supraorbital bar from the frontal squama.

Opinion (of the International Commission on Zoological Nomenclature) A (binding) decision by the Commission on a problem of nomenclature.

opisthocranion The posterior point on the braincase.

originator internodon An internodon (q.v.) along which an autapomorphic state arises and becomes fixed.

OTU Operational taxonomic unit (q.v.).

outgroup A taxon, or group of taxa, that is not part of the group under study (the ingroup), and so a potential source of polarity (q.v.) determination.

paleogene The first half of the Cenozoic Era: Paleocene, Eocene, and Oligocene.

paralectotype A specimen serving as a kind of "deputy" to the lectotype, from which new type material can be selected if the lectotype is lost.

parallelism The development of similar structures independently but from similar bases.

parapatric Of two or more populations or taxa, inhabiting adjoining geographic areas.

paraphyletic Of a taxonomic group, descended from a common ancestor that was itself a member of that taxonomic group, but not including *all* the descendants of that ancestor.

paraspecies A species that is known to be nonmonophyletic.

paratype A specimen designated as a kind of "deputy" to the holotype, from which it would be appropriate to select a new type specimen if the holotype is lost.

parsimony In cladistics, the principle that fewer evolutionary changes are more probable than many (i.e., that homoplasy is less probable than homology).

PCA Principal components analysis (q.v.).

phaeomelanin A form of melanin producing red tones when concentrated and yellow when dilute.

phenetics (numerical taxonomy) A taxonomic school relying on phenetic relationships, without taking character state polarity into account.

phylogenetic (phyletic) tree A representation of evolutionary relationships.

phylogenetic species concept (PSC) The concept that a species is "the smallest diagnosable cluster of individual organisms within which there is a parental pattern of ancestry and descent" (Cracraft 1983:170). The key concept is diagnosability. *See* diagnosable.

phylogeny Evolutionary history.

phylogeography "The study of the principles and processes governing the geographic distributions of genealogical lineages" (Avise 1994:233).

plesiomorphic, -ous Primitive (but without pejorative connotations).

plesion A "dummy rank" for a fossil group of limited diversity and time depth, and preferably without autapomorphies, inserted at any convenient place in a taxonomic scheme to prevent proliferation of taxonomic ranks and to avoid giving a short-lived fossil group the same rank as a long-lived, diverse sister group.

polarity The phylogenetic status of a character state, whether primitive (q.v.) or derived (q.v.).

polyphyletic Of a taxonomic group, not descended from a common ancestor that was itself a member of that taxonomic group.

polytypic Of a species, having subspecies.

primary homonym A specific name that was a homonym when the species was first described.

primitive (plesiomorphic) Of a character state, unchanged from an ancestral state.

principal components analysis A form of multivariate analysis (q.v.) in which the individual specimens are the OTUs, and there is no a priori grouping of specimens.

principle of least violence The principle that taxonomic modification should be as conservative as is consistent with monophyly or rank/time association.

priority In nomenclature, the principle that the first available name given to a taxon should be the one that is thereafter applied to it.

prosthion The most anterior point on the cranium, between the upper central incisor alveoli.

PSC Phylogenetic species concept (q.v.).

pterygoid plates The bony plates behind the choanae; usually two pairs, internal and external.

recognition species concept The concept that a species is "that most inclusive population of biparental organisms which share a common fertilization system" (Pa-

terson 1986:63). The key concept is the specific mate recognition system (q.v.).

reproductive isolating mechanisms (RIMs) Mechanisms that keep species from interbreeding under natural conditions (e.g., different courtship behaviors).

reversal The loss of a derived condition, reverting to the primitive.

revision A new taxonomic scheme of a group of organisms.

secondary homonym A specific name that becomes a homonym only when it is transferred to a genus other than the one in which the species was first described.

senior synonym The earliest available name for a taxon.

shared derived (synapomorphic) Of a character state, shared between two or more OTUs by virtue of being changed from its original ancestral state.

shared primitive (symplesiomorphic) Of a character state, shared between two or more OTUs by virtue of being retained unchanged from a distant common ancestor.

similum (plural, simila) A group of organisms (or study specimens) that share one or more key similarities.

sister groups Two taxa resulting from a single branching of an evolutionary lineage.

SMRS Specific mate recognition system (q.v.).

specific mate recognition system (SMRS) The mechanism whereby a member of a species recognizes another, of the opposite sex, as a potential mate.

stem-group The shorter lineages along the stem from the ancestor of the crown-group back to the next major branch.

subjective synonyms Two or more names that, in the opinion of a taxonomist, refer to the same taxon.

submetacentric Metacentric (q.v.), but with one of the two arms very short.

subspecies A geographic segment of a species, distinguishable from other such segments by strong gene frequency differences (traditionally, 75% of individuals are different from members of all other subspecies).

sympatric Of two or more populations or taxa, inhabiting the same geographic area.

symplesiomorphic, -ous Shared primitive (q.v.).

synapomorphic, -ous Shared derived (q.v.).

synonym One of several names that are thought (by a taxonomist) to refer to the same taxon.

syntypes Two or more specimens simultaneously designated as type specimens of a taxon at the time of its original description.

taxon (plural, taxa) "A group of real organisms recognised as a formal unit at any level of a hierarchic classification" (Simpson 1961:19).

taxonomy The classification of organisms.

termination The end point of a branch in a cladogram or phylogenetic tree (indicating either an extinction or arrival at the present day).

topotype A specimen from the type locality.

total group Crown-group plus stem-group.

transformed or pattern cladism An attempt to free cladistics from its evolutionary basis.

trinomial A three-word scientific name for a subspecies, consisting of a binomial (q.v.) plus a subspecific name.

type locality The locality at which a type specimen was obtained.

type species The species on which a genus (or subgenus) was originally founded and to which the name is forever indissolubly attached.

type specimen The specimen on the evidence of which a species or subspecies was originally named and to which the name is forever indissolubly attached. *See* holotype, lectotype, and neotype.

unavailable Of a scientific name, improperly published or a homonym, and so not even potentially usable in nomenclature.

uniquely derived (autapomorphic) Of a character state, changed from an ancestral state and unique to a given OTU.

valid Of a taxon, distinct at some level.

Literature Cited

Agrawal, V. C. 1974. The taxonomic status of Barbe's leaf-monkey, *Presbytis barbei* Blyth. *Primates* 15:235–239.

Ahmed, M. M. U., and R. Kanagasuntheram. 1965. A note on the mammary glands in the lesser bush baby (*Galago senegalensis senegalensis*). *Acta Anatomica* 60:253–261.

Aimi, M., and A. Bakar. 1992. Taxonomy and distribution of *Presbytis melalophos* group in Sumatera, Indonesia. *Primates* 33:191–206.

———. 1996. Distribution and deployment of *Presbytis melalophos* group in Sumatera, Indonesia. *Primates* 37:399–409.

Aimi, M., H. S. Hardjasasmita, A. Sjarmidi, and D.Yuri. 1986. Geographical distribution of aygula-group of the genus *Presbytis* in Sumatra. *Kyoto University Overseas Research Report of Studies on Asian Non-Human Primates* 5:45–58.

Albrecht, G. H., and J. M. A. Miller. 1993. Geographic variation in Primates: A review with implications for interpreting fossils. In *Species, species concepts, and Primate evolution*, W. H. Kimbel and L. B. Martin, eds., 123–161. New York: Plenum Press.

Albrecht, G., P. D. Jenkins, and L. R. Godfrey. 1990. Eco-geographic size variation among the living and subfossil prosimians of Madagascar. *American Journal of Primatology* 22:1–50.

Allen, G. M. 1939a. A checklist of African mammals. *Bulletin of the Museum of Comparative Zoology at Harvard College* 83:1–763.

———. 1939b. *Mammals of China and Mongolia.* Vol. 1. New York: American Museum of Natural History.

Allen, J. A. 1884. Are trinomials necessary? *Auk* 1:102–104.

———. 1916. Mammals collected on the Roosevelt Brazilian Expedition, with field notes by Leo E. Miller. *Bulletin of the American Museum of Natural History* 35:559–610.

———. 1925. Primates collected by the American Museum Congo Expedition. *Bulletin of the American Museum of Natural History* 47:283–499.

Alperin, R. 1993. *Callithrix argentata* (Linnaeus, 1771): Considerações taxonômicas e descriçao de subespécie nova. *Boletim del Museo Paraense Emilio Goeldi Série Zoologia* 9:317–328.

Anderson, J. 1902. *Zoology of Egypt: Mammalia.* London: Hugh Rees.

Andrews, P. 1992. Evolution and environment in the Hominoidea. *Nature* 360:641–646.

Andrews, P., and D. Pilbeam. 1996. The nature of the evidence. *Nature* 379:123–124.

Ansell, W. F. H. 1960. Contributions to the mammalogy of Northern Rhodesia. *Occasional Papers of the National Museum of Southern Rhodesia* 24B:351–398.

———. 1963. Additional breeding data on Northern Rhodesian mammals. *Puku* 1:9–28.

———. 1978. *The mammals of Zambia.* Chilanga: National Parks and Wildlife Service.

———. 1989. *African mammals 1938–1988.* St. Ives: Trendrine Press.

Ashlock, P. D. 1971. Monophyly and associated terms. *Systematic Zoology* 20:63–69.

———. 1979. An evolutionary systematist's view of classification. *Systematic Zoology* 28:441–450.

Avila, J. Morphological variation between two subspecies of *Cebus apella* Erxleben, 1777: A preliminary report. (in press.)

Avila-Pires, F. D. de. 1985. On the validity and geographical distribution of *Callithrix argenta emiliae* Thomas, 1920 (Primates, Callithrichidae). *A Primatologia no Brasil* 2:319–322.

Avise, J. C. 1994. *Molecular marks, natural history, and evolution*. London: Chapman and Hall.

Avise, J. C., and R. M. Ball. 1990. Principles of genealogical concordance in species concepts and biological taxonomy. *Oxford Surveys in Evolutionary Biology* 7:45–67.

Ayres, J. M. 1985. On a new species of squirrel monkey, genus *Saimiri*, from Brazilian Amazonia (Primates, Cebidae). *Papéis Avulsos de Zoologia* 36:147–164.

Baba, M. L., L. G. Darga, and M. Goodman. 1979. Immunodiffusion systematics of the Primates. V. The Platyrrhini. *Folia Primatologica* 32:207–238.

Barroso, C. M. L. 1995. Molecular phylogeny of the Callitrichinae. *Neotropical Primates* 3:186.

Bartlett, E. 1871. Notes on the monkeys of eastern Peru. *Proceedings of the Zoological Society of London* 1871:217–220.

Bauer, K., and A. Schreiber. 1997. Double invasion of Tertiary island South America by ancestral New World monkeys? *Biological Journal of the Linnean Society* 60:1–20.

Baum, D. 1992. Phylogenetic species concepts. *Trends in Ecology and Evolution* 7:1–2.

Bearder, S. K., P. E. Honess, and L. Ambrose. 1995. Species diversity among galagos with special reference to mate recognition. In *Creatures of the dark: The nocturnal prosimians*, L. Alterman, G. A. Doyle, and M. K. Izard, eds., 1–22. New York: Plenum Press.

Bender, M. A., and E. H. Y. Chu. 1963. The chromosomes of primates. In *Evolutionary and genetic biology of the Primates*, Vol.1, J. Buettner-Janusch, ed., 261–310. London: Academic Press.

Benefit, B. R., and M. McCrossin. 1993. The lacrimal fossa of Cercopithecoidea, with special reference to cladistic analysis of Old World monkey relationships. *Folia Primatologica* 60:133–145.

———. 1997. Earliest known Old World monkey skull. *Nature* 388:368–371.

Bernstein, I. S. 1970. Some behavioral elements of the Cercopithecoidea. In *Old World monkeys*, J. R. and P. H. Napier, eds., 263–295. New York: Academic Press.

Bigelow, R. S. 1965. Hybrid zones and reproductive isolation. *Evolution* 19:449–458.

Bishop, M. J., and A. E. Friday. 1987. Tetrapod relationships: The molecular evidence. In *Molecules and morphology in evolution: Conflict or compromise?*, C. Patterson, ed., 123–139. Cambridge: Cambridge University Press.

Blyth, E. 1847. Supplementary report of the Curator of the Zoological Department. *Journal of the Asiatic Society of Bengal* (Calcutta) 16:728–737.

———. 1863. *Catalogue of the Mammalia in the Museum of the Asiatic Society of Bengal*. Calcutta: Savielle and Cranenburgh.

Bodini, R., and R. Pérez-Hernàndez. 1987. Distribution of the species and subspecies of cebids in Venezuela. *Fieldiana Zoology*, n. s. 39:231–244.

Bonvicino, C. R., A. Langguth, and R. A. Mittermeier. 1989. A study of pelage color and geographic distribution in *Alouatta belzebul* (Primates: Cebidae). *Revista Nordestina de Biologia* 6:139–148.

Bonvicino, C. R., M. E. B. Fernandes, and H. N. Seuánez. 1995. Morphological analysis of *Alouatta seniculus* species group (Primates, Cebidae). A comparison with biochemical and karyological data. *Human Evolution* 10:169–176.

Booth, A. H. 1955. Speciation in the mona monkeys. *Journal of Mammalogy* 36:434–449.

———. 1956. The Cercopithecidae of the Gold and Ivory Coasts: Geographic and systematic observations. *Annals and Magazine of Natural History* (12) 9:476–480.

Booth, C. P. 1968. Taxonomic studies of *Cercopithecus mitis* Wolf (East Africa). *Research Reports of the National Geographic Society* 1963:37–51.

Bory de St. Vincent, G. 1825. Homme, *Homo*. In *Dictionnaire classique d'histoire naturelle*. Vol. 8, I. B. Andouin, ed., 269–346. Paris: Rey and Gromier.

Braga, J. 1998. Chimpanzee variation facilitates the interpretation of the incisive suture closure in South African Plio-Pleistocene hominids. *American Journal of Physical Anthropology* 105:121–135.

Brandon-Jones, C. 1995. Long gone and forgotten: Reassessing the life and career of Edward Blyth, zoologist. *Archives of Natural History* 22:91–95.

Brandon-Jones, D. 1978. Comment on the proposal to conserve Colobidae Blyth, 1875, as the family-group name for the leaf-eating monkeys (Mammalia, Primates). *Bulletin of Zoological Nomenclature* 35:69–70.

———. 1984. Colobus and leaf-monkeys. In *The encyclo-*

paedia of mammals. Vol. 1, D. Macdonald, ed., 398–408. London: George Allen and Unwin.

———. 1993. The taxonomic affinities of the Mentawai Islands sureli, *Presbytis potenziani* (Bonaparte, 1856) (Mammalia: Primates: Cercopithecidae). *Raffles Bulletin of Zoology* 41:331–357.

———. 1995a. *Presbytis fredericae* (Sody, 1930), an endangered colobine species endemic to central Java, Indonesia. *Primate Conservation* 16:68–70.

———. 1995b. A revision of the Asian pied leaf-monkeys (Mammalia: Cercopithecidae: superspecies *Semnopithecus auratus*), with a description of a new subspecies. *Raffles Bulletin of Zoology* 43:3–43.

———. 1996. The zoogeography of sexual dichromatism in the Bornean grizzled sureli, *Presbytis comata* (Desmarest, 1822). *Sarawak Museum Journal* 50:177–200.

Brockelman, W. Y. 1985. A gibbon pelt *(Hylobates lar entelloides)* from Khao Yai National Park, Saraburi Province, Thailand. *Natural History Bulletin of the Siam Society* 33:55–57.

———. 1991. Book review: A theory of Primate evolution *[sic]. Australian Primatology* 5 (4): 5–7.

Brooks, D. M., and L. Pando-Vasquez. 1997. Crossing the great barrier: *Callicebus cupreus discolor* north of the Napo River. *Neotropical Primates* 5:11.

Butchart, S. H. M., R. Barnes, C. W. N. Davies, M. Fernandez, and N. Seddon. 1995. Observations of two threatened primates in the Peruvian Andes. *Primate Conservation* 16:15–19.

Bynum, E. L., D. Z. Bynum, and J. Supriatna. 1997. Confirmation and location of the hybrid zone between wild populations of *Macaca tonkeana* and *Macaca hecki* in central Sulawesi, Indonesia. *American Journal of Primatology* 43:181–209.

Cabrera, A. 1940. Los nombres cientificos de algunos monos Americanos. *Ciencia Méxicana* 1:402–405.

———. 1956. Sobre la identificacion de *Simia leonina* Humboldt (Mammalia, Primates). *Neotropica* 2:49–53.

———. 1957, 1961. *Catalogo de los mamíferos de America del Sur* (1957, i–xxii, 1–308; 1961, 309–732). Buenos Aires: Editora Coni.

Cain, A. J. 1954. *Animal species and their evolution.* London: Hutchinson's University Library.

Cameron, D. W. 1998. Anatomical variability and systematic status of the hominoids currently allocated to the African Dryopithecinae. *Homo* 49:101–137.

Campbell, B. G. 1965. The nomenclature of the Hominidae. *Royal Anthropological Institute Occasional Paper* No. 22. London: Royal Anthropological Institute.

Canavez, F., G. Alves, T. G. Fanning, and H. N. Seuánez.

1996. Comparative karyology and evolution of the Amazonian *Callithrix* (Platyrrhini, Primates). *Chromosoma* 104:348–357.

Carr, S. M., S. W. Ballinger, J. N. Derr, L. H. Blankenship, and J. W. Bickham. 1986. Mitochondrial DNA analysis of hybridization between sympatric white-tailed deer and mule deer in West Texas. *Proceedings of the National Academy of Sciences of the United States of America* 83:9576–9580.

Cartelle, C., and W. C. Hartwig. 1996. A new extinct primate among the Pleistocene megafauna of Bahia, Brazil. *Proceedings of the National Academy of Sciences of the United States of America* 93:6405–6409.

Cartmill, M., and K. Milton. 1977. The lorisiform wrist joint and the evolution of "brachiating" adaptations in the Hominoidea. *American Journal of Physical Anthropology* 47:249–272.

Charig, A. J. 1983. Systematics in biology: A fundamental comparison of some major schools of thought. In *Problems of phylogenetic reconstruction,* K. A. Joysey and A. E. Friday, eds., 363–440. London: Academic Press.

Chasen, F. N. 1940. A handlist of Malaysian mammals. *Bulletin of the Raffles Museum* (Singapore) 15:1–209.

Chasen, F. N., and C. B. Kloss. 1927. Spolia Mentaweiensia. Mammals. *Proceedings of the Zoological Society of London* 4:797–840.

Cheverud, J. M., and A. J. Moore. 1990. Subspecific morphological variation in the saddle-back tamarin *(Saguinus fuscicollis), American Journal of Primatology* 21:1–15.

Chiarelli, B. 1963. Comparative morphometrics of primate chromosomes. III. The chromosomes of the genera *Hylobates, Colobus,* and *Presbytis. Caryologia* 16:637–648.

Choudhury, A. 1992. Golden langur—distribution confusion. *Oryx* 26:172–173.

———. 1997. *Checklist of the mammals of Assam.* Guwahati, Assam, India: Gibbon Books, with Assam Science Technology and Environment Council.

———. 1998. Père David's macaque discovered in India. *The Rhino Foundation Newsletter* 2:7.

Christoffersen, M. L. 1995. Cladistic taxonomy, phylogenetic systematics, and evolutionary ranking. *Systematic Biology* 44:440–454.

Coimbra-Filho, A. F. 1985. Especies ameaçades de extinção: Sagui-de-Wied. *Fundaçao Brasileira para a Conservaçao da Natureza Boletim Informativo* (Rio de Janeiro) 9 (4, October/December): 5.

———. 1990. Sistematicá, distribuiçâo geográfica, e situaçâo atual dos simios brasileiros (Platyrrhini-Primates). *Revista Brasileira Biologica* 50:1063–1079.

Coimbra-Filho, A. F., and R. A. Mittermeier. 1973. New data on the taxonomy of the Brazilian marmosets of the genus *Callithrix* Erxleben. *Folia Primatologica* 20:241–264.

Coimbra-Filho, A. F., A. Pissinatti, and A. B. Rylands. 1993. Experimental multiple hybridism and natural hybrids among *Callithrix* species from eastern Brazil. In *Marmosets and tamarins: Systematics, behaviour, and ecology*, A. B. Rylands, ed., 95–120. Oxford: Oxford University Press.

Coimbra-Filho, A. F., I. de Gusmão Câmara, and A. B. Rylands. 1995. On the geographic distribution of the red-handed howling monkey, *Alouatta belzebul*, in North-East Brazil. *Neotropical Primates* 3:176–179.

Coimbra-Filho, A. F., A. B. Rylands, A. Pissinatti, and I. B. Santos. 1995. The distribution and status of the buff-headed capuchin monkey, *Cebus xanthosternos* Wied 1820, in the Atlantic Forest region of eastern Brazil. *Primate Conservation* 12–13 (for 1991–1992): 24–30.

Coimbra-Filho, A. F., A. Pissinatti, and A. B. Rylands. 1997. A simulacrum of *Saguinus bicolor ochraceus* Hershkovitz, 1966, obtained through hybridising *S. b. martinsi* and *S. b. bicolor* (Callitrichidae, Primates). *A Primatologia no Brasil* 5:179–184.

Colyn, M. 1988. Distribution of guenons in the Zaïre-Lualaba-Lomami river system. In *A Primate radiation: Evolutionary biology of the African guenons*, A. Gautier-Hion, F. Bourlière, J.-P. Gautier, and J. Kingdon, eds., 104–124. Cambridge: Cambridge University Press.

———. 1991. L'importance zoogéographique du bassin du Fleuve Zaïre pour la spéciation: La cas des Primates Simiens. *Koninklijk Museum voor Midden-Afrika, Tervuren, België, Annalen, Zoologische Wetenschappen* 264:i–ix, 1–250.

———. 1993. Coat colour polymorphism of red colobus monkeys (*Colobus badius*, Primates, Colobinae) in eastern Zaire: Taxonomic and biogeographic implications. *Journal of African Zoology* 107:301–320.

———. 1999. Étude populationnelle de la super-espèce *Cercopithecus cephus* habitant l'enclave forestière Sangha-Oubangui (République Centrafricaine). Description de *C. cephus ngottoensis* subsp. nov. *Mammalia* 63:137–147.

Colyn, M., and U. Rahm. 1987. *Cercopithecus hamlyni kahuziensis* (Primates, Cercopithecidae): Une nouvelle sous-espèce de la forêt de bambous du Parc National "Kahuzi-Biega" (Zaire). *Folia Primatologica* 49:203–208.

Colyn, M. M., and W. N. Verheyen. 1987a. *Colobus rufomitratus parmentieri*, une nouvelle sous-espèce du Zaïre (Primates, Cercopithecidae). *Revue de Zoologie Africaine* 101:125–132.

———. 1987b. Considérations sur la provenance de l'holotype de *Cercopithecus mitis maesi* Lönnberg, 1919 (Primates, Cercopithecidae) et description d'une nouvelle sous-espèce: *Cercopithecus mitis heymansi*. *Mammalia* 51:271–281.

Colyn, M., A. Gautier-Hion, and D. Thys van den Audenaerde. 1991. *Cercopithecus dryas* Schwarz 1932 and *C. salongo* Thys van den Audenaerde 1977 are the same species with an age-related coat pattern. *Folia Primatologica* 56:167–170.

Conroy, G. C. 1981. Cranial asymmetry in ceboid Primates: The emissary foramina. *American Journal of Physical Anthropology* 55:187–194.

Consigliere, S., R. Stanyon, U. Koehler, G. Agoramoorthy, and J. Wienberg. 1996. Chromosome painting defines genomic rearrangements between red howler monkey subspecies. *Chromosome Research* 4:264–270.

Cope, D. A., and M. G. Lacy. 1992. Falsification of a single species hypothesis using the coefficient of variation: A simulation approach. *American Journal of Physical Anthropology* 89:359–378.

Coppens, Y., V. J. Maglio, C. T. Madden, and M. Beden. 1978. Proboscidea. In *Evolution of African mammals*, V. J. Maglio and H. B. S. Cooke, eds., 336–367. Cambridge, Mass.: Harvard University Press.

Coryndon, S. C. 1977. The taxonomy and nomenclature of the Hippopotamidae (Mammalia, Artiodactyla) and a description of two new fossil species. Parts I and II. *Proceedings of the Koninklijke Nederlandse Akademie van Wetenschappen* 80B:61–88.

Costello, R. K., C. Dickinson, A. L. Rosenberger, S. Boinski, and F. S. Szalay. 1993. Squirrel monkey (genus *Saimiri*) taxonomy: A multidisciplinary study of the biology of species. In *Species, species concepts, and Primate evolution*, W. H. Kimbel and L. B. Martin, eds., 177–210. New York: Plenum Press.

Coues, E. 1884. Trinomials are necessary. *Auk* 1:197–198.

Courtenay, D. O., and S. K. Bearder. 1989. The taxonomic status and distribution of bushbabies in Malawi with emphasis on the significance of vocalizations. *International Journal of Primatology* 10:17–34.

Couturier, J., and J.-M. Lernould. 1991. Karyotypic study of four gibbon forms previously considered as subspecies of *Hylobates (Nomascus) concolor* (Primates, Pongidae, Hylobatidae). *Folia Primatologica* 56:95–104.

Cracraft, J. 1983. Species concepts and speciation analysis.

In *Current ornithology*. Vol. 1, R. F. Johnston, ed., 159–187. New York: Plenum Press.

———. 1992. The species of the birds-of-paradise (Paradiseidae): Applying the phylogenetic species concept to a complex pattern of diversification. *Cladistics* 8:1–43.

Crisp, M. D., and G. T. Chandler. 1996. Paraphyletic species. *Telopea* 6:813–844.

Cronin, J. E., and V. M. Sarich. 1975. Molecular systematics of the New World monkeys. *Journal of Human Evolution* 4:357–375.

Cropp, S. J., A. Larson, and J. M. Cheverud. 1999. Historical biogeography of tamarins, genus *Saguinus:* The molecular phylogenetic evidence. *American Journal of Physical Anthropology* 108:65–89.

Crovella, S., D. Montagnon, and Y. Rumpler. 1992. Highly repeated DNA analysis and systematics of the Lemuridae, a family of Malagasy prosimians. *Primates* 34:61–69.

Crovella, S., J. C. Masters, and Y. Rumpler. 1994. Highly repeated DNA sequences as phylogenetic markers among the Galaginae. *American Journal of Primatology* 32:177–185.

Cruz Lima, E. da. 1945. *Mammals of Amazonia, 1: General introduction and Primates.* Belém do Pará: Museu Paraense E. Goeldi.

Dandelot, P. 1959. Note sur la classification des Cercopithèques du groupe *aethiops*. *Mammalia* 23:357–368.

———. 1962. Note sur la position taxonomique de *Cercopithecus albogularis* Sykes. *Mammalia* 26:447–450.

———. 1971. Order Primates. In *The mammals of Africa: An identification manual,* Part 3, 1–45. Washington, D.C.: Smithsonian Institution.

Dandelot, P., and J. Prévost. 1972. Contribution à l'étude des primates d'Éthiopie (simiens). *Mammalia* 36:607–633.

Dao V. T. 1983. On the North Indochinese gibbons *(Hylobates concolor)* (Primates, Hylobatidae) in North Vietnam. *Journal of Human Evolution* 12:367–372.

———. 1985. *Khao Sát Thú o miên bác Viêt Nam* [Scientific results of some mammal surveys in North Vietnam], *1957–1971.* Hanoi: Scientific and Technical Publishing House.

Dekeyser, P. L., and J. Derivot. 1960. Sur de nouveaux spécimens de cynocéphales du Tibesti. *Bulletin de l'Institut Français d'Afrique Noire, Série A, Sciences Naturelles* 22:1453–1456.

Delson, E. 1976. The family-group name of the leaf-eating monkeys (Mammalia, Primates): A proposal to give Colobidae Blyth, 1875, precedence over Semnopithecidae Owen, 1843, and Presbytina Gray, 1825. *Bulletin of Zoological Nomenclature* 33:85–89.

———. 1980. Fossil macaques, phyletic relationships and a scenario of deployment. In *The macaques: Studies in ecology, behavior, and evolution,* D. G. Lindburg, ed., 10–30. New York: Van Nostrand Reinhold Co.

Delson, E., and P. H. Napier. 1976. Request for the determination of the generic names of the baboon and the mandrill (Mammalia: Primates, Cercopithecidae). *Bulletin of Zoological Nomenclature* 33:46–60.

———. 1977. Correction to the request to determine the generic names of the baboon and the mandrill. *Bulletin of Zoological Nomenclature* 33:149.

Dene, H., M. Goodman, W. Prychodko, and G. W. Moore. 1976. Immunodiffusion systematics of the Primates. III. The Strepsirhini. *Folia Primatologica* 25:35–61.

Deraniyagala, P. E. P. 1955. *Some extinct elephants, their relatives, and the two living species.* Colombo: Ceylon National Museums Publication.

Disotell, T. 2000. Molecular systematics of the Cercopithecoidea. In *Old World monkeys,* P. Whitehead and C. J. Jolly, eds., 29–56. Cambridge: Cambridge University Press.

Dobroruka, L. J., and J. Badalec. 1966. Zur Artbildung der Mangaben Gattung *Cercocebus* (Cercopithecidae, Primates). *Revue de Zoologie et de Botanique Africaines* 73:345–350.

Donoghue, M. J. 1985. A critique of the biological species concept and recommendations for a phylogenetic alternative. *Bryologist* 88:172–181.

Dunbar, R. I. M., and P. Dunbar. 1974. On hybridization between *Theropithecus gelada* and *Papio anubis* in the wild. *Journal of Human Evolution* 3:187–192.

Dutrillaux, B. 1988. New interpretation of the presumed common ancestral karyotype of platyrrhine monkeys. *Folia Primatologica* 50:226–229.

Dutrillaux, B., and J. Couturier. 1981. The ancestral karyotype of platyrrhine monkeys. *Cytogenetics and Cell Genetics* 30:232–242.

Dutrillaux, B., and Y. Rumpler. 1995. Phylogenetic relations among Prosimii with special reference to Lemuriformes and Malagasy nocturnals. In *Creatures of the dark: The nocturnal prosimians,* L. Alterman, G. A. Doyle, and M. K. Izard, eds., 141–150. New York: Plenum Press.

Dutrillaux, B., J. Couturier, and E. Viegas-Péquignot. 1986. Évolution chromosomique des Platyrhiniens. *Mammalia* 50 (supplement): 56–81.

Dutrillaux, B., M. Muleris, and J. Couturier. 1988. Chromosomal evolution of Cercopithecinae. In *A Primate radiation: Evolutionary biology of the African guenons*, A. Gautier-Hion, F. Bourlière, J.-P. Gautier, and J. Kingdon, eds., 150–159. Cambridge: Cambridge University Press.

Easteal, S., and G. Herbert. 1997. Molecular evidence from the nuclear genome for the time frame of human evolution. *Journal of Molecular Evolution* 44 (supplement 1): S121–132.

Ellerman, J. R., and T. C. S. Morrison-Scott. 1951. *Checklist of Palaearctic and Indian mammals, 1758–1947*. London: British Museum (Natural History).

Ellerman, J. R., T. C. S. Morrison-Scott, and R. W. Hayman. 1953. *Southern African mammals 1758 to 1951: A reclassification*. London: Trustees of the British Museum.

Elliot, D. G. 1913. *A review of the Primates*. 3 vols. New York: American Museum of Natural History.

Eschscholtz, F. 1821. Description of a new species of monkey *(Presbytis mitrata)*. In O. von Kotzebue, *A voyage of discovery in the South Sea, and to Behring's Straits: In search of a north-east passage, undertaken in the years 1815, 16, 17, and 18, in the ship Rurick*, 353–356. London: Printed for Sir Richard Phillips and Co.

Evans, B. J., J. C. Morales, J. Supriatna, and D. J. Melnick. 1999. Origin of Sulawesi macaques (Cercopithecidae: *Macaca*) as suggested by mitochondrial DNA phylogeny. *Biological Journal of the Linnean Society* 66:539–560.

Evans, M. I., P. M. Thompson, and A. Wilson. 1995. A survey of the lemurs of Ambatovaky Special Reserve, Madagascar. *Primate Conservation* 14–15 (for 1993–1994): 13–21.

Faith, D. P. 1989. Homoplasy as pattern: Multivariate analysis of morphological convergence in Anseriformes. *Cladistics* 5:1–24.

Faith, D. P., and P. S. Cranston. 1990. Could a cladogram this short have arisen by chance alone? On permutation tests for cladistic structure. *Cladistics* 7:1–28.

Falk, D. 1979. Cladistic analysis of New World monkey sulcal patterns: Methodological implications for Primate brain studies. *Journal of Human Evolution* 8:637–645.

Farris, J. S. 1976. Phylogenetic classification of fossils with recent species. *Systematic Zoology* 25:271–282.

Felsenstein, J. 1981. Evolutionary trees from DNA sequences: A maximum-likelihood approach. *Journal of Molecular Evolution* 17:368–376.

———. 1985. Confidence limits on phylogenies: An approach using the bootstrap. *Evolution* 39:783–791.

———. 1991. *PHYLIP: Phylogeny inference package*. Version 3. 4. Seattle: University of Washington.

Feng Z., Cai G., and Zheng C. 1986. *The mammals of Xizang*. Beijing: Science Press [in Chinese].

Ferrari, S. F., and M. A. Lopes. 1992. A new species of marmoset, genus *Callithrix* Erxleben, 1777 (Callitrichidae, Primates), from western Brazilian Amazon. *Goeldiana Zoologia* 12:1–13.

Fitter, R., and P. Scott. 1978. *The penitent butchers*. London: Collins.

Fooden, J. 1963. A revision of the woolly monkeys (genus *Lagothrix*). *Journal of Mammalogy* 44:213–247.

———. 1966. Identification of the type specimen of *Simia apedia* Linnaeus, 1758. *Mammalia* 30:507–508.

———. 1971. Report on Primates collected in western Thailand, January–April, 1967. *Fieldiana Zoology*, n. s. 59:1–62.

———. 1975. Taxonomy and evolution of liontail and pigtail macaques (Primates: Cercopithecidae). *Fieldiana Zoology* 67:1–169.

———. 1976. Provisional classification and key to living species of macaques (Primates: *Macaca*). *Folia Primatologica* 25:225–236.

———. 1979. Taxonomy and evolution of the *sinica* group of macaques: 1. Species and subspecies accounts of *Macaca sinica*. *Primates* 20:109–140.

———. 1981. Taxonomy and evolution of the *sinica* group of macaques: 2. Species and subspecies accounts of the Indian bonnet macaque, *Macaca radiata*. *Fieldiana Zoology*, n.s. 9:v–vii, 2–52.

———. 1982. Taxonomy and evolution of the *sinica* group of macaques: 3. Species and subspecies accounts of *Macaca assamensis*. *Fieldiana Zoology*, n. s. 10:1–52.

———. 1983. Taxonomy and evolution of the *sinica* group of macaques: 4. Species and subspecies accounts of *Macaca thibetana*. *Fieldiana Zoology*, n.s. 17:1–20.

———. 1987. Type locality of *Hylobates concolor leucogenys*. *American Journal of Primatology* 12:107–110.

———. 1988. Taxonomy and evolution of the *sinica* group of macaques: 6. Interspecific comparisons and synthesis. *Fieldiana Zoology*, n.s. 45:v–vi, 1–44.

———. 1989. Classification, distribution, and ecology of Indian macaques. In *Perspectives in Primate biology*. Vol. 2, P. K. Seth and S. Seth, eds., 33–46. New Delhi: Today and Tomorrow's Publishers.

———. 1990. The bear macaque, *Macaca arctoides*: A systematic review. *Journal of Human Evolution* 19:607–686.

———. 1991. Systematic review of Philippine macaques

(Primates, Cercopithecidae: *Macaca fascicularis* subspp.). *Fieldiana Zoology,* n.s. 64:iii–iv, 1–44.

———. 1995. Systematic review of Southeast Asian long-tail macaques, *Macaca fascicularis* (Raffles, [1821]). *Fieldiana Zoology,* n.s. 81:iii–vi, 1–206.

———. 1996. Zoogeography of Vietnamese primates. *International Journal of Primatology* 17:845–899.

———. 1997. Tail length variation in *Macaca fascicularis* and *M. mulatta. Primates* 38:221–231.

Fooden, J., and S. M. Lanyon. 1989. Blood-protein allele frequencies and phylogenetic relationships in *Macaca:* A review. *American Journal of Primatology* 17:209–241.

Fooden, J., Quan G., Zhang Y., Wu M., and Liang M. 1994. Southward extension of the range of *Macaca thibetana. International Journal of Primatology* 15:623–627.

Ford, S. M. 1986. Systematics of New World monkeys. In *Comparative Primate biology. 1: Systematics, evolution, and anatomy,* D. R. Swindler and J. Erwin, eds., 73–135. New York: Alan R. Liss.

———. 1994a. Primitive platyrrhines? Perspectives on anthropoid origins from platyrrhine, parapithecid, and preanthropoid postcrania. In *Anthropoid origins,* J. G. Fleagle and R. F. Kay, eds., 595–673. New York: Plenum Press.

———. 1994b. Taxonomy and distribution of the owl monkey. In *Aotus: The owl monkey,* 1–57. London: Academic Press.

Forey, P. L., C. J. Humphries, I. J. Kitching, R. W. Scotland, D. J. Siebert, and D. M. Williams. 1992. *Cladistics: A practical course in systematics.* Oxford: Clarendon Press.

Forsyth Major, C. I. 1894. Über die Malagassischen Lemuriden-Gattungen *Microcebus, Opolemur,* und *Chirogale. Novitates Zoologicae* 1:2–39.

Freudenstein, J. V. 1998. Paraphyly, ancestors, and classification—a response to Sosef and Brummitt. *Taxon* 47:95–104.

Frisch, J. E. 1965. *Trends in the evolution of the hominoid dentition. Bibliotheca Primatologica* 5.

Froehlich, J. W., and P. H. Froehlich. 1987. The status of Panama's endemic howling monkeys. *Primate Conservation* 8:58–62.

Froechlich, J. W., J. Supriatna, and P. H. Froehlich. 1991. Morphometric analyses of *Ateles:* Systematic and biogeographic implications. *American Journal of Primatology* 25:1–22.

Froehlich, J. W., N. Babo, S. Akbar, and J. Supriatna. 1998. A new species of Sulawesi monkey: The Balantak macaque. *Abstracts of the 17th Congress of the International Primatological Society,* no. 038.

Garner, K. J., and O. A. Ryder. 1996. Mitochondrial DNA diversity in gorillas. *Molecular Phylogenetics and Evolution* 6:39–48.

Garza, J. C., and D. S. Woodruff. 1992. A phylogenetic study of the gibbons *(Hylobates)* using DNA obtained noninvasively from hair. *Molecular Phylogenetics and Evolution* 1:202–210.

Gautier, J.-P. 1988. Interspecific affinities among guenons as deduced from vocalizations. In *A Primate radiation: Evolutionary biology of the African guenons,* A. Gautier-Hion, F. Bourlière, J.-P. Gautier, and J. Kingdon, eds., 194–226. Cambridge: Cambridge University Press.

Gebo, D. L. 1985. The nature of the Primate grasping foot. *American Journal of Physical Anthropology* 67:269–277.

Gebo, D. L., and M. Dagosto. 1988. Foot anatomy, climbing, and the origin of the Indriidae. *Journal of Human Evolution* 17:135–154.

Gebo, D. L., L. MacLatchy, R. Kityo, A. Deino, J. Kingston, and D. Pilbeam. 1997. A hominoid genus from the early Miocene of Uganda. *Science* 276:401–404.

Geissmann, T. 1989. A female black gibbon, *Hylobates concolor* ssp., from northeastern Vietnam. *International Journal of Primatology* 10:455–476.

———. 1991. Sympatry between white-handed gibbons *(Hylobates lar)* and pileated gibbons *(H. pileatus)* in southeastern Thailand. *Primates* 32:357–363.

———. 1994. Systematik der Gibbons. *Zeitschrift des Kölner Zoo* 37:65–77.

Gentry, A. W. 1978. Bovidae. In *Evolution of African mammals,* V. J. Maglio and H. B. S. Cooke, eds., 540–572. Cambridge, Mass.: Harvard University Press.

———. 1994. The Miocene differentiation of Old World pecora (Mammalia). *Historical Biology* 7:115–158.

Gentry, A. W., and A. Gentry. 1978. Fossil Bovidae (Mammalia) of Olduvai Gorge, Tanzania. Part I. *Bulletin of the British Museum (Natural History), Geology* 29:289–446.

Geoffroy-Saint-Hilaire, E. 1812. Notes sur trios dessins de Commerçon, représentant des Quadrumanes d'un genre inconnu. *Annales du Muséum d'Histoire Naturelle, Paris* 19:171–175.

Geoffroy-Saint-Hilaire, I. 1851. *Catalogue méthodique de la collection des mammifées, de la collection des oiseaux et des collections annexes.* Paris.

Ghiselin, M. T. 1966. On psychologism in the logic of taxonomic controversies. *Systematic Zoology* 15:207–215.

———. 1974. A radical solution to the species problem. *Systematic Zoology* 23:536–544.

Godfrey, L. R., E. L. Simons, P. S. Chatrath, and B. Rako-tosamimanana. 1990. A new fossil lemur (*Babakotia*, Primates) from northern Madagascar. *Comptes Rendus de l'Académie des Sciences, Série II* (Paris) 310:81–87.

Godinot, M. 1998. A summary of adapiform systematics and phylogeny. *Folia Primatologica* 69 (supplement 1): 218–249.

Goldman, E. A. 1914. Descriptions of five new mammals from Panama. *Smithsonian Miscellaneous Collections 63*, no.5: 1–7.

Gonder, M. K., J. F. Oates, T. R. Disotell, M. R. J. Forst-ner, J. C. Morales, and D. J. Melnick. 1997. A new West African chimpanzee subspecies? *Nature* 388:337.

Goodman, M., C. A. Porter, J. Czelusniak, S. L. Page, H. Schneider, J. Shoshani, G. Gunnell, and C. P. Groves. 1998. Toward a phylogenetic classification of Primates based on DNA evidence complemented by fossil evidence. *Molecular Phylogenetics and Evolution* 9:585–598.

Goodman, S. M., and O. Langrand. 1996. A high moun-tain population of the ring-tailed lemur *Lemur catta* on the Andringitra Massif, Madagascar. *Oryx* 30:259–268.

Goonan, P. M., C. P. Groves, and R. D. Smith. 1995. Karyotype polymorphism in the slender loris *(Loris tar-digradus)*. *Folia Primatologica* 65:100–109.

Gray, J. E. 1821. On the natural arrangement of verte-brose animals. *London Medical Repository* 15:296–310.

———. 1825. An outline of an attempt at the disposition of Mammalia into tribes and families, with a list of the genera apparently appertaining to each tribe. *Annals of Philosophy,* n. s. 10:337–344.

———. 1870. *Catalogue of monkeys, lemurs, and fruit-eating bats in the collection of the British Museum.* London: Brit-ish Museum Trustees.

———. 1872. On the varieties of *Indris* and *Propithecus*. *Annals and Magazine of Natural History* (4) 10:474.

Grine, F. E., L. B. Martin, and J. G. Fleagle. 1986. Enamel structure in platyrrhine primates: What are the impli-cations? *American Journal of Physical Anthropology* 69:208.

Groves, C. P. 1971. Systematics of the genus *Nycticebus*. In *Proceedings of the Third International Congress of Primatol-ogy* (Zürich, 1970).Vol. 1, 44–53. Basel: S. Karger.

———. 1972. Systematics and phylogeny of gibbons. In *Gibbon and siamang*. Vol. 1, D. M. Rumbaugh, ed., 1–89. Basel: S. Karger.

———. 1974. Taxonomy and phylogeny of prosimians. In *Prosimian biology,* R. D. Martin, G. A. Doyle, and A. C. Walker, eds., 449–473. London: Duckworth.

———. 1976. The origin of the mammalian fauna of Su-lawesi (Celebes). *Zeitschrift für Säugetierkunde* 41:201–216.

———. 1978. Phylogenetic and population systematics of the mangabeys (Primates: Cercopithecoidea). *Primates* 19:1–34.

———. 1980. Species in *Macaca:* The view from Sulawesi. In *The macaques: Studies in ecology, behavior and evolu-tion,* D. G. Lindburg, ed., 84–124. New York: Van Nos-trand Reinhold Co.

———. 1986. Systematics of the Great Apes. In *Compara-tive Primate biology. 1: Systematics, evolution, and anat-omy,* D. R. Swindler and J. Erwin, eds., 187–217. New York: Alan R. Liss.

———. 1988. Gentle lemurs: New species, and how they are formed. *Australian Primatology* 3 (2/3): 9–12.

———. 1989. *A theory of human and Primate evolution*. Ox-ford: Clarendon Press.

———. 1992. Book review: Titis, New World monkeys of the genus *Callicebus* (Cebidae, Platyrrhini): A prelimi-nary taxonomic review, by P. Hershkovitz. *International Journal of Primatology* 13:111–112.

———. 1993. Speciation in living hominoid primates. In *Species, species concepts, and Primate evolution,* W. H. Kimbel and L. B. Martin, eds., 109–121. New York: Plenum Press.

———. 1996a. Taxonomic diversity in Arabian gazelles: The state of the art. In *Conservation of Arabian gazelles,* A. Greth, C. Magin, and M. Ancrenaz, eds., 8–39. Ri-yadh: National Commission for Wildlife Conservation and Development.

———. 1996b. The nomenclature of the Tanzanian man-gabey and the Siberut macaque. *Australian Primatology* 10 (4): 2–5.

———. 1997. Species concept in palaeoanthropology. *Per-spectives in Human Biology* 3:13–20.

———. 1998. Systematics of tarsiers and lorises. *Primates* 39:13–27.

———. 2000a. Phylogeny of the Cercopithecoidea. In *Old World monkeys,* P. Whitehead and C. J. Jolly, eds., 77–98. Cambridge: Cambridge University Press.

———. 2000b. The genus *Cheirogaleus:* Unrecognised bio-diversity in dwarf lemurs. *International Journal of Prima-tology* 21:943–962.

Groves, C. P., and R. H. Eaglen. 1988. Systematics of the Lemuridae. *Journal of Human Evolution* 17:513–538.

Groves, C. P., and L. B. Holthuis. 1985. The nomencla-ture of the orang utan. *Zoologische Mededelingen* (Leiden) 59:411–417.

Groves, C. P., and I. Tattersall. 1991. Geographical variation in the fork-marked lemur, *Phaner furcifer* (Primates, Cheirogaleidae). *Folia Primatologica* 56:39–49.

Groves, C. P., and R. W. Thorington. 1970. An annotated classification of the Cercopithecoidea. In *Old World monkeys*, J. R. and P. H. Napier, eds., 629–647. New York: Academic Press.

Groves, C. P., and J. W. H. Trueman. 1995. Lemurid systematics revisited. *Journal of Human Evolution* 28:427–437.

Groves, C. P., and Wang Y. 1990. The gibbons of the subgenus *Nomascus* (Primates, Mammalia). *Zoological Research* 11:147–154.

Groves, C. P., C. Westwood, and B. T. Shea. 1992. Unfinished business: Mahalanobis and a clockwork orang. *Journal of Human Evolution* 22:327–340.

Groves, C. P., R. Angst, and C. Westwood. 1993. The status of *Colobus polykomos dollmani* Schwarz. *International Journal of Primatology* 14:573–586.

Grubb, P. 1973. Distribution, divergence, and speciation of the drill and mandrill. *Folia Primatologica* 20:161–177.

Grubb, P., and C. B. Powell. 1999. Discovery of red colobus monkeys *(Procolobus badius)* in the Niger delta with the description of a new and geographically isolated subspecies. *Journal of Zoology* (London) 248:67–73.

Grubb, P., J.-M. Lernould, and J. F. Oates. 1999. Validation of *Cercopithecus erythrogaster pococki* as the name for the Nigerian white-throated guenon. *Mammalia* 63:389–392.

Hafen, T., H. Neveu, Y. Rumpler, I. Wilden, and E. Zimmermann. 1998. Acoustically dimorphic advertisement calls separate morphologically and genetically homogeneous populations of the grey mouse lemur *(Microcebus murinus)*. *Folia Primatologica* 69 (supplement 1): 342–356.

Hagen, H.-O. von. 1978. Zur Verwandtschaft der "Echten Makis" (Prosimii, Gattung *Lemur*). *Zoologische Beiträge* 24:91–122.

Haimoff, E. H., D. J. Chivers, S. P. Gittins, and A. Whitten. 1982. A phylogeny of gibbons (*Hylobates* spp.) based on morphological and behavioural characters. *Folia Primatologica* 39:213–237.

Hall, L. M., D. S. Jones, and B. A. Wood. 1998. Evolution of gibbon subgenera inferred from Cytochrome *b* DNA sequence data. *Molecular Phylogenetics and Evolution* 10:281–286.

Hall-Craggs, E. C. B. 1965. An osteometric study of the hind limb of the Galagidae. *Journal of Anatomy* 99:119–126.

Hanihara, T., and M. Natori. 1989. Evolutionary trends of the hairy-face *Saguinus* in terms of the dental and cranial morphology. *Primates* 30:531–541.

Harper, F. 1940. The nomenclature and type localities of certain Old World mammals. *Journal of Mammalogy* 21:191–203.

Harrison, M. J. S. 1988. A new species of guenon *(Cercopithecus)* from Gabon. *Journal of Zoology* (London) 215:561–575.

Hartwig, W. C., and C. Cartelle. 1996. A complete skeleton of the giant South American primate *Protopithecus*. *Nature* 381:307–311.

Hartwig, W. C., A. L. Rosenberger, P. W. Garber, and M. A. Norconk. 1996. On atelines. In *Adaptive radiations of neotropical Primates,* M. A. Norconk, A. L. Rosenberger, and P. A. Garber, eds., 427–431. New York: Plenum Press.

Hayashi, S., K. Hayasaka, O. Takenaka, and S. Horai. 1995. Molecular phylogeny of gibbons inferred from mitochondrial DNA sequences: Preliminary report. *Journal of Molecular Evolution* 41:359–365.

Hayes, V. J., L. Freedman, and C. E. Oxnard. 1990. The taxonomy of savannah baboons: An odontomorphometric analysis. *American Journal of Primatology* 22:171–190.

Hayman, R. W. 1937. A note on *Galago senegalensis inustus* Schwarz. *Annals and Magazine of Natural History* (10) 20:149–151.

Heltne, P. G., and L. M. Kunkel. 1975. Taxonomic notes on the pelage of *Ateles paniscus paniscus, A. p. chamek* (*sensu* Kellogg and Goldman, 1944), and *A. fusciceps rufiventris* (= *A. f. robustus* Kellogg and Goldman, 1944). *Journal of Medical Primatology* 4:83–102.

Henneberg, M. 1997. The problem of species in hominid evolution. *Perspectives in Human Biology* 3:21–31.

Henneberg, M., and G. Brush. 1994. Similum, a concept of flexible synchronous classification replacing rigid species in evolutionary thinking. *Evolutionary Theory* 10:278.

Hennig, W. 1950. *Grundzüge einer Theorie der phylogenetischen Systematik.* Berlin: Deutsches Zentralverlagen.

———. 1966. *Phylogenetic systematics.* Translated by D. D. Davis and R. Zangerl. Urbana: University of Illinois Press.

———. 1969. *Die Stammesgeschichte der Insekten.* Frankfurt: E. Kramer.

Hernández-Camacho, J., and R. W. Cooper. 1976. The nonhuman Primates of Colombia. In *Neotropical Primates: Field studies and conservation,* R. W. Thorington and P. G. Heltne, eds., 35–69. Washington: National Academy of Sciences.

Hershkovitz, P. 1949. Mammals of Northern Colombia. Preliminary report no. 4: Monkeys (Primates), with taxonomic revisions of some forms. *Proceedings of the United States National Museum* 98:323–427.

———. 1958. Type localities and nomenclature of some American Primates, with remarks on secondary homonyms. *Proceedings of the Biological Society of Washington* 71:53–56.

———. 1963. A systematic and zoogeographic account of the monkeys of the genus *Callicebus* (Cebidae) of the Amazonas and Orinoco River basins. *Mammalia* 27:1–80.

———. 1966. Taxonomic notes on tamarins, genus *Saguinus* (Callithricidae, Primates), with descriptions of four new forms. *Folia Primatologica* 4:381–395.

———. 1968. Metachromism or the principle of evolutionary change in mammalian tegumentary colors. *Evolution* 22:556–575.

———. 1970. Cerebral fissural patterns in platyrrhine monkeys. *Folia Primatologica* 13:213–240.

———. 1977. *Living New World monkeys (Platyrrhini), with an introduction to Primates.* Vol. 1. Chicago: University of Chicago Press.

———. 1979. Races of the emperor tamarin, *Saguinus imperator* Goeldi (Callitrichidae, Primates). *Primates* 20:277–287.

———. 1983. Two new species of night monkeys, genus *Aotus* (Cebidae, Platyrrhini): A preliminary report on *Aotus* taxonomy. *American Journal of Primatology* 4:209–243.

———. 1984. Taxonomy of squirrel monkeys, genus *Saimiri* (Cebidae, Platyrrhini): A preliminary report with description of a hitherto unnamed form. *American Journal of Primatology* 7:155–210.

———. 1985. A preliminary taxonomic review of the South American bearded saki monkeys, genus *Chiropotes* (Cebidae, Platyrrhini), with the description of a new subspecies. *Fieldiana Zoology,* n. s. 27: iii + 46 pp.

———. 1987a. Uacaries, New World monkeys of the genus *Cacajao* (Cebidae, Platyrrhini): A preliminary taxonomic review with the description of a new subspecies. *American Journal of Primatology* 12:1–53.

———. 1987b. The taxonomy of South American sakis, genus *Pithecia* (Cebidae, Platyrrhini): A preliminary report and critical review with the description of a new species and a new subspecies. *American Journal of Primatology* 12:387–468.

———. 1990. Titis, New World monkeys of the genus *Callicebus* (Cebidae, Platyrrhini): A preliminary taxonomic review. *Fieldiana Zoology,* n. s. 55:v + 109 pp.

Hill, J. E., and T. D. Carter. 1939. The mammals of Angola, Africa. *Bulletin of the American Museum of Natural History* 78:3–211.

Hill, W. C. O. 1934. A monograph on the purple-faced leaf-monkeys *(Pithecus vetulus).* Ceylon Journal of Science, Section B, Zoology 19:23–88.

———. 1936. Supplementary observations on purple-faced leaf-monkeys (genus *Kasi*). *Ceylon Journal of Science, Section B, Zoology* 20:115–133.

———. 1939. An annotated systematic list of the leaf-monkeys. *Ceylon Journal of Science* 21:277–305.

———. 1952. *Primates: Comparative anatomy and taxonomy. I. Strepsirhini.* Edinburgh: Edinburgh University Press.

———. 1953. Notes on the taxonomy of the genus *Tarsius. Proceedings of the Zoological Society of London* 123:13–16.

———. 1955. *Primates: Comparative anatomy and taxonomy. II. Haplorhini: Tarsioidea.* Edinburgh: Edinburgh University Press.

———. 1957. *Primates: Comparative anatomy and taxonomy. III. Hapalidae.* Edinburgh: Edinburgh University Press.

———. 1960. *Primates: Comparative anatomy and taxonomy. IV. Cebidae, Part A.* Edinburgh: Edinburgh University Press.

———. 1962. *Primates: Comparative anatomy and taxonomy. V. Cebidae, Part B.* Edinburgh: Edinburgh University Press.

———. 1966. *Primates: Comparative anatomy and taxonomy. VI. Catarrhini, Cercopithecoidea, Cercopithecinae.* Edinburgh: Edinburgh University Press.

———. 1970. *Primates: Comparative anatomy and taxonomy. VIII. Cynopithecinae: Papio, Mandrillus, Theropithecus.* Edinburgh: Edinburgh University Press.

———. 1974. *Primates: Comparative anatomy and taxonomy. VII. Cynopithecinae: Cecocebus, Macaca, Cynopithecus.* Edinburgh: Edinburgh University Press.

Hinton, M. A. C. 1929. Mr. M. R. Oldfield Thomas, F.R.S. *Nature* 124:101–102.

Hofer, H. O. 1976. Preliminary study of the comparative anatomy of the external nose of South American monkeys. *Folia Primatologica* 23:193–214.

Hohmann, G. 1988. Analysis of loud calls provides new evidence for hybridization between two Asian leaf monkeys *(Presbytis johnii, Presbytis entellus). Folia Primatologica* 51:209–213.

Hohmann, G., and M. Herzog. 1985. Die braunen Languren Südindiens. *Zeitschrift des Kölner Zoo* 28:37–41.

Honess, P. E. 1996. Speciation among galagos (Primates, Galagidae) in Tanzanian forests. Ph.D. thesis, Oxford Brookes University, United Kingdom.

Hooijer, D. A. 1962. Quaternary langurs and macaques from the Malay archipelago. *Zoologische Verhandelingen, Leiden* 55:1–64.

———. 1972. A late Pliocene rhinoceros from Langebaanweg, Cape Province. *Annals of the South African Museum* 59:151–191.

Horovitz, I., R. Zardoya, and A. Meyer. 1998. Platyrrhine systematics: A simultaneous analysis of molecular and morphological data. *American Journal of Physical Anthropology* 106:261–282.

Hrdy, S. B. 1977. *The langurs of Abu.* Cambridge, Mass.: Harvard University Press.

Hull, D. B. 1979. A craniometric study of the black and white *Colobus* Illiger 1811 (Primates: Cercopithecoidea). *American Journal of Physical Anthropology* 51:163–181.

Husson, A. M. 1957. Notes on the Primates of Suriname. *Studies of the Fauna of Suriname and Other Guyanas* 1:13–40.

———. 1978. *The mammals of Suriname.* Zoological Monographs of the Rijksmuseum van Natuurlijke Historie, Leiden, no. 2. Leiden: E. J. Brill.

Huxley, J. S. (ed.) 1940. *The new systematics.* Oxford: Clarendon Press.

Huxley, J. S. 1942. *Evolution: The modern synthesis.* London: Allen and Unwin.

Inagaki, H. 1996. Some hair characteristics of *Macaca* monkeys and an attempt to group them based on those features. In *Variations in the Asian macaques,* T. Shotake and K. Wada, eds., 89–96. Tokyo: Tokai University Press.

International Commission on Zoological Nomenclature. 1929. Opinion 114. Under suspension *Simia, Simia satyrus,* and *Pithecus* are suppressed. *Smithsonian Miscellaneous Collections* 73:423–424.

———. 1954a. Opinion 238. Validation, under the plenary powers, of the generic name *Cercopithecus* as from Linnaeus, 1758 (Class Mammalia). *Opinions and Declarations of the International Commission on Zoological Nomenclature* 4 (28): 25–26.

———. 1954b. Opinion 257. Rejection for nomenclatorial purposes of the work by Zimmermann (A. E. W. von [sic]) published in 1777 under the title Specimen Zoologicae Geographicae, Quadrupedum Domicilia et Migrationes Sistens and acceptance for the same purposes of the work by the same author published in the period 1778–1783 under the title Geographische Geschichte des Menschen, und der Allgemein Verbreiteten Vierfüssigen Thiere. *Opinions and Declarations of the International Commission on Zoological Nomenclature* 5 (18): 231–244.

———. 1956. Rejection for nomenclatorial purposes of volume 3 (Zoologie) of the work by Lorenz Oken entitled *Okens Lehrbuch der Naturgeschichte* published in 1815–1816. *Opinions and Declarations Rendered by the International Commission on Zoological Nomenclature* 14:1–42.

———. 1970. Opinion 935. *Simia leucophaea* F. Cuvier, 1807 (Mammalia): Validated under the plenary powers. *Bulletin of Zoological Nomenclature* 27:171–172.

———. 1982a. Opinion 1199. *Papio* Erxleben, 1777, and *Mandrillus* Ritgen, 1824 (Mammalia, Primates): Designation of type species. *Bulletin of Zoological Nomenclature* 39:15–18.

———. 1982b. Opinion 1219. *Homo lar* Linnaeus, 1771, neotype designated; *Hylobates entelloides* I. Geoffroy St. Hilaire, 1842 and *Simia hoolock* Harlan, 1834 (Mammalia, Primates): Placed on the official list. *Bulletin of Zoological Nomenclature* 39:168–171.

———. 1982c. Opinion 1224. *Simia syndactyla* Raffles, 1821 (Mammalia, Hylobatidae): Given precedence over *Simia gibbon* C. Miller, 1779. *Bulletin of Zoological Nomenclature* 39:183–185.

———. 1985a. Opinion 1329. *Galago crassicaudatus* E. Geoffroy, 1812 (Primates, Galagidae [sic]): Neotype designated. *Bulletin of Zoological Nomenclature* 42:226–227.

———. 1985b. Opinion 1368. The generic names *Pan* and *Panthera* (Mammalia, Carnivora): Available as from Oken, 1816. *Bulletin of Zoological Nomenclature* 42:365–370.

———. 1986. Opinion 1400. *Simia fascicularis* Raffles, 1821 (Mammalia, Primates): Conserved. *Bulletin of Zoological Nomenclature* 43:229–230.

Izawa, K., and G. Bejarano. 1981. Distribution ranges and patterns of nonhuman Primates in Western Pando, Bolivia. *Kyoto University Overseas Research Reports of New World Monkeys* 2:1–11.

Jablonski, N. G., and Peng Y. 1993. The phylogenetic relationships and classification of the doucs and snub-nosed langurs of China and Vietnam. *Folia Primatologica* 60:36–55.

Janczewski, D. N., D. Goldman, and S. J. O'Brien. 1990. Molecular genetic divergence of orang utan *(Pongo pygmaeus)* subspecies based on isozyme and two-dimensional gel electrophoresis. *Journal of Heredity* 81:375–387.

Jenkins, P. D. 1987. *Catalogue of Primates in the British Museum (Natural History) and elsewhere in the British Isles.*

Part IV: Suborder Strepsirrhini, including the subfossil Madagascan lemurs and family Tarsiidae. London: British Museum (Natural History).

———. 1990. *Catalogue of Primates in the British Museum (Natural History) and elsewhere in the British Isles. Part V: The apes, superfamily Hominoidea.* London: British Museum (Natural History).

Jiang X., Wang Y., and Ma S. 1991. Taxonomic revision and distribution of subspecies of rhesus monkey *(Macaca mulatta)* in China. *Zoological Research* 12:241–247 [in Chinese].

Jolly, A. 1966. *Lemur behavior.* Chicago: University of Chicago Press.

Jolly, C. J. 1972. The classification and natural history of *Theropithecus (Simopithecus)* (Andrews, 1916), baboons of the African Plio-Pleistocene. *Bulletin of the British Museum (Natural History) Geology* 22:1–123.

———. 1993. Species, subspecies, and baboon systematics. In *Species, species concepts, and Primate evolution,* W. H. Kimbel and L. B. Martin, eds., 67–107. New York: Plenum Press.

Jolly, C. J., and P. J. Ucko. 1969. The riddle of the Sphinx-monkey. In *Man in Africa,* M. Douglas and P. M. Kaberry, eds., 319–335. London: Tavistock Publications.

Jolly, C. J., T. Woolley-Barker, S. Beyene, T. R. Disotell, and J. E. Phillips-Conroy. 1997. Intergeneric hybrid baboons. *International Journal of Primatology* 18:597–627.

Jung, K. Y., S. Crovella, and Y. Rumpler. 1992. Phylogenetic relationships among lemuriform species determined from restriction genomic DNA banding patterns. *Folia Primatologica* 58:224–229.

Jungers, W. L., and Y. Rumpler. 1976. Craniometric corroboration of the specific status of *Lepilemur septentrionalis,* an endemic lemur from the north of Madagascar. *Journal of Human Evolution* 5:317–321.

Jungers, W. L., L. R. Godfrey, E. L. Simons, P. S. Chatrath, and B. Rakotosamimanana. 1991. Phylogenetic and functional affinities of *Babakotia* (Primates), a fossil lemur from northern Madagascar. *Proceedings of the National Academy of Sciences of the United States of America* 88:9082–9086.

Kawamura, S. 1984. Distribution and vocalization of *Presbytis melalophos* and *P. femoralis* varieties in West Central Sumatra—A summarized report. *Kyoto University Overseas Research Report of Studies on Asian Non-Human Primates* 3:37–44.

Kay, R. F. 1990. The phyletic relationships of extant and fossil Pitheciinae (Anthropoidea, Platyrrhini). *Journal of Human Evolution* 19:175–208.

Kay, R. F., and B. A. Williams. 1994. Dental evidence for anthropoid origins. In *Anthropoid origins,* J. G. Fleagle and R. F. Kay, eds., 361–445. New York: Plenum Press.

Kellogg, R., and E. A. Goldman. 1944. Review of the spider monkeys. *Proceedings of the United States National Museum* 96:1–45.

Khajuria, H., and V. C. Agrawal. 1979. On the types of *Presbytis barbei* Blyth. *Primates* 20:317–319.

Kingdon, J. 1971. *Atlas of East African mammals.* Vol. 1. London: Academic Press.

———. 1980. Role of visual signals and face patterns in African forest monkeys (guenons) of the genus *Cercopithecus. Transactions of the Zoological Society of London* 35:425–475.

———. 1988. What are face patterns and do they contribute to reproductive isolation in guenons? In *A Primate radiation: Evolutionary biology of the African guenons,* A. Gautier-Hion, F. Bourlière, J.-P. Gautier, and J. Kingdon, eds., 227–245. Cambridge: Cambridge University Press.

———. 1997. *The Kingdon field guide to African mammals.* London: Academic Press.

Kloss, C. B. 1929. Some remarks on the gibbons, with the description of a new subspecies. *Proceedings of the Zoological Society of London* 1929:113–127.

Kobayashi, S. 1995. A phylogenetic study of titi monkeys, genus *Callicebus,* based on cranial measurements: I. Phyletic groups of *Callicebus. Primates* 36:101–120.

Kobayashi, S., and A. Langguth. 1999. A new species of titi monkey, *Callicebus* Thomas, from north-eastern Brazil (Primates, Cebidae). *Revista Brasileira de Zoologia* 16:531–551.

Kornet, D. J., and J. W. McAllister. 1993. The composite species concept. In D. J. Kornet, *Reconstructing species: Demarcations in genealogical networks,* 61–89. Leiden: Instituut voor Theoretische Biologie, Rijksherbarium.

Kuhn, T. S. 1962. *The structure of scientific revolutions.* Chicago: University of Chicago Press.

Lawrence, B. 1933. Howler monkeys of the *palliata* group. *Bulletin of the Museum of Comparative Zoology* (Harvard) 75:314–354.

Lee, M. S. Y. 1997. Species concepts and the recognition of ancestors. *Historical Biology* 10:329–339.

Leigh, S. R., and W. L. Jungers. 1994. Brief communication: A re-evaluation of subspecific variation and canine dimorphism in woolly spider monkeys *(Brachyteles arachnoides). American Journal of Physical Anthropology* 95:435–442.

Lemos de Sá, R. M., T. R. Pope, K. E. Glander, T. T. Struhsaker, and G. A. B. de Fonseca. 1990. A pilot study of genetic and morphological variation in the muriqui (*Brachyteles arachnoides*). *Primate Conservation* 11:26–30.

Lemos de Sá, R. M., T. R. Pope, T. T. Struhsaker, and K. E. Glander. 1993. Sexual dimorphism in canine length of woolly spider monkeys (*Brachyteles arachnoides*, E. Geoffroy 1806). *International Journal of Primatology* 14:755–763.

Leo Luna, M. 1987. Primate conservation in Peru: A case study of the yellow-tailed woolly monkey. *Primate Conservation* 8:122–123.

Lernould, J.-M. 1988. Classification and geographical distribution of guenons: A review. In *A Primate radiation: Evolutionary biology of the African guenons*, A. Gautier-Hion, F. Bourlière, J.-P. Gautier, and J. Kingdon, eds., 54–78. Cambridge: Cambridge University Press.

Lesson, R. P. 1840. *Species des mammifères bimanes et quadrumanes.* Paris: J. B. Baillière.

Li, Z., and Z. Lin. 1983. Classification and distribution of living Primates in Yunnan China. *Zoological Research* 4:111–120.

Li, Z., and S. Ma. 1980. A revision of the white-headed langur. *Acta Zootaxonomica Sinica* 5:440–442.

Lima, M. M. C. de, and H. N. Seuánez. 1989. Cytogenetic characterization of *Alouatta belzebul* with atypical pelage coloration. *Folia Primatologica* 52:97–101.

Lima, M. M. C. de, M. I. C. Sampaio, M. P. C. Schneider, W. Scheffrahn, H. Schneider, and F. M. Salzano. 1990. Chromosome and protein variation in red howler monkeys. *Brazilian Journal of Genetics* 4:789–802.

Lima, M., I. Shalqueiro, M. Pinheiro, and E. H. de Oliveira. 1997. Chromosome variability in four species of *Alouatta* (Primates, Cebidae). *A Primatologia no Brasil* 5:362–363.

Linnaeus, C. 1758. *Systema Naturae per Regna Tria Naturae, Secundum Classes, Ordines, Genera, Species, cum Characteribus, Differentiis, Synonymis, Locis,* 10th ed. Stockholm: Laurentius Salvius.

———. 1766. *Systema Naturae per Regna Tria Naturae, Secundum Classes, Ordines, Genera, Species, cum Characteribus, Differentiis, Synonymis, Locis,* 12th ed. Stockholm: Laurentius Salvius.

Lönnberg, E. 1939. Remarks on some members of the genus *Cebus. Arkiv för Zoologi* 31A, 23:1–24.

———. 1940. Notes on some members of the genus *Saimiri. Arkiv för Zoologi* 32A, 21:1–18.

Lorenz von Liburnau, L. 1917. Beitrag zur Kenntnis der Affen und Halbaffen von Zentralafrika. *Annalen von der naturhistorisches Hofmus, Wien* 31:169–241.

Lorini, M. L., and V. G. Persson. 1990. Nove espécie de *Leontopithecus* Lesson, 1840, do sul do Brasil (Primates, Callitrichidae). *Boletim do Museu Nacional,* n. s., *Zoologia* (Rio de Janeiro) 338:1–14.

Lowther, F. de L. 1940. A study of the activities of a pair of *Galago senegalensis moholi* in captivity. *Zoologica* 25:433–462.

Loy, J. 1987. The sexual behavior of African monkeys and the question of estrus. In *Comparative behavior of African monkeys,* E. L. Zucker, ed., 197–234. New York: Alan R. Liss.

Luckett, W. P. 1993. Developmental evidence from the fetal membranes for assessing Archontan relationships. In *Primates and their relatives in phylogenetic perspective,* R. D. E. MacPhee, ed., 149–186. New York: Plenum Press.

Luckett, W. P., and W. Maier. 1986. Developmental evidence for anterior tooth homologies in the aye-aye, *Daubentonia. American Journal of Physical Anthropology* 69:233.

Ma S., and Wang Y. 1986. The taxonomy and distribution of the gibbons in southern China and its adjacent region—with description of three new subspecies. *Zoological Research* 7:393–410.

Ma, S., Y. Wang, and F. E. Poirier. 1988. Taxonomy, distribution, and status of gibbons (*Hylobates*) in southern China and adjacent areas. *Primates* 29:277–286.

Macedonia, J. M., and K. F. Stanger. 1994. Phylogeny of the Lemuridae revisited: Evidence from communication signals. *Folia Primatologica* 63:1–43.

Machado, A. de B. 1969. Mamíferos de Angola ainda não citados ou pouco conhecidos. *Publicações Culturais da Companhia de Diamantes de Angola* 46:93–232.

Machida, H., E. Perkins, and L. Giacometti. 1966. The skin of Primates, XXIX. The skin of the pigmy bushbaby (*Galago demidovii*). *American Journal of Physical Anthropology* 24:199–203.

Maddison, W. P. 1997. Gene trees in species trees. *Systematic Biology* 46:523–536.

Maddison, W. P., and D. R. Maddison. 1992. *MacClade: Analysis of phylogeny and character evolution.* Version 3. Sunderland, Mass.: Sinauer Associates.

Maples, W. R. 1972. Systematic reconsideration and a revision of the nomenclature of Kenya baboons. *American Journal of Physical Anthropology* 36:9–19.

Marroig, G. 1995. Espécies ou subespécies em *Callithrix? Neotropical Primates* 3:10–13.

Marshall, J. T., and E. Marshall. 1976. Gibbons and their territorial songs. *Science* 193:235–237.

Marshall, J. T., and J. Sugardjito. 1986. Gibbon systematics. In *Comparative Primate biology. 1: Systematics,* D. R. Swindler and J. Erwin, eds., 137–185. New York: Alan R. Liss.

Martin, L. D. 1989. Fossil history of the terrestrial Carnivora. In *Carnivore behavior, ecology, and evolution,* J. L. Gittleman, ed., 536–568. Ithaca, N.Y.: Comstock Publishing Associates.

Martin, R. D. 1990. *Primate origins and evolution: A phylogenetic reconstruction.* Princeton: Princeton University Press.

———. 1995. Prosimians: From obscurity to extinction? In *Creatures of the dark: The nocturnal prosimians,* L. Alterman, G. A. Doyle, and M. K. Izard, eds., 535–563. New York: Plenum Press.

Martin, R. D., and A. M. MacLarnon. 1988. Quantitative comparisons of the skull and teeth in guenons. In *A Primate radiation: Evolutionary biology of the African guenons,* A. Gautier-Hion, F. Bourlière, J.-P. Gautier, and J. Kingdon, eds., 160–183. Cambridge: Cambridge University Press.

Masters, J. C. 1986. Geographic distributions of karyotypes and morphotypes within the greater galagos. *Folia Primatologica* 46:127–141.

———. 1988. Speciation in the greater galagos (Prosimii: Galaginae): A review and synthesis. *Biological Journal of the Linnean Society* 34:149–174.

———. 1991. Loud calls of *Galago crassicaudatus* and *G. garnettii* and their relation to habitat structure. *Primates* 32:153–167.

———. 1993. Primates and paradigms: Problems with the identification of genetic species. In *Species, species concepts, and Primate evolution,* W. H. Kimbel and L. B. Martin, eds., 43–64. New York: Plenum Press.

Masters, J. C., W. H. R. Lumsden, and D. A. Young. 1988. Reproductive and dietary parameters in wild greater galago populations. *International Journal of Primatology* 9:573–592.

Masters, J. C., R. J. Rayner, H. Ludewick, E. Zimmermann, N. Molez-Verrière, F. Vincent, and L. T. Nash. 1994. Phylogenetic relationships among the Galaginae as indicated by erythrocytic allozymes. *Primates* 35:177–190.

Materson, T. J. 1995. Morphological relationships between the Ka'apor capuchin (*Cebus kaapori* Queiroz, 1992) and other male *Cebus* crania: A preliminary report. *Neotropical Primates* 3:165–169.

Matschie, P. 1904. Bemerkungen über die Gattung *Gorilla.* *Sitzungsberichte der Gesellschaft der naturforschenden Freunde* (Berlin) 1904:45–53.

Mayr, E. 1940. Speciation phenomena in birds. *American Naturalist* 74:249–278.

———. 1942. *Systematics and the origin of species.* New York: Columbia University Press.

———. 1963. *Animal species and evolution.* Cambridge, Mass.: Harvard University Press, Belknap Press.

———. 1969. *Principles of systematic zoology.* New York: McGraw-Hill.

Mayr, E., and P. D. Ashlock. 1991. *Principles of systematic zoology,* 2d ed. New York: McGraw-Hill.

Mayr, E., E. G. Linsley, and R. L. Usinger. 1953. *Methods and principles of systematic zoology.* New York: McGraw-Hill.

Medeiros, M. A. A., M. Ponsá, M. Garcia, F. Garcia, J. C. Pieczarka, C. Y. Nagamachi, J. Egozcue, and R. M. S. Barros. 1997. Radiation and speciation of the genus *Ateles* analysed chromosomally. *A Primatologia no Brasil* 5:361–362.

Meester, J. A. J., I. L. Rautenbach, N.J. Dippenaar, and C. M. Baker. 1986. *Classification of southern African mammals.* Monograph No. 5. Pretoria: Transvaal Museum.

Meier, B., and R. Albignac. 1991. Rediscovery of *Allocebus trichotis* Günther 1875 (Primates) in Northeast Madagascar. *Folia Primatologica* 56:57–63.

Meier, B., R. Albignac, A. Peyriéras, Y. Rumpler, and P. Wright. 1987. A new species of *Hapalemur* (Primates) from South East Madagascar. *Folia Primatologica* 48:211–215.

Melnick, D. J., G. A. Hoelzer, R. Absher, and M. V. Ashley. 1993. mtDNA diversity in rhesus monkeys reveals overestimates of divergence time and paraphyly with neighbouring species. *Molecular Biology and Evolution* 10:282–295.

Mendes, S. L. 1997. Hybridization in free-ranging *Callithrix flaviceps* and the taxonomy of the Atlantic Forest marmosets. *Neotropical Primates* 5:6–8.

Miller, G. S. 1903. Seventy new Malayan mammals. *Smithsonian Miscellaneous Collections* 45:1–73.

———. 1934. The langurs of the *Presbytis femoralis* group. *Journal of Mammalogy* 15:124–137.

Milne-Edwards, A., and A. Grandidier. 1890. *Histoire physique, naturelle et politique de Madagascar. 10. Histoire Naturelle des Mammifères. 5* (Atlas) 2. Paris.

Minezawa, M., M. Harada, O. C. Jordan C., and C. J. V. Borda. 1985. Cytogenetics of Bolivian endemic red howler monkeys (*Alouatta seniculus sara*): Accessory chromosomes and Y-autosome translocation related numerical variation. *Kyoto Overseas Research Reports of New World Monkeys* 5:7–16.

Mittermeier, R. A., A. B. Rylands, and A. F. Coimbra-Filho. 1988. Systematics: Species and subspecies—an update. In *Ecology and behavior of neotropical Primates.* Vol. 2, R. A. Mittermeier, A. B. Rylands, A. F. Coimbra-Filho, and G. A. B. da Fonseca, eds., 13–75. Washington, D.C.: WWF.

Mittermeier, R. A., M. Schwarz, and J. M. Ayres. 1992. A new species of marmoset, genus *Callithrix* Erxleben, 1777 (Callitrichidae, Primates) from the Rio Maués Region, state of Amazonas, central Brazilian Amazonia. *Goeldiana Zoologia* 14:1–17.

Mittermeier, R. A., I. Tattersall, W. R. Konstant, D. M. Meyers, and R. B. Mast. 1994. *Lemurs of Madagascar.* Washington, D.C.: Conservation International.

Montagna, W., and J. S. Yun. 1962. The skin of Primates, VII. The skin of the great bushbaby *(Galago crassicaudatus). American Journal of Physical Anthropology* 20:149–166.

Montagnon, D., S. Crovella, and Y. Rumpler. 1993. Comparison of highly repeated DNA sequences in some Lemuridae and taxonomic implications. *Cytogenetics and Cell Genetics* 63:131–134.

Morales, J. C., and D. J. Melnick. 1998. Phylogenetic relationships of the macaques (Cercopithecidae: *Macaca),* as revealed by high resolution restriction site mapping of mitochondrial ribosomal genes. *Journal of Human Evolution* 34:1–23.

Morbeck, M. E., and A. L. Zihlman. 1989. Body size and proportions in chimpanzees, with special reference to *Pan troglodytes schweinfurthii* from Gombe National Park, Tanzania. *Primates* 30:369–382.

Morin, P. A., J. J. Moore, R. Chakraborty, L. Jin, J. Goodall, and D. S. Woodruff. 1994. Kin selection, social structure, gene flow, and the evolution of chimpanzees. *Science* 265:1193–1201.

Mouri, T. 1994. Distribution of lacrimal fossa in cercopithecids. *Anthropological Science* 102:395–407.

Moyà-Solà, S., and M. Köhler. 1996. A *Dryopithecus* skeleton and the origins of great-ape locomotion. *Nature* 379:156–159.

Mudry, M. D., M. Rahn, M. Gorostiaga, A. Hick, M. S. Merani, and A. J. Solari. 1998. Revised karyotype of *Alouatta caraya* (Primates: Platyrrhini) based on synaptonemal complex and banding analyses. *Hereditas* 128:9–16.

Müller, S. 1839. Die zoogdieren van den Indischen Archipel. In *Verhandelingen over de natuurlijke geschiednis der Nederlandsche overzeesche bezittingen, Zoologie (Mammalia),* C. J. Temminck, ed., 1–57. Leiden: E. J. Brill.

Musser, G. G., and M. Dagosto. 1987. The identity of *Tarsius pumilus,* a pygmy species endemic to the montane mossy forests of central Sulawesi. *American Museum Novitates* 2867:1–53.

Nadler, T. 1996. Verbreitung und Status von Delacour-, Tonkin-, und Goldschopf-languren *(Trachypithecus delacouri, Trachypithecus francoisi,* und *Trachypithecus poliocephalus)* in Vietnam. *Zoologische Garten,* n. f. 66:1–12.

———. 1997. A new subspecies of douc langur, *Pygathrix nemaeus cinereus* ssp. nov. *Zoologische Garten,* n. f. 67:165–176.

———. 1998. Black langur rediscovered. *Asian Primates* 6:10–12 (for December 1996/March 1997).

Nagamachi, C. Y., J. C. Pieczarka, R. M. S. Barros, M. Schwarz, J. A. P. C. Muniz, and M. S. Mattevi. 1996. Chromosomal relationships and phylogenetic and clustering analyses on genus *Callithrix,* group *argentata* (Callitrichidae, Primates). *Cytogenetics and Cell Genetics* 72:331–338.

Napier, J. R., and P. H. Napier. 1967. *A handbook of living Primates.* London: Academic Press.

Napier, P. H. 1976. *Catalogue of Primates in the British Museum (Natural History). Part I: Families Callitrichidae and Cebidae.* London: British Museum (Natural History).

———. 1981. *Catalogue of Primates in the British Museum (Natural History) and elsewhere in the British Isles. Part II: Family Cercopithecidae, subfamily Cercopithecinae.* London: British Museum (Natural History).

———. 1985. *Catalogue of Primates in the British Museum (Natural History) and elsewhere in the British Isles. Part III: Family Cercopithecidae, subfamily Colobinae.* London: British Museum (Natural History).

Nash, L. T., S. K. Bearder, and T. R. Olson. 1989. Synopsis of *Galago* species characteristics. *International Journal of Primatology* 10:57–80.

Natori, M. 1986. Interspecific relationships of *Callithrix* based on dental characters. *Primates* 27:321–336.

———. 1988. A cladistic analysis of interspecific relationships of *Saguinus. Primates* 29:263–276.

———. 1990. Numerical analysis of the taxonomical status of *Callithrix kuhlii* based on measurements of the postcanine dentition. *Primates* 31:555–562.

Natori, M., and T. Hanihara. 1988. An analysis of interspecific relationships of *Saguinus* based on cranial measurements. *Primates* 29:255–262.

Neff, N. A. 1991. The cats and how they came to be. In *Great cats: Majestic creatures of the wild,* J. Seidensticker and S. Lumpkin, eds., 15–27. Sydney: Weldon Own Pty Ltd.

Nelson, G. J. 1972. Phylogenetic relationships and classification. *Systematic Zoology* 21:227–230.

Nelson, G. J., and N. I. Platnick. 1981. *Systematics and biogeography.* New York: Columbia University Press.

Nhat, P., D. Tuoc, and T. V. La. 1998. Preliminary survey for Hatinh langur in North Central Vietnam. *Asian Primates* 6:13–17 (for December 1996/March 1997).

Niemitz, C. 1984 Taxonomy and distribution of the genus *Tarsius* Storr, 1780. In *Biology of tarsiers,* C. Niemitz, ed., 1–16. Stuttgart: Gustav Fischer Verlag.

Niemitz, C., A. Nietsch, S. Warter, and Y. Rumpler. 1991. *Tarsius dianae:* A new primate species from central Sulawesi (Indonesia). *Folia Primatologica* 56:105–116.

Nijman, V. 1997. Geographical variation in pelage characteristics in *Presbytis comata* (Desmarest, 1822) (Mammalia, Primates, Cercopithecidae). *Zeitschrift für Säugetierkunde* 62:257–264.

Nishida, T., J. Itani, M. Hiraiwa, and T. Hasegawa. 1981. A newly-discovered population of *Colobus angolensis* in East Africa. *Primates* 22:557–563.

Nixon, K. C., and Q. D. Wheeler. 1990. An amplification of the phylogenetic species concept. *Cladistics* 6:211–223.

Nogami, Y., and M. Yoneda. 1983. Structural patterns of enamel in the superfamily Ceboidea. *Primates* 24:567–574.

Nozawa, K., T. Shotake, M. Minezawa, Y. Kawamoto, K. Hayasaka, and S. Kawamoto. 1996. Population-genetic studies of the Japanese macaque, *Macaca fuscata.* In *Variations in the Asian macaques,* T. Shotake and K. Wada, eds., 1–36. Tokyo: Tokai University Press.

Oates, J. F. 1981. Mapping the distribution of West African rain-forest monkeys: Issues, methods, and preliminary results. *Annals of the New York Academy of Sciences* 376:53–64.

———. 1985. The Nigerian guenon, *Cercopithecus erythrogaster:* Ecological, behavioral, systematic and historical observations. *Folia Primatologica* 45:25–43.

———. 1988. The distribution of *Cercopithecus* monkeys in West African forests. In *A Primate radiation: Evolutionary biology of the African guenons,* A. Gautier-Hion, F. Bourlière, J.-P. Gautier, and J. Kingdon, eds., 79–103. Cambridge: Cambridge University Press.

———. 1994. The Niger delta's red colobus monkey: A new subspecies? *African Wildlife Update* March–April 1994: 4.

Oates, J. F., and P. A. Anadu. 1989. A field observation of Sclater's guenon (*Cercopithecus sclateri* Pocock, 1904). *Folia Primatologica* 52:93–96.

Oates, J. F., and T. F. Trocco. 1983. Taxonomy and phylog-

eny of black-and-white colobus monkeys. *Folia Primatologica* 40:83–113.

Oates, J. F., P. A. Anadu, E. L. Gadsby, and L. O. Werre. 1992. Sclater's guenon. *Research and Exploration* 8:476–491.

Oates, J. F., A. G. Davies, and E. Delson. 1996. The diversity of living colobines. In *Colobine monkeys: Their ecology, behaviour, and evolution,* A. G. Davies and J. F. Oates, eds., 45–73. Cambridge: Cambridge University Press.

Oliveira, E. H. de, M. M. C. de Lima, and I. J. Shalqueiro. 1995. Chromosomal variation in *Alouatta fusca. Neotropical Primates* 3:181–183.

Olson, T. R. 1979. Studies on aspects of the morphology and systematics of the genus *Otolemur.* Ph.D. thesis, University of London.

———. 1980. *Galago crassicaudatus* E. Geoffroy, 1812 (Primates: Galagidae): Proposed use of the plenary powers to suppress the holotype and to designate a neotype. *Bulletin of Zoological Nomenclature* 37:176–185.

———. 1981. Systematics and zoogeography of the greater galagos. *American Journal of Physical Anthropology* 54:259.

Oxnard, C. E. 1981. The uniqueness of *Daubentonia. American Journal of Physical Anthropology* 54:1–21.

Pan R., Peng Y., Ye Z., Wang H., and Yu F. 1992. Classification and relationships of the macaque population on Hainan Island, China. *Folia Primatologica* 59:39–43.

Pariente, G. 1970. Rétinographies comparées des Lémuriens malgaches. *Comptes Rendus de l'Académie des Sciences* (Paris) 270:1404–1407.

Passamani, M., L. M. S. Aguiar, R. B. Machado, and E. Figueiro. 1997. Hybridization between *Callithrix geoffroyi* and *C. penicillata* in southeastern Minas Gerais, Brazil. *Neotropical Primates* 5:9–10.

Paterson, H. E. H. 1986. Environment and species. *South African Journal of Science* 82:62–65.

Patterson, C. 1980. Cladistics. *Biologist* 27:234–240.

Patterson, C., and D. E. Rosen. 1977. Review of ichthyodectiform and other Mesozoic teleost fishes and the theory and practice of classifying fossils. *Bulletin of the American Museum of Natural History* 158:81–172.

Peng Y., Pan R., Yu F., Ye Z., and Wang H. 1993. Cranial comparison between the populations of rhesus monkeys (*Macaca mulatta*) distributing China and India. *Acta Theriologica Sinica* 13:1–10.

Peres, C. A. 1993. Notes on the Primates of the Juruá River, western Brazilian Amazonia. *Folia Primatologica* 61:97–103.

Peres, C. A., J. L. Patton, and M. N. F. da Silva. 1996. Riverine barriers and gene flow in Amazonian saddle-back tamarins. *Folia Primatologica* 67:113–124.

Perkins, E. M. 1975. Phylogenetic significance of the skin of New World monkeys (Order Primates, Infraorder Platyrrhini). *American Journal of Physical Anthropology* 42:395–424.

Pero, M. del, S. Crovella, P. Cervella, G. Ardito, and Y. Rumpler. 1995. Phylogenetic relationships among Malagasy lemurs as revealed by mitochondrial DNA sequence analysis. *Primates* 36:431–440.

Petter, J.-J., and A. Petter-Rousseaux. 1960. Remarques sur la systématique du genre *Lepilemur. Mammalia* 24:76–86.

Petter, J.-J., R. Albignac, and Y. Rumpler. 1977. *Faune de Madagascar. 44. Mammifères lémuriens (Primates prosimiens).* Paris: ORSTOM/CNRS.

Petter-Rousseaux, A., and J.-J. Petter. 1967. Contribution à la systématique des *Cheirogaleinae* (Lémuriens malgaches). *Allocebus,* gen. nov., pour *Cheirogaleus trichotis* Günther 1875. *Mammalia* 31:574–582.

Phillips, W. W. A. 1935. *Mammals of Ceylon.* Colombo: Colombo Museum.

Pieczarka, J. C., and C. Y. Nagamachi. 1988. Cytogenetic studies of *Aotus* from eastern Amazonia: Y/autosome rearrangement. *American Journal of Primatology* 14:255–263.

Pieczarka, J. C., R. M. de Souza Barros, F. M. de Faria Jr., and C. Y. Nagamachi. 1993. *Aotus* from the southwestern Amazon region is geographically and chromosomally intermediate between *A. azarae boliviensis* and *A. infulatus. Primates* 34:197–204.

Pieczarka, J. C., C. Y. Nagamachi, R. M. S. Barros, and M. S. Mattevi. 1996. Analysis of constitutive heterochromatin by fluorochromes and in situ digestion with restriction enzymes in species of the group *Callithrix argentata* (Callitrichidae, Primates). *Cytogenetics and Cell Genetics* 72:325–330.

Plavcan, J. M. 1993. Catarrhine dental variability and species recognition in the fossil record. In *Species, species concepts, and Primate evolution,* W. H. Kimbel and L. B. Martin, eds., 239–263. New York: Plenum Press.

Pocock, R. I. 1907. A monographic revision of monkeys of the genus *Cercopithecus. Proceedings of the Zoological Society of London* 1907:677–746.

———. 1918. On the external characters of the lemurs and of *Tarsius. Proceedings of the Zoological Society of London* 1918:19–53.

———. 1927. The gibbons of the genus *Hylobates. Proceedings of the Zoological Society of London* 3:719–741.

———. 1928. The langurs, or leaf monkeys, of British In-

dia. *Journal of the Bombay Natural History Society* 32:472–504, 660–677.

———. 1929. Michael Rogers Oldfield Thomas—1858–1929. *Proceedings of the Royal Society of London* 100:108–113.

———. 1932. The rhesus macaques *(Macaca mulatta). Journal of the Bombay Natural History Society* 35:530–551.

———. 1935. The monkeys of the genus *Pithecus* (or *Presbytis*) and *Pygathrix* found to the east of the Bay of Bengal. *Proceedings of the Zoological Society of London* 1934:895–961.

———. 1939. *The fauna of British India including Ceylon and Burma: Mammalia.* Vol. 1. *Primates and Carnivora (in part), families Felidae and Viverridae.* London: Taylor and Francis.

Poorman, A. P. 1983. The banded chromosomes of Coquerel's sifaka, *Propithecus verreauxi coquereli* (Primates, Indriidae). *International Journal of Primatology* 4:419–425.

Poorman-Allen, P., and M. K. Izard. 1990. Chromosome banding patterns of the aye-aye, *Daubentonia madagascariensis* (Primates, Daubentoniidae). *International Journal of Primatology* 11:401–410.

Porter, C. A., S. L. Page, J. Czelusniak, H. Schneider, M. C. Schneider, I. Sampaio, and M. Goodman. 1997. Phylogeny and evolution of selected Primates as determined by sequences of the epsilon-globin locus and 5' flanking regions. *International Journal of Primatology* 18:261–295.

Pusch, B. von. 1941. Der Arten der Gattung *Cebus. Zeitschrift für Säugetierkunde* 16:183–237.

Qi T., and K. C. Beard. 1998. Late Eocene sivaladapid primate from Guangxi Zhuang Autonomous Region, People's Republic of China. *Journal of Human Evolution* 35:211–220.

Quammen, D. 1997. You looking for me? *Sports Illustrated* 86 (4): 66–76.

Queiroz, H. L. 1992. A new species of capuchin monkey, genus *Cebus* Erxleben, 1777 (Cebidae: Primates) from eastern Brazilian Amazonia. *Goeldiana Zoologia* 15:1–13.

Queiroz, K. de, and J. Gauthier. 1992. Phylogenetic taxonomy. *Annual Review of Ecology and Systematics* 23:449–480.

———. 1994. Toward a phylogenetic system of biological nomenclature. *Trends in Ecology and Evolution* 9:27–31.

Rahm, U. 1966. Les mammifères de la forêt equatoriale de l'est du Congo. *Annales du Musée Royale de l'Afrique Centrale,* Série 8, 149:39–121.

Ramirez-Cerquera, J. 1983. Reporte de una nueva especie

de Primates del genero *Aotus* de Colombia. In *Symposio sobre primatologia en Latinoamerica (IX Congreso Latinoamericano de Zoologia, Arequipa, Peru)*, 146. Lima: Pacific Press.

Reed, C. A. 1960. Polyphyletic or monophyletic ancestry of mammals, or: What is a class? *Evolution* 14:31–322.

Ridley, M. 1989. The cladistic solution to the species problem. *Biology and Philosophy* 4:1–16.

Rijksen, H. D., and E. Meijaard. 1999. *Our vanishing relative*. The Hague: Kluwer.

Rodgers, W. A., T. T. Struhsaker, and C. C. West. 1984. Observations on the red colobus *(Colobus badius tephrosceles)* of Mbisi forest, south-west Tanzania. *African Journal of Ecology* 22:187–194.

Röhrer-Ertl, O. 1984. *Orang-utan studien*. Neuried: Hieronymus Verlag.

Roonwal, M. L. 1981. Intraspecific variation in size, proportion of body parts, and weight in the hanuman langur, *Presbytis entellus* (Primates), in South Asia, with remarks on subspeciation. *Reviews of the Zoological Survey of India* 79:125–158.

Roonwal, M. L., R. S. Prite, and S. S. Saha. 1984. Geographical boundary between the northern and southern tail styles in the common South Asian langur, *Presbytis entellus* (Primates). *Journal of the Zoological Society of India* 36:15–26.

Roosmalen, M. G. M. van, and T. van Roosmalen. 1997. An eastern extension of the geographical range of the pygmy marmoset, *Cebuella pygmaea. Neotropical Primates* 5:3–6.

Roosmalen, M. G. M. van, T. van Roosmalen, R. A. Mittermeier, and G. A. B. da Fonseca. 1998. A new and distinctive species of marmoset (Callitrichidae, Primates) from the Lower Rio Aripuanã, state of Amazonas, central Brazilian Amazonia. *Goeldiana Zoologia* 22:1–27.

Rosenberger, A. L. 1977. *Xenothrix* and ceboid phylogeny. *Journal of Human Evolution* 6:461–481.

———. 1981. Systematics: The higher taxa. In *Ecology and behavior of neotropical Primates*. Vol. 1, A. F. Coimbra-Filho and R. A. Mittermeier, eds., 9–27. Rio de Janeiro: Academia Brasileira de Ciências.

———. 1983a. Tale of tails: Parallelism and prehensility. *American Journal of Physical Anthropology* 60:103–107.

———. 1983b. Aspects of the systematics and evolution of the marmosets. *A Primatologia no Brasil,* Anuario de 1st Congreso Brasileira do Primatologia, 159–180. Belo Horizonte: Primatological Society of Brazil.

Rosenberger, A., and A. F. Coimbra-Filho. 1984. Morphology, taxonomic status, and affinities of the lion tamarins, *Leontopithecus* (Callitrichinae, Cebidae). *Folia Primatologica* 42:149–179.

Rosenberger, A. L., and W. G. Kinzey. 1976. Functional patterns of molar occlusion in platyrrhine Primates. *American Journal of Physical Anthropology* 45:284–298.

Rosenblum, L. L., J. Supriatna, M. N. Hasan, and D. J. Melnick. 1997. High mitochondrial DNA diversity with little structure within and among leaf monkey populations *(Trachypithecus cristatus* and *Trachypithecus auratus). International Journal of Primatology* 18:1005–1028.

Ross, C., B. Williams, and R. F. Kay. 1998. Phylogenetic analysis of anthropoid relationships. *Journal of Human Evolution* 35:221–306.

Rothschild, M. 1983. *Dear Lord Rothschild*. London: Hutchinson.

Rowell, T. E. 1988. The social system of guenons, compared with baboons, macaques, and mangabeys. In *A Primate radiation: Evolutionary biology of the African guenons,* A. Gautier-Hion, F. Bourlière, J.-P. Gautier, and J. Kingdon, eds., 439–451. Cambridge: Cambridge University Press.

Rumpler, Y. 1975. The significance of chromosomal studies in the systematics of the Malagasy lemurs. In *Lemur biology,* I. Tattersall and R. W. Sussman, eds., 25–40. New York: Plenum Press.

Rumpler, Y., and R. Albignac. 1973. Cytogenetic study of the endemic Malagasy lemur: *Hapalemur,* I. Geoffroy, 1851. *Journal of Human Evolution* 2:267–270.

———. 1975. Intraspecific chromosome variability in a lemur from the north of Madagascar: *Lepilemur septentrionalis,* species nova. *American Journal of Physical Anthropology* 42:425–429.

Rumpler, Y., and B. R. Rakotosamimanana. 1972. Coussinets palmo-plantaires et dermatoglyphes des représentants des lémuriformes malgaches. *Bulletin de l'Association des Anatomistes* 154:1127–1143.

Rumpler, Y., S. Warter, B. Meier, H. Preuschoft, and B. Dutrillaux. 1987. Chromosomal phylogeny of three Lorisidae: *Loris tardigradus, Nycticebus coucang,* and *Perodicticus potto. Folia Primatologica* 48:216–220.

Rumpler, Y., S. Warter, J.-J. Petter, R. Albignac, and B. Dutrillaux. 1988. Chromosomal evolution of Malagasy lemurs. *Folia Primatologica* 50:124–129.

Rumpler, Y., S. Warter, B. Ishak, and B. Dutrillaux. 1989. Chromosomal evolution in Primates. *Human Evolution* 4:157–170.

Rumpler, Y., S. Warter, C. Rabarivola, J.-J. Petter, and B. Dutrillaux. 1990. Chromosomal evolution in Malagasy lemurs: XII. Chromosomal banding study of *Avahi laniger occidentalis* (syn.: *Lichanotus laniger occidentalis*) and cytogenetic data in favour of its classification in a species apart—*Avahi occidentalis*. *American Journal of Primatology* 21:307–316.

Rumpler, Y., S. Warter, M. Hauwy, V. Randrianasolo, and B. Dutrillaux. 1991. Brief report: Cytogenetic study of *Hapalemur aureus*. *American Journal of Physical Anthropology* 86:81–84.

Ruvolo, M. 1988. Genetic evolution in the African guenons. In *A Primate radiation: Evolutionary biology of the African guenons*, A. Gautier-Hion, F. Bourlière, J.-P. Gautier, and J. Kingdon, eds., 127–139. Cambridge: Cambridge University Press.

Ryder, O. A., and L. G. Chemnick. 1993. Chromosomal and mitochondrial DNA variation in orang utans. *Journal of Heredity* 84:405–409.

Rylands, A. B., and D. Brandon-Jones. 1998. Scientific nomenclature of the red howlers from the northeastern Amazon in Brazil, Venezuela, and the Guianas. *International Journal of Primatology* 19:879–905.

Rylands, A. B., A. F. Coimbra-Filho, and R. A. Mittermeier. 1993. Systematics, geographic distribution, and some notes on the conservation status of the Callitrichidae. In *Marmosets and tamarins: Systematics, behaviour, and ecology*, A. B. Rylands, ed., 11–77. Oxford: Oxford University Press.

Rylands, A. B., R. A. Mittermeier, and E. R. Luna. 1995. A species list for the New World Primates (Platyrrhini): Distribution by country, endemism, and conservation status according to the Mace-Lande system. *Neotropical Primates* 3 (supplement): 113–160.

Sampaio, M. I. C., M. P. C. Schneider, C. M. L. Barroso, B. T. F. Silva, H. Schneider, F. Encarnación, E. Montoya, and F. M. Salzano. 1991. Carbonic anhydrase II in New World monkeys. *International Journal of Primatology* 12:389–402.

Sampaio, I., M. P. C. Schneider, and H. Schneider. 1996. Taxonomy of the *Alouatta seniculus* group: Biochemical and chromosome data. *Primates* 37:65–73.

Sampaio, I., M. P. C. Schneider, A. Pissinatti, A. F. Coimbra-Filho, M. Goodman, and H. Schneider. 1997. A molecular approach to the adaptive radiation of the atelids: Parallels and controversies. *A Primatologia no Brasil* 5:352.

Sarmiento, E. E., T. M. Butynski, and J. Kalina. 1996. Gorillas of Bwindi–Impenetrable Forest and the Virunga Volcanoes: Taxonomic implications of morphological and ecological differences. *American Journal of Primatology* 40:1–21.

Sauer, E. G. F., and E. M. Sauer. 1963. The South-West African bushbaby of the *Galago senegalensis* group. *Journal South West African Scientific Society* 16:5–35.

Schilling, D. 1984. Gibbons in European zoos, with notes on the identification of subspecies of concolor gibbon. In *The lesser apes*, H. Preuschoft, D. J. Chivers, W. Y. Brockelman, and N. Creel, eds., 51–60. Edinburgh: Edinburgh University Press.

Schlegel, H. 1876. *Les singes. Simiae.* Leiden: E. J. Brill.

Schmid, J., and P. M. Kappeler. 1994. Sympatric mouse lemurs (*Microcebus* spp.) in western Madagascar. *Folia Primatologica* 63:162–170.

Schneider, H., I. Sampaio, M. L. Harada, C. M. L. Barroso, M. P. C. Schneider, J. Czelusniak, and M. Goodman. 1996. Molecular phylogeny of the New World monkeys (Platyrrhini, Primates) based on two unlinked nuclear genes: IRBP Intron 1 and epsilon-globin sequences. *American Journal of Physical Anthropology* 100:153–179.

Schneider, H., I. Sampaio, and M. P. C. Schneider. 1997. Systematics of the platyrrhines. *A Primatologia no Brasil* 5:315–324.

Schneider, M. P. C., H. Schneider, M. I. C. Sampaio, N. M. Carvalho-Filho, F. Encarnación, E. Montoya, and F. M. Salzano. 1995. Biochemical diversity and genetic distances in the Pitheciinae subfamily (Primates, Platyrrhini). *Primates* 36:129–134.

Schouteden, H. 1953. Un lémurien noir de Ruanda. *Revue de Zoologie et de Botanique Africaines* 48:111–114.

Schreiber, A., and K. Bauer. 1997. Strepsirhine dichotomy from a human perspective (Primates: Lorisoidea, Lemuroidea). *Journal of Zoological Systematics and Evolution Research* 35:121–129.

Schultz, A. H. 1948. Number of young at birth and the number of nipples in Primates. *American Journal of Physical Anthropology* 6:1–23.

Schultze, H., and B. Meier. 1995. The subspecies of *Loris tardigradus* and their conservation status: A review. In *Creatures of the dark: The nocturnal prosimians*, L. Alterman, G. A. Doyle, and M. K. Izard, eds., 193–209. New York: Plenum Press.

Schwartz, J. H. 1996. *Pseudopotto martini:* A new genus and species of extant lorisiform primate. *Anthropological Papers of the American Museum of Natural History* 78:1–14.

Schwartz, J. H., and J. C. Beutel. 1995. Species diversity in lorisids: A preliminary analysis of *Arctocebus, Perodicticus,* and *Nycticebus.* In *Creatures of the dark: The nocturnal prosimians*, L. Alterman, G. A. Doyle, and M. K. Izard, eds., 171–192. New York: Plenum Press.

Schwartz, J. H., and I. Tattersall. 1985. Evolutionary relationships of living lemurs and lorises (Mammalia, Primates) and their potential affinities with European Eocene Adapidae. *Anthropological Papers of the American Museum of Natural History* 60:1–100.

Schwarz, E. 1927. Paul Matschie. *Journal of Mammalogy* 8:292–295.

———. 1928a. Note on the classification of the African monkeys in the genus *Cercopithecus* Erxleben. *Annals and Magazine of Natural History* (10) 1:649–663.

———. 1928b. Bemerkungen über die roten Stummelaffen. *Zeitschrift für Säugetierkunde* 3:92–97.

———. 1929. On the local races and distribution of the black and white colobus monkeys. *Proceedings of the Zoological Society of London* 1929:585–598.

———. 1931a. On the African long-tailed lemurs or galagos. *Annals and Magazine of Natural History* (10) 7:41–66.

———. 1931b. On the African short-tailed lemurs or pottos. *Annals and Magazine of Natural History* (10) 8:249–256.

———. 1931c. A revision of the genera and species of Madagascar Lemuridae. *Proceedings of the Zoological Society of London* 1931:399–426.

———. 1934a. On the local races of the chimpanzee. *Annals and Magazine of Natural History* (10) 13:576–585.

———. 1934b. Notes on the nomenclature and systematic position of some African mammals. *Annals and Magazine of Natural History* (10) 14:258–261.

Sclater, P. L. 1872. On the Quadrumana found in America north of Panama. *Proceedings of the Zoological Society of London* 1872:2–9.

Scott, K. M., and C. M. Janis. 1987. Phylogenetic relationships of the Cervidae, and the case for a superfamily "Cervoidea." In *Biology and management of the Cervidae,* C. M. Wemmer, ed., 3–20. Washington, D.C.: Smithsonian Institution Press.

Sereno, P. 1999. Definitions in phylogenetic taxonomy: Critique and rationale. *Systematic Biology* 48:329–351.

Seuánez, H. N., J. L. Armada, L. Freitas, R. Rocha e Silva, A. Pissinatti, and A. Coimbra-Filho. 1986. Intraspecific chromosome variation in *Cebus apella* (Cebidae, Platyrrhini): The chromosomes of the yellow breasted capuchin *Cebus apella xanthosternos* Wied, 1820. *American Journal of Primatology* 10:237–247.

Seuánez, H. N., G. Alves, M. M. C. de Lima, R. de Souza Barros, C. M. L. Barroso, and J. A. P. C. Muniz. 1992. Chromosome studies in *Chiropotes satanas utahicki* Hershkovitz, 1985 (Cebidae, Platyrrhini): A comparison with *Chiropotes satanas chiropotes*. *American Journal of Primatology* 28:213–222.

Shea, B. T., S. R. Leigh, and C. P. Groves. 1993. Multivariate craniometric variation in chimpanzees: Implications for species identification. In *Species, species concepts, and Primate evolution,* W. H. Kimbel and L. B. Martin, eds., 265–296. New York: Plenum Press.

Shedd, D. H., and J. M. Macedonia. 1991. Metachromism and its phylogenetic implications for the genus *Eulemur* (Prosimii: Lemuridae). *Folia Primatologica* 57:221–231.

Shekelle, M., S. M. Leksono, L. L. S. Ichwan, and Y. Masala. 1997. The natural history of the tarsiers of North and central Sulawesi. *Sulawesi Primate Newsletter* 4 (2): 4–11.

Shortridge, G. C. 1942. Field notes of the first and second expeditions of the Cape Museum's Mammal Survey of the Cape Province, and descriptions of some new subgenera and subspecies. *Annals of the South African Museum* 36:27–100.

Shoshani, J., C. P. Groves, E. L. Simons, and G. F. Gunnell. 1996. Primate phylogeny: Morphological vs molecular results. *Molecular Phylogenetics and Evolution* 5:102–154.

Simons, E. L. 1988. A new species of *Propithecus* (Primates) from Northeast Madagascar. *Folia Primatologica* 50:143–151.

———. 1994. The giant aye-aye *Daubentonia robusta*. *Folia Primatologica* 62:14–21.

———. 1998. The prosimian fauna of the Fayum Eocene/Oligocene deposits of Egypt. *Folia Primatologica* 69 (supplement 1): 286–294.

Simons, E. L., and Y. Rumpler. 1988. *Eulemur*: New generic name for species of *Lemur* other than *Lemur catta*. *Comptes Rendus de l'Académie des Sciences* (Paris) 307:547–551.

Simpson, G. G. 1945. The principles of classification and a classification of mammals. *Bulletin of the American Museum of Natural History* 85:i–xvi, 1–350.

———. 1961. *Principles of animal taxonomy*. New York: Columbia University Press.

Snowdon, C. T. 1993. A vocal taxonomy of the callitrichids. In *Marmosets and tamarins: Systematics, behaviour, and ecology,* A. B. Rylands, ed., 78–94. Oxford: Oxford University Press.

Sokal, R. R., and T. J. Crovello. 1970. The biological species concept: A critical evaluation. *American Naturalist* 104:127–153.

Sokal, R. R., and P. H. A. Sneath. 1963. *Principles of numerical taxonomy*. San Francisco: Freeman.

Soligo, C., and A. E. Müller. 1999. Nails and claws in primate evolution. *Journal of Human Evolution* 36:97–114.

Sonnerat, M. 1781. *Voyage aux Indes Orientales et à la Chine.* Paris: Froulé.

Souza Barros, R. M. de, C. Y. Nagamachi, and J. C. Pieczarka. 1990. Chromosomal evolution in *Callithrix emiliae. Chromosoma* 99:440–447.

Stanger-Hall, K. F. 1997. Phylogenetic affinities among the extant Malagasy lemurs (Lemuriformes) based on morphology and behavior. *Journal of Mammalian Evolution* 4:163–194.

Stanyon, R., J. Wienberg, E. L. Simons, and M. K. Izard. 1992. A third karyotype for *Galago demidovii* suggests the existence of multiple species. *Folia Primatologica* 59:33–38.

Stanyon, R., S. Tofanelli, M. A. Morescalchi, G. Agoramoorthy, O. A. Ryder, and J. Wienberg. 1995. Cytogenetic analysis shows extensive genomic rearrangements between red howler (*Alouatta seniculus* Linnaeus) subspecies. *American Journal of Primatology* 35:171–183.

Stewart, C.-B., and T. R. Disotell. 1998. Primate evolution—in and out of Africa. *Current Biology* 8:R582–R588.

Stiles, C. W., and M. B. Orleman. 1927. The nomenclature for man, the chimpanzee, the orang-utan, and the Barbary ape. *Bulletin of the Hygienic Laboratory, U.S. Public Health Service* (Washington, D.C.) 145:1–60.

Strasser, E. 1988. Pedal evidence for the origin and diversification of cercopithecid clades. *Journal of Human Evolution* 17:225–245.

———. 1994. Relative development of the hallux and pedal digit formulae in Cercopithecidae. *Journal of Human Evolution* 26:413–440.

Strasser, E., and E. Delson. 1987. Cladistic analysis of cercopithecoid relationships. *Journal of Human Evolution* 16:81–99.

Struhsaker, T. T. 1970. Phylogenetic implications of some vocalizations of *Cercopithecus* monkeys. In *Old World Monkeys*, J. R. Napier and P. H. Napier, eds., 365–444. New York: Academic Press.

———. 1975. *The red colobus monkey*. Chicago: University of Chicago Press.

———. 1981. Vocalizations, phylogeny, and paleogeography of red colobus monkeys *(Colobus badius). African Journal of Ecology* 19:265–283.

Struhsaker, T. T., and L. Leland. 1980. Observations on two rare and endangered populations of red colobus monkeys in East Africa: *Colobus badius gordonorum* and *Colobus badius kirki. African Journal of Ecology* 18:191–216.

Su B., Wang W., and Zhang Y. 1998. Protein polymorphism and genetic divergence in slow loris (genus *Nycticebus*). *Primates* 39:79–84.

Swofford, D. L. 1993. *PAUP: Phylogenetic analysis using parsimony*. Version 3. 1. 1. Washington, D.C.: Smithsonian Institution.

Szalay, F. S., and C. C. Katz. 1974. Phylogeny of lemurs, galagos, and lorises. *Folia Primatologica* 19:88–103.

Takahata, N., and Y. Satta. 1997. Evolution of the primate lineage leading to modern humans: Phylogenetic and demographic inferences from DNA sequences. *Proceedings of the National Academy of Sciences of the United States of America* 94:4811–4815.

Tan, B. 1985. The status of primates in China. *Primate Conservation* 5:63–81.

Tate, G. H. H. 1939. The mammals of the Guiana region. *Bulletin of the American Museum of Natural History* 76:151–229.

Tattersall, I. 1979. Another note on nomenclature and taxonomy in the Lemuridae. *Mammalia* 43:256–257.

———. 1982. *The Primates of Madagascar*. New York: Columbia University Press.

———. 1986a. Systematics of the Malagasy strepsirhine Primates. In *Comparative Primate biology. 1: Systematics, evolution, and anatomy*, D. R. Swindler and J. Erwin, eds., 43–72. New York: Alan R. Liss.

———. 1986b. Notes on the distribution and taxonomic status of some subspecies of *Propithecus* in Madagascar. *Folia Primatologica* 46:51–63.

———. 1988. A note on nomenclature in the Lemuridae. *Physical Anthropology Newsletter* 7:14.

———. 1992. Systematic versus ecological diversity: The example of the Malagasy Primates. In *Systematics, ecology, and the biodiversity crisis*, N. Eldredge, ed., 25–39. New York: Columbia University Press.

———. 1993. Speciation and morphological differentiation in the genus *Lemur*. In *Species, species concepts, and Primate evolution*, W. H. Kimbel and L. B. Martin, eds., 163–176. New York: Plenum Press.

Tattersall, I., and J. H. Schwartz. 1974. Craniodental morphology and the systematics of the Malagasy lemurs (Primates, Prosimii). *Anthropological Papers of the American Museum of Natural History* 52:141–192.

———. 1991. Phylogeny and nomenclature in the *Lemur*-group of Malagasy strepsirhine primates. *Anthropological Papers of the American Museum of Natural History* 69:3–18.

Templeton, A. R. 1989. The meaning of species and speciation: A genetic perspective. In *Speciation and its consequences*, D. Otte and J. A. Endler, eds., 3–27. Sunderland, Mass.: Sinauer Associates.

Thalmann, U., and T. Geissmann. 2000. Distribution and geographic variation in the western woolly lemur *(Avahi occidentalis)*, with description of a new species *(A. unicolor)*. *International Journal of Primatology* 21:915–941.

Thalmann, U., and N. Rakotoarison. 1994. Distribution of lemurs in central western Madagascar, with a regional distribution hypothesis. *Folia Primatologica* 63:156–161.

Thalmann, U., T. Geissmann, A. Simona, and T. Mutschler. 1993. The indris of Anjanaharibe-Sud, northeastern Madagascar. *International Journal of Primatology* 14:357–381.

Thomas, O. 1895. On some mammals collected by D. E. Modigliani in Sipora, Mentawei Islands. *Annali del Museo Civico di Storia Naturale di Genova* (2a), 14 (34): 1–13.

———. 1903. Notes on South American monkeys, bats, carnivores and rodents with descriptions of new species. *Annals and Magazine of Natural History* (7) 12:455–464.

———. 1911. The mammals of the tenth edition of Linnaeus: An attempt to fix the types of the genera and the exact bases and localities of the species. *Proceedings of the Zoological Society of London* 1911:20–158.

———. 1917. The geographical races of *Galago crassicaudatus*. *Annals and Magazine of Natural History* (8) 20:47–50.

Thorington, R. W. 1985. The taxonomy and distribution of squirrel monkeys *(Saimiri)*. In *Handbook of squirrel monkey research*, L. A. Rosenblum and C. L. Coe, eds., 1–33. New York: Plenum Publishing Corp.

———. 1988. Taxonomic status of *Saguinus tripartitus* (Milne-Edwards, 1878). *American Journal of Primatology* 15:367–371.

Tjio, H., and A. Levan. 1956. The chromosome number of man. *Hereditas* 42:1–6.

Trouessart, E. L. 1878. Catalogue des Mammifères vivants et fossils. Ordo 2. Prosimiae, Haeckel, 1866. *Revue et Magasin de Zoologie* (3) 6:162–169.

Tuinen, P. van, and D. H. Ledbetter. 1983. Cytogenetic comparison and phylogeny of three species of Hylobatidae. *American Journal of Physical Anthropology* 61:453–466.

Uchida, A. 1998a. Variation in tooth morphology of *Gorilla gorilla*. *Journal of Human Evolution* 34:55–70.

———. 1998b. Variation in tooth morphology of *Pongo pygmaeus*. *Journal of Human Evolution* 34:71–79.

Vandebroek, G. 1969. *Évolution des vertébrés de leur origine à l'homme*. Paris: Masson and Cie.

Van Valen, L. 1976. Ecological species, multispecies, and oaks. *Taxon* 25:233–239.

———. 1988. Species, sets, and the derivative nature of philosophy. *Biology and Philosophy* 3:49–66.

Vassart, M., A. Guédant, J. C. Vié, J. Kéravec, A. Séguéla, and V. T. Volobuev. 1996. Chromosomes of *Alouatta seniculus* (Platyrrhini, Primates) from French Guiana. *Journal of Heredity* 87:331–334.

Verheyen, W. 1962. Contribution à la craniologie comparée des primates. *Annales du Musée Royale de l'Afrique Centrale,* Série 8, Zoologie 105:1–256.

Vermeer, J. 1998. New information about the distribution of *Presbytis* on Sumatra. *Asian Primates* 6:9–10 (for December 1996/March 1997).

Vieira, C. da C. 1944. Os simios do estado da Sâo Paulo. *Papeis Avulsos de Zoologia* (Sao Paulo) 4:1–31.

Villalba, J. S., C. M. Prigiogini, and A. C. Sappa. 1995. Sobre la posible presencia de *Alouatta caraya* en Uruguay. *Neotropical Primates* 3:173–174.

Vincent, F. 1969. Contribution à l'étude des prosimiens africaines: Le galago de Demidoff. Thèse du Docteur ès Sciences Naturelles, Faculté des Sciences de Paris.

Vivo, M. de. 1991. *Taxonomia de Callithrix* Erxleben, 1777 (Callitrichidae, Primates). Belo Horizonte: Fundação Biodiversitas para Conservação da Diversidade Biológica.

Vuillaume-Randriamanantena, M., L. R. Godfrey, and M. R. Sutherland. 1985. Revision of *Hapalemur (Prohapalemur) gallieni* (Standing 1905). *Folia Primatologica* 45:89–116.

Wang, W., M. R. J. Forstner, Y. Zhang, Z. Liu, Y. Wei, H. Huang, H. Hu, Y. Xie, D. Wu, and D. J. Melnick. 1997. A phylogeny of Chinese leaf monkeys using mitochondrial ND3-ND4 gene sequences. *International Journal of Primatology* 18:305–320.

Wang Y., Jiang X., and Li D. 1998. Classification and distribution of the extant subspecies of golden snub-nosed monkey *(Rhinopithecus roxellana)*. In *The natural history of the doucs and snub-nosed monkeys*, N. G. Jablonski, ed., 53–54. Singapore: World Scientific.

Watanabe, K., and S. Matsumura. 1991. The borderlands and possible hybrids between three species of macaques, *M. nigra, M. nigrescens,* and *M. hecki,* in the northern peninsula of Sulawesi. *Primates* 32:365–369.

Watanabe, K., H. Lapasere, and R. Tantu. 1991a. External characteristics and associated developmental changes in two species of Sulawesi macaques, *Macaca tonkeana* and *M. hecki,* with special reference to hybrids and the borderland between the species. *Primates* 32:61–76.

Watanabe, K., S. Matsumura, T. Watanabe, and Y. Hamada. 1991b. Distribution and possible intergradation between *Macaca tonkeana* and *M. ochreata* at the borderland of the species in Sulawesi. *Primates* 32:385–389.

Watanabe, T. 1982. Mandible/basihyal relationships in red howling monkeys *(Alouatta seniculus):* A craniometrical approach. *Primates* 23:105–129.

Wayne, R. K., R. E. Benveniste, D. N. Janczewski, and S. J. O'Brien. 1989. Molecular and biochemical evolution of the Carnivora. In *Carnivore behavior, ecology, and evolution,* J. L. Gittleman, ed., 465–494. Ithaca, N.Y.: Comstock Publishing Associates.

Weitzel, V., and C. P. Groves, 1984. The nomenclature and taxonomy of the colobine monkeys of Java. *International Journal of Primatology* 6:399–409.

Weitzel, V., C. M. Yang, and C. P. Groves. 1988. A catalogue of Primates in the Singapore Zoological Reference Collection, Department of Zoology, National University of Singapore (formerly the Zoological Collection of the Raffles Museum). *Raffles Bulletin of Zoology* 36:1–166.

Whitten, A. J., and J. E. J. Whitten. 1982. Preliminary observations of the Mentawai macaque on Siberut Island, Indonesia. *International Journal of Primatology* 3:445–459.

Wiley, E. O. 1980. Phylogenetic systematics and vicariance biogeography. *Systematic Botany* 5:194–220.

Willermet, C., and B. Hill. 1997. Fuzzy set theory and its implications for speciation models. In *Conceptual issues in modern human origins research,* G. A. Clark and C. M. Willermet, eds., 77–88. New York: Aldine de Gruyter.

Wilson, C. C., and W. L. Wilson. 1977. Behavioral and morphological variation among Primate populations in Sumatra. *Yearbook of Physical Anthropology* [for 1976] 20:207–233.

Wilson, E. O., and W. L. Brown. 1953. The subspecies concept and its taxonomic application. *Systematic Zoology* 2:97–111.

Wolfheim, J. H. 1983. *Primates of the world.* Seattle: University of Washington Press.

Wyner, Y., R. Absher, G. Amato, E. Sterling, R. Stumpf, Y. Rumpler, and R. Desalle. 1999. Species concepts and the determination of historic gene flow patterns in the *Eulemur fulvus* (brown lemur) complex. *Biological Journal of the Linnean Society* 66:39–56.

Xu, X., and U. Arnason. 1996. The mitochondrial DNA molecule of Sumatran orangutan and a molecular proposal for two (Bornean and Sumatran) species of orangutan. *Journal of Molecular Evolution* 43:431–437.

Yasuda, K., T. Aoki, and W. Montagna. 1961. The skin of Primates, IV. The skin of the lesser bushbaby. *American Journal of Physical Anthropology* 19:23–34.

Yoder, A. D. 1994. Relative position of the Cheirogaleidae in strepsirhine phylogeny: A comparison of morphological and molecular methods and results. *American Journal of Physical Anthropology* 94:25–46.

———. 1997. Back to the future: A synthesis of strepsirrhine systematics. *Evolutionary Anthropology* 6:11–22.

Yoder, A. D., B. Rakotosamimanana, and T. J. Parsons. 1999. Ancient DNA in subfossil lemurs: Methodological challenges and their solutions. In *New directions in lemur studies,* H. Rasaminanana, B. Rakotosamimanana, S. Goodman, and J. Ganzhorn, eds., 1–17. New York: Plenum Press.

Zhang Y. 1997. Mitochondrial DNA sequence evolution and phylogenetic relationships of gibbons. *Acta Genetica Sinica* 24:231–237.

Zhang, Y., and O. A. Ryder. 1998. Mitochondrial cytochrome *b* gene sequences of Old World monkeys: With special reference on evolution of Asian colobines. *Primates* 39:39–49.

Zhang Y., and Shi L. 1993. Phylogenetic relationships of macaques as inferred from restriction endonuclease analysis of mitochondrial DNA. *Folia Primatologica* 60:7–17.

Zhang, Y., S. Jin, G. Quan, S. Li, Z. Ye, F. Wang, and M. Zhang. 1997. *Distribution of mammalian species in China.* Beijing: China Forestry Publishing House.

Zimmermann, E. 1990. Differentiation of vocalizations in bushbabies (Galaginae, Prosimiae, Primates) and the significance for assessing phylogenetic relationships. *Zeitschrift für Zoologische Systematik und Evolutionsforschung* 28:217–239.

Zimmermann, E., S. K. Bearder, G. A. Doyle, and A. B. Andersson. 1988. Variations in vocal patterns of Senegal and South African lesser bushbabies and their implications for taxonomic relationships. *Folia Primatologica* 51:87–105.

Zimmermann, E., P. Ehresmann, V. Zietemann, U. Radespiel, B. Randrianambinina, and N. Rakotoarison.

1997. A new primate species in northwestern Madagascar: The golden-brown mouse lemur *(Microcebus ravelobensis)*. *Primate Eye* 63:26–27.

Zimmermann, E., S. Cepok, N. Rakotoarison, and V. Zietemann. 1998. Sympatric mouse lemurs in North-West Madagascar: A new rufous mouse lemur species *(Microcebus ravelobensis)*. *Folia Primatologica* 69:106–114.

Zingeser, M. R. 1973. Dentition of *Brachyteles arachnoides* with reference to alouattine and ateline affinities. *Folia Primatologica* 20:351–390.

Index